History of Science and Technology In Ancient India

— The Beginnings —

Debiprasad Chattopadhyaya

with a foreword by
JOSEPH NEEDHAM

FIRMA KLM PVT. LTD. CALCUTTA 1986

Published by
FIRMA KLM PRIVATE LIMITED
257-B, B. B. Ganguly Street
Calcutta 700 012
INDIA

First Edition : Calcutta, 1986

Printed in India by :
Sreema Mudran,
8B, Shib Narayan Das Lane,
Calcutta 700 006

Rs. 325/

A project sponsored by

NATIONAL INSTITUTE OF SCIENCE TECHNOLOGY
& DEVELOPMENT STUDIES
A Constituent Establishment of C.S.I.R.

General Advisers

Abdur Rahman
Ashok Jain
Sushil Kumar Mukherjee

Special Advisers

Astronomy and Mathematics
Apurba Kumar Chakrabarty
Ramatosh Sarkar
with the assistance of
Subinoy Ray

Archaeology
Amita Ray
with the assistance of
Santanu Maity

Sanskritist
Mrinal Kanti Gangopadhyaya

Aid to the reader

1. *Notes and References*

The references in the foot-notes are to the works given in the Bibliography. Where only one book by an author is included in the Bibliography only his name occurs in the foot-note ; where more than one work is included in the Bibliography by the same author, the name of the author is followed by the initials of the title of the work in Roman capitals. Contributions to journal-articles and general reference works are indicated by the word 'in' preceding these.

2. *On the Indian Words*

Diacritical marks for Indian words are avoided in the body of the book. But all Indian words occurring in the text are included in the Index with diacritical marks to indicate their phonetic value.

3. *Invited Contributions*

Invited contributions from the following are included in the book. *Articles* : D. P. Agrawal, Navjyoti Singh and Santanu & Sujata Maiti. *Appendices* : Apurba Kumar Chakrabarty, Navjyoti Singh, Ramatosh Sarkar and Subinoy Ray. *Brief Comments* : Dilip Kumar Chakrabarti and Partha Ghosh.

Foreword for Debiprasad Chattopadhyaya's "History of Science and Technology in Ancient India"

JOSEPH NEEDHAM

It is almost too much of an honour for me to be asked to contribute a foreword to this new book of Chattopadhyaya and the team of excellent scholars which he has gathered together to help him in the enterprise. When I was younger I thought I knew something about the history and the philosophies of India, but now I realise how little it ever was. Yet it is quite clear that the history of science and technology in India will bear comparison with that of all the other ancient civilisations, and I would like to congratulate the main author and all his colleagues warmly on this endeavour, which they have brought to such a successful fruition.

Debiprasad Chattopadhyaya made his name in the world of learning some thirty years ago, with his book "Lokayata" in which he showed how much theoretical materialism there had been in ancient India, and how it had been systematically obscured and vilified by the theologians of all the Indian religions. He has never ceased to uphold the banner of the naturalists of India, and some twenty years later, in his book on "Science and Society...." he showed in detail how the medical men had to struggle against the religious theorists. The former were searching for the naturalistic causes of disease—a point of view entirely justified by modern medical science—but the theologians always wanted to attribute diseases to the bad *karma* incurred in previous existences. All this could be demonstrated particularly by the nature and fate of the ancient medical book *Caraka-samhita*.

If there is one thing more than anything else which has characterised the work of Chattopadhyaya from the beginning, it has been his conviction of the importance of relating the history of science, technology and medicine to the social conditions which surrounded their growth. This principle will alone enable us to understand in depth the story of their slow development. For example, take the grand question which looms behind all the volumes of "Science and Civilisation in China", why, in spite of so many wonderful discoveries and inventions during sixteen or seventeen centuries before the Scientific Revolution, did modern science not develop in China but only in Europe? The answer can only be stated in social and economic terms. Only when one knows that China was characterised by bureaucratic feudalism, while Europe had military-aristocratic feudalism, seemingly stronger but in fact much weaker, and so exposed to overthrow when the time came for the rise of the bourgeoisie; then only can one begin to see why modern science, along with capitalism and the Reformation, originated in Europe and in Europe alone. How things went in India I could not attempt to say, but I would expect that apart from wars and colonialism, some concrete social and economic factors will in the end account for the fact that, in spite of wonderful past achievements, modern science did not originate there either.

Here in this present book we have the beginning of the story, and most exciting it is. Chattopadhyaya and his colleagues speak of two urbanisations, the first associated with the Indus Valley culture which produced such splendid cities as Mohenjo-daro and Harappa—roughly speaking corresponding to the Shang-Yin civilisation of China or rather earlier, and the beginning of the 2nd. millennium BC. The reasons for their decline and fall are not yet fully understood, and the subject is discussed here, but it is sure that they were followed by the Aryan invasions and the Vedic Age. Then came the second urbanisation in the 6th century BC. Although the Indus Valley script is not yet fully interpreted, we can see from ocular aspection (as Sir Thomas Browne would have said) the wonders of their hydraulic engineering systems, and the great harbours that they built for their maritime commerce, presumably with the Middle East.

Perhaps the most illuminating correlation which this book contains—or so it was for me—is that the first beginnings of mathematical geometry in India, later preserved in the *Sulva-sutras*, was a direct result of the baked brick industry of the Indus Valley cities. This was rather earlier than either Rome or China, where we do not seem to find baked bricks much before the Warring States period in the 1st. millennium BC. But in any case, the industry was a very early example of mass-production, and since sizes of all shapes had to be exactly specified, their geometrical relationships shown in building would naturally follow. But throughout the book, for example in metallurgy and in ceramics, there is no lack of other examples of practice coming first, and then theory arising out of it afterwards.

Yet another interesting question taken up by Chattopadhyaya is to what extent there was genuine curiosity and nature-study among the Indus Valley people ; in astronomy for example. Did they trace the path of the moon through the stars, and did they originate the *nakshatra* moon-stations, which we call *hsiu* in China ? There has long been controversy about what civilisation it was that initiated them ; perhaps this book will help to settle it.

Again, all that has been written here about *rta*, that ancient Indian concept of the Order of Nature, its pattern and organisation, self-originating and underlying all that happens, is well worth reading. This concept is somewhat analogous with what in Chinese we call the *Tao,* or *li,* also self-originating, *tzu-jan*. Apparently it was not characteristic of the Indus Valley civilisation but rather to be found in the *Rgveda* and such works. As a recognition of the regularity and uniformity of Nature it was certainly wisdom, but it had to be fleshed out with specific theories about natural phenomena, and these to a large extent arose out of technological practice. Of course it was the ancestor too of what today we call "laws of Nature", those laws which Westerners once thought of as due to the will of a transcendent creator deity, but which are now regarded as descriptive rather than prescriptive.

Finally I should like to say that I sympathise very much with the attempt to "de-mystify" ancient science, and to destroy the arguments which primitive theology brought against it. But we must beware of "pouring out the baby with the bath-water" (as we say in my country). Today ethics is needed more than ever, whatever one's attitude may be to developed religion. One of the most striking experiences of my life has been connected with the ethical value to be attached to science. When I was young, in the thirties of the present century, I was one of the "Science and Society" movement among the "younger scientists" of the time. It never occurred to any of us that science was something inimical to mankind, or dangerous ; the whole complaint of colleagues like J. D. Bernal was that capitalism prevented the full employment of science for human benefit, and that was quite reason enough for opposing it. But the whole situation has utterly changed since the Second World War. People in general are now suspicious and fearful of science, only too conscious of its dread possibilities—whether in atomic and chemical weapons, nuclear power, acid rain, environmental pollution, genetic engineering, loss of privacy due to information banks, etc. etc. The fact that so many of these things are due to the machinations of evil men, engaged in the struggle for wealth and power, is forgotten. Capitalism, the very incarnation of this struggle, still stands condemned. But now people are desperately afraid of what modern science and industry are capable of, as in the recent case at Bhopal. This has been one of the greatest changes which I have seen in my not too short life, and it seems to me that ethics and morality are more necessary than ever for the human condition.

But now it only remains to salute Debiprasad Chattopadhyaya's new book, and to wish every success to it and to the volumes which will follow it.

Cambridge
June 15, 1986

Joseph Needham

AUTHOR'S PREFACE

A study in science and technology in Indian history is much more than a matter of mere academic exercise. It has profound significance for our national requirements, specially in these grim days through which we are passing. We shall try to explain this, beginning with some accredited admissions.

In the inaugural address to the fiftyfifth session of the Indian Science Congress (Varanasi), our prime minister observed : "We must transform an ancient tradition-bound people into a modern nation." For this purpose, she naturally looked for aid from science : "The quicker way is that of science. What do we expect of science ? The immediate answer is, generally, that we seek for more advanced technologies and their application to bring material benefit and to take knowledge and training within the reach of different sections of our people, thus enabling them to produce wealth in their fields and factories and to exploit our vast untapped resources. While this must remain a primary objective of scientific endeavour in any country, we are equally aware of the importance of other aspects and of basic science."

Hence she felt anxious to report on how much her government had already done for the spread of science in this country : "It is a measure of our resolve to give science and technology an important place in our scheme of things that India has made considerable investment in stimulating their growth. The awareness of science and technology is part of our national policy and we have made strenuous efforts to give practical shape and content to this ideology in the form of institutions. We have today some thirty national research laboratories. We have more than a dozen major agricultural and medical research centres. We have seventy universities and a sophisticated atomic energy programme."

Still, the prime minister felt that there was something wrong somewhere, and hence added : "And yet we must admit that all these developments have not made a significant impression on the consciousness even of our scientists, educationists and policy planners. We do still continue to lead a somewhat schizophrenic existence—one half of our individual self pays homage to science and the scientific approach, while the other half remains deeply rooted in the past. Paradoxically enough, this applies even to some who work in science!" There was thus something that resisted the scientific approach, without overcoming which scientific education itself could not serve the purpose of developing the scientific outlook proper.

What, then, is this force of resistance?

This led the prime minister to re-examine the question of traditional thought. And she observed: "Much of what is called tradition in our country is no more than a fossilization of thought and habit. These layers of superstition and dead habit have no meaning in our times or relevance to our needs. They impede the growth of science and the scientific temper. This dichotomy between our social life and scientific needs has to be overcome. Our science should not only be an effort of individuals and institutions, but should develop general social significance and movement."

All this was said in 1968. In the meanwhile, the number of our universities has gone up, science laboratories and research centres have gained both in quantity and quality: these are now better financed and better equipped. Sophisticated research in various branches of science is increasingly gaining global prestige. How much, indeed, is being done for science and technology in India today!

But that is only one side of the picture.

Also happened in the country events that are absolutely stunning—assasinations, murders, caste malevolence and communal carnage. Factors operating behind all this are complex no doubt. What specially concern us here are those that may broadly be called ideological. How much of the scientific temper is actually cultivated among the people—a temper that has a good deal to do in changing their consciousness so that they can rise above the false lure of casteism, communalism and regional chauvinism? Not surely in a scale that can make us feel proud. Even the working scientists cannot absolve themselves of the responsibility of this failure; they cannot seek convenient evasion of their responsibility under the shelter of sophistication and specialisation.

Science is not a mere marvel, or better, that which makes

it really marvellous is the basic attitude that makes it possible. It is, in short, the scientific temper, of which the scientists themselves are expected to be the best custodians.

An antidote to the malevolence with which we are being confronted today is the spread of the scientific temper. And one of the special problems created in the country is the illusion fomented by the regional chauvinists, communalists and fundamentalists is their claim to be the real custodians of our national cultural heritage. The claim is a fiction—in fact the most dangerous fiction. And it has got to be debunked. But it cannot be debunked with mere demagogy. We have to lead our people to meet the technicians, engineers and scientists in our own history and to show how they were defending the scientific temper in their own way, defying the dark forces that threatened it. This had indeed been a very significant aspect of our national cultural heritage. We have also to try to lead our people to see what, in the past, inhibited our scientists—with all their personal gifts—to move forward to what is ordinarily called modern science, i.e. science in the sense that developed in Europe from the days of Galileo and others. When we do this, we are confronted with an unexpected situation. The factors that inhibited the development of modern science in Indian history are inclusive of those that are still creating the zeal for casteism and communalism, murder and malevolence. In other words, we meet the same monster from whom inspiration is still being drawn, often surreptitiously though also often overtly.

That is why, a study of science in Indian history is more than a mere academic exercise. It is linked up also with the question of our very survival. At least that is how we have tried to define our task. Not that we want to flatter ourselves with the idea that the task is successfully executed in our own study. We are aware of its limitations and shortcomings. Nevertheless, nothing would be more rewarding for our labour if our own limitations irritate better scholars to carry on the work with superior abilities and in sounder lines.

I am anxious here to add only one more point. Without the moral and material support received from the National Institute of Science Technology and Development Studies and also the untiring labour of my colleagues who worked with me,

I could not write this book, or at least could not have completed it within a rather brief period. However, the views expressed here are mine and so also are the errors. In any case, the views expressed are by no means to be considered as those of NISTADS.

Calcutta
May 24, 1985.

Debiprasad Chattopadhyaya

ACKNOWLEDGEMENTS

I am profoundly grateful to Professor A. Rahman, then Director NISTADS, for initiating the project on the history of science and technology in India and for giving me the responsibility of looking after the ancient period of the work. Every help that I required and asked for was readily given to me. Besides, his otherwise busy programme notwithstanding, he never hesitated to spare sufficient time to discuss with me certain academic questions on which I required clarification.

Professor A. Rahman retired from the post of the Director of the NISTADS when the work on this book was under progress. He was succeeded by Dr. Ashok Jain. I am happy to mention that in spite of this change in the administrative set-up, the moral and material help received for this work from NISTADS remained unaffected. Dr. Ashok Jain have been helping in all possible ways to work on the present history.

Besides A. Rahman and A. Jain I am most grateful to Dr. Sushil Kumar Mukherjee, to whose unpretentious and characteristic modest way of helping the present generation of our scientists is well-known. The active help and guidance I personally received from him has been most encouraging.

Sense of gratitude demands that I mention here the following. On a personal invitation of Dr. Joseph Needham—and with the financial support received from NISTADS and the British Council—I had recently the opportunity of spending some weeks at Cambridge. Dr. Needham spared long hours with me to discuss the general plan of the book and the methodology required for working on it. It has been exceedingly kind of him to contribute a *Foreword* to the book and this in spite of his present age and the incredible pressure of work of his own. I do not honestly know how to express my gratitude to him adequately.

The Cambridge visit also gave me the opportunity to discuss with Bridget and Raymond Allchin many questions concerning Indian archaeology, which proved most valuable for me.

As for the general understanding of ancient Indian history, I had never any hesitation to approach Professor R. S. Sharma, freely exploiting his personal affection for me.

Three chapters in the book and the *Appendices* are contributed by other scholars, whose names are mentioned in proper places, though I must add that I was truly surprised by the readiness with which a scientist of the stature of D. P. Agrawal agreed to associate himself with my humble work.

I am anxious to mention here the names of two other scientists—Partha Ghosh, the physicist, and D.K. Chakrabarti, the archaeologist. The former made a critical assessment of the recent claim of some (advertised in a rather big way) that the modern theory of unified field is already to be found in the Vedas. His assessment of the claim is to be found in pp. 400-402 of the present book. D.K. Chakrabarti recently told me in Delhi that he was preparing a paper on the agricultural technology in the Harappan period. Since he evidently needs a longish time to complete the paper, he readily agreed to my request to prepare a synopsis of his main points which is to be found in pp. 110-11 of the present book.

Talking of archaeology I must mention here the long discussions I had the opportunity to have with N. C. Ghosh (now of the Visvabharati University) which proved very useful for my own clarifications.

As for the young colleagues on whom I have throughout depended I am anxious to mention here specially the following: Santanu Maity, Subinoy Kumar Ray, Sujata Maity, Ramkrishna Bhattacharyya, G. Ramakrishna, Sanjay Biswas and Ramakrishna Maitra.

I am thankful to Hrishikesh Chakraborty for typing (and often retyping) the entire manuscript.

Debiprasad Chattopadhyaya

BACKGROUND NOTE

ASHOK JAIN

Director NISTADS

Influences of science and technology, some pleasant and some not so pleasant, are seen in the shades and textures of different social fabrics. One of the central areas of enquiry of the National Institute of Science Technology & Development Studies (NISTADS) has been to understand the nexus between science, technology and society. In this context, it was felt that an elucidation of the relationship between science and society in different periods of Indian history was essential. We were conscious of the fact that when the professional historians of India were still debating on many questions concerning the economic, political and social history of the country, it would be premature to attempt any final version of the history of scientific and technological activities in India. What was nevertheless possible—and considered necessary—was to try to make some probe into it, howsoever tentative and sometimes even controversial may be the result reached by it.

This was the guiding idea of Professor Nurul Hasan, then the Vice-President of the Council of Scientific & Industrial Research (of which NISTADS is a constituent establishment) when he desired initiation of a project on the History of Science & Technology in India. Professor A. Rahman who was then the Director of NISTADS set up three teams in Calcutta, Lucknow and NISTADS to work on the project to cover the ancient, medieval & modern periods respectively.

Professor Debiprasad Chattopadhyaya in Calcutta kindly agreed to guide the work covering the ancient period and set up a team for it. The present book embodies the results reached by this team for the pre-historic and proto-historic periods.

It needs to be specially emphasised that the present elucidation, tentative as it may be, reflects the understanding and viewpoint of the scholar entrusted with the work. Professor Debiprasad Chattopadhyaya requests me (true to the spirit of scholarship) to state very clearly and categorically that the views expressed by him in this book are his own and he takes full responsibility of the authenticity of the materials on the basis of which he has attempted to reach here certain results. Nevertheless, I would like to add that though NISTADS as an institution has not formulated views on the subject or arrived at conclusions, we are happy that our support had enabled Chattopadhyaya to initiate a probe into the history of science and technology in ancient India. This book, we feel, is a good beginning. Amendments, corrections and elaborations may be required in the future ; even the controversies it may provoke are likely to stimulate a deeper probing into the subject by other scholars. As Professor Debiprasad Chattopadhyaya unhesitatingly acknowledges his indebtendness to his predecessors—inclusive of those with whom he sometimes sharply differs—so also the future scholars may positively or negatively be indebted to him. But then if the book generates scholarly interest in the history of science and technology in ancient India, NISTADS would have justified its role.

New Delhi *Ashok Jain*
September 12, 1986.

CONTENTS

Foreword
JOSEPH NEEDHAM

AUTHOR'S PREFACE

Acknowledgements

Background Note : ASHOK JAIN

CHAPTER 1

INTRODUCTORY

1. Preliminary Remarks 2
2. P. C. Ray 2
3. Ray's Place among the Pioneers 4
4. Ray and the Scientific Temper 7
5. Rejection of the 'Internalist Hypothesis' .. 8
6. On Head and Hand 11
7. Back to Bernal 13
8. Science and Philosophy 14
9. M. N. Saha 17
10. The Scientific Method 20
11. Ideological Retreat 22
12. Hindu Revivalism 24
13. B. N. Seal on Scientific Method 26
14. Limitations of B. N. Seal 27
15. Scientific Method and the Working Scientists .. 30
16. S. N. Dasgupta 33
17. Mainstream of Global Science 35
18. Europe-Centrism 37
19. Apparent Anomaly 38
20. Slavery and the Decline of Greek Science .. 39
21. The Middle Ages 40
22. The Renaissance 40
23. Indebtedness to the East 42
24. 'Arrogant Ignorance' 44
25. Co-respondent to Conservatism 46
26. Not a mere Catalogue of Scientific Achievements 48
27. Pioneering Work 49
28. Science and Society : J. Needham 52
29. Concluding Remarks 53

Chapter 2

ON PERIODISATION

1. Indian Studies : Past & Present 55
2. Veda-Centrism and Indian History 62
3. Archaeology : New Light on Ancient India .. 64
4. Later Archaeological Work and Literature .. 66
5. Archaeology and Ancient Technology 68
6. Problem of Periodisation Reopened 70
7. Two Urbanisations and the "Dark Age" .. 72

Chapter 3

EXACT SCIENCE AND THE URBAN REVOLUTION

1. Preliminary Remarks 76
2. Script and Exact Science 77
3. "The Urban Revolution" 78
4. Mathematics, Astronomy and the Urban Revolution 82

Chapter 4

THE FIRST URBANISATION

1. Preliminary Remarks 86
2. Extent and Population 86
3. Agricultural Surplus 89
4. Postulate of Centralised Political Power .. 91
5. Problem of Origin 94
6. Agriculture and the Agricultural Surplus .. 98
7. Brick Technology and the Harappan Culture .. 100
8. Bricks in the First Urbanisation 104
9. Brief Appendix on 'Agricultural Technology and Harappan Culture' by D. K. Chakrabarti .. 110

Chapter 5

MATHEMATICS IN ITS MAKING

1. Preliminary Remarks 112
2. Origin of Geometry : Herodotus and the recent Corrections of his view 113
3. R. S. Sharma and Sulva-Geometry 116
4. Brick Technology and Magico-Religious Beliefs 119

5. An Apparent Archaeological Anomaly .. 123
6. R. S. Sharma's Theory of Mud Bricks 128
7. Burrow on 'Arma' and 'Armaka' 130
8. Evidence of the *Satapatha Brahmana* .. 137
9. The Question Reopened 139
10. Prehistory of Sulva Geometry 143

CHAPTER 6

TECHNICIANS AND THE VEDIC PRIESTS

1. Preliminary Remarks 147
2. Origin of Geometry 147
3. 'What is to be Done' & 'How is it to be Done'? 150
4. 'Thus we are told' 155
5. An Example: Baudhayana's Procedure .. 157
6. Mathematics to meet 'Theological Twaddle' .. 161
7. B. B. Datta's Analysis of a Problem 164
8. Role of the Technicians 168
9. Admission in the Vedic Tradition 170
10. Evading the Entire Problem of the Physical Construction of the Brick-structures 172
11. Question of Vocabulary and Terminology .. 176
12. Bricks in the *Yajurveda* : *Taittiriya Samhita* .. 178
13. Bricks in the *Satapatha Brahmana* 183
14. Bricks in the *Sulva-sutras* : General Observations 184
15. Brick Types: Some Examples 187
16. Masonry and Architecture 197
17. An Example of Mathematical Excellence .. 199
18. Terminological Precision 201
19. Political Philosophy of the *Satapatha Brahmana* 203
20. General Theoretical Temper 209
21. Unsolved Problems and Pointers to Further Research 214
22. Future of Sulva Mathematics 217

CHAPTER 7

SCIENCE IN FIRST URBANISATION

1. Preliminary Remarks: Mathematics 223
2. The Archaeological Data 224
3. Scales of Length Measure 225

4. Brick Technology and Mathematics in First Urbanisation 232
5. System of Weights 237
6. Mathematical Instruments? 240
7. Astronomy in First Urbanisation 241
8. Specimen of Wild Conjecture: "Great Bath" an Astronomical Observatory? 245
9. Astronomy and the Indus "Seals" 248
10. Method of Retrospective Probing: Chronological Pointer 253
11. Arbitrariness of the Interest in Astronomy of Vedic Priests 263
12. Method of Retrospective Probing: Geographical Pointer 266

CHAPTER 8

POTTERY, TRANSPORT, TEXTILE AND OTHER TECHNOLOGIES

(Contributed by Santanu Maity and Sujata Maity)

1. Preliminary Remarks 274
2. Pottery and Ceramic 275
3. Terracotta 285
4. Textile 288
5. Transport 290
6. Technology of Some Stone Objects 294
7. Seal-cutting and Engraving 299
8. Beads 303

Corrections to Tables I & III 312

CHAPTER 9

METAL TECHNOLOGY OF THE HARAPPA CULTURE

(Contributed by D. P. Agrawal)

1. Introduction 316
2. The Problems 318
3. Chemical Analysis 318
4. Alloying 327

CHAPTER 10

POSSIBILITY OF "CONSCIOUS" NATURE-SCIENCE IN FIRST URBANISATION

1. Preliminary Remarks 334

2. Ancient Techniques and Magic 335
3. Channelising Social Surplus 336
4. Nature-Science and Urban Revolution .. 339
5. Superstition and Nature-Science in Egypt : Plato 341
6. Possible Role of Superstition in Indus Administration 345
7. Conscious Nature-Science in Harappan Culture? 350

CHAPTER 11

END OF THE FIRST URBANISATION

1. Preliminary Remarks 352
2. Decline of the Indus Civilisation 353
3. Aryans and the End of Indus Civilisation .. 360
4. R. P. Chanda and the Theory of Aryan Invasion 363
5. Archaeology an Aid to Vedic Studies .. 366
6. D. D. Kosambi and the Vrtra Myth .. 367
7. Summing up 370

CHAPTER 12

BETWEEN THE TWO URBANISATIONS

1. Preliminary Remarks 372
2. Recent Archaeology and the "Dark Age" .. 373
3. Painted Grey Ware and the Vedic People .. 374
4. The *Rgveda-samhita* and Science in India .. 378
5. Predominantly Pastoral Economy 381
6. *Rta* : The Primordial Complex of Natural Law and Moral Law 383
7. Wrong Way of reading Science in the Vedas .. 398
8. Right way of reading Science in the Vedas .. 404

CHAPTER 13

LINGUISTICS AND ORAL TRADITION IN THE PERIOD BETWEEN THE DECLINE OF HARAPPAN CULTURE AND THE RISE OF MAGADHAN CULTURE

(Contributed by Navjyoti Singh)

1.0. Preliminaries 406
1.1. A Historical Paradox 406
1.2. Way Out of the Paradox 408
2.0. Non-literate Oral Phase and Fixation of Long Compositions 409

2.1. Conjecture on the Inability of Harappan Script to Fix Long Compositions 410
2.2. Beginning of Long Vedic Compositions and Oral Approach to Fix them 412
2.3. Institutionalisation of Priestly Communes for the Purpose 416
3.0. Standardisation and Fixing of the Vedic Texts and the Exact Science of Language 417
3.1. Problem of Structurisation of *Rgveda* : Origin of Indices, Concordances and Lexicons .. 418
3.2. Theory of Metres Developed to Articulate Structure of *Rksamhita* 420
3.3. Theory of Pronunciation Developed to Fix Incantation of *Samhita* 425
3.4. Significance of Meaning for Fixation and the Importance of Words 428
3.5. Invention of Device of *Padapatha* and Development of the theory of *Sandhi* of Sounds and Accents 429
3.6. Invention of the Device of *Karmapatha* and Internalisation of the knowledge-body of *Sandhi* in Recitation Strategy 433
3.7. Scientific Strategy for Orally Fixing *Rgveda* Became Model for Fixing Other Long Texts .. 435
3.8. Fixing of *Samaveda Samhita* and Failure to Fix Musical Aspect of Samans 436
3.9. Fixing *Yajurveda Samhita* and Invention of Complexly Knitted Recitation Strategy .. 437
3.10. Fixing of the Text of *Atharvaveda Samhita* .. 440
3.11. Linguistics : A Unique Feature of the Exact Science in India 440
4.0. Ancient Literature Dealing with the knowledge-body of Linguistics 441
5.0. Conclusion .. ..

APPENDICES

I. Basic geometrical propositions in the Sulva-sutras. By Subinoy Roy .. 457

II. Some observations relating to the lunar asterisms Krttika. By Ramatosh Sarkar .. 481

III. Some observations relating to the longest day of the year. By Ramatosh Sarkar .. 487

IV. The Asterisms. By Apurba Kumar Chakrabarty 495

V. Further notes on the Krttikas. By Ramatosh Sarkar .. 502

VI. Some remarks on Brij Bhusan Vij's paper on linear standard in the Indus Civilisation. By Apurba Kumar Chakrabarty 504

VII. Illustrations of various kinds of recitations of *Rgveda* which were devised to preserve long compositions orally. By Navjoyti Singh .. 506

BIBLIOGRAPHY .. 515

INDEX

CHAPTER I

INTRODUCTORY

In science, more than in any other human institution, it is necessary to search out the past in order to understand the present and to control the future.

Such an assertion would, at least until recently, have received scant support from working scientists. In natural science, and especially in the physical sciences, the idea is firmly held that current knowledge takes the place of and supersedes all the knowledge of the past. It is admitted that future knowledge will in turn make present knowledge obsolete, but for the moment it is the best available knowledge. All useful earlier knowledge is absorbed in that of the present ; what has been left out are only the mistakes of ignorance. Briefly, in the words of Henry Ford, 'History is bunk.'

Fortunately more and more scientists in our time are beginning to see the consequences of this attitude of neglect of history, and with it, necessarily, of any intelligent appreciation of the place of science in society. It is only this knowledge that can prevent the scientists, for all the prestige they enjoy, being blind and helpless pawns in the great contemporary drama of the use and misuse of science. It is true that in the recent past scientists and people at large got on very nicely in the comfortable belief that the application of science led automatically to a steady improvement in human welfare. The idea is not a very old one. It was a revolutionary and dangerous speculation in the days of Roger Bacon and was first confidently asserted by Francis Bacon 300 years later. It was only the immense and progressive changes in science and manufacture that came about with the Industrial Revolution that were to make this idea of progress an assured and lasting truth—almost a platitude—in Victorian times. It is certainly not so now, in these grim anxious days, when the power that science can give is seen to be more immediately capable of wiping out civilization and even life itself from the planet than of assuring an uninterrupted progress in the arts of peace. Though even here doubt has crept in and some neo-Malthusians fear that even curing disease is dangerous on an overcrowded planet.

Whether for good or ill the importance of science today needs no emphasizing, but it does, just because of that importance, need understanding. Science is the means by which the whole of our civilization is rapidly being transformed. And science is growing ; not, as in the past, steadily and imperceptibly, but rapidly, by leaps and bounds, for all to see. The fabric of our civilization has already changed enormously in our own lifetimes and is changing more and more rapidly from year to year. To understand how this is taking place it is not sufficient to know what science is doing now. It is also essential to be aware of how it came to be what it is, how it has responded in the past to the successive forms of society, and how in its turn it has served to mould them.

—J. D. Bernal, *Science in History*

1. PRELIMINARY REMARKS

We have quoted J. D. Bernal not because we have the intention—and certainly not the ability—to treat the question of science in history in the vast global canvas which he so inspiringly does. Our own scope is, in comparison, extremely limited. Nevertheless, it is necessary for our purpose to have a valid general perspective and also an acceptable methodology for the work. What we have quoted from Bernal provides us with both—better than many other works on the subject we are aware of. But this does not mean that all of our own scientists were totally unaware of the need of basically the same perspective and the same methodology, howsoever different might have been their modes of expression and how much fundamentally different might have been the general intellectual climate and socio-economic conditions in which they worked. We have chosen to illustrate this specially with two of the makers of modern science in India—P. C. Ray and M. N. Saha—whose credentials as scientists it would be too churlish to question though whose approach to science in Indian history is not sufficiently emphasised by most of our writers on the history of science in India.

2. P. C. RAY

We shall begin with a few words on *acarya* Prafulla Chandra Ray (1861-1944), whose name became somewhat lagendary even during his lifetime. Of the manifold activities with which he was connected we have the scope here to mention only those that have immediate bearing on our present attempt to sketch the outlines of the history of science and technology in ancient India.

P. C. Ray was the first *notable working scientist* in India to have realised the importance of an intelligent appreciation of the place of science and scientific activities in Indian history.

It is true that his *History of Hindu Chemistry* (first published : vol. i, 1902/03 ; vol ii. 1908) was inspired to a considerable extent by the writings of M. Berthelot (1827-1909), about which Ray himself drew our attention in his *Autobiography.*[1] However, what specially interest us in the present context are some of the points on which Ray's *History* differs from that of Berthelot.

Berthelot, who succeeded Louis Pasteur as Secretary of the *Academie des Sciences,* Paris, belonged to an intellectual climate in which science as an intrinsic value was on the whole generally accepted. These were the days of great optimism about science in Europe. There was thus no need for any general defence of science and of the methodology of science, specially against the alleged claim of real knowledge to be found only in scriptural revelation. However, the intellectual climate to which Ray belonged was different. The defence of science—of the importance of observation and experiment on which it is based—was still frowned upon.

Among his predecessors Akshay Kumar Datta (1820-1886)—literally some kind of science-intoxicated man about whom much more requires to be written in our times[2],—was virtually ostracised and sacked from his job because of uncompromising enthusiasm for modern science and its implications. It may be incidentally mentioned here that, towards the end of his life, Datta himself was seeking the roots of science and scientific temper in the Indian history, so that it is not easily denounced as alien to our tradition, though this work was left unfinished by his illness and untimely death. Among the successors of P. C. Ray, the outstanding scientist Megh Nad Saha had to waste much of his valuable time for arguing against the orthodox champions of Indian spiritualism claiming that the wisdom of our ancient sages was inclusive of everything worthwhile in modern science, and hence the defence and cultivation of

1. P. C. Ray, *Auto* 93ff. The second volume of his *History of Hindu Chemistry,* first published in 1908, was dedicated to the memory of Berthelot.
2. A brief but very able summary of his life and struggle for the introduction of modern science in India is to be found in *Rationalist Annual,* 1962. pp. 20-30 *Akshay Dutt, Pioneer of Indian Rationalism* by A. K. Bhattacharyya.

modern science in India is nothing more than an index to our slavish mentality.[3] We shall have to see if, in spite of Saha's brilliant essays, the tendency continues in our times.

It was in such an intellectual climate that P. C. Ray had to work. Evidently, at least one factor that made him the first important working scientist in India to have worked on the history of science in India was to show that the tradition of science and scientific activity did form an important dimension of our national heritage. This made him, moreover, the first to realise the need for this purpose to be clear about the methodology of science. What was no less remarkable about him is that he drew our attention to the fact that the defence of this methodology was a matter of active struggle—a struggle not merely theoretical but also against the prevailing social conditions.

3. RAY'S PLACE AMONG THE PIONEERS

Others before P. C. Ray made no doubt the most wonderful pioneering work in rediscovering scientific aspects of Indian cultural heritage, though for this purpose they had to go against the stream—against the assumption generally prevailing among the European historians of science that science is an essentially European phenomenon. Leaving for the time being the discussion of this, and leaving also the account of the earlier visiting scientists like I-Tsing (643-713) and more particularly al-Biruni (973-1048), we cannot but recall here the names of a number of outstanding European scholars, without whose pioneering work our knowledge of science in Indian history would not perhaps have been what it is today. To mention only a few of them : H. T. Colebrooke (1765-1837), Alexander Csoma de Koros (1784-1842), A. Weber (1825-1901), E. Burgess (1805-1807), A.F.R. Hoernle (1841-1918), P. Cordier (1870-1914), F.G.W. Thibaut (1848-1914), J. Jolly (1848-1932), Th. Stcherbatsky (1866-1942),—to which many more names may indeed be added, though it is our misfortune that the Indian pandits under whom they learned the language and studied

3. We shall see how a scientist like M. N. Saha had to fight against this tendency, which unfortunately still persists.

their subjects are, generally speaking, not known to us.[4] But the fact remains that they did invaluable work in collecting the manuscripts, settling their reading and interpreting their science-potentials—works judged in the standard of sheer textual study have often been simply stupenduous. Specially for knowing the contributions to mathematics, astronomy and medicine in ancient India, contemporary investigators are often obliged to depend on them.[5]

But howsoever important their contributions might have been, the fact remains that they were antiquarians after all. Professionally speaking, most of them had little to do with natural science.[6]

It is thus one of the significant points on which P. C. Ray differed from most of them. He was above all a working scientist—in fact one that earned considerable reputation as a chemist even in Europe, having been elected a Vice-President in 1887-88 session of Edinburgh University Chemical Society, where, in the absence of the President-elect, he was required to preside over the session.[7]

Indeed, many aspects of his activities in introducing modern science to India are rather well-known. Under the most adverse circumstances and against the resistance of the colonial policy of the British rule, he succeeded in setting up the first full-fledged chemical laboratory in Calcutta in the University College of Science.[8] Among his "brilliant students" he mentions Jnan Chandra Ghosh, Jnanendra Nath Mukherjee, Makhanlal Dey, Satyendra Nath Bose, Pulin Behari Sarkar, Rasik Lal Datta, Nilratan Dhar and Megh Nad Saha.[9] Realising the importance

4. Little work is so far done on this, though undoubtedly it is in need of more intensive investigation.
5. For a selection of such pioneering works see, D. Chattopadhyaya (ed.) SHSI.
6. There were few exceptions to this : P. Cordier, and Buchanan-Hamilton e.g., were physicians by profession.
7. P. C. Ray, *Auto*. 56.
8. *Ibid* Chap. XV.
9. For the Bengalee readers, the most notable name in this connection is that of Rajsekhar Bose (See, Ray *Auto*. 404-5 and 89), who is better remembered as by far the most prominent writer of wit and humour in Bengali literature ; his lifelong devotion to build up the Bengal Chemical and Pharmaceutical Works Limited, is not generally remembered.

of science being actually applied to industry, he worked for building up the first Indian chemical industry, starting with the savings of his personal paltry salary, which, thanks to the support he received from some of his close Indian friends, eventually flourished as the Bengal Chemical and Pharmaceutical Works Limited.

Among the varied activities of this multi-dimensional personality what we want here to focus on, however, is one point. We are not aware of any other working scientist—none surely of the stature of P. C. Ray—to have developed an absorbing interest in science and scientific activities in Indian history. Understandably, much of this enthusiasm to seek roots of science in the Indian tradition was inspired by the spirit of national awakening of his time. As he himself says in the preface to the second volume of the *History of Hindu Chemistry*, "The Hindu nation with its glorious past and vast latent potentialities may yet look forward to a still more glorious future, and if the perusal of these pages will have the effect of stimulating my countrymen to strive for regaining their own position in the intellectual hierarchy of nations, I shall not have laboured in vain."[10]

We are aware, of course, of much more passionate expressions of the patriotic sentiment in those days, as we have also learnt the rather unfortunate results of the use of the word "Hindu" for Indians, as was somewhat customary in P. C. Ray's time.[11] But all this is besides the point of our present discussion. The point rather is that when a working scientist of P. C. Ray's stature wants to look back at the scientific activities in ancient and medieval India, the understanding of science itself acquires a dimension far more serious than that of mere antiquarian curiosity, as was more or less the case with Colebrooke and others. Thus while most of the other scholars were on the whole contributing to the preparation of some kind of inventory of the achievements of the ancient Indians in different departments of science, P. C. Ray raised and tried to answer questions far more important for our

10. P. C. Ray, HHC. Vol. 2, Pref. p.D.
11. That Ray himself was furthest from any communal understanding of the word "Hindu" will be seen from section 12 of the present introduction.

understanding of the history of science. We shall discuss mainly two of these. First, what it was that infused real vitality to the scientific activities in ancient India. Secondly, what it was that inhibited their growth and eventual decline.

4. RAY AND THE SCIENTIFIC TEMPER

The two questions are inter-related, and it remains for us to see how in answering these—specially the second one—P. C. Ray was in a sense ahead of his own times.

But let us begin with some of his general observations. In a lecture delivered in 1918, Ray observed :

> I shall endeavour to unfold before you to-day a forgotten chapter in the history of the intellectual development of the Indian people, namely the cultivation of the Experimental Sciences. It is generally taken for granted that the Hindus were a dreamy, mystical people given to metaphysical speculation and spiritual contemplation. Due credit is, no doubt, assigned to them for the production of such priceless treasures as the Upanishads, the Six Systems of Philosophy, including the abstruse Samkhya and the Gita, with their transcendental teachings. But the fact that the Hindus had very large hand in the cultivation of the experimental sciences is hardly known in these days.
>
> It should, however, be borne in mind that Experimental Sciences such as we now understand them are of very recent origin and growth, even in Europe.
>
> The controversies of the Schoolmen in the Middle Ages lend colour to the theory that in approaching the discussion of the most evident truths of nature the learned men of Europe always avoided the test of appealing to experiments. As some of you are aware, a solemn discussion arose among the foundation members of the Royal Society as to whether a dead fish weighed more than a live one, though it never occurred to them that the solution of the problems lay in directly weighing a fish—live and dead. When the Royal Society was founded in 1662 by Boyle, Hooke, Christopher Wren and other students of Nature, Hobbes sneered at them as 'experimentarians.' If such was the respect for accurate knowledge even in England in the 17th century, we should not be justified in applying a rigid test to the knowledge of India in the past ages.
>
> Experiments and observations constitute the fundamental bases of Sciences. It is naturally a relief to come across such dicta as laid down by two standard works on Hindu Chemistry, namely, *Rasendra-cintamani* by Ramacandra and *Rasa-prakasa-sudhakara* by Yasodhara, both belonging to the 13th or 14th century A.D.
>
> Says the former : 'That which I have heard of learned men and have read in the Sastra-s but have not been able to verify by experiment

I have discarded. On the other hand those operations which I have, according to the directions of my sage teachers, been able to perform with my own hands—those alone I am committing to writing.

'Those are to be regarded as real teachers who can verify by experiments what they teach—those are to be regarded as laudable disciples who can perform what they have learned—teachers and pupils, other than these are mere actors on the stage.'

Yasodhara, the author of the latter, observes: 'All the chemical operations described in my book have been performed with my own hands—I am not writing from mere hearsay. Everything related is based upon my own conviction and observations'.[12]

In the modern writings on science in India, we have here for the first time a clear emphasis on the methodology of science—specially on the importance of observation and experiment—as of decisive importance for the making of science. This emphasis on direct observation and experiment, went strongly against the dominant philosophical view in India and also against favourite priestly dictum that evidently sensed danger in science and the scientific method—in the attempt to know nature as it is without any alien addition. This must have been one of the factors that went against the emphasis on the need of interrogating nature by direct observation and experiment from which scientific activity drew its nourishment. P. C. Ray had the hindsight to note it. In his *History of Hindu Chemistry,* he found it necessary to have a special chapter on "Knowledge of Technical Arts and Decline of Scientific Spirit"—a theme which none before Ray felt the need of discussion, and which, among his notable scientist-successors only M. N. Saha cared to consider seriously. Though brief, there is much for the historians of science in India to draw from this chapter of P. C. Ray.

5. REJECTION OF THE INTERNALIST HYPOTHESIS

The first point to be noted about his discussion is that he wanted to see the real cause of the decline of the scientific spirit in India *not within* the general framework of science itself but *outside* it, i.e. mainly in the social conditions that developed in this country. This means, without being aware of the recent controversies over the hypothesis of the "inter-

12. P. C. Ray, in SHSI Vol. I, 344-5.

nalists" or "autonomists"—the hypothesis wanting to explain the history of science "solely in terms of the internal or autonomous filiation of ideas, theories, mental or mathematical techniques and practical discoveries, handed on like torches from one man to another", P. C. Ray did in fact reject it in his own way. The main cause of the decline of the scientific spirit in India was the entrenchment of caste society, with its disastrous degradation of the social status of the technicians, craftsmen and other manual workers. This, he thought, took place "when the Brahmins reasserted their supremacy on the decline and expulsion of Buddhism", because people in "the Vedic age did not form an exclusive caste of their own but followed different professions according to their convenience or natural taste."[13] How much the contemporary historians of India have to amend such a view of the triumph of the caste-society is evidently a different question.[14] Our point rather is, how much they have to learn from Ray's analysis of the effect of the caste-structure of society on the progress and development of science in India. As he put it :[15]

> The drift of Manu and of the later Puranas is in the direction of glorifying the priestly class, which set up most arrogant and outrageous pretensions. According to Susruta, the dissection of dead bodies is a *sine qua non* to the student of surgery and this high authority lays particular stress on knowledge gained from experiment and observation. But Manu would have none of it. The very touch of a corpse, according to Manu, is enough to bring contamination to the sacred person of Brahmin. Thus we find that shortly after the time of Vagbhata, the handling of a lancet was discouraged and Anatomy and Surgery fell into disuse and became to all intents and purposes lost

13. P. C. Ray HHC Vol. I, 192. A Vedic scholar would perhaps want to add that this was true only of the early Rigvedic period.
14. The literature on the origin of the caste system in India is already enormous and this is still being added to. Some idea of it may be obtained from Suvira Jaiswal in IHR Vol. VI, No. 1-2, pp. 1-63. It may however be mentioned in this connection that the contempt for manual work characterizing the caste system was unknown to the early Vedic Poets. See D. Chattopadhyaya, WLWDIP, 144 ff. It is generally believed that in the Rigveda the concept of the four castes is first foreshadowed in a very late hymn called the *Purusa-sukta*.
15. P. C. Ray HHC Vol. I, 192-197.

sciences to the Hindus. It was considered equally undignified to sweat away at the forge like a Cyclops. Hence the cultivation of the *Kalas* by the more refined classes of the society of which we get such vivid pictures in the ancient Sanskrit literature has survived only in traditions since a very long time past.

The arts being thus relegated to the low castes and the professions made hereditary, a certain degree of fineness, delicacy and deftness in manipulation was no doubt secured but this was done at a terrible cost. The intellectual portion of the community being thus withdrawn from active participation in the arts, the how and why of phenomena—the coordination of cause and effect—were lost sight of—the spirit of enquiry gradually died out among a nation naturally prone to speculation and metaphysical subtleties and India for once bade adieu to experimental and inductive sciences. Her soil was rendered morally unfit for the birth of a Boyle, a Des Cartes or a Newton and her very name was all but expunged from the map of the scientific world.

In this land of intellectual torpor and stagnation the artizan classes, left very much to themselves and guided solely by their mother wit and sound commonsense, which is their only heritage in this world, have kept up the old traditions. In their own way they display marvellous skill in damescening, making ornamental designs on metals, carving on ivory, enamelling, weaving, dyeing, lace-making, goldsmith's and jeweller's works, etc.

There was nothing new, of course, about censoring the evils of the caste-system. Many of our social reformers—inclusive of some of the contemporaries of P. C. Ray—did it, some perhaps with greater passion and eloquence than Ray showed. What was nevertheless striking about P. C. Ray is that he saw in the caste structure of society something that made science a prey to creeping paralysis. In contemporary terminology, the caste society ushered in and entrenched the ruinous separation of theory from practice—of mental work from manual work—as a consequence of the condemnation and degredation of actual craftsmen and technicians. But they alone possessed the tools and apparatus of interrogating nature, without which the "how and why" of phenomena cannot be known. The craftsmen and technicians were thus left only with their craftlores, improving these to the best of their abilities, but without enabling science proper to draw upon these and get enriched.

Since P. C. Ray was the first historian of Indian science to have clearly realised this point and to have boldly asserted it, we may as well try to understand some of its implications.

6. ON HEAD AND HAND

First, if the real clue to the decline of science is to be sought in the social degradation of the craftsmen and the technicians, it obviously follows that science draws its ultimate nourishment from the techniques. This does not mean, of course, that science is to be equated to mere technology. What it means nevertheless is that science is implicit in the techniques and hence inconceivable without the latter. It may be noted that P. C. Ray realised it already in the first decade of our century, i.e. long before the publication of E. Zilsel's famous paper on *The Genesis of the Concept of Physical Law*[16] which appeared in 1942, of B. Farrington's *Greek Science,* first published in 1944 and J. D. Bernal's *Science in History* which was first published in 1954—works that have given a new turn to the historiography of science and in which basically the same theme is worked out, though understandably with more historical data.

Farrington's *Greek Science* and also his subsequent brief but brilliant book *Head and Hand in Ancient Greece,* first published in 1947, lead us to see the second significant point sought to be emphasised by P. C. Ray. Though we do not expect of Ray the wide range of Greek studies as that of Farrington, it would be wrong to overlook the fact that Ray wanted to understand in his own way that this calamitous consequence for science resulting from the degradation of manual work was some kind of universal phenomenon, as was evident *also by what happened in ancient Greece.* Farrington himself emphasises this point as a correction to the lop-sided importance often attached to purely theoretical accomplishments of the ancient Greeks usually alleged to account for their scientific achievements. "Many moderns", says he, "misled no doubt by some of the ancient Greeks themselves, have combined pride in the theoretical brilliance of Greek science with a wish to ignore or deny its practical triumphs."[17] Among other things what such a tendency fails to take note of is the adverse effect on Greek science of Aristotle's attempted justification of slavery,

16. This paper forms a landmark in the contemporary approach to the history of science from the social point of view. We shall see more of it later.
17. B. Farrington, GS 18.

notwithstanding the personal genius of Aristotle himself. We have a brilliant analysis of this in Farrington's *Greek Science*.[18] What appears to be remarkable about P. C. Ray is that decades before the appearance of Farrington's work and depending only on comparatively meagre information about the Greek situation available to him, he wanted to draw our attention to basically the same phenomenon as illustrating the ruinous consequence for science resulting from the contempt and social degradation of the manual workers, as it happened in India. Thus he observed :

> Similar dangers have threatened Europe from time to time but her sturdy sons have proved better of them in the long run. Thus 'Aristotle's opinion that industrial work tends to lower the standard of thought was certainly of influence here. In accordance with this dictum the educated Greeks held aloof from the observation and practice of technical chemical processes ; a theoretical explanation of the reactions involved in these lay outside their circle of interest.'
>
> Paracelsus flings a sneer at the physicians of his time and compares them with the alchemists in the following terms : 'For they are not given to idleness nor go in a proud habit, or plush and velvet garments, often showing their rings upon their fingers or wearing swords with silver hilts by their sides, or fine and gay gloves upon their hands, but diligently follow their labours, sweating whole days and nights by their furnaces. They do not spend their time abroad for recreations but take delight in their laboratory. They wear leather garments with a pouch, and an apron wherewith they wipe their hands. They put their fingers amongst coals, into clay, and filth, not into gold rings. They are sooty and black like smith and colliers, and do not pride themselves upon clean and beautiful faces.'
>
> Even so late as the middle of the last century, the pursuit of Chemistry in England was not regarded in a serious light and 'chemists were ashamed to call themselves so because the apothecaries had appropriated the name'—a circumstance which led Liebig in 1837 to declare 'that England was not the land of science'.[19]

Thus when the physicians of Paracelsus' time became only fashionable social parasites,[20] representing science only in its

18. *Ibid* 112 ff.
19. P. C. Ray HHC Vol. I, 194 note.
20. It may be noted here that his real name was P.A. T.B. von Hohenheim (1493-1541) ; but he called himself Paracelsus to show contempt for Celsus, the great doctor of antiquity. To prove the supremacy of direct experience over any authority, he went to the extent of burning books of Galen and Avicenna in market-place. See J. D. Bernal, SH 398.

degeneration, Paracelsus himself saw in the alchemists' devotion to manual work the real hope of the rejuvenation of natural science. And it is here that we have the clue to one of the most significant factors that ushered in the spirit of modern science in Renaissance Europe—a phenomenon referred to by Ray as 'the sturdy sons of Europe eventually overcoming the dangers' created for science by this undesirable social phenomenon. Since P. C. Ray referred to this phenomenon—though cryptically and in his own way—we may as well turn to J. D. Bernal who has worked out the point in greater details, because the point itself is of crucial significance for our understanding of science in Indian history.

7. BACK TO BERNAL

Bernal wants us to note that one important factor accounting for the revival of science in modern Europe is that "the Renaissance healed, though only partially, the breach between aristocratic theory and plebeian practice." Elaborating the point he observes[21] :

> What was really new, however, was the respect given to the practical arts of spinning, weaving, pottery, glass-making, and, most of all, to the arts that provided for the twin needs of wealth and war—those of the miners and the metal-workers. The techniques of the arts were ot more account in the Renaissance than in classical times because they were no longer in the hands of slaves but of free men, and these were not, as they had been in the Middle Ages, far removed socially and economically from the rulers of the new society. In medieval Florence, for instance, the artists had been subordinate members of the major guild of doctors and spice-dealers, *Medicie Speciali ;* the sculptors were lower down with the minor guild of the masons and bricklayers. By the beginning of the sixteenth century, however, individual painters and sculptors could command the favours of popes and kings, though they often had to press hard to obtain payment for their work.
>
> The enhancement of the status of the craftsman made it possible to renew the link between his traditions and those of the scholars that had been broken almost since the beginning of the early civilizations. Both had a great contribution to make : the craftsman could add to the old techniques of classical antiquity the new devices that had arisen during the Middle Ages ; the scholar could contribute the world views, the ideas, and possibly most of all, the logical methods of argument derived from the Greeks by way of Arabic and scholastic

21. J. D. Bernal SH 386.

philosophy, and the newly evolved methods of computation. The combination of the two approaches took some time to work out, and spread rather gradually at first through the different parts of knowledge and action. But once the constituents had been brought together there was no stopping the combination—it was an explosive one. The intellectual task of the Renaissance was essentially the rediscovery and mastery of the world of art and Nature.

8. SCIENCE AND PHILOSOPHY

But let us return to P. C. Ray. If he was the first working scientist in India to have rejected in his own way the "internalist hypothesis" by looking at the cause of the decline of scientific spirit in India not *within* science but *outside* it—in the development of a society requiring condemnation of manual work—let us not overlook the fact that he was also aware of the ideological or philosophical factors that worked as correspondents to the downfall of the scientific temper in Indian history. Hence he felt that a critical review of the Indian philosophical situation was necessary in this connection. As he realised, among the Indian philosophical views, there were some that favoured science or the scientific spirit, just as there were others that were basically hostile to the requirements of natural science. Though not specialising in philosophy, Ray had evidently enough grasp of the Indian philosophical situation to differentiate between the two. Thus he could see that the philosophy of atomism, associated in the popular mind with the name of Kanada as its founder, had significant science-potential in ancient Indian context. He could also see that the world-denying metaphysics (*maya-vada*), generally known as Samkara's Vedanta, could not but be inimical to science. In P. C. Ray's judgement, therefore, Samkara, too, stood accused for the decline of science in Indian history—a judgment too courageous to pronounce in the Indian context, where Samkara's name often carries the epithet of being an incarnation of God. But Ray had the courage and he observed :

> The Vedanta philosophy, as modified and expanded by Samkara, which teaches the unreality of the material world, is also to a large extent responsible for bringing the study of physical science into disrepute. Samkara is unsparing in his strictures on Kanada and his system. One or two extracts from Samkara's Commentary on the *Vedanta Sutras*, will make the point clear : [Observed Samkara] 'It thus appears that the atomic doctrine is supported by very weak arguments only,

is opposed to those scriptural passages which declare the Lord to be the general cause, and is not accepted by any of the authorities taking their stand on scripture, such as Manu and others. Hence it is to be altogether disregarded by highminded men who have a regard for their own spiritual welfare.' [Again :] 'The reasons on account of which the doctrine of the Vaisesikas cannot be accepted have been stated above. That doctrine may be called semi-destructive (or semi-nihilistic)'.[22]

To this Ray added :

Among a people ridden by caste and hide-bound by the authorities and injunctions of the Vedas, Puranas, and Smrtis and having their intellect thus cramped and paralysed, no Boyle could arise to lay down such sound principles for guidance as : '...I saw that several chymists had, by a laudable diligence obtained various productions, hit upon many more phenomena, considerable in their kind, than could well be expected from their narrow principles : but finding the generality of those addicted to chymistry, to have had scarce any view, but to the preparation of medicines, or, to the improving of metals, I was tempted to consider the art, not as a physician or an alchymist, but a philosopher. And, with this view, I once drew up a scheme for a chymical philosophy ; which I shou'd be glad that any experiments or observations of mine might any way contribute to complete.

'.....And, truly, if men were willing to regard the advancement of philosophy ; more than their own reputations, it were easy to make them sensible, that one of the most considerable services they could do the world is, to set themselves diligently to make experiments, and collect observations, without attempting to establish theories upon them, before they have taken notice of all the phenomena that are to be solved'.[23]

Here, again, we come across a working scientist, while enquiring into the history of science in India, realising the need of analysing the interaction of science and philosophy. It cannot but be reminiscent of what J. D. Bernal observes only a few decades later, though Bernal—equipped as he is with scientific sociology—connects the philosophical controversy also with sociological interests. He observes[24]:

The general character of the theoretical controversy inside science is, however, not new. As will emerge clearly from a study of its history, a sometimes latent, sometimes active struggle has been going on ever since the dawn of science between two main opposing ten-

22. P. C. Ray HHC I. 195-6 note.
23. *Ibid* I. 196-7 note.
24. J. D. Bernal, SH 53-54.

dencies : one, formal and idealistic ; the other, practical and materialistic. We shall see this conflict as the dominant one in Greek philosophy, but it must have originated much earlier, indeed from the first formation of class societies, for the general social affinities of the two sides in the conflict have never been in doubt.

The idealist side is the side of 'order, the aristocracy, and established religion ; its most persuasive champion is Plato. The objective of science, in its view, is to explain why things are as they are and how impossible, as well as impious, it is to hope to change them in essentials. In Plato's mind all that is necessary is to remove a few blemishes, such as democracy, for the republic to be established safely for ever under the care of the guardians, the men of gold. As the perfections of this state of affairs may not be at once apparent to inferior ranks, it is necessary to prove to them the illusoriness of the material world and consequently the unreality of evil in it. In this imagined world, change is evil ; the ideal, the good, the true, and the beautiful are eternal and beyond question ; and as they are palpably not very prevalent on earth they must be sought for in a perfect heaven. This view has had a profound effect on the development of science, particularly in astronomy and physics, and even today, in more elaborate and sophisticated forms, there is again a strong tendency to enforce it on science.

The materialist view, partly because of its practical nature and even more because of its revolutionary implications, did not for centuries find much support in literate circles and rarely formed part of official philosophy. One expression of it, however, survives in Lucretius' Epicurean poem *De Rerum Natura* (on the nature of things), which shows both its power and its danger to established order. It is essentially a philosophy of objects and their movements, and explanation of Nature and society from below and not above. It emphasizes the inexhaustible stability of the ever-moving material world and man's power to change it by learning its rules. The classical materialists could go no further because, as we shall see, of their divorce from the manual arts ; nor could, in later days, the great reformulator of materialism, Francis Bacon. Once the Industrial Revolution was under way, science became in practice materialist, though continuing to give, for political and religious reasons, some lip service to idealism. Up to the middle of the nineteenth century materialism remained philosophically inadequate because it did not concern itself with society and its transformation, and was thus unable to account for politics and religion. The extension and transformation of materialism to include these was the work of Marx and his followers. First effective in the political and economic field, the new dialectical materialism is only now beginning to enter the sphere of the natural sciences.

The struggle between idealist and materialist tendencies in science has been a persistent feature in its history from earliest times. The idealism of Plato is in some sense an answer to the materialism of

Democritus, the founder of the atomic theory. In the Middle Ages, Roger Bacon attacked the prevailing Platonic-Aristotelian philosophy and preached a science aimed at practical utility and was imprisoned for his pains. In the great struggle of the Renaissance to create modern experimental science the prime enemy was formal Aristotelianism backed by the Church. The same opposition was to be found in the last century in the warfare between science and religion over Darwinian evolution. The very persistence of the struggle, despite the successive victories won by materialist science, shows that it is not essentially a philosophic or a scientific one, but a reflection of political struggles in scientific terms. At every stage idealist philosophy has been invoked to pretend that present discontents are illusory and to justify an existing state of affairs. At every stage materialist philosophy has relied on the practical test of reality and on the necessity of change.

To a section of readers, Bernal's way of connecting philosophical controversy not merely with theoretical issues *within* science but also with socio-economic factors apparently *outside* science may appear to be but a mark of his political preference and partisanism. It remains for us to see, however, whether the Indian situation really justifies it or not.[25]

For the present, let us return again to P. C. Ray and the tradition he wanted to set up.

9. M. N. SAHA

Among the "talented students" of the Presidency College who are indebted to P. C. Ray for their early initiation in modern science, M. N. Saha appears to be the most prominent one. He carried forward not only Ray's defence of science as a value in itself but also the superb insight of Ray into the relation of science and society. For this, Saha had to confront literally a barrage of attack from the champions of Vedic orthodoxy.[26] He was accused of showing only a slavish mentality in defending modern science flourishing in Europe, overlooking the fact that everything worthwhile in modern science is already to be found in the Veda-centric culture of ancient India, which in many ways was alleged to have been far ahead of modern "European" science for example, in developing the

25. I have, in my book SSAI, tried to show that this has also been the case in Indian history. We shall return to the point in the next volume of the present work.
26. See *Meghnad Racana Sankalan* (in Bengali) 117 ff.

caste system which was supposed to impart a kind of stability to society and hence enabled the Indians to evade the social turmoil of capitalist Europe. Saha had to waste much of his valuable time to clean up such rubbish.[27] We do not unfortunately have the scope here to quote the whole of it. But we shall mention here a few of his points, because of their bearing on what we are going to discuss later.

A point of exceeding importance raised and briefly discussed by Saha is that the champions of Veda-centric or Aryan culture somehow feel obliged to suppress or by-pass the recent archaeological work, which proves that while the Vedic people—apart from their literary work that often proves highly obscure for us today—left practically nothing worth mentioning as material achievements, there developed already at least a millenium before their coming to India an imposing civilization in the Indus valley which—at least according to a section of serious archaeologists—was finally destroyed by the invading Aryans. There are controversies, of course, about the decline and final destruction of the Indus Valley Civilization. Without entering into these, Saha raises a simple but significant question : Why do his opponents—the champions of the Vedas—prefer to remain silent about this glorious past of Indian civilization ? Is it simply because of the fact that the makers of the Indus Valley Civilization were pre-Aryans and pre-Vedic ?[28] It is understandable that P. C. Ray in his *History of Hindu Chemistry* did not raise this question, because the Indus Valley Civilization was discovered after its publication. But it is worth mentioning that outside the archaeologists, M. N. Saha is about the only notable scientist who, already in 1939, found it worthwhile to write a longish article in *Science and Culture* on "The Indus Valley 5000 years Ago".[29] It remains for us to see why all this is of material importance for our own understanding of the history of science and technology in Indian history.

27. *Ib*. 108-189. These writings still remain in Bengali and it seems highly desirable that these should be rendered into English for readers outside Bengal.
28. *Ib*. 128.
29. M. N. Saha in SC Vol. V No. 1 July 1939 and Vol. V No. 2, August, 1939.

No less important is Saha's observation on the causes that inhibited the development of modern science in India—an observation in which he carries forward the view already expressed by P. C. Ray. We quote him, though inevitably missing much of the force of his argument in our rough English translation. Refuting the alleged benefits of the caste system, Saha observes :

"But I have looked at the matter from a different angle. In my view the caste system has completely snapped the connecting link between the hand and the brain, and this is why the material culture of India is lagging far behind that of Europe and America. One who belongs to the intelligentsia is perpetually busy with bookish knowledge, commentaries, glosses and grammatical debates ; the ideal for the medieval Indian scholars was to create awe among the people with the extent of their learning. They had little connection with real life. They never cared for the development of industry and commerce, which perhaps carried the risk of being degraded in caste hierarchy".[30] By contrast, adds Saha, Hargreaves was an illiterate labourer, Arkwright a pennybarber, Cartwright a village clergy ; James Watt was a smith running a repairing-shop—it was because he came in contact with Professor Black of Glasgow University that he was able to invent steam engine.

As against all this, his orthodox critic—hitting as it were below the belt—argued that neither Megh Nad Saha nor Rabindranath Tagore is a craftsman ; will then the social status of a skilled shoemaker or weaver be higher than them ? Saha's answer to this is a bitter one. What he is arguing is a different point altogether. "Why", he asks, "should be the social prestige of an illiterate priest—who, without really knowing the meaning of the Sanskrit verses, make others mutter these during the marriage or *Sraddha* ritual—be higher than a weaver or shoemaker ? After all, the weaver or the shoe-maker serves the society with his labour ; but what else can you possibly say about the illiterate priest than a social cheat ?"[31]

Saha's passionate defence of manual work and his emphasis on the need of the unity of head and hand as an essential pre-

30. *Meghnad Racana Sankalan* (in Bengali) 132.
31. *Ib.* 133.

condition for scientific development cannot but be reminiscent of the teachings of P. C. Ray. So also in a sense, his way of looking back at the Indian philosophical tradition and its impact on science. Thus, for example, when Ray argued that the overwhelming influence of Samkara's Vedanta, with its attempted suppression of the atomic hypothesis, had only an inhibitory influence on the scientific activities in India, Saha wants us not to forget the fact that this Vedic philosophy had not been the only one in the general fund of Indian philosophical thought, evidently implying that the genuine national sentiment may as well look to other directions in philosophy that could better cater to the requirements of our national development. the focal point of which in Saha's view is of course the development of modern science and technology. Addressing his orthodox critic who insists on the exclusive glory of the Vedas, Saha asks : "Is he not aware of the fact that both Buddhism and Jainism, which ushered in the most glorious periods of Indian civilization have completely rejected the Vedas as but a bundle of errors ? Is he not aware of the fact that, according to the Lokayata view, 'These makers of the Vedas are but cunning cheats and thieves ?' All this means that sometimes before Christ, there was a group of rationalists who could realise that it was extremely difficult to comprehend the actual meaning of the Vedas ; only a few hypocrites propagate erroneous views with the alleged sanction of the Vedas. So to seek the roots of Hindu religion and philosophy only in the Vedas is about ninety per cent erroneous and it is this error that makes the essay of the critic full of mistakes".[32]

10. SCIENTIFIC METHOD

To quote the authority of the Lokayata philosophy with approval tacit or otherwise—is perhaps the limit of heresy from the orthodox point of view.[33] Nevertheless, from the standpoint

32. *Ib.* 130.
33. How the Indian philosophical literature is almost saturated with the contempt for Lokayata is too well-known to be mentioned here. See D. Chattopadhyaya *Lokayata : A Study in Ancient Indian Materialism.* Incidentally, in a Bengali article written by P. C. Ray also there seems to be a tacit approval of the Lokayata view : reprinted in *Utsa Manus* (Bengali), Sept. 1984, pp. 229-30.

of science there appears to be some justification for it. Among the philosophical views of traditional India, Lokayata is about the only one that puts an uncompromising emphasis on direct observations as the primary way of knowing, so much so that the other philosophers usually depict it as accepting immediate perception as the only source of right knowledge and therefore reject even inference as a way of knowing—a position that creates obvious difficulties in philosophical investigation. It remains for us to see whether the opponents of Lokayata are justified in imputing to it such an exclusive emphasis on direct observation, i.e. to the exclusion of the validity of worldly or normal inference, for there are enough indications in the Indian philosophical literature to think that the representatives of Lokayata were only insisting on the primacy of perception and were prepared to accept the validity of inference, though only in so far as it was based on perception.[34]

For the present, however, our point is a different one. To the scientists and also to the historians of science in India, there is obvious reason to look back at the Lokayata emphasis on the primacy of perception or direct observation with some enthusiasm, specially in the general intellectual climate of a country characterised by an almost suffocating scriputure-mongering. In direct contrast to this tendency of accepting the scriptures as embodying ultimate wisdom, science demands that the starting point of its method is the interrogation of nature by direct observation. This is why, as we have already seen, P.C. Ray himself enthusiastically quoted two alchemists—Ramacandra and Yasodhara—because of their claim of putting very strong emphasis on direct observation, adding : "Experiments and observations constitute the fundamental bases of sciences."

All this leads us to consider the question of scientific method in Indian tradition. From P.C. Ray's *Autobiography*[35] we learn that while working on the *History of Hindu Chemistry* he felt the need of the cooperation of B. N. Seal (1864-1938) : certain basic questions pertaining to the history of science in India remain embodied in texts that are viewed as philosophical ones in the restricted sense. In spite of a very wide range of his

34. D. Chattopadhyaya *Lokayata* 22 ff.
35. P. C. Ray *Auto.* 132.

own studies, these philosophical texts are often too technical to be tackled by Ray himself, specially because of his own multifarious activities. During his time, B.N. Seal had the reputation of an encyclopaedic mind, specially in matters concerning the Indian philosophical texts. It was, therefore, only natural for Ray to seek Seal's cooperation and get two chapters to contribute to Ray's *History*. Seal wrote these two chapters, which first appeared as incorporated in Ray's *History* and seem to have formed the starting point of Seal's own book, which was eventually published with the title *The Positive Sciences of the Ancient Hindus* (London 1915). As B.N. Seal put it in his foreword to it : "The chapter on the 'Mechanical, Physical and Chemical theories of the Ancient Hindus' appeared in P. C. Ray's *Hindu Chemistry* and that on the 'Scientific Method of the Hindus' as an appendix to the same work".[36]

We have mentioned this specially to emphasise one point. Ray evidently felt that a history of science, in order to be scientific, was in need of a survey of the scientific method. It remains for us to see that Seal's own understanding of the methodology of science—like his reading of various aspects of scientific development in ancient India—seems to be in need of serious revision, notwithstanding its historical importance as the first attempt at a comprehensive reconstruction of the scientific method from the Indian sources. Before we pass on to it, however, there remains for us to note another point, which appears to be rather sad, if not a positive misfortune.

11. IDEOLOGICAL RETREAT

Thanks to the initiative of S.S. Bhatnagar—whom Ray proudly calls "my chemical grand pupil"[37]—and with the active cooperation of J.C. Ghosh and J.N. Mukherjee, there came into being in 1924, The Indian Chemical Society, with P. C. Ray as its first president.[38] In view of the fact that Ray's *History of Hindu Chemistry* has become somewhat dated in 1948 the Indian Chemical Society resolved to publish "a revised edition of the book". "It was further resolved that the new publication should

36. B. N. Seal PSAH Preface iii.
37. P. C. Ray *Auto.* 130, also note.
38. *Ib.* 152-53 & 157-8.

incorporate all important additional materials that had since been brought to light, and that its name should consequently be changed to *History of Chemistry in Ancient and Medieval India*".[39] The task of the preparation of this new version was assigned to Priyadaranjan Ray—a direct disciple of P.C. Ray, about whom P. C. Ray himself expressed high hopes as a young chemist with much promise.[40] As edited by him this new book appeared in 1956, claiming in its title page to "incorporate" P. C. Ray's *History of Hindu Chemistry*.

It would be wrong, of course, to take a totally negative attitude to Priyadaranjan Ray's new version of the history of chemistry in India. When, for example, P.C. Ray wrote his *History*, nothing was known about the imposing Indus Valley Civilization, the material remains of which are being unearthed only from the third decade of the 20th century. These are indicative of high technological achievements, inclusive of some that are significant chemical operations. Descriptive accounts of some of these are rightly given in the new book. So also are incorporated in it some alchemical texts which P.C. Ray could not consult. Specially interesting among these are the works that are irrevocably lost in the Indian originals and survive only in Tibetan version.[41]

However, such important additions notwithstanding, the new book gives us the sad impression of some kind of ideological retreat—retreat specially from the scientific temper on which P.C. Ray puts so much emphasis. It is, for example, strange to see in this book allegedly incorporating P.C. Ray's *History*, the entire chapter on Scientific Method by B.N. Seal summarily scrapped. Even some of P.C. Ray's own observations on the methodology of the alchemists are just brushed aside. Of the two alchemists so enthusiastically quoted by P.C. Ray for their scientific temper, Ramacandra does not receive even a cursory mention in the new book. The other alchemist quoted by P.C. Ray, Yasodhara, has some place in the new book. We find him being mentioned five times (and even his *Rasaprakasa-sudhakara* is given in extracts), but *nowhere is mentioned the*

39. P. Ray HCAMI. Foreword p. E.
40. P. C. Ray *Auto.* 155.
41. P. Ray HCAMI 449 ff.

strong emphasis he put on direct observation and experiment or, according to the alchemists' way of putting the latter, on personally performing an operation as a precondition for accepting a chemical or proto-chemical proposition.

The new book in fact shows some worse features from the viewpoint of the scientific temper. In the general fund of Indian philosophical thought, there are views going strongly against the requirements of science. P.C. Ray found the most prominent of these in the world-denying philosophy (*maya-vada*) of Samkara's Vedanta. Hence he sharply came out against it, arguing that the great prestige and popularity of this philosophy was an important factor accounting for the decline of the scientific spirit in India. Peculiarly enough, Priyadaranjan Ray, in his allegedly revised version of P.C. Ray's *History,* chooses to expunge the entire comment of P.C. Ray on the adverse effect of Samkara's Vedanta on the development of science in India. That is hardly the way of editing a book—specially one that has acquired the status of some kind of classic among the recent works on Indian science.

All this, however, seems to be a matter more serious than that of mere editorial norm. The conscious suppression of P.C. Ray's observations of Samkara's Vedanta seems to be indicative of an ideological retreat from the bold defence of science and the scientific temper of P.C. Ray. Following Samkara's own claim, this philosophy is proclaimed in the country for centuries as having the highest scriptural sanction. Many of our contemporary philosophers continue to be under its spell, producing tons of books in defence of the philosophy. To be critical of it amounts to many as some kind of sacrilege, if not a sin.

12. HINDU REVIVALISM

We can thus see where exactly is the rub. If, in defence of science, P.C. Ray found it necessary to come sharply against the world-denying philosophy of Samkara, the editor of his book seems to seek safety in maintaining a discrete silence about the whole thing. The result is that, though giving us a somewhat enlarged account of the chemical and proto-chemical techniques in ancient and medieval India, Priyadaranjan Ray prefers to avoid the troublesome question of the scripture-oriented philosophy coming in open confrontation with the basic requirements of natural science. Does it not amount to aiding and abetting Hindu revivalism, which P.C. Ray himself considered fatal for the country? In his *Autobiography*, he found it necessary to write a special section on "The Revival of Hindu Orthodoxy Fatal to the Progress of India".[42] We quote an extract from it:

> Our excellent friends, the Hindu revivalists, will talk by the hour on the transcendental truths and sublime thoughts in the *Gita* and deliver elaborate discourses on the catholicity of Hinduism and its superiority over all other existing creeds; will condemn untouchability in unmeasured terms and so on. When, however, it comes to carrying the precepts into practice, they are the first to show the white feathers; as Professor Wadia puts it :—
>
> 'Our Vedantist will flood you with quotations to show how catholic Hinduism is, but woe to the man who dared to take this seriously and ever acted upon them. Quotations are for show, not for action. In fact, I believe so much precious time has been wasted in proving that Hinduism is cosmopolitan, is catholic, that no time has been spared for the practice of it. Fear leads to repressions and without its conquest no man can find himself or rise to his full stature'—*Presidential Address at the Philosophical Congress* (*Dec. 1930*).
>
> No wonder that in spite of the empty vapourings of the Hindu Sabhas and Sangathans, conversion to Islam goes on day by day as ever. And why should it not? Islam knows no distinction of race, colour, or creed as far as social matters are concerned. Untouchability it is a stranger to; it is a 'perfect equaliser of men', as Carlyle has it..... No wonder our friends the *Namasudras* are tired of the hypocritical assertions of the Hindu leaders and are sometimes eager to seek shelter in the bosom of Islam.[43]

Whether Islamic fundamentalism is having in the world today an adverse effect on the progress of science, is, of course, a different question. Fundamentalism in any form is indeed anti-science, and in any case what Ray referred here was to an altogether different aspect of the message of Islam. What proved specially difficult for the development of science in India in his time was Hindu revivalism. That is evidently one of the reasons why he put special emphasis on the need of scientific method which could be an effective antidote to religious revivalism in all forms.

42. P. C. Ray *Auto.* 434.
43. *Ib.*

13. B. N. SEAL ON SCIENTIFIC METHOD

As already mentioned, P.C. Ray considered it desirable to depend on B.N. Seal for a comprehensive survey of the scientific method in Indian tradition and we are not aware of any other scholar before Seal attempting to treat the subject seriously. But this pioneering study in the scientific method seems itself to be in need of a scientific scrutiny.

B.N. Seal opened his own discussion of the Scientific Method in Indian thought with a strong defence of it, and, what is more important, he wanted to draw our attention to the fact that, according to some Indian thinkers at any rate, the main point about it is that practice is the criterion of truth. As he put it :

A study of the Hindu Methodology of Science is absolutely essential to a right understanding of Hindu positive Science, its strength and its weakness, its range and its limitations. Apart from this rigorous scientific method, Hindu Chemistry, for example, would be all practical recipe, or all unverified speculation. This, however, would be a very inadequate and indeed erroneous view of this early achievement of the human mind. That the whole movement was genuinely and positively scientific, though arrested at an early stage will appear from the following brief synopsis of the Hindu Methodology of Science.

Criterion or Test of Truth, after the Buddhists :—The ultimate criterion of Truth is found, not in mere cognitive presentation, but in the correspondence between the cognitive and the practical activity of the Self, which together are supposed to form the circuit of consciousness. That knowledge is valid which prompts an activity ending in fruition. (Cf. the distinction between *samvadi* and *visamvadi jnana*. Also compare *pravrtti-samarthyat arthavat pramanam*—Vatsyayana). Truth, the Buddhists contend, is not self-evidence, not the agreement between ideas, not the agreement of the idea with the reality beyond, if any, for this cannot be attained direct, but the harmony of experience (*samvada*), which is implied when the volitional reaction, that is prompted by a cognition and that completes the circuit of consciousness, meets with fruition, i.e. realises its immediate end (with this compare Sriharsa, *Khandana Khanda Khadya,* on the relation of *prama* to *loka-vyavahara*). This is the material aspect of Truth. The formal aspect is given in a principle which governs all presentations in consciousness, and which combines the three moments of Identity, non-Contradiction, and Excluded Middle in every individual cognitive operation [*taduktam tat paricchinatti* (identity), *anyad vyavacchinatti* (non-contradiction), *trtiya-prakarabhavam ca sucayati* (excluded middle) *iti ekapramana-vyaparah*—Jayanta, *Nyayamanjari, pramana-dvaividhya-khandanam*].[44]

We have quoted all this for the purpose of focussing on one point. At least as understood by the earlier generation of our scientists and thinkers like P.C. Ray and B.N. Seal, any attempt to write a history of science in India along with a total neglect of the scientific method cannot be legitimate. That, in other words, would be following a wrong model for the history of science. From this point of view, therefore, it is unfortunate that such a false model has become the dominant one for the later writers on the history of science in India. An example of this is the imposing volume sponsored by the Indian National Science Academy and published under the title *A Concise History of Science in India.*[45] It professes to give us an impressive account of the achievements of the ancient Indians in mathematics, astronomy, medicine, and so on, without anywhere seriously raising the question concerning the scientific way of knowing or of the methodology followed by the ancient Indians for moving in the field of science proper. The result is worse than indiscriminately jumbling up technology with science. The indifference to scientific method allows even patently a-scientific and anti-scientific ideas and attitudes to enjoy the same status as that of science, as is evident, for example, in the tendency to read in this book science in all sorts of predominantly religious texts.[46] Such a danger could have been easily avoided if only the editors and contributors to this book cared to remember the elementary point already emphasised by P.C. Ray and B.N. Seal that there is no real science without the methodology of science.

14. LIMITATIONS OF B. N. SEAL

But let us return to B.N. Seal's study of the scientific method in ancient India. As against the prevailing but eclectic tendency

44. B. N. Seal PSAH 244-45.
45. Practically the same model was followed by Th. Stcherbatsky in "Scientific Achievements of Ancient India" (See SHSIA 3-22) and B. K. Sarkar in his *Hindu Achievements in the Exact Sciences,* New York and Calcutta 1918.
46. It needs to be acknowledged, however, that as a compilation of useful information about scientific activities in India the volume continues to have its importance.

to discuss the history of science in India without taking note of the scientific method, the importance of the main point on which he put special emphasis can hardly be exaggerated. So also is the significance of many of his observations on the scientific method in ancient India, which are perhaps only to be expected of the encyclopaedic erudition of B.N. Seal. However, specially for the purpose of our understanding of the problem of the *emergence of science* in ancient India, it may be useful for us to begin with a few points which, though extremely relevant to the methodology of science, escaped his attention.

Here is a preliminary enumeration of some of these :

First, the scientific method—like science itself—is a growing process, and hence it would be wrong to expect it as fully formulated at any stage of the history of ideas.

Secondly, the commitments to the scientific method—like the commitment to science itself—is often a matter of active struggle, because there is also some such thing as anti-science with its own methodological prerequisites.

Thirdly, the struggle between the scientific method and its opposite, though manifesting itself mainly at the level of theoretical considerations, is ultimately rooted in concrete social conditions, without noting which even the most erudite compilation of bare textual data runs the risk of petering out in puerile pedantry.

Let us begin with some clarification of the first of these points. Scientific method does not mean any royal road to Scientific Truth, which is discovered once for all, leaving the scientists only with the task of properly following it. As J.D. Bernal[47] has very lucidly put the point :

> There is a danger of considering it [i.e. the scientific method] as a kind of ideal Platonic form, as if there were one proper way of finding the Truth about Nature or Man, and the scientists' only task was to find this way and abide in it. Such an absolute conception is belied by the whole history of science, with its continual development of a multiplicity of new methods. The method of science is not a fixed thing, it is a growing process. Nor can it be considered without bringing out its closer relations with the social, and particularly the class, character of science. Consequently scientific method,

47. Bernal SH 35.

like science itself, defies definition. It is made up of a number of operations, some mental, some manual. Each of these in its time, has been found useful, first in the formulation of questions that seem urgent at any stage and then in the finding, testing and using the answers to them.

In view of the refinements and ramifications of the methodology in the subsequent course of the history of science, it may be useful to add to the above another point. Specially for understanding science in the ancient world—though also partly for disentangling scientific method from the dregs of some metaphysical views with which science is intertwined in the writings of many modern writers—it is useful and in a sense even crucial—to analyse the first step taken from pre-science to conscious science. When we look back at it and wrongly judge it in the standard of modern knowledge, we may have the impression of this having been a trivial one. But such an impression would be most misleading. Historically speaking, the first step to science is indeed the most prodigious one, for it requires the profoundest transformation in the totality of ideas and attitudes.

With all his emphatic assertion of the importance of the scientific method, B. N. Seal does not take note of these points. For his own understanding of the scientific method in ancient India, he began straightway with a class of texts—inclusive of much later ones—which, in default of a better descriptive epithet, are usually called texts on "logic". Some of these were written by philosophers belonging to the Nyaya school, some others by a section of philosophers professing the Buddhist creed.[48] Depending mainly on these texts, B. N. Seal proposed to construct his own account of the scientific method in ancient India.

There was some real ground no doubt for his procedure. The writings of these "logicians" are indeed for us the most outstanding works discussing many questions considered essen-

48. Seal's somewhat omnibus use of the word "Buddhist" cannot but be confusing. Among the numerous Buddhist sects, only the one represented by Dignaga and his followers took a positive interest in "logic" and "scientific method", though for this purpose they had to suspend, as it were, their enthusiasm for extreme subjective idealism. See, D. Chattopadhyaya WLWDIP 56 ff.

tial for the methodology of science in the nineteenth century Europe. Such, for example, was the question concerning the establishment of materially true universal real propositions which, in its turn, called for a good deal of discussion about the technique of determining the causal connection. The Indian philosophers went into much details of all these which, B. N. Seal energetically argued, made their position much more advanced than that of the bare formal logic of Aristotle and in fact anticipated the Inductive Methods of J. S. Mill.

15. SCIENTIFIC METHOD AND THE WORKING SCIENTISTS

However, admitting all that is really substantial in B. N. Seal's discussion, he leads us to an apparently peculiar situation. Notwithstanding all their interest in epistemology, the writers on whom Seal depended were philosophers after all. There is nothing improbable, of course, about some philosophers taking interest in the scientific method, as many of the contemporary philosophers are ostensibly doing. What is odd, however, is to think that the methodology of science could develop *totally outside the circle of the working scientists* or irrespective of the experience of the scientists themselves. How, then, could the scientists themselves possibly work? Could they work without the need of any scientific method—a possibility that makes the very concept of scientific method a misnomer? Alternatively, we have to imagine that the scientists had first of all to learn their methodology from the philosophers before engaging themselves to their own research. Evidently, there can be no ground to take such a possibility seriously. Besides, it is not easily conceivable that the philosophers could be taking an interest in the methodology of science already *before* the emergence of science or without there being a general theoretical climate created by the development of science at least upto a certain stage. Evidently again, the methodology of science cannot be the product of pure reason or sheer speculative considerations, which the philosophers are supposed to be left with in default of the development of natural science. In short, one can hardly be serious about the very concept of the scientific method, ignoring at the same time the possibility of the working scientists participating in its making.

The obvious way out of this difficulty is to think that the emergence of science took place in ancient India already *before* the "logicians". This means that the real pioneers of science had an understanding of their own methodology, though it was left for the "logicians" to elucidate, elaborate and systematise the theoretical implications of this methodology. It remains for us to see how far historically speaking such could possibly be the real situation. The very structure of the earliest work on Indian "logic" (*Nyaya-sutra*) seems to presuppose a theoretical climate in which natural science was developed enough to knock, as it were, for new doors to be opened in philosophy for more adequate understanding of the scientific method.[49] But B. N. Seal did not concede to this possibility. Therefore, with all his emphatic assertion that the scientific activities in ancient India cannot be understood without the appreciation of the scientific method, he was virtually led to the assumption that the scientific method somehow or other developed *before* natural science among the philosophers, from whom the natural scientists borrowed it. Let us see how he committed himself to such a position.

Though nowhere facing the historical question of the actual formation of the scientific method, B. N. Seal was obviously aware of the fact that the *Caraka-samhita*—which comes down to us as the earliest compilation of medical science in India—contains a fairly elaborate discussion of the scientific method. How is this to be explained? Does it mean that the ancient Indian physicians—in the capacity of working scientists—developed a methodology that suited their purpose? Or, does it mean that they borrowed it from some other quarter? B. N. Seal argued in favour of the second possibility. To reconcile the evidence of the *Caraka-samhita* with his preconceived notion that the scientific method was actually worked out by the philosophers or logicians, he wanted us to believe that the physicians themselves found this method already worked out, awaiting only to be *applied* to their own field of work. He introduced two terms to explain this: *Logic* and *Applied logic*. The first meant the scientific method in its most general

49. S. N. Dasgupta's hypothesis that the Nyaya logic might have originated from the methodology of ancient Indian medicine may give us a useful clue to the primary stock of empirical logic presupposed by Nyaya.

sense or having a universal applicability. The second meant the same method as used by a section of scientists to meet their specific requirements. With this terminological innovation, B. N. Seal argued[50] :

I will conclude with a few observations on Applied Logic, i.e. the logic of the special sciences which is such a characteristic feature of Hindu Scientific investigation. What is characteristic of the Hindu scientific mind is that, without being content with the general concepts of science and a general Methodology, it elaborated the fundamental categories and concepts of such of the special sciences as it cultivated with assiduity, and systematically adopted the general principles of Scientific Method to the requirements of the subject-matter in each case. The most signal example of applied logic (or Scientific Method) worked out with systematic carefulness is the Logic of Therapeutics in Caraka, a logic which adopts the general concepts of cause, effect, energy, operation, etc. and the general methodology of science, to the special problems presented in the study of diseases, their causes, symptoms and remedies.

The argument is peculiar indeed, for it amounts to the claim that some philosophers in ancient India could somehow discover some royal road to Truth without being bothered by the rigor of systematic investigation of any department of nature. Even admitting this, what could be the justification of equating it to the methodology of science?

From the standpoint of natural science, B. N. Seal in fact wanted us to believe in something more strange than this. Assuming, as he did, that the scientific method was the product of the philosophers' brain, he had no hesitation in viewing it as some kind of universal solvent of all problems, i.e. not merely of the problems of natural science but even that of attaining the highest metaphysical illumination. As he put it[51] :

This doctrine of Scientific Method, in Hindu Logic, is only a subsidiary discipline, being comprehended under the wider conception of Methodology, which aims at the ascertainment of Truth whether scientific (*vijnana*) or philosophical (*jnana*) ; the latter being the ulterior aim......Now the various *Pramana*-s, proofs, i.e. sources of valid knowledge, in Hindu Logic, viz., Perception, Inference, Testimony, Mathematical Reasoning (*Sambhava,* including Probability in one

50. B. N. Seal PSAH 290-1.
51. *Ib.* 289-90.

view), are only operations subsidiary to the ascertainment of Truth (*tattvanirnaya*). And the Scientific Methods are merely ancillary to these *pramana*-s themselves.

All this appears to be worse than considering scientific method "as a kind of ideal Platonic form", against the possible danger of which we have already quoted the warning of J. D. Bernal.

16. S. N. DASGUPTA

But this does not mean that others among the recent interpreters of Indian philosophical tradition simply echoed Seal's views. In the second volume of *A History of Indian Philosophy,* S. N. Dasgupta proposed a hypothesis which in fact amounts to the very opposite of that of B. N. Seal. While, according to the latter, the original architects of the scientific method in India were the logicians from whom the physicians borrowed it, Dasgupta raised the rather unusual question: "Did logic originate in the discussions of the Ayurveda physicians ?"[52] The answer to this, according to him, is presumably in the affirmative. Before passing on to quote him at some length, let us have a few preliminary clarifications.

The earliest extant work on Indian "logic" is the *Nyaya-sutra.* It is attributed to a certain Aksapada (also called Caranaksa by Samkara), literally "One with eyes on the feet". About him—some legends apart—we known practically nothing. Even the very name Aksapada could be a sarcastic innovation[53] of the pure spiritualists, because of his essentially mundane interest: one with eyes on the feet could hardly have any vision of the lofty metaphysical wisdom. Nor do we know the exact date of the *Nyaya-sutra,* which is usually conjectured to be roughly the second century A.D.[54] Though the modern scholars have various views about the history of the formation

52. S. N. Dasgupta HIP ii. 373 ff.
53. D. Chattopadhyaya *Bharatiya-darsana* (in Bengali) 25.
54. It must be remembered, however, that the actual formation of the Nyaya logic must have been much earlier than its codification in the form of the *Nyaya-sutra.* Besides, the *Nyaya-sutra,* in the form in which it reaches us, is full of later grafts—inclusive of grafts of palpably alien ideas. See D. Chattopadhyaya, Introduction to *"Indian Philosophy in its Sources, Vol. I, Nyaya",* Calcutta 1982.

of the extant *Nyaya-sutra,* the fact remains that we have no positive or definite knowledge about the origin of Nyaya logic. S. N. Dasgupta, however, points to certain exceedingly interesting internal evidences of the Nyaya literature, the possible implication of which cannot be overlooked. As he put it, "Incidentally it may be mentioned that Jayanta, in his *Nyaya-manjari,* discussing the probable sources from which Aksapada drew his materials, suggests that he probably elaborated his work from what he may have gathered from some other science (*sastrantarabhyasat*) ; but it is difficult to say whether by *sastrantara* Jayanta meant Ayur-veda. The *Nyaya-sutra,* however, expressly justifies the validity of the Vedas on the analogy of the validity of Ayur-veda".[55]

Though belonging to a somewhat late period (*c.* A.D. 900), Jayanta is regarded as a highly authoritative exponent of the Nyaya view. His suggestion that Aksapada might have received the fundamentals of the Nyaya view from some other science (*sastrantara*) cannot be easily dismissed, specially when we remember that the very structure of Nyaya logic presupposes the accumulation of a good deal of empirical knowledge and can by no means be the product of pure reason. In ancient India, as we shall later see more fully, medicine or Ayurveda was about the only science that showed the most absorbing interest in a vast amount of empirical knowledge. Thus the possibility cannot be easily dismissed that Jayanta might have had Ayurveda in mind when he said that Aksapada could draw his basic materials from *sastrantara* or "other science". All this seems to be all the more plausible, because we read in the *Caraka-samhita* extensive discussions concerning the problems of scientific method—"logic" or "proto-logic".

But more of this later. For the present we have mentioned all this only to emphasise the point that one of the foremost historians of Indian philosophy suggests the possibility of the scientific method in India developing among working scientists rather than among the metaphysicians searching for a royal road to ultimate truth.

On this point, therefore, B. N. Seal's exposition of "The Scientific Method of the Hindus" appears to be in need of serious amendment, notwithstanding the circumstance of Seal

55. S. N. Dasgupta HIP ii. 399.

having been the pioneer in discussing it in modern times and in putting very strong emphasis on its importance for our understanding of the history of science in India. Indeed, he tacitly assumed that the Scientific Method of the Indian thinkers was something more than was needed to meet the requirement of science, inasmuch as it was, in his view, some royal road to absolute Truth. As we have already quoted him claiming, "The doctrine of Scientific Method, in Hindu Logic, is only a subsidiary discipline, being comprehended under the wider conception of Methodology which aims at the ascertainment of Truth, whether scientific (*Vijnana*) or philosophical (*Jnana*)." Such an absolutistic or quasi-absolutistic view of the Scientific Method creates a number of difficulties in the understanding of the history of science, some of which may be noted here.

If the ancient Indians did in fact develop a method for ascertaining Truth (with a capital T), there remained little or nothing for them to learn from the scientific activities of the other nations, i.e. they remained cut off from the mainstream of international science to which various nations in various periods did contribute and from which all nations are required to—and, as a matter of fact, historically did—acquire elements for development and nourishment. In other words, Seal's understanding of Scientific Method in India carried on its heels the risk of making Indian science a rather closed body of Truth and thus lose the perspective of the current of international science.

17. MAINSTREAM OF GLOBAL SCIENCE

From this point of view, Seal seemed to have overlooked P. C. Ray's position who, in his own way, visualised the mainstream of global science into which the contributions of various countries flowed in various periods. Here are some of his observations, from which our chauvinists have much to learn.[56]

> That vigorous and robust thinking which characterised the days when the six systems of Hindu philosophy had been elaborated and which has very aptly been styled the Rationalistic Age had been for ages a thing of the past. Hindu intellect came to be under the domination of scholastic philosophy and revelled in the dialectics of

56. P. C. Ray *Auto.* 114-17.

the schoolmen and a sort of learning was in vogue under which, to quote the happy words of Buckle, the more learned the votaries were the more ignorant they grew......

In the history of nations it is often found that contact of one civilisation with another brings about strange and on the whole beneficial results. Proud Rome did not disdain to learn at the feet of vanquished Greece. Alexandria [was] favoured by her position as the meeting-place of the Eastern and Western nations ..... besides the exchange of commercial products there was also an interchange of thoughts.

Borrowing does not always mean slavish imitation or lack of originality. As Emerson says: 'The greatest genius is the most indebted man'......

The development and enrichment of Arabic literature may be cited here as a notable illustration. The orthodox and Umayyad Caliphates are from the intellectual point of view barren. It was however during the Abbasid rule that the many sided life of the Moslems found full expression and vigour in a copious literature which was enriched by wholesale borrowing from Greece. Under Caliphs Mansur and Mamun, Hellenic culture found full scope. The works of Aristotle, Plato, Galen, Ptolemy as also of the neo-platonists Plotinus and Porphyry were translated often from the Syraic versions as also direct from the Greek text. Among the Falasifa school (i.e. those who read in the original Greek) the names of Al-Kindi, Al Forabi, Ibn Sina (Avicenna), Al Razi (Rhazes) as also the Spanish philosopher Ibn Rhshd (Averroes) who flourished in the 12th Century A.D., stand conspicuous.

'This material expansion (in trade) was accompanied by an outburst of intellectual activity such as the East had never witnessed before. ...In quest of knowledge men travelled over three continents and returned home, like bees laden with honey, to impart the precious stores which they had accumulated to crowds of eager disciples, and to compile with incredible industry those works of encyclopaedic range and erudition from which modern science, in the widest sense of the word, has derived far more than is generally supposed'. The contributions of the Arabs in the domain of philosophy and science in the middle ages need not be dwelt upon here, nor it is necessary to mention that in mathematics and medicine they are deeply in debt to India.

The Arabs in their palmy days were in turn the bringers of light to mediaeval Europe and wielded an enormous influence on Latin scholasticism. A separate chapter may be written on the reciprocity of intellectual debt between Asia and Europe.

It remained for M. N. Saha, the able pupil of P. C. Ray, to carry forward the main argument of Ray. As against the orthodox chauvinists claiming to read everything worthwhile in contemporary science in the Indian scriptures—specially the Vedas—Saha shows how much of the scientific activities in India was actually enriched by those of abroad. One way Saha shows this is by concentrating on the field of his own specialization, namely astronomy. Remaining fully aware of the achievements of the Indian astronomers specially as judged in their own historical context, Saha shows that from the time of the Sakas (*c.* A.D. 100) to that of *pandit* Jagannath, the court astronomer of Jay Singh in the 17th century A.D., there was repeated effort in India to improve astronomical knowledge by way of absorbing the best results in the field achieved abroad. We shall see some detail of all this in our own discussion of the development of astronomy in ancient India.

18. EUROPE-CENTRISM

If on the authority mainly of P. C. Ray and M. N. Saha we have so far tried to argue how wrong are the Indian chauvinists refusing to see the history of science as an international stream to which different nations contributed in different periods, let us not forget the other side of the picture, namely the somewhat arrogant claim of most of the European historians of science, namely that science itself is an essentially European phenomenon. We begin with a brief outline of their position.

With the first foreshadowing of modern science in Europe in the sixteenth century, there also grew the tendency of working backward and trace the beginnings of scientific thought to the achievements of Mediterranean antiquity. This direct linking of modern science with the ancient Greek tradition is increasingly utilised by a flourishing literature in the cause of a rather simplified understanding of the history of global science.

The story, we are told, began with some kind of Greek "miracle" resulting in the dawn of science. Not that the facts of the older civilisations—of Egypt and Mesopotamia, and, in recent years, also of the Indus—are denied. But these are mentioned cursorily and mainly for the purpose of showing why there could be no real science before the Greeks. As Arnold Reymond[57] says, "Compared to the empirical and frag-

57. A Reymond, *Science in Greco-Roman Antiquity*. Quoted by B. Farrington, GS 16.

mentary knowledge which the peoples of the East had laboriously gathered together during long centuries, Greek science constitutes a veritable miracle."

After an exciting career of about seven hundred years in Greece, Alexandria and Greco-Roman world, science is said to have suffered an eclipse, resulting in the darkness of the middle ages. The darkness prevailed over a thousand years. Then there was the illumination of Renaissance Europe, when the old Greek tradition was taken up again and conditions were created for the rise of modern science, with an ever-increasing rate of progress since then.

This, in brief, is supposed to be the history of science. What is true in it is, of course, often emphasised. What is fallacious about it is discussed only by a minority of conscientious scholars. The fallacy in short lies in the tacit equation of global science with science in European history. We have the scope here to discuss it mainly in so far as it has a bearing on the need of understanding the contributions of the Asian countries to the mainstream of science, which are usually ignored or at best desultorily mentioned.

That such an understanding of the history of science is not sufficiently scientific becomes obvious when we consider only one point. Even for the restricted purpose of understanding scientific developments in Europe, it is essential to take note of the contributions of the Asian countries. Let us see why.

19. APPARENT ANOMALY

To begin with, there is something apparently odd about the story as usually told. If the Greek tradition had so much to bequeath to modern science, how was it that among the Greeks and Romans themselves it suffered a creeping paralysis? The question becomes all the more perplexing when we remember that the essential intellectual tools for the making of modern science were worked out by the Greeks long before Galen who died in A. D. 199. As Farrington[58] sums up: "Before the end of the third century B.C. Theophrastus, Strato, Herophilus and Erasistratus, Ctesibius and Archimedes had done their work. In the Lyceum and the Museum the prose-

58. B. Farrington, GS 302.

cution of research had reached a high degree of efficiency. The capacity to organise knowledge logically was great. The range of positive information was impressive, the rate of its acquisition more impressive still. The theory of experiment had been grasped. Applications of science to various ingenious mechanisms were not lacking. It was not then only with Ptolemy and Galen that the ancients stood on the threshold of the modern world. By that late date they had already been loitering on the threshold for four hundred years. They had indeed demonstrated conclusively their inability to cross it."

20. SLAVERY AND THE DECLINE OF GREEK SCIENCE

Why, then, was this inability? We shall quote Farrington again, who has answered the question in his masterly survey of *Greek Science* : "The failure was a social one and the remedy lay in public policies that were beyond the grasp of the age. The ancients rigorously organised the logical aspects of science, lifted them out of the body of technical activity in which they had grown or in which they should have found their application, and set them apart from the world of practice and above it. This mischievous separation of the logic from the practice of science was the result of the universal cleavage of society into freeman and slave. This was not good either for practice or for theory. As Francis Bacon put it, if you make a vestal virgin of science you must not expect her to bear fruit." [59]

When the muscles of the slaves were the only recognised source of power, science became increasingly irrelevant as a means of transforming the conditions of life. It became a relaxation and a pastime for a handful of social parasites. Cut off from the actual process of interrogating nature, the tools and implements for which were left exclusively to the slaves with no social status at all, science ceased to be a knowledge of nature and a power over it. Such was the blind alley into which science was pushed in the Greco-Roman world. Nothing short of a social revolution could rescue it. Between A.D. 400 and 800 the revolution took place, though as the work of the northern barbarians. "Even if in the end," says Engels, "we find almost the same main classes as in the beginning,

59. *Ibid.* 303.

still, the people who constituted these classes had changed. The ancient slavery had disappeared ; gone were also the beggared poor freemen, who had despised work as slavish. Between the Roman *colonus* and the new serf there had been the free Frankish peasant. The 'useless reminiscences and vain strife' of doomed Romanism were dead and buried. The social classes of the ninth century had taken shape not in the bog of a declining civilisation, but in the travail of a new."[60]

21. THE MIDDLE AGES

It is not the place for us to go into the detail of the transition from slavery to feudalism, or, what is more important for the history of modern science, from feudalism to capitalism. But it is of material importance to note the slow operation of historical process that allowed the intellectual movements of the Middle Ages to bring modern science into being in the European cultural area. The most important clue to this is to be sought in the series of technological innovations witnessed by Europe roughly from the ninth century, which gradually transformed the economic basis of society. Quoting an inventory of these (from the IXth century harness of the saddle-horse down to the XVth century printing) prepared by Des Noettes, Farrington observes :

> In another of his writings, a masterpiece of research and of historical analysis, Des Noettes discusses the social consequences of this series of inventions. He is not wrong when he insists that 'by fundamentally transforming the means of production they fundamentally transformed the social organism'. Nor is his conclusion lessened in importance when we understand that one of the transformations of the social organism involved was the disappearance of the last vestige of slavery and the possibility of undertaking immense constructional works with free labour—works of a kind which had normally been performed in antiquity by the forced labour of slaves. This implied an immense improvement in the consciousness of the modern world over the ancient.[61]

22. THE RENAISSANCE

It was with this improved consciousness that Renaissance Europe looked back at Greek science and tried to understand its message, which was lost to the Graeco-Roman world itself.

60. F. Engels, OF 253-54.
61. B. Farrington, GS 306-7.

Graeco-Roman science was good seed, but it could not grow on the stony ground of ancient slave society. The technical revolution of the Middle Ages was necessary to prepare the soil of Western Europe to receive the seed and the technical device of printing was necessary to multiply and broadcast the seed before the ancient wisdom could raise a wholesome crop.[62]

We shall presently return to the question of these technological innovations. Before that, let us try to be clearer about the nature of the inheritance of Greek science by modern Europe. J. D. Bernal warns us against a naive understanding of it :

It would be a mistake, natural enough in the time of the Renaissance but unpardonable now, to assume that all that happened then was the taking up again of classical culture where it left off, or even where it was at its best. What happened was something different and far more important. *The civilisations that took over the classical heritage of science had a hard task to prevent themselves from being stifled by it......* There was still, however, the vast store of knowledge to be found in books available to any with the desire or skill to read them. The Syrians and Arabs, and after them the medieval schoolmen and the humanists of the Renaissance, had to trace that store step by step back to its Greek originals....... That they managed to absorb and transform it at all was by virtue of their own vigorous cultural developments. The very rediscovery of the works of the Ancients was the effect, far more than the cause, of the spurts of intellectual activity that characterised the beginning of Islamic science in the ninth century, of medieval science in the twelfth, and of Renaissance science in the fifteenth century.

....Late classical culture was limited both socially and geographically. Socially it had become an almost exclusively upper-class preserve and was consequently abstract and literary, for ingrained intellectual snobbery had barred the learned from access to the enormous wealth of practical knowledge that was locked in the traditions of almost illiterate craftsmen. One of the greatest achievements of the new movement which culminated in the Renaissance was to raise the dignity of the crafts and to break down the barriers between them and the learned world.

The geographical range of classical culture had largely been limited to the countries of the Mediterranean and the Near East. Its very completeness formed a barrier to the use of the common stock of techniques and ideas of the other ancient cultures of India and China.

62. *Ibid,* 307-8.

With the breakdown of the Roman Empire the way was open to much wider exchange and influence.[63]

23. INDEBTEDNESS TO THE EAST

We are now nearing the point we have been trying to make. The point is that even for the limited purpose of understanding the history of science in Western Europe, it is not enough to rely exclusively on information of European cultural area. The enormous importance of the contributions of Central and Western Asia to the first foreshadowing of modern science in Europe is now being increasingly realised, though a good deal of more research remains to be done on the subject. But let us leave that point for the present. Let us concentrate instead *only* on the technological innovations of the Middle Ages, without which, as we have just seen, the new enthusiasm for the Greek heritage of Renaissance Europe cannot be understood. How are we to understand these technological innovations? Joseph Needham has boldly answered the question :

> In case after case it can be shown with overwhelming probability that the fundamental discoveries and inventions made in China were transmitted to Europe, for example, magnetic science, equatorial celestial coordinates and the equatorial mounting of observational astronomical instruments, quantitative cartography, the technology of cast iron, essential components of the reciprocating steam-engine such as the double-acting principle and the standard interconversion of rotary and longitudinal motion, the mechanical clock, the boot stirrup and the efficient equine harnesses, to say nothing of gunpowder and all that followed therefrom. These many diverse discoveries and inventions had earth-shaking effects in Europe, but in China the social order of bureaucratic feudalism was very little disturbed by them.[64]

The importance of the last point mentioned by Needham is surely not to be overlooked. It is connected, as he elsewhere says, with "what I believe is one of the greatest problems in the history of culture and civilisation—namely the great problem of why modern science and technology developed in Europe and not in Asia." [65] For the historian of science in India there is no escape from the problem. One of the ques-

63. J. D. Bernal, SH 266-7. Emphasis added. *Cf.* also p. 243.
64. J. Needham, GT 213.
65. *Ibid,* 154.

tions he is obliged to face is what inhibited the development of modern science in India, in spite of its brilliant early promise.[66]

For the present, what we are trying to understand, however, is a different question. How are we to understand the emergence of modern science in Europe, or, more specifically, the technological stimulants required by it? Here is how Needham sums up the results of his research: "The more you know about Chinese technology in the medieval period, the more you realise that, not only in the case of certain things very wellknown, such as the invention of gunpowder, the invention of paper, printing, and the magnetic compass, but in many other cases, inventions and technological discoveries were made in China which changed the course of Western civilisation, and indeed that of the whole world." [67]

We may be yet far from a thorough and systematic exploration of the other important cultural areas of Asia—notably India and Central and Western Asia. But we have before us the stupendous volumes of Joseph Needham's *Science and Civilisation in China,* and we are expecting from him more volumes of the work.[68] Enough is contained in these to be considered as the most massive verdict on the facile claim that science is an essentially European phenomenon. One man has indeed exploded the myth nourished by generations.

It is mainly on the strength of the results reached in this great work that J.D. Bernal has come out with the following observation on the main point we have been trying to discuss:

66. J.D. Bernal mentions in this connection a possibility which is favoured by some contemporary Indian historians: "In the East, once the earlier stimulus to economic progress failed, the intellectual stimulus also vanished. Both might have revived later, but by the time they showed signs of this, as in India under the Moguls, their development was cut short by the superior commercial and military achievements of early European capitalism." SH 284.
67. Needham, GT 154.
68. "I have just turned 81, and although eleven volumes of the *SCC* series are already out, there are nine more to be finalised and issued before everything planned is completed": Joseph Needham, personal communication dated Dec. 13, 1981.

The technical advances of the Middle Ages were made possible by the exploitation and development of inventions and discoveries which, taken together, were to give Europeans greater powers of controlling and ultimately of understanding the world than they could get from the classical heritage. Significantly, the major inventions....were not themselves developed in feudal Europe. All seem to have come from the East, and most of them ultimately from China...Already enough is known to show that the whole concept of the superiority of Western Christian civilisation is one based on an *arrogant ignorance of the rest of the world.*[69]

24. 'ARROGANT IGNORANCE'

"Arrogant ignorance" is an exasperated expression indeed. But it is hitting the nail on the head and hitting it hard. The old prejudice that science cannot but be an essentially European phenomenon sometimes goes to the extent of flouting obvious facts. Reviewing [70] the papers of a Symposium held in Delhi in November 1950 on *History of Sciences in South Asia,* Needham quotes Filliozat for a rather glaring example of this.

We quote Filliozat over again:

The greatest historians of science have not always escaped from the inconvenience of knowing only one side of the matter. Paul Tannery, so famous for his studies on ancient mathematics, is an example. We know that the trigonometric sine is not mentioned by Greek mathematicians and astronomers, that it was used in India from the Gupta period onwards (third century A.D.), that the *Surya-siddhanta* (fourth or fifth century A. D.) gives a table of sines, that the Arab astronomers knew them from their Indian contacts and passed them on to Europe in the twelfth century A. D., when the work of al-Battani was translated into Latin. The only conclusion possible is that the use of sines was an Indian development and not a Greek one. But Tannery, persuaded that the Indians could not have made any mathematical inventions, preferred to assume that the sine was a Greek idea not adopted by Hipparchus, who gave only a table of chords. For Tannery, the fact that the Indians knew of sines was sufficient proof that they must have heard about them from the Greeks.

If this is the way we are to argue, there was never any science other than Greek science, and the question whether science has any origins other than the Greek 'miracle' is solved in advance. Only a profound study of Indian scientific developments in parallel with those

69. J. D. Bernal, *op. cit.*, 311. Emphasis added.
70. J. Needham in *Nature,* Vol. 168, pp. 64 ff.

which took place elsewhere about the same times can reveal the degree of originality of that science, and hence enable us to understand the role which India played in the history of the growth of man's knowledge of nature.[71]

It is not difficult to mention other examples. Amazed by the discussion of the preparation of alkalies in the *Susruta-samhita*, the eminent chemist and historian of chemistry M. Berthelot suggested that this portion of the *Susruta* could only be a later interpolation inserted into the text after the Indians had contact with European chemists. He had no patience for some elementary chronological considerations which make such a claim palpably absurd and to which P. C. Ray draws our attention.[72]

More examples are perhaps not necessary. But this tendency to flout or ignore facts in defence of the idea of science being a monopoly of the Europeans cannot but lead to the suspicion of racialism, however disguised and even unconscious it may be. In recent years it is passionately argued by some Asian scholars that the whole concept is used for inducing submissiveness among the Asians to the scientifically and technologically superior Western races, *i.e.* for colonial domination and colonial exploitation. "The political purpose behind this was to create a sense of inferiority amongst Asians and use science and technology as an instrument both of intellectual domination as well as exploitation."[73] Significantly, before Europe entered the career of colonial expansion, there was no such zeal to deny or undermine Indian contribution to the mainstream of science. Here is what a Spanish Muslim scholar wrote in A. D. 1068 : "Among the nations, during the course of centuries and throughout the passage of time, India was known as the mine of wisdom and the fountainhead of justice and good government and the Indians were credited with excellent intellects, exalted ideas, universal maxims, rare inven-

71. J. Filliozat CDIM pp. xix-xx. We have quoted the passage as translated by Needham.
72. P.C. Ray in SHSI 369.
73. A. Rahman, Introduction to *Science and Technology in Medieval India : A Bibliography of Source Material in Sanskrit, Arabic and Persian,* p. vi. I have also before me the manuscript of A. Rahman's *Intellectual Colonisation : Science and Technology in East West Relations,* where the same point is more vigorously argued.

tions and wonderful talents. They have studied arithmetic and geometry. They have also acquired copious and abundant knowledge of the movements of the stars, the secrets of the celestial sphere and all other kinds of mathematical sciences. Moreover, of all the peoples they are the most learned in the science of medicine and thoroughly informed about the properties of drugs, the nature of composite elements and peculiarities of the existing things."[74] If, in view of the complexities of Indian history we are being increasingly aware of, such an observation of about a thousand years back appears today to be rather naive, it is also refreshing if for no other reason than the complete absence of racialism—conscious or unconscious.

25. CO-RESPONDENT TO CONSERVATISM

The view of science being a monopoly of Western Europe has other undesirable consequences. It serves the forces of conservatism within the Asian countries,[75] which, in defence of stagnation and *status quo,* prefer to cut off Asian culture from the mainstream of global science. This makes it convenient to project the irrational religious-mystical trends of the past as representing the quintessence of Asian culture. In India at any rate we are painfully aware of where this leads to. In its cruder form, it debauches people's minds by accustoming them to ignore science in favour of obscurantism, which is required for caste hatred and communalism sanctifying malevolence and murder. In its sophisticated form, it inflates our ego and wants us to be convinced that, compared to the inferior ideal of science and rationalism, Indian sages discovered the secret of some mysterious supra-scientific knowledge. S. Radhakrishnan, for example, goes to the extent of regretting the modern fascination for science and rationalism.[76]

74. Abu'l-Qasim Sa'id bin 'Abdur-Rahman bin Muhammad bin Sa'id al-Andalusi's comments on India in *Tabaqat al-Umam* (*Categories of Nations*), A.D. 1068/460 A.H. Quoted by M. Saber Khan, "India in Hispano-Arabic Literature : An Eleventh Century Hispano-Arabic Source for Ancient Indian Sciences and Culture", in *Studies in the Foreign Relations of India* (Professor H. K. Sherwani Felicitation Volume), Hyderabad 1975, p. 359.
75. See S. Nurul Hasan, Introduction to *Ibn Sina* : *His Life and Contributions* by S. M. Ibrahim, New Delhi 1981.
76. S. Radhakrishnan IVL 127-33.

> The Western mind lays great stress on science, logic and humanism. Hindu thinkers as a class hold with great conviction that we possess a power more interior than intellect by which we become aware of the real in its intimate individuality. . . . Intuitive realisation is the means to salvation. . . . 'He who knows that supreme *brahman* becomes that *brahman* itself' . . . . While the dominant feature of Eastern thought is its insistence on creative intuition, the Western systems are generally characterised by a greater adherence to critical intelligence. . . . . From the Socratic insistence on the concept to Russell's mathematical logic, the history of Western thought has been a supreme illustration of the primacy of the logical. Rationalism is deep in our bones, and we feel secure about scientific knowledge and sceptical about religious faith.

Perhaps the only point of any historical significance about this breath-taking generalisation is that only one among the many philosophical trends in India, namely Vedanta, was keen on denying logic and rationalism in order to make room for an abject faith in the scriptures, declaring the scriptures as the repositories of direct spiritual realisation, and all this as sharply contrasted with other philosophical trends strongly defending logic and rationalism.[77] That the philosophical trend glorified by Radhakrishnan received strong support of the Indian law-makers is a pointer not only to its immense prestige among the Indian elites but also to its social function, for obviously enough the law-makers would not boost a philosophy that did not serve their main purpose and the main purpose of the law-makers was the defence of the caste-structure of society.[78] Nevertheless, because of the inflated importance attached to this philosophical trend by scholars like Radhakrishnan and others, considerable confusion is created even among our working scientists, some of whom—with admirable scientific skill in their professional life—are inclined to nourish obscurantist views as their private convictions, perhaps under the delusion that this is the way of seeking sanction of the national heritage. Hence are the well-known cases of "split personality" of our scientists.[79] This, to say the least, is undesirable and self-defeating. However patriotic it may seem, it does create an impediment in the way of the formation of the scientific atti-

77. D. Chattopadhyaya WLWDIP ch. 1.
78. *Ibid.* ch. 5.
79. The point is already mentioned in the *Preface*.

tude, without which the present socio-economic set-up can never be radically changed.

26. NOT A MERE CATALOGUE OF SCIENTIFIC ACHIEVEMENTS

But the generalised claim that Indian culture is essentially spiritual is as much a myth as the one with which it is in open collusion, namely that science is something essentially European. We have to scrap both and the right way of doing it is to work for the reconstruction of the origin and development of science and scientific thought in Indian history, as Needham has done in the case of Chinese history. This does not surely mean that we are equipped today to achieve comparable results. What it means is that the work must have priority while we think of the areas of our research. It is important not only for a better understanding of our own cultural heritage but also for correcting the prevailing imbalance in the story of global science. And the work itself, as Needham observed, "remains enthralling".[80]

The work is not easily done. It is necessary for the purpose to seek answers to a considerable number of questions. What did India contribute to the general fund of science and scientific thought specially in the ancient and medieval periods ? What were the special areas of these contributions and how are we to account for the importance attached to these ? How in different ages did science respond—or was prevented from responding—to the technological experience and the store of empirical knowledge locked up in the craft-lores ? How was the literary tradition in science related to the folk tradition, which, as is evident, for example, in the case of medicine, has always been very strong in India ? What was the nature of interaction between science and other dimensions of Indian culture, like religion, philosophy and jurisprudence ? What was the nature of scientific exchange of India with other countries—with China and Tibet, with Central and Western Asia, with Greece and Rome, and in the still earlier period with Egypt and Mesopotamia ? What role did foreign trade and commerce play in this interchange ? Above all, how was science related to society in the different stages of Indian history ? How

80. J. Needham in *Nature,* Vol. 168 pp. 64ff.

far, in this relation, we are to seek clues to the periods of outbursts in scientific activities alternating with periods of stagnation and decay? Lastly, what were the inhibiting factors that prevented the rise of modern science in India in spite of its early promise and prolonged continuity?[81]

Evidently enough, it is desirable to have a team of historians, scientists, philologists, philosophers and specialists in other branches to tackle such a wide range of questions. There is today some talk in the country of forming such a team. In the meanwhile, something remains to be done. We have to consolidate the results already reached by the earlier generations of scholars.

27. PIONEERING WORK

It is true that compared to the tons of books written on Indian metaphysics, religion and mysticism, there has been a sad neglect of what B. N. Seal[82] called "the work of constructing scientific concepts and methods in the investigation of physical phenomena". But this does not mean that we have to start today from mere scratch. Though in a minority, some of the scholars—both Eastern and Western—went against the stream and took an absorbing interest in the scientific activi-

81. Continuation of the tradition of astronomy specially among the scientists of Kerala may be mentioned here as an example. K. V. Sarma of Hoshiarpur, to whom we are indebted for a good deal of work on the subject, writes (in a personal communication dated 6.1.82 to my friend Sri Ramkrishna Bhattacharya) : "To be sure, there has been steady and rather reverential 'continuity in astronomical science [after Bhaskara II] but streaks of 'progress' through a rationalistic questioning mind had been there, as evidenced by works like *Rasi-gola-sphuta-niti* of Madhava of the 14th century, *Jyotirmimamsa* of Nilakantha (born 1444) and *Ganitayuktayah,* being short rationales of mathematics and astronomy by several astronomers of Kerala. But, even I am at a loss to visualise realistically how they made this progress. The possible explanation could be the nature of Indian tradition of throwing away, *i.e.,* not keeping, the record of the intermediate steps and arguments of derivation, once the resultant formulae have been reached—a tradition in distinct contrast of the Western tradition from early periods."

82. B. N. Seal, PSAH p.iv.

ties in India. The tradition they respresent is not to be slighted. They include the early visiting scientists like I-Tsing (whose medical background is sometimes overshadowed by his image of being a Buddhist monk) and al-Biruni, whose vision of science as an international endeavour led him not only to make Indian works on astronomy available in Arabic translation but also to make Greek works on science available to the Indians translated into Indian language, i.e. Sanskrit.[83]

They include some of the profoundest scholars of the eighteenth, nineteenth and twentieth centuries doing a great deal of pioneering work—searching for the manuscripts, settling their reading, interpreting their science potentials and trying to solve the most difficult chronological questions. The contemporary historian of science in India cannot but depend on their work. But their contributions remain buried often in the brittle pages of rare journals, often in books gone long out of print, and thus becoming increasingly inaccessible to us.

It remains for us to add only one point. Admirable though the contributions of the pioneers are, indispensable though these may be for the contemporary historians of science in India, the presentation of these in the form of a convenient handbook can in no way be claimed as an adequate account

83. In the Introduction to *Ghuraat al-Zigat or Karana Tilaka* (A Handbook of Astronomy by Bijayanand [sic] of Benaras), Translated from Original into Arabic...by...al-Biruni [prepared for publication by N. A. Baloch], Sind (Pakistan), Institute of Sindhology, 1973, are mentiond the following books translated into Sanskrit by al-Biruni :
(a) *Elements* of Euclid (No. 30).
(b) *Almagest* of Ptolemy (No. 31).
(c) Book on Astrobale by al-Biruni (No. 32).
(d) Key to the Science of Astronomy by al-Biruni (No. 8).
It seems however that these translations were not preserved, because from the *Rekhaganitam* of Jagannatha we learn that under the direction of Jayasimha II (of Jaipur, ruled A. D. 1699-1743) Ptolemy's *Almagest* (*Majisti*) and the *Elements* of Euclid (from the Arabic versions of al-Tusi) were translated again into Sans-Krit : see S. N. Sen, *A Bibliography of Sanskrit Works on Astronomy and Mathematics,* New Delhi 1966, pp. 89-99. Incidentally, as a reminder to the norm of internationalism in science, it may be mentioned here that Jagannatha's *Siddhanta-samraj* repeatedly refers to the work of Ulugh Beg of Samarkand.

of science in Indian history. As is perhaps evident from some of the questions just mentioned, science in Indian history is an enormously complicated subject and is surely not to be confused with some kind of a catalogue of the prominent achievements of Indian scientists. We have, for example, the model of such a catalogue as prepared by Th. Stcherbatsky in 1923. In spite of perhaps what is inevitable, namely that some of his observations are in need of correction in the light of later researches, the usefulness of such a catalogue when it was prepared is not to be undermined. It was then some kind of a novelty and it was in response to a necessity then keenly felt. We can judge this from what Stcherbatsky said only a few years before (1916) :

> The Indian thought on the whole still remained enveloped in the mist of oriental fantasy and the orderly forms of its consistent logical theories were hidden from the keen sight of the historians of philosophy owing first to the inadequacy of the material available to them and second to the lack of any systematic methods for its scientific study. Besides this stage of scientific knowledge, there could be discerned, in the wider circles of reading public, a morbid interest in Indian philosophy caused by the hazy state of our knowledge of the subject and the various fables of supernatural powers rampant therein.[84]

It was in such a situation that Stcherbatsky wanted his readers to meet the logicians and the atomists, the astronomers and the mathematicians, the physicians and the chemists, the technicians and the engineers of ancient India. A mere list of them and of some of their achievements could be and were indeed of much significance for the earlier stage of historical research, though the fact is that many earlier conclusions have been rejected and corrected.

We have no doubt outgrown this stage. This is not merely because the different areas of scientific activities in India are more intensively explored and still being explored, though only by a minority of scholars. It is more particularly because of the profound change in the scientific historiography of science that has in the meanwhile taken place. But basically the same model of catalogue-making without any reference to the social and economic factors sometimes persists, as is evidenced by

84. Th. Stcherbatsky, Introtuction to the Russian translation of Dharmakirti's *Santanatara-siddhi.* Petrograd 1916. See *Papers of Stcherbatsky,* Calcutta 1969, 73.

Binoy Kumar Sarkar's *Hindu Achievements in Exact Sciences* (1918) and the much more enlarged version of practically the same model in *A Brief History of Science in India* (1971) edited by D. M. Bose, S. N. Sen and B. V. Subbarayappa.

Its other limitations apart, the model of catalogue-making remains exposed to a number of risks. The most serious of these seems to be that it may encourage the tendency to look at science as some kind of an autonomous discipline without being basically influenced by society. That is not helpful for understanding science either in Europe or in Asia. Let us end by quoting Joseph Needham.

28. SCIENCE AND SOCIETY : J. NEEDHAM

In recent decades much interest has been aroused in the history of science and technology in the great non-European civilisations, specially China and India, interest, that is, on the part of scientists, engineers, philosophers, and orientalists, but not, on the whole, among historians. Why, one may ask, has the history of Chinese and Indian science been unpopular among them ? Lack of the necessary linguistic and cultural tools for approaching the original sources has naturally been an inhibition, and of course if one is primarily attracted by+18th and+19th century science European developments will monopolise one's interest. But I believe there is a deeper reason.

The study of great civilisations in which *modern* science and technology did not spontaneously develop obviously tends to raise the causal problem of how modern science did come into being at the European end of the Old World, and it does so in acute form. Indeed, the more brilliant the achievements of the ancient and medieval Asian civilisations turn out to have been the more discomforting the problem becomes. During the past thirty years historians of science in Western countries have tended to reject the sociological theories of the origin of modern science which had a considerable innings earlier in this century. The forms in which such hypotheses had then been presented were doubtless relatively crude, but that was surely no reason why they should not have been refined. Perhaps also the hypotheses themselves were felt to be too unsettling for a period during which the history of science was establishing itself as a factual academic discipline....

"The study of other civilsations therefore places traditional historical thought in a serious intellectual difficulty. For the most obvious and necessary kind of explanation which it demands is one which would demonstrate the fundamental differences in social and economic structure and mutability between Europe on the one hand and the great Asian civilisations on the other, differences which would account not only for the development of modern science in Europe alone, but also of capitalism in Europe alone, together with its typical accompaniments of protestantism, nationalism, etc., not paralleled in any other part of the globe......

But if you reject the validity or even the relevance of sociological accounts of the 'scientific revolution' of the late Renaissance, which brought modern science into being, if you renounce them as too revolutionary for that revolution, and if at the same time you wish to explain why Europeans were able to do what Chinese and Indians were not, then you are driven back upon an inescapable dilemma. One of its horns is called pure chance, the other is racialism however disguised. To attribute the origin of modern science entirely to chance is to declare the bankruptcy of history as a form of enlightenment of the human mind. To dwell upon geography and harp upon climate as chance factors will not save the situation, for it brings you straight into the question of city-states, maritime commerce, agriculture and the like, concrete factors with which autonomism declines to have anything to do. The 'Greek miracle', like the scientific revolution itself, is then doomed to remain miraculous. But what is the alternative to chance? Only the doctrine that one particular group of peoples, in this case the European 'race', possessed some intrinsic superiority to all other groups of peoples. Against the scientific study of human races, physical anthropology, comparative haematology, and the like, there can of course be no objection, but the doctrine of European superiority is racialism in the political sense and has nothing in common with science. For the European autonomist, I fear, 'we are the people, and wisdom was born with us'. However, since racialism (at least in its explicit forms) is neither intellectually respectable nor internationally acceptable, the autonomists are in a quandary which may be expected to become more obvious as time goes on. I confidently anticipate therefore a great revival of interest in the relations of science and society during the crucial European centuries, as well as a study ever more intense of the social structures of all the civilisations, and the delineation of how they differed in glory, one from another.

In sum, I believe that the analysable differences in social and economic pattern between China and Western Europe will in the end illuminate, as far as anything can ever throw light on it, both the earlier predominance of Chinese science and technology and also the later rise of modern science in Europe alone.[85]

29. CONCLUDING REMARKS

We have quoted Needham at some length, as we have quoted Bernal at the very beginning for we have in these the main

85. J. Needham GT 214ff.

guidelines for our own work. Not that we have in our humble team anybody of Needham or Bernal's stature nor do we delude ourselves with the idea that we are going to achieve anything even remotely comparable to what they have done. Nevertheless, we shall consider our attempt amply rewarded if it is considered as a step—howsoever faltering and tentative it may be—towards a history of science in India and if the errors or inadequacies in it may negatively stimulate more competent scholars to work on a better study of the subject.

CHAPTER 2

ON PERIODISATION

1. INDIAN STUDIES : PAST & PRESENT

The story of what Gordon Childe[1] describes as "the dramatic entry of India on the stage of Oriental history with the excavation of Harappa and Mohenjodaro", though well-known, may be briefly recapitulated. We follow B. K. Thapar[2] for the purpose.

In spite of "an undignified scramble for archaeological loot", the excavations undertaken during the latter half of the nineteenth century at Nineveh, Nimrud, Nippur and Lagash initiated the process of the dazzling revelation of the brilliance of the ancient Mesopotamian civilization.

> Strangely enough, an equally ancient civilization of the Indus, with comparable environmental and economic pattern to that of Mesopotamia, had remained unrecognized and even unsuspected during these eventful years. And this despite the fact that the type-site, Harappa, had been extensively despoiled by brick robbing, and been excavated several times by Alexander Cunningham who had obtained various antiquities including a typical Indus seal (collected by a Major Clark), bearing pictographic characters. India thus continued to be referred to as a country where 'there was no twilight before dark'. It was not until 1921 and 1922 when preliminary trial diggings at Harappa by Daya Ram Sahni, and at Mohenjo-daro by R. D. Banerjee, had yielded identical finds, including exotic seals, that the potential of the sites came to be realized and the elements of a forgotten civilization identified. After examining the collection of antiquities from these two widely-separated sites and being convinced that they were totally distinct from anything previously known in India, Marshall announced the discovery in a London Weekly [in 1924]. This took the archaeological world by surprise, for the excavations at Ur by Leonard Woolley, almost during the same period, had already created a great sensation among Old World archaeologists. Writing elsewhere, Marshall had averred that 'the discoveries had at a single bound taken our knowledge of Indian civilization some 3000 years earlier'.

1. Childe NLMAE 2.
2. Thapar in FIC 1. This may be contrasted with the simplified statement of Lamberg-Karlovsky 189. "In 1921, Sir John Marshall excavated the major metropolises of the pre-aryan Indus civilization—Harappa and Mohenjo-daro."

What specially interests us for our present study may at once be noted. The excavation, in Childe's[3] words, necessitated the admission that "India confronts Egypt and Babylonia by the third millennium with a thoroughly individual and independent civilization of her own, *technically the peer of the rest*". One result of the profound revolution in this knowledge of technology is the liberating influence it had for serious scholars from the earlier but somewhat obligatory limitation of accepting the Vedic literature as virtually the only source of information for the earliest chapter of the story of India's progress to civilization.

Vincent Smith's article[4] in *Indian Antiquary,* "The Copper Age and Prehistoric Bronze Implements of India"—and even the contribution of John Marshall[5] (then already the Director General of Archaeology in India) to the first volume of *The Cambridge History of India* (published in 1922) on "The Monuments of Ancient India"—show how desultory had then been the knowledge of Indian archaeology. In default of archaeological data, the scholars had recourse mainly to literary sources. The Vedic literature having been the earliest of these had to be taken as the starting point for understanding Indian history.

This explains the earlier tendency of trying to trace the beginnings of everything about Indian achievements to the Vedas. Understandably, technology and science were not exceptions to this. What, indeed, could the scholars do when nothing serious was known to them outside the Vedic literature as earliest available records of Indian achievements?

Examples of this are indeed numerous. It may be enough for our purpose to mention here only two.

I have before me a book containing over 350 pages by Binode Behari Dutt bearing the title *Town Planning in Ancient India.* The book was published in 1925, and was evidently written before anything substantial was known about the Indus Civilization. Though containing valuable information from later

3. Childe NLMAE 183. Emphasis added.
4. IA xxxiv. 229ff & xxxvi. 53ff. Vincent Smith remarks here that the history of India began with the Maurya, or still better with Alexander's invasion of India.
5. Marshall in CHI i. 612-648.

Indian architectural texts, the author also felt the need of discussing the origin of the technique in India. And he had then nowhere else to look for it than the Vedic sources. Hence he spoke of "The Vedic Origin of Town-planning", and remarked : "the Vedic Aryans had certainly developed a far more advanced knowledge in the science of building than a mere inchoate and crude sciolism. The plan of the towns and their denominations were identical with those of the geometrical figures that had to be, and are even now, drawn on the sacrificial altars. These figures suggested the plans and the names. It is sure that the Vedic civilization had long ago immensely outgrown the primitive stage and still glows with innumerable and irrefragable evidences of its high water-mark".[6]

If one discerns a spirit of patriotism in such an observation, it needs also to be added in the light of the new knowledge we have today that the patriotism is somewhat misplaced. Reserving for the present the question of the Vedic sacrificial altars and of their possible connection with house building—a question which we shall have to discuss later in some detail—it is relevant to mention here only one point. Town planning—and, for that matter, an amazingly advanced form of it—is actually witnessed by Indian history at a very ancient period. This was already over a thousand years before the making of the Vedas and the elaboration of the Vedic sacrificial cult connected with the fire altars.

In the first comprehensive report on the excavations at Mohenjo-daro and Harappa, Marshall[7] observes, "Any one walking for the first time through Mohenjo-daro might fancy himself surrounded by the ruins of some present-day working town at Lancashire. That is the impression produced by the wide expanse of bare red-brick structures devoid of any semblance of ornament, and bearing in every feature the mark of stark utilitarianism. And the illusion is helped out, or rather the comparison is prompted, by the fact that the bricks themselves of which these buildings are composed are made much of a size with modern English bricks, but differ conspicuously from any during the historic period in India". Mackay adds,[8]

6. Dutt TPAI 7.
7. Marshall MIC 15.
8. Mackay EIC 18.

"and it is interesting to note that these ancient cities of the Indus plains are the earliest sites yet discovered where a scheme of town-planning existed. There is no evidence of such a scheme at Ur as late as 2000 B. C., though there are traces of one at Babylon at about that date ; and also at the Twelfth Dynasty town at Kahum in Egypt". In 1978, depending also on later archaeological work, S. P. Gupta and Shashi Asthana[9] give us a more detailed idea of town-planning in the ancient Harappan civilization which we quote at some length :

From their foundation the cities of the Indus system appear to have been laid out in accordance with some pre-arranged scheme of things. It seems that building regulations were strictly enforced at Mohenjo-daro, Kalibangan, Lothal and other places for many centuries and the greatest care were taken to prevent any structure from encroaching upon the roads and lanes. The major roads and lanes in the cities ran in straight lines and were crossed by others at right angles. The secondary streets were dog-legged. Within this planning it was seen by the planners that the major roads were aligned from east to west and from north to south, since the prevailing winds always came from the latter quarters. These roads had differing width, ranging from 30 feet to about 9 feet, which could accommodate several lines of wheeled traffic. The entire city was thus divided into a number of blocks arranged in a more or less chessboard pattern. One could easily reach the buildings of different blocks quickly and smoothly even on wheeled-carts...

The Harappan towns are well-known for their elaborate drainage system. Covered drains from the lanes discharged dirty water into the major drains which ultimately emptied it into tanks, jars, etc., or open areas outside the city. There are many other features both in town-planning and architectural devises.... which are well known and need not be detailed here.

S. R. Rao,[10] writing for a popular journal, somewhat dramatically observes :

What is common between Mohenjo Daro and New York ? A glance at the road maps of the two cities reveals that the ancient metropolis was as well laid out as the modern one with roads running parallel in cardinal directions and crossed at right angles by streets leading to the individual buildings. And not just Mohenjo Daro, most Indus cities were as well planned—obviously one of the pointers of a highly evolved culture ... The orderliness and the consequent urban discipline of the Indus

9. S.P. Gupta and Sashi Asthana in ME ii. 47-8.
10. S. R. Rao in ST June 1982, 13-14.

civilization are reflected in its uniform system of weights and measures, in the planning of cities and the trade regulations which the Harappans were able to enforce throughout the vast empire. The municipal laws were strict. For instance, no encroachment was allowed on public streets. Over a period of four centuries (2300 B. C. and 1900 B. C.) the width of the Bazar street remained the same (4. 5 metres) even though the flanking houses had to be reconstructed thrice due to heavy damage caused by floods. Similarly, inspection chambers in private drains were compulsory, so that solids could be removed before liquid waste entered the public drains...The Harappans were the first town planners of the world. In the earlier urban civilizations elsewhere such as the Sumerian, the cities were not planned ; the streets of Ur, Kish and Brak were tortuously winding. The meticulous planning of the Harappans extended also to production, storage and distribution of food-grains as suggested by remains of the granaries seen at Harappa and Mohenjo Daro. The interlinking by river of towns in the interior with ports was another great advantage which ensured exchange of goods. For instance, the port at Lothal was connected by the Sabarmati with the north Gujarat hinterland. Harappa and Mohenjo Daro were interconnected by the Indus system, and so on.

We have quoted these to emphasise only one point. After the discovery of the Indus civilization, the history of town planning in India is certainly in need of being rewritten, as certainly again its starting point cannot be the Vedic literature.

I have also before me the well-known work by the late Radha Kumud Mookerjee on *Indian Shipping : A History of the Sea-borne Trade and Maritime Activity of the Indians from the Earliest Times.* For many years it was considered about the only full-length study of the subject attempted by the modern scholars, and its importance for the students of Indian history is generally admitted. Where, however, it reads strange today is about the maritime activity of the Indians of "the earliest times". Written in 1910 and first published in 1912, the author had nowhere else to look for it than the earliest literary sources available. Accordingly, he observes : "The oldest evidence on record is supplied by the *Rgveda*, which contains several references to sea voyages undertaken for commercial and other purposes".[11] In substantiation of this, Mookerji mentions five passages of the *Rgveda,* to which he could have added many more from the *Vedic Index* by

11. Mookerji 37. Following passages of the Rgveda are cited i. 25.7 ; i. 48.3 ; i. 56.2 ; i. 116. 3 ; vii. 88. 3-4.

Macdonell and Keith,[12] who argue against those that denied the Vedic people any actual knowledge of the sea.[13]

As for "sea-borne trade and maritime activity", however, there remained some problem. There are legends presumably coming down from a hoary antiquity and recorded in the *Jataka*-s indicating ancient Indian maritime trade with Mesopotamia. What could possibly be the basis of such legends? R. K. Mookerji[14] observes :

The *Baveru Jataka* without doubt points to the existence of commercial intercourse between India and Babylon in pre-Asokan days. The full significance of this important *Jataka* is thus expressed by the late Professor Buhler : 'The now well-known *Baveru-Jataka* to which Professor Minayef first drew attention, narrates that Hindu merchants exported peacocks to Baveru. The identification of Baveru with Babiru or Babylon is not doubtful', and considering the 'age of the materials of the *Jatakas,* the story indicates that the Vanias of Western India undertook trading voyages to the shores of the Persian Gulf and of its rivers in the 5th, perhaps even in the 6th century B. C. just as in our days'.

Macdonell and Keith[15] observe, "That there was any sea-trade with Babylon in Vedic times cannot be proved...... There is, besides, little reason to assume an early date for the trade that no doubt developed later, perhaps about 700 B. C."

We have quoted these, because these are specimens of best scholarship on ancient Indian maritime activity that could be possible before the archaeological work of the last few decades. Indeed when the mental horizon of the historians could not be extended backward beyond the Vedic literature, the account of the maritime activity and sea-trade of the ancient Indians had to remain limited and circumscribed.

Today, however, the situation is quite changed. One has only to go through the brief paper on *Indus-Mesopotamian Trade* (1979) by Sashi Asthana[16] or glance through some of

12. A. A. Macdonell and A. B. Keith ii. 432f.
13. Interestingly enough, it was A. B. Keith, who earlier doubted the knowledge of the sea on the part of the Vedic Aryans. Keith in CHI i. 79.
14. Mookerji 51.
15. A. A. Macdonell and A. B. Keith ii. 432.
16. Sashi Asthana, in EIP 31-43.

the writings on the subject by S. R. Rao and others to realise how profound are the changes in our knowledge of it, thanks mainly to recent archaeological work. We quote here only one observation[17] on overseas trade of the ancient Indus period :

> The sea-borne trade of the Bronze Age cities was unbelievably far-flung. As a predominantly-mercantile community the Lothal merchants sent their ships to distant ports in the Persian Gulf and even beyond. The excavations at Ras Shamra have revealed that ivory rods, now suspected to be of Indus workmanship, reached as far as the north Syrian coast. Archaeological evidence alone is sufficient to show that overseas trade was well-organized in the third millennium B. C. The merchants had established colonies outside their homeland and used specific types of seals prescribed by the local rulers or merchant guilds. For example, cylinder seals were brought into vogue in the Euphratis-Tigris valley, circular stamp seals in the Persian Gulf islands and square or rectangular stamp seals in the Indus Empire. It is thus possible to know the source of goods from the seals and seal-impressions recovered at different sites. The establishment of colonies of Indus merchants in the Bahrein islands, the Euphratis-Tigris valley and the Diyala region is attested to by the seals bearing Indus motifs and script found at Ras-al-Qala, Ur, Kish Asmar and the Diyala sites. Other evidences of trade between the Indus and Sumerian cities are provided by Indus weights and beads in south Mesopotamia and gold beads and painted pottery of Mesopotamian origin at Lothal and other Harappan sites.

We shall see more of overseas trade later.[18] For the present the point is that in the light of the new knowledge we have

17. S. R. Rao LIC 117.
18. S. R. Rao, "Shipping and Maritime Trade of the Indus People" in *Expedition* Vol. 7, No. 3. Philadelphia 1965, S. R. Rao, "Shipping in Ancient India, India's contribution to World thought and culture," Madras : 1970, 89-107 ; Mallowan, "The Mechanism of Ancient Trade in Western Asia", *Iraq,* Vol. III (1965). pp. 1-7 ; D. K. Chakrabarti, "The external trade during Harappan period : evidence and hypothesis", in *50 years of Harappan Discovery,* ed. B. B. Lal and S. P. Gupta (in press) ; D. K. Chakrabarti, "Gujrat Harappan connection with West Africa : A reconstruction of the evidence" in JESHO 18, 3 337-42 ; E.C.L. During-Caspers, "Harappan trade in the Arabina Gulf in third millenium B. C." in *Mesopotamia* VII. (1972) 167-191 ;—'Etched carnelian beads' in Bulletin of the Institute of Archaeology X. (1972). 83-98 ;—"New Archaeological Evidence for maritime trade in the Persian Gulf during the late Proto-literate Period" in EW Vol. 21-(1-2). 1971

today, what is earlier viewed about maritime trade—like that of town planning—is certainly in need of being rewritten, as certainly again its starting point is not the Vedic literature.

2. VEDA-CENTRISM AND INDIAN HISTORY

But let us first be clear about one point. What we are trying to emphasise here is not the mere fact of the accumulation of greater information about the earlier period of Indian history resulting from the archaeological work. What we want to emphasise instead is the *profound change in the very orientation of Indian studies* and this mainly because of the excavations of Harappa, Mohenjo-daro and other sites of the ancient Indus civilization. For the purpose of a proper appreciation of this, we begin with a few words on the limitations of the earlier approach.

The Indian tradition strongly insists that the Vedic literature is essentially religious. So also are the other literatures of India seriously studied by the earlier generation of Indian historians. The result has been the periodisation of Indian history mainly in religious terms.

This is easily seen from the first volume of *The Cambridge History of India*, which for many years virtually remained the model sketching of ancient Indian history. Here are some of its prominent chapter headings : "The Age of the *Rgveda*", "The Period of the later Samhitas, the Brahmanas, the Aranyakas and the Upanishads", "The History of the Jainas", "The Early History of the Buddhists"—and so on, until the historians felt somewhat relieved of being obsessed by religious

21-44 ;—"Sumer, coastal Arabia and the Indus Valley in protoliterate and Early Dynastic eras. Supporting evidence for a cultural linkage in JESHO 22,2 (1979). 121-35 ; S. Parpola, A. Parpola and R. H. Brunswig, Jr. "The Meluha village : evidence of acculturation of Harappan traders in late 3rd millenium Mesopotamia" in JESHO 20,2 (1977). 129-65 ; S. R. Rao, LIC 114 ff. S. Ratnagar, "Long distance trade of the Harappan civilization" in "50 years of Harappan discovery," eds. B. B. Lal and S. P. Gupta (in Press) ; I. J. Gelb, "Makkan and Meluha in Early Mesopotamian Sources" *Revue d' Assyriologie* 64 (1970). 1-8 ; S. N. Kramer, "The Indus civilization and Dilmun, the Sumerian Paradise land", in *Expedition* 6. No. 3 (1964) 44-52.

preoccupation when they reached the age of the Mauryas or of the Persian domination of northern India, and, above all the time of Alexander's invasion.

This religious bias apart, what then influenced the understanding of ancient Indian history was some kind of a racial bias, tacit though it might have been. This was also, at least to a considerable extent, the result of accepting the Vedas as the only valid starting point of Indian studies. Those who composed the Vedas called themselves the *arya*-s or Aryans, literally "the nobles". The language of this literature, as was first shown by the famous address of William Jones in 1786, was closely connected with the languages of the Persians, Greeks, Romans, Celts, Germans and Slavs. It was thus thought that all these developed out of some root-language, for which the term often used was Indo-European. "But the study of this family of languages has from the beginning been beset with a subtle fallacy. There has been throughout an almost constant confusion between the languages and the persons who spoke them."[19] Thus came into being the theory of the Aryans—or of the Indo-Aryans—who gave to Indian culture the Vedic literature. Ancient Indian history was accordingly conceived largely in terms of the colonization of India by the invading Aryans. The Aryans, it was assumed, brought civilization to the local peoples, the most prominent of whom—largely on linguistic considerations again—were called the Dravidians. How far such a model of ancient Indian history also suited the general temper of the European colonisers is a different question, into which we need not at present digress. In fairness, however, it must be admitted that some of the European scholars were in the forefront in breaking away from this understanding of ancient Indian history, while some of the Indian scholars still remain consciously or unconsciously exposed to its influence.[20]

What concerns our present discussion is to note how the archaeologists' spade shattered this earlier model of ancient

19. P. Giles in CHI i. 64-65.
20. Even a scientist like Meghnad Saha, as already noted, had to enter into a prolonged controversy with the traditionalists claiming that practically everything worthwhile in contemporary science was already known to the Vedic Aryans. See Chapter 1.

Indian history. The discovery of Harappa and Mohenjo-daro, in other words, had for the serious scholars a liberating effect from the dual limitations of the earlier understanding, inasmuch as—following what was firmly argued by R. P. Chanda,[21] B. S. Guha[22] and others[23]—the builders of the Indus civilization were viewed as pre-Aryans and non-Aryans. This is a point on which the later archaeologists are on the whole agreed. T. N. Ramachandran[24] and some others[25] have, of course, a dissenting note to this, for they are inclined to think that the Vedic Aryans were themselves the builders of the Indus civilization. Before putting too much confidence on their view, the readers may as well go through K. C. Chattopadhyaya's *Studies in Vedic and Indo-Iranian Religion and Literature*[26] : practically everything decisive that needs to be said from the viewpoint of Vedic scholarship against the possible Vedic origin of the Indus Civilization is already said in this.

3. ARCHAEOLOGY : NEW LIGHT ON ANCIENT INDIA

We shall quote here only two examples of the first flush of enthusiasm created for the reinterpretation of ancient Indian culture by the excavations of Harappa and Mohenjo-daro. One of these is concerning the new way of looking at Hindu religion, often considered as the basic plank of Indian culture. The other is more immediately relevant for the main theme of our present discussion, because it indicates the possibility of tracing the roots of classical Indian science to the achievements of the Indus period.

In his first comprehensive report *Mohenjo-daro and the Indus Civilization,* published in 1931, Marshall wrote a longish chapter on the religion of the Indus civilization. Though written about five decades back the chapter is still considered

21. R. P. Chanda, in MA SI No. 31.
22. B. S. Guha, in Marshall's MIC ii. chap. xxx, 599-644.
23. Childe NLMAE 185.
24. T. N. Ramachandran, *Presidential Address, Section I* Indian History Congress, xix session, Agra : 1956, 1-14.
25. B. K. Chattopadhyaya, "Mohenjo-daro Civilization" in CR vol. 139 : 121-6 ; vol. 141 : 252-60 ; vol. 144 : 127-33.
26. K. C. Chattopadhyaya, SVIIRL ii. 40-50. cf. also R. S. Sharma in MCSFAI 171.

"brilliant"[27] by some the leading archaeologists of our time. Marshall[28] observes :

> Many of the basic features of Hinduism are not traceable to an Indo-Aryan source at all. They come into view, not in the earliest Vedic literature, which represents the more or less pure Indo-Aryan tradition, but either in the later Vedas or in the still later *Brahmanas*, Upanishads and Epics, when the Vedic Aryans had long since amalgamated with the older races and absorbed some measure of their culture and teachings.... Whence these various elements were derived and when they found their way into the fabric of the national religion has never yet been satisfactorily explained....A few of these features, it has been conceded, may have been taken over from the pre-Aryans, but only such primitive ones as the worship of trees and animals and stones, which are common to the majority of uncivilized races. Those who have championed this view (and they include the chief writers on the subject) knew little, of course, of the great pre-Aryan civilization that has now been revealed. They pictured the pre-Aryans as nothing more than untutored savages, whom it would have been grotesque to credit with any reasoned scheme of religion or philosophy. Now that our knowledge of them has been revolutionized and we are constrained to recognize them as no less highly civilized—in some respect, indeed, more highly civilized—than the contemporary Sumerians or Egyptians, it behoves us to redraw the picture afresh and revise existing misconceptions regarding their religion as well as their material culture..... In view of these facts, is it not reasonable to presume that the peoples who contributed so much to the cultural and material side of Hinduism, contributed also some of the essential metaphysical and theological ideas so intimately associated with it ?

Gordon Childe seems to go a step further and expect new light to be eventually thrown by this archaeological discovery even on the possible indebtendness of occidental science to the scientific achievements in the ancient Indus civilization. The Indus script, which survives for us mainly on the "seals", argues Childe on Sumerian and Cretan analogies, could have been primarily devised for documents of accounts keeping, though such documents must "have perished with the unknown

27. B. & R. Allchin, RCIP 213.
28. Marshall MIC 77-78. Childe NLMAE 185 sums up Marshall's observations on the religion of the Indus, and adds : "for the above reasons alone the Indus civilization will be regarded as non-Aryan and pre-Aryan".

material on which they were written".[29] Referring to this script, he observes[30] :

> With this equipment the Bronze Age citizens of the Indus valley could have—and, indeed, must have—developed exact science as well as Sumerians and Egyptians, and for the same imperious reasons. For instance, a free use in decorative art of squares inscribed in compass-drawn intersecting circles suggests a study of geometry. But the results of such sciences are not directly known.....
>
> The imposing civilization perished utterly as a result of internal decay accelerated by the shock of barbarian raids. Only since 1920 have its dumb outlines been rescued from complete oblivion by archaeologists....
>
> Nevertheless, since Indus manufactures were imported into Sumer and Akkad, and Indus cults were actually celebrated there, the forgotten civilization must have made direct if undefinable contributions to the cultural tradition we inherit through Mesopotamia. Moreover, the technical traditions of the Bronze Age craftsmen, at least of potters and wainwrights, persist locally until today. Fashions of dress, established in the Indus cities, are still observed in contemporary India. Hindu rituals and deities have roots in the cults depicted in the prehistoric art. *So classical Hindu science, too, and through it occidental science, may be indebted to the prehistoric to an unexpected degree.* From this standpoint the Bronze Age civilization of India has not utterly perished ; 'for its work continueth far beyond our knowledge.'

4. LATER ARCHAEOLOGICAL WORK AND LITERATURE

The observations just quoted were published decades back—that of Marshall in 1931 and of Childe in 1942. Since then considerable archaeological work is being done on the cultural frontiers of ancient India, inclusive of significant work done by archaeologists in Pakistan since 1947. Though still considered inadequate in certain respects (specially as restricted mainly to vertical excavations)—and though the delay in publishing the reports on works already done is often regretted—an enormous amount of new materials is available today. Apart from what is known from the better explorations and excavations of the early stone age sites, we now have extensive information about the pre-Harappan, Harappan and post-Harappan settlements in the broader Indus region, about the Neolithic-Chalcolithic

29. Childe WHH 128.
30. *Ibid* 128-129. Emphasis added.

settlements beyond the Indus system, and about the iron age and early historical sites.

So also we have before us a very imposing literature on the subject. What is written on the Harappan Culture alone, as evidenced by the *Bibliography*[31] prepared by B. M. Pande and K. S. Ramachandran and published in 1971, is quite substantial. But this is only one example of the literature being stimulated by recent archaeological work on the cultural frontiers of ancient India and it is already somewhat dated. Besides, excavations and explorations are going on, adding to the volumes of the reports on these.

The vast literature, it needs to be noted, does not necessarily restrict itself to what is sometimes called cold archaeological data. All sorts of questions are also raised in these writings concerning the possible deductions from such data. In answering these questions, the writers—inclusive of very eminent archaeologists—are actually far from being unanimous. There are often sharp controversies on extremely vital issues among scholars, none of whose authority it is easy to dismiss. Thus, for example, the archaeologists are yet far from being agreed on the origin and formation of the Harappan civilization, the nature of the socio-political organization and intellectual climate in it, the cause of decline, degeneration and end of the civilization, of its legacy in later Indian culture, and so on, though thanks to the advent of the technique of radiocarbon dating the chronological controversies are now on the whole abated, notwithstanding the storm still going on about relating the archaeological data with literary evidences.

In such circumstances, any tendency to accept or reject the earlier observations of Marshall and others without taking note of the later archaeological literature is liable to be oversimplified. But so also would be the hope of following some safer course, steering clear of the recent controversies, inclusive of those with which the names of eminent archaeologists are associated. Lest, however, we get lost into the maze of all

31. B. M. Pande, and K. S. Ramachandran BHC. D. K. Chakrabarti in *Puratattva* No. 6 (1972-73) p. 97-98 reviews it, pointing to some lapses and errors in it. It may be added that the bibliography given at the end of Possehl's ACI contains 1607 entries.

these, it is necessary for us first of all to be clear about the exact scope of our present discussion and of the special relevance for it of archaeological evidences.

5. ARCHAEOLOGY AND ANCIENT TECHNOLOGY

Our purpose is not to survey Indian archaeology in the sense in which it is done by many eminent scholars, both in India and abroad. But we are obliged to begin with some account of recent archaeological researches because of a number of important considerations. We shall mention here a few of these.

One reason for beginning with archaeology is quite on the surface. Any attempt to understand technology specially of the ancient period without depending on the archaeologists is just absurd. For the pre-historic and proto-historic periods, archaeology—and archaeology alone—provides us with the surest information about technology and its development. In the contexts in which very ancient literary sources are also available—as in the case of India where we have the extensive Vedic literature—the material relics of extinct societies often prove to be of surprising significance for interpreting certain archaic words and passages than the method of pure philological analysis, though of course the literary sources in their turn may sometimes be of much help to the archaeologists for understanding their data. But the question of co-relating archaeology with literature is a very complex one and let us not digress into it for the present.

Instead of that, we may briefly mention here another point, namely that of the relation of technology with science. The difficulties about defining science are, of course, well known. J. D. Bernal questions the very possibility of cramping science into any neat definition : "Science is so old, it has undergone so many changes in its history, it is so linked at every point with other social activities, that any attempted definition, and there have been many, can only express more or less inadequately one of the aspects, often a minor one, that it has had at some period of its growth".[32] It will be observed no doubt that even such an observation has to presuppose some basic understanding of science. What is it that is so old and that

32. J. D. Bernal SH 30.

has undergone so many changes in history that it frustrates the very possibility of a clear-cut definition? In other words, in default of any exact definition of science, a study in its history cannot but begin—tacitly at any rate—with some understanding of it.

The basic understanding of science which Bernal himself has in mind is not difficult to judge. He speaks of "the generalised techniques from which science arose and to which it is still attached".[33] Science, he observes, "does not appear in the first place in a recognizable form.... It is necessary to search for its hidden sources in the histories of human arts and institutions".[34] "The tool-using and fire-using animal is well on the way to scientific humanity. Just as the tool is the basis of physical and mechanical science, so is fire the basis of chemical science".[35] "It was in the ways of extracting and fashioning materials so that they could be used as tools to satisfy the prime needs of man that first techniques and then science arose. A technique is an individually acquired and socially secured way of doing something; a science is a way of understanding how to do it in order to do it better. When we come to examine in greater detail...the first appearance of distinct sciences and the stages of their development it will become increasingly plain that they evolve and grow only when they are in close and living contact with the mechanism of production".[36]

More quotations are not necessary. It is already obvious that in Bernal's understanding at any rate any attempt to view science without relating it to technology will be fallacious. This becomes increasingly obvious as we move backward to human history. Not that this means that we may equate science to technology. What it means, however, is that science cannot be viewed as a purely disinterested search for truth or the product of pure reason, as it was and is being still viewed by many. On the contrary, the assumption on the basis of which we propose to work—an assumption endorsed by Joseph Needham, Benjamin Farrington, Gordon Childe, George Thomson and

33. *Ibid* 49.
34. *Ibid* 61.
35. *Ibid* 70.
36. *Ibid* 47.

many others, in whose writings the history of science has become sufficiently scientific—is that science is *implicit* in technology and hence can at best be very inadequately understood—if not positively misunderstood—without the broader perspective of technology. This is the second major justification for beginning with archaeology, for, as already observed, archaeology provides us with the most dependable information about technology of the ancient period.

6. PROBLEM OF PERIODISATION REOPENED

But there is another consideration that leads us to begin with archaeology specially in the Indian context. The importance of this consideration can hardly be exaggerated. Archaeological work on the cultural frontiers of ancient India during roughly the last six decades has in fact thoroughly revolutionized the mode of periodising ancient Indian history.

We have already noted how the discovery of Mohenjo-daro and Harappa has relieved the serious scholars from the earlier but somewhat necessary obligation of beginning with the Vedic literature as virtually the only starting point for understanding the earliest chapter of India's progress to civilization. But the implications of the recent archaeological works are much more profound than this.

The discovery of the Harappan civilization took place in the third decade of our century. Since then the archaeologists have worked not merely on the frontiers of the Indus civilization but have excavated and explored—as they are still doing—very large areas of India, and of course also of areas beyond the present political boundaries of India. Spectacular results are often reached by them, of which we shall mention here only those that have direct bearings on the periodisation of ancient Indian history.

Notwithstanding the debate still going on about the origin of the Harappan civilization, the archaeologists are fully agreed that, judged specially in the ancient context, it was a very advanced form of civilization, embracing within its orbit a number of significant urban centres like Mohenjo-daro, Harappa, Kalibangan and Lothal. In short, there is no scope to doubt that the Harappan civilization was a highly urban one ; it is in fact for us the evidence of First Urbanization so far known about ancient Indian history.

How exactly it came to its "end" is, of course, a very thorny question. There is literally a wilderness of conjectures concerning the cause of its decline and eventual destruction—ranging from malarial epidemic to the shock of Aryan attack. There are also controversies on the question whether the "end" of the Harappan civilization was a total one, leaving no trace of it whatsoever in later Indian culture. We shall have to return to some of these questions because of their obvious bearing on the main subject of our discussion. For the present the point is that after a flourishing period of many centuries, the civilization did come to an "end" and—thanks to the advent of radiocarbon dating—at least the majority of serious archaeologists are now agreed to accept the date of this event roughly as 1750 B.C.

Our question naturally is : What happened in Indian history specially in northern India after the decline or destruction of the Harappan civilization? In 1942, Gordon Childe has observed, "In Egypt, Mesopotamia and India, eras of prosperity that have left a vivid impression in the archaeological record were succeeded by Dark Ages form which few buildings and inscriptions survive. *In India civilization itself seems to have been extinguished*".[37]

We are concerned here with what happened in India. After the end of the first urbanization, it took a period of about a thousand years or more for civilization in its true sense to take shape or for the reintroduction of urban life. It is the usual practice of the archaeologists to refer to this reintroduction of urban life as indicative of the period of Second Urbanization. The period intervening the two urbanizations—which from the archaeological point of view is often called the "Dark Age" or the "Dark Period"—persisted for a thousand years or more.

We shall have to see what archaeology has to tell us about this intervening period between the two urbanizations. We shall also have to see what data other than the archaeological ones we have that throw light on this period. For the present, the point is that mainly as the result of the archaeological work of about the last six decades, our understanding of the periodisation of ancient Indian history is—or requires to be—basically changed. We can better understand it if divided into

37. Childe WHH 151. Emphasis added.

three main periods, namely those of 1) the First Urbanization, (2) an intervening "Dark Period" and (3) the Second Urbanization.

7. TWO URBANIZATIONS AND "THE DARK AGE"

Thus in Indian history, we do not have the story of a unilinear process of urbanization beginning in some hoary antiquity and continuing down to a much later period, which is usually characterised as the "historical" one. What we have instead are two distinct processes of urbanization with a period of over a thousand years intervening between the end of the former and the beginnings of the latter.

But what is meant by "the two urbanizations"? What happened during the period intervening the two? These, we are going to see, are questions crucial for our understanding of the history of technology and science in ancient India. We begin with brief descriptive answers to these.

Of the two urbanizations, only the first answers to what Gordon Childe calls the "urban revolution". This, as we shall presently see, is in his view a precondition for the making of mathematics and astronomy as "exact and predictive" sciences. In the archaeologists' view, this is exemplified by the Indus valley civilization, for which some of the recent writers prefer to use the expression Harappan Culture or Mature Harappan Culture—Harappa being taken as the typesite of this culture, which extended far beyond geographical borders of the Indus Valley.[38]

Earlier archaeologists like Marshall and others tried to estimate its date, depending mainly on evidences that may be called circumstantial. In recent years, however, it has become possible to be fairly certain about it. As the Allchins[39] observe :

> The advent of radiocarbon dating has provided a welcome new source of information on what must otherwise have remained a very vague position, and may well necessitate a revision of the earlier views.

38. See, M. R. Mughal, in Possehl's ACI 91 and B. K. Thapar in FIC 1 ff. for these terminological differences. Such differences, however, are not considered essential for our own discussion and hence are used interchangeably.
39. B. & R. Allchin RCIP 218.

By 1956 Fairservis had seen in the (as yet uncalibrated) radiocarbon dates of his excavations at Quetta valley a need to bring down the dating of the Harappan culture to between 2000 and 1500 B. C. In 1964 D. P. Agrawal, of the radiocarbon laboratory attached to the Tata Institute of Fundamental Research in Bombay, was able to plot some two dozen dates, including those for Kot Diji, Kalibangan and Lothal, and to draw the conclusion that the total span of the culture should be between 2300 and 1750 B. C. (based on uncalibrated dates). This evidence still appears to be most plausible.

We may be yet far from being certain about the actual cause of the final decline or destruction of the Harappan cities. Whatever it had been, however, there are grounds to think that this took place sometimes near 1750 B.C. We have thus an idea of the time when the First Urbanization in ancient India came to its end.

And what, approximately, is the date of the beginnings of the Second Urbanization? We quote A. Ghosh,[40] to whom we owe an important monograph on the subject: "For her next cities, her 'second urbanization', India had to wait for over a thousand years after the disappearance of the Indus cities—till the middle of the sixth century B.C., which saw simultaneously the beginnings of her historical period."

And how do the archaeologists want us to view the period intervening the two urbanizations? Subbarao[41] says, it "is dramatically called the Dark Period". B. B. Lal[42] also speaks of "the Dark Age between the end of the Indus civilization and the beginning of the early historical period."

Following this practice of the archaeologists, therefore, we may sum up the periodisation of ancient Indian history as follows, though adding that the scene of this was confined mainly to northern India:

(a) Period culminating in First Urbanization as evidenced by the Mature Harappan Culture, which came to its end sometime near 1750 B.C.

(b) A Dark Period or Dark Age of over a thousand years following the end of the First Urbanization.

40. A. Ghosh, CEHI 2.
41. B. Subbarao, 100.
42. B. B. Lal, in CF 22.

(c) Period of Second Urbanization, first foreshadowed roughly in the sixth century B.C. and leading eventually to the formation of the early historical cities.

Before passing on to further discussion, we may have a brief note on the second of these three. The archaeologists called this the Dark Period or the Dark Age, because nothing substantial was known about it even a few decades back. Thanks to the brisk field-work of the Indian archaeologists in recent years, however, we have ampler information about this period than we had even at the time of Indian independence.

But, in spite of this, specially from the viewpoint of technology and science, the period continues to be "dark" after all. It was—as compared to the highly urban achievements of the ancient Indus cities—a period of reverting again to the stage of pre-literate peasant communities. The technique of writing—or, more strictly, of the use of script—which is an important trait of the First Urbanization—is lost during this intervening period, notwithstanding the attempt to see in the graffiti marks on the pottery of Rangpur some kind of survival of the Indus scripts.[43] The technique of using kiln-burnt bricks for construction of houses and other monumental structures in the Indus cities is totally forgotten and, in the intervening period, we come across only humble habitations made of mud or mud-bricks. And so on.[44] Archaeologically, the main index to this period is a pottery type, known as the Painted Grey Ware.[45]

43. S. R. Rao, in AI Nos. 18-19. 1963, 5-207. See also, A. Ghosh CEHI 75, and B. B. Lal in AI No. 16, 4-24.

44. But all this, as we shall see later, does not substantiate A. Ghosh's thesis CEHI 73ff of the complete and absolute extinction of the Harappan culture in Indian history. In spite of the intervening dark age, later Indian culture is characterised by strong and important survivals of the Harappan culture.

45. About 700 PGW sites are located so far mainly in the region associated with Vedic settlements and there is the general tendency among the scholars to associate this pottery with the Vedic people. But R. S. Sharma in MCSFAI 57 wants us to note the following—"However there is nothing like an exclusive PGW culture because other wares such as black-and-red ware, black-slipped ware, red ware and plain grey ware are also associated with them. Although very distinctive, the PGW sherds are not numerically predominant at any place. At Atranjikhera, where the PGW covers an area of about 650 sq. m., its incidence ranges between three and ten per

Though as pottery it is considered very fine, its sites are not indicative of advanced material culture. In this context of practically all-round technological regression, it is evidently useless to expect much of scientific activities, outside perhaps certain formal disciplines arising from the technique of orally transmitting a vast literature.

The evidence of this "dark age" intervening the periods of two urbanizations is important for our understanding of the history of technology and science in ancient India, because it is this that defines the chronological horizon within which we are to place all that could have been the distinct contributions of the Indo-European speaking peoples who called themselves the Aryans and who gave to Indian culture a vast body of orally composed songs and hymns that were eventually compiled in the form of the *Rgveda-samhita*. We shall later see that it is impermissible to imagine these people entering the Indian subcontinent much before the end of the First Urbanization. Nor is it permissible to imagine that these people could have retained much of their original identity towards the beginnings of the process of Second Urbanization, because by this time they got inextricably mixed up with the local peoples among whom they advanced, adapting and thriving mainly on the material culture that had locally developed. As Ghosh[47] puts it, "In short, the early Aryan society had made room for the Indian Society, in which it is difficult to isolate Aryan and non-Aryan elements." Thus whatever could have been distinctive of the contributions of these Aryans is to be sought in the period intervening the two urbanizations. For the main subject of our study, all these warn us against the extravagant enthusiasm of some of our scholars to read in the Vedas a great deal of achievement in science and technology. But more of this later.

cent of the total pottery complex. Even where their number is fairly large, the PGW sherds may not exceed fifteen per cent of the total pottery recovered from the PGW layers. Thus the PGW horizon represents a composite culture, just as the culture revealed by the later Vedic texts represents an amalgam of Sanskritic and non-Sanskritic, Aryan and non-Aryan elements."

46. This is corroborated by the internal evidences of the *Rgveda*. For example, RV. I. 38. 14.
47. A. Ghosh, CEHI 4.

CHAPTER 3

EXACT SCIENCE AND THE URBAN REVOLUTION

1. PRELIMINARY REMARKS

It is perhaps expected that we should now pass on straight-way to the account of the two urbanizations—and also of the period intervening between the two—and take up the question of science and technology in these periods. However, it may be a methodological advantage if we begin with some preliminary clarification of the general question concerning the relation between urbanization and the making of exact science.

We have already had some words on the relation between technology and science and we have referred to the view according to which any attempt to try to understand science without its relation to technology is liable to be arbitrary. Following this understanding, Farrington quotes J. G. Crowther, who describes science as "the system of behaviour by which man acquires mastery of his environment". [1] Farrington makes this the basis of his extremely meaningful survey of *Greek Science,* inclusive of the tendency of some of the later Greek philosophers to view science as a purely disinterested search for Truth. As he[2] briefly puts it :

> Science whatever be its ultimate developments, has its origin in techniques, in arts and crafts, in the various activities by which man keeps body and soul together. Science arises in contact with things, it is dependent on the evidence of the senses, and however far it seems to move from them, must always come back to them. It requires logic and the elaboration of theory, but its strict logic and choicest theory must be proved in practice. Science in the practical sense is the necessary basis for abstract and speculative science.

There is thus an obvious advantage in viewing science as a system of behaviour by which man acquires mastery of his environment. Taken in an absolute sense, however, it has the risk of making the denotation of science too wide, inasmuch as it makes the emergence of science coextensive with that

1. Farrington GS 18.
2. *Ibid.*

of man as a tool-making animal. Not that there is nothing in such a view. As Gordon Childe[3] says :

> Even the simplest tool made out of a broken bough or a chipped stone is the fruit of long experience—of trials and errors, impressions noticed, remembered and compared. It may seem an exaggeration, but it is yet true to say that any tool is an embodiment of *science.* For it is a practical application of remembered, compared, and collected experiences of the same kind as are systematized and summarized in scientific formulas, descriptions and prescriptions.

Notwithstanding what is evidently true in this, the fact remains that, ordinarily at any rate, when we speak of science we do not use the word in so wide a sense. In Gordon Childe's own masterly survey of global archaeology, it was only after the prolonged acccumulation of quantitative changes in the technological accomplishments of man that a stage is reached when a qualitative leap is taken in man's ability to control and understand the world in which he lives. Hence is the need felt for a new terminology to refer to the new dimensions assumed basically by the same ability. The term Gordon Childe himself uses for it is "exact and predictive science." It knocks as it were for the gates to be first opened for the coming of what we are ordinarily accustomed to call 'science' today.

2. SCRIPT AND EXACT SCIENCE

The decisive invention that makes science in this sense possible is that of writing or of the script. In the pre-literate communities, the technological experience in which science is only implicit is transmitted as craftlore, in the form of oral precepts and examples. Such a mode of the transmission of knowledge is essentially imitative with little or no "opportunity of introducing a variation which might be beneficial".[4] Hence there is no scope for developing an understanding of how to do something in order to do it better.[5] Craftlore, in short, is circumscribed by severe conservatism, various aspects of which we shall later see.

With the invention of the script and the possibility of committing to writing in general terms the knowledge derived

3. Childe WHH 9.
4. *Ibid* 79.
5. Bernal SH 47.

through the interrogation of nature, the conservatism characteristic of craftlore is largely broken, and, along with it, many a fetter for technology to develop into science. The profound importance of the invention of writing for the historical development of science can hardly be exaggerated. This, as Gordon Childe observes, "was destined to revolutionize the transmission of human knowledge. By its means a man can immortalize his experience and transmit it directly to contemporaries living far off and to generations yet unborn. It is the first step to raising science above the limits of space and time."[6]

It remains for us to see, of course, if there is another side of this tremendous step forward in the development of knowledge. But let us not complicate our present discussion with such questions.

3. "THE URBAN REVOLUTION"

Our understanding of the invention of writing—and therefore also of the first formation of exact science as leaning on this—would be very inadequate if viewed without their appropriate socio-economic background. These are rather to be understood as features of a profoundly significant socio-economic transformation for which Gordon Childe uses the expression "the urban revolution."

In 1936, he formulates this concept in the book *Man Makes Himself*. In 1942, he reiterates his main points about it and carries forward the subsequent story of man upto the decline and fall of the ancient world in the book *What Happened in History*. In 1950, he sums up his main results in a brief article published with the title *The Urban Revolution*.[7]

In this article—surveying mainly the civilizations of the Old World, namely Egypt, Mesopotamia and India, and comparing these also with the Mayas of America—he formulates "ten rather abstract criteria, all deducible from the archaeologist's data" which "serve to distinguish even the earliest cities from any older or contemporary village."[8]

6. Childe MMH 186.
7. Childe 'The Urban Revolution' first published in *Town Planning Review*, Vol. 21 No. 1, 3-17. Liverpool Univ. Press. Reprinted in Possehl's ACI 12-17. Our references to the article are to the latter.
8. Childe 'The Urban Revolution' in Possehl's ACI 15.

This article has provoked a great deal of discussion among the archaeologists, sociologists and historians, though some of the critics of Gordon Childe appear to have overlooked his main thesis specially as worked out in his earlier books in greater details. In any case, as it is rightly observed, there is no better categorization of the urban traits than offered in this brief essay.[9] We may conveniently begin with it.

Here is how Gordon Childe[10] sums up the most essential point for understanding the urban revolution :

> About 5000 years ago irrigation cultivation (combined with stock-breeding and fishing) in the valleys of the Nile, the Tigris-Euphrates and the Indus had begun to yield a social surplus, large enough to support a number of resident specialists who were themselves released from food-production. Water-transport, supplemented in Mesopotamia and the Indus Valley by wheeled vehicles and even in Egypt by packed animals, made it easy to gather food stuffs at a few centres. At the same time dependence on river water for the irrigation of the crops restricted the cultivable areas while the necessity of canalising the waters and protecting habitations against annual floods encouraged the aggregation of population. Thus arose the first cities—units of settlement ten times as great as any known neolithic village. It can be argued that *all cities in the old world are offshoots of those of Egypt, Mesopotamia and the Indus basin.* So the latter need not be taken into account if a minimum definition of civilization is to be inferred from a comparison of its independent manifestations.

Before any attempt is made at a critical evaluation of Gordon Childe's view, it is necessary to be clear about the view itself. Two points need to be specially noted in this connection. First, what he calls "the urban revolution" is not to be equated to any and every process of urbanization, revolutionary though such a process may be. In other words, what he is really discussing is the first momentous socio-economic transformation in human history that leads to the formation of the earliest cities. This takes place in the old world in three regions, namely Egypt, Mesopotamia and the Indus basin, which are accordingly viewed by him as the three "primary centers" of the urban revolution. From Gordon Childe's viewpoint, all

9. D.K. Chakrabarti, and A. Ghosh in *Puratattva* No. 6, 1972-73. 27, 34. See also, A. Ghosh in Possehl's HC p. 20ff.
10. Childe 'The Urban Revolution' in Possehl's ACI 14-15. Emphasis added.

the ten distinguishing features of the urban revolution are not to be sought in the later centres of urbanization, though all these later centres are historically indebted to the three primary centres of the urban revolution. This point is worked out at some length in the book *What Happened in History,*[11] though in the brief article on *The Urban Revolution* also, Gordon Childe wants to be very clear about it :[12]

> .... all later civilizations in the Old World may in a sense be regarded as lineal descendants of those of Egypt, Mesopotamia or the Indus. But this was not a case of like producing like. The maritime civilizations of Bronze Age Crete or classical Greece for example, to say nothing of our own, differ more from their reputed ancestors than these did among themselves. But the urban revolutions that gave them birth did not start from scratch. They could and probably did draw upon the capital accumulated in the three allegedly primary centres. That is most obvious in the case of cultural capital. Even today we use the Egyptians' calendar and the Sumerians' divisions of the day and the hour. Our European ancestors did not have to invent for themselves these divisions of time nor repeat the observations on which they are based ; they took over—and very slightly improved systems elaborated 5,000 years ago !

Secondly, the ten criteria mentioned by Gordon Childe for distinguishing the earliest cities from the neolithic villages are not to be understood as just a compilation or an inventory of certain empirically observed features of the urban centres. From his point of view, at any rate, these are rather internally related with each other, and hence no one of these is to be understood as detached from the rest. Adams[13] observes that these traits are but "loosely associated features", which is precisely what Gordon Childe does not intend his view to be understood. If, therefore, one or other of these traits is not palpable or obvious in the case of any of the three primary centres of the urban revolution, from Gordon Childe's viewpoint it would be worthwhile to seek for circumstantial evidences that may be pointers to it, instead of hastily concluding that in this particular centre such a trait is just missing.

11. Childe WHH Chap. vii. 130 ff.
12. Childe 'The Urban Revolution' in Possehl's ACI 17.
13. R.M. Adams 11.

With these two clarifications, let us proceed to understand his main criteria of the urban revolution. We shall begin with brief notes on the first few of these and then pass on to consider the traits that interest us most for our own discussion. Thus :

(1) The first cities are "more extensive and more densely populated than any previous settlements". (2) These accommodate—perhaps over and above a small minority who are still peasants—"classes who did not themselves procure their own food by agriculture, stock-breeding, fishing or collecting, full-time specialist craftsmen, transport workers, merchants, officials and priests"—all these supported by the surplus produced by the peasants living mainly in the dependent villages. (3) Each primary producer paid over the tiny surplus he could wring from the soil...as tithe or tax to an imaginary deity or a divine king who thus concentrated the surplus. (4) "Truly monumental public buildings" in the cities, symbolizing "the concentration of the social surplus." (5) Though in the cities all those not engaged in food-production were supported by the surplus accumulated in the temple or royal granaries, yet "naturally priests, civil and military leaders and officials absorbed a major share of the concentrated surplus", and thus formed the "ruling class", which "did confer substantial benefits upon their subjects in the way of planning and organisation." All this helps us to understand the sixth and seventh traits of the urban revolution which, because of their immediate relevance for our discussions, we quote[14] in full from his essay :

> (6) They (the members of ruling classes) were in fact compelled to invent systems of recording and exact, but practically useful sciences. The mere administration of the vast revenues of a Sumerian temple or an Egyptian pharaoh by a perpetual corporation of priests or officials obliged its members to devise conventional methods of recording that should be intelligible to all their colleagues and successors, that is, to invent systems of writing and numeral notation. Writing is thus a significant, as well as a convenient, mark of civilization. But while writing is a trait common to Egypt, Mesopotamia, the Indus Valley and Central America, the characters themselves were different in each region and so were the normal writing materials.

14. Childe 'The Urban Revolution' in Possehl's ACI 16.

(7) The invention of writing—or shall we say the inventions of scripts—enabled the leisured clerks to proceed to the elaboration of exact and predictive sciences—arithmetic, geometry and astronomy. Obviously beneficial and explicitly attested by the Egyptian and Maya documents was the correct determination of the tropic year and the creation of a calendar. For it enabled the rulers to regulate successfully the cycle of agricultural operations. But once more the Egyptian, Maya and Babylonian calendars were as different as any systems based on a single natural unit could be. Calendrical and mathematical sciences are common features of the earliest civilizations and they too are corollaries of the archaeologists' criterion, writing.

4. MATHEMATICS, ASTRONOMY AND THE URBAN REVOLUTION

We shall return later to discuss Gordon Childe's formulation of some other traits of the urban revolution. For the present let us have some clarifications about what is already quoted.

We propose specially to note three points in this connection.

First, the emergence of mathematics and astronomy as exact sciences is viewed here not as a mere concomitance of the urban revolution. It is understood, on the contrary, as an essential feature of a highly complex and profoundly significant socio-economic transformation, which Gordon Childe has been discussing. As he puts it elsewhere, "The synchronism is not accidental. The practical needs of the new economy had, in fact, evoked the innovations."[15]

Secondly, while discussing the first creation of exact and predictive sciences, Gordon Childe mentions only mathematics (arithmetic and geometry) and astronomy (inclusive of the calendrical science). He does not mention natural sciences well-known to us in other forms, like chemistry and physics, botany and zoology—and not even any system of rationalist medicine. Evidently, there is no evidence for the making of these natural sciences in the earliest centres of civilization. This is quite striking, because the *potentials* of these must have been there in the spectacular technological achievements presupposed by the urban revolution. "Between 6000 and 3000 B.C.", as he says, "man has learnt to harness the force of oxen and of winds, he invents the plough, the wheeled cart

15. Childe MMH 179.

and the sailing boat, he discovers the chemical processes involved in smelting copper ores and the physical properties of metals, and he begins to work out an accurate solar calendar... In no period of history till the days of Galileo was progress in knowledge so rapid or far-reaching discoveries so frequent".[16] However, when it came to the question of processing this knowledge and of giving it the form of exact science, steps were taken for the creation of mathematics and astronomy, and not the physical and biological sciences so familiar to us. The techniques in which the potentials of these natural sciences are implicit continued to be transmitted in the form of oral precepts and examples, allowing knowledge in other forms to be committed to writing as exact sciences. Apparently, there was something about the urban revolution that required the making of mathematics and astronomy as apparently again there was also something about the profound socio-economic transformation that discouraged the creation of natural sciences in other forms. We shall see how Gordon Childe wants us to understand this when we return to discuss his tenth trait of the urban revolution—a theme discussed by him in fuller details in the last two chapters of his *Man Makes Himself*.

Thirdly, his view of the urban revolution is based on a review of mainly three primary centres of the earliest civilizations in the Old World, namely Mesopotamia, Egypt and India. But his observation just quoted about the emergence of exact science is silent about India, notwithstanding his assertion that from the standpoint of technological accomplishments the Indian centre was "the peer of the rest". Could it, then, be that in spite of the splendid technological achievements, there was no step taken in the ancient Indus civilization towards the creation of mathematics and astronomy? This would certainly be farthest from his understanding. The ten traits of the urban revolution are for him not a mere jumble of loosely connected or unconnected features empirically observed in the earliest civilizations; these are—though discernible—rather internally related traits in terms of which it may be convenient for us to understand the momentous step taken by man towards civilization. The making of mathematics and astronomy being one of these traits is only to be expected also in the

16. *Ibid* 105.

Indus civilization. As a matter of fact, as we have already seen, Gordon Childe, in his book *What Happened in History*, proposes to seek the roots of classical Indian science in the achievements of the Indus civilization.

However, the difficulties about a straight-forward assertion of this are well-known. We have nothing of the nature of a direct document to prove mathematics and astronomy in the Indus period. The Indus civilization is equipped with the technique of writing no doubt. But, in spite of there being a rather imposing literature on the Indus script,[17] the fact is that it still remains undeciphered. It is even argued that the script "is virtually undecipherable".[18] Even admitting the possibility of it being deciphered in the future, there is really little or no hope of the Indus inscriptions—the total number of which so far known is in the region of 2,500[19]—throwing light on the scientific activities of the time, though there are some recent efforts to reconstruct from these elements of astronomical and calendrical knowledge.[20] These inscriptions are usually found on what are called the "seals", and a few also on the reverse side of small flat copper tablets. But the archaeologists are yet to be exact about what these "seals" were really meant for. "Though all these objects have been classed as seals, there is no evidence of their having been used to seal anything, whereas in Mesopotamia and Egypt sealings are far commoner than seals... The term 'seal' is therefore conventional, and the objects thus denoted must be classed as ritual—a learned way of saying that we have no notion what they were for".[21]

17. We may have some idea of this from the references given by Asko Parpola in EIP 180-186.
18. Arlene R.K. Zide, in Possehl's ACI 259.
19. B. & R. Allchin. RCIP 212. Mahadevan IS 6 catalogues 2906 legible inscriptions.
20. Asko Parpola in EIP 179. cf. also Bongard Levin, *The Image of India* 190 ff and Bongard-Levin in SUSSR 1981. 71 ff. However, as we shall later see, it seems to be premature to be exact about the value of such a claim.
21. Childe NLMAE 182. But is is recently claimed by some Indian archaeologists that there are evidences of these being actually used for sealing purposes. S.R. Rao LIC 119, e.g., asserts: "Lothal has the unique distinction of proving conclusively that

In any case, there is no ground to imagine that the inscriptions on the seals were intended to document scientific activities, beyond perhaps what is implied by the bare use of certain signs often viewed as numerals. The inscriptions are in fact too brief—consisting of five or six signs on an average—to suggest such documentation in spite of the increasing importance being attached to the hypothesis of these inscriptions embodying some calendrical information.

Thus, in short, we have nothing among the archaeological relics of the Indus civilization directly documenting the making of mathematics and astronomy as exact sciences. We have nothing comparable to the Egyptian papyri or Mesopotamian clay tablets witnessing to the same in the other two primary centres of the urban revolution. Of course, the possibility of there having been documents written on unknown but perishable materials cannot be absolutely ruled out, particularly because it is *prima facie* difficult to imagine that the script was invented by the makers of the Indus civilization only for the purpose of using these on the 'seals' on which these survive. But it is somewhat precarious to build up any dependable hypothesis mainly on the strength of the unknown, though from the presumption that the seal-cutters were themselves not the scribes, the use of perishable materials originally used by the scribes cannot be easily dismissed.

At the same time, as we are going to argue, there are indirect or circumstantial evidences justifying Gordon Childe's presumption of classical Indian mathematics and astronomy having roots in the ancient Harappan achievements.

the Indus seals were used for a commercial purpose and not as amulates or ornaments". The question, still appears to remain an open one.

CHAPTER 4

THE FIRST URBANIZATION

1. PRELIMINARY REMARKS

Before we proceed to the question of the possibility of the making of exact science—specially of mathematics and astronomy—in our First Urbanization, it is desirable to have some idea of the First Urbanization itself.

After the discovery of Harappa and Mohenjo-daro, excavations and explorations of about the last six decades—inclusive of the work done in Pakistan—have considerably extended our idea of the frontiers of the Indus civilization. So also have increased the nature and number of controversies about it. An important reason for the controversies is already noted. The Indus script remains undeciphered, notwithstanding many claims to the contrary. Compared to Egypt and Mesopotamia, we are not in possession of any literary document giving direct knowladge of such features of the Harappan culture as its socio-political organization or general intellectual climate. As the Allchins rightly put it, "for any society lacking written records, or whose script is still undeciphered, evidence of such matters as political conditions is clearly hard to come by, and is at best inferential".[1] On such matters, therefore, we are left with the choice between dependable inferences and less dependable ones, or between sound and comparatively less sound presumptions based on materials unearthed, combined often with circumstantial evidences.

In any case, our starting point is the body of the material evidences, about which there is no scope for conjecture.[2]

2. EXTENT AND POPULATION

In 1972, M. R. Mughal[3] observes :

> The evidence shows that the area covered by the Indus Civilization was larger than any of the known civilizations of the ancient world.

1. B & R. Allchin, BIC 129n.
2. Though, in this connection we have to note the observation of D.K. Chakrabarti in EIP 206 that there is sometiems discripancy even between the actual data and published reports about some of the sites.
3. M.R. Mughal, in Possehl's ACI 91.

Starting from the borders of Afghanistan, in northern Baluchistan (at Periano Ghundai), and the Iranian border on the Makran coast (at Sutkegan-Dor), it extended east and south-east and covered the entire Makran coast, the Greater Indus Valley and Gujarat. Beyond the vast plain of the Punjab, including that part which was formerly drained by the Ghaggar-Hakra river, remains of the Indus culture have also been found near Delhi in the Ganges-Yamuna doab. This enormous area could not have been limited to only 144 sites (the number so far securely identified) and, indeed, claims have been made for the discovery of twentyseven sites of Harappan affinity in east Punjab and the Doab near Saharanpur.

In 1983, the Allchins give us an approximate measurement of the area : "The area enclosed by a line joining the outermost sites at which the material culture of this civilization has been discovered is little less than half a million square miles, considerably larger than modern Pakistan".[4]

It is true that within this vast area of about 500,000 square miles, some of the sites where the remains of the Indus culture have been found "were probably ports or trading posts situated in an otherwise separate culture region"[5] while some others like those "in the Kutch and Saurashtra area, and those which penetrated into the doab represent later movements of the people from the Indus Valley proper".[6] Admitting all this, the area covered by the civilization has still to be considered as quite vast.

In 1979, D. K. Chakrabarti wants to go into some detail of this vast area : "There is no detailed gazetteer of the Harappan sites available and most of the sites are only perfunctorily reported. Pande and Ramachandran listed 258 sites in 1971. While some sites . . . should now be added to the list there is no assurance that all of these 258 sites should strictly come under the category of the Harappan".[7] Such difficulties notwithstanding, Chakravarti—depending on published data—gives us a list of some of the major Harappan settlements along with their size estimated in square metres. We give below a synopsis of his list :[8]

4. B. & R. Allchin, RCIP 167.
5. *Ibid* 167.
6. M.R. Mughal, in Possehl's ACI 91.
7. D.K. Chakrabarti in EIP 205.
8. *Ibid* 210-212.

Site	*Size in Square Metres*	*Site*	*Size in Square Metres*
Mohenjo-Daro	848058 (or roughly 850000)	Rahman Dheri	189070 (or 200000)
Ali Murad	101840 (or 100000)	Harappa	850000
Chanhu-Daro	64752 (or 65000)	Vainiwal	73872 (or 75000)
Karchat	47196 (or 50000)	Ahmad Khan Dheri	20748 (or 20000)
Lohum-jo-Daro	49886 (or 50000)	Dewali-na Ther	215364 (or 222000)
Nokjo-Shadinzai	75076 (or 75000)	Lure-wala	819200 (or 820000)
Mehi	99029 (or 100000)	Trekoe	417195 (or 420000)
Pandi Wahi	14522 (or 15000)	Qadar-wali Ther	53144 (or 55000)
Thano-Bula-Khan	14934 (or 15000)	Sandhan-wala Ther	53144 (or 55000)
Kotasur	91790 (or 90000)	Kali-bangan	140210 (or 150000)
Naru-Waro-Dharo	347234 (or 350000)	Lothal	46938 (or 47500)
Theri Bahadur Shah	38120 (or 40000)	Kanjetar	23104 (or 20000)
Judeir-jo-Daro	250436 (or 250000)	Kindar-khera	8281 (or 10000)
Kot Diji	22022 (or 22000)	Devaliyo	318384 (or 320000)
Sutka-gendor	17819 (or 20000)	Bhim-patal	11011 (or 10000)
Dabar-kot	187827 (or 200000)	Akru	40200 (or 40000)
		Malva	18291 (or 20000)
		Surkotada	26000
		Kotadi	140000

Attempt has also been made to estimate the possible population of these. Here is how Chakrabarti[9] sums it up :

> Taking the population estimates for Mohenjo-daro as an index one may speculate on the number of sites which were likely to contain at least more than 5000 people. There are three estimates for the population of Mohenjo-daro... On the basis of his estimate of the

9. *Ibid* 206. Compare population estimate in major Mesopotamian sites in Childe WHH 94.

quantity of grain stored in the granaries, which, he thought, was used only for the general civic consumption, Datta ... [gives] a total population of 33,469 for Mohenjo-daro ... Allowing a ratio of 800 square feet per person, Fairservis put the total population of Mohenjo-daro at 41,250 and considered this figure to be on the conservative side. ... On the basis of a nineteenth century statistics for Shikarpur in north-west Sind, which he thought, closely resembled Mohenjo-daro both in dimensions and lay-out, Lambrick estimated the total population of Mohenjo-daro to be 35,000.

Accepting Lambrick's estimate of 35,000 people in Mohenjo-daro with an area of 850,000 square metres, and "assuming that the density of population in other settlements was more or less the same, one feels that the settlements having an area of 1,25,000 square feet or more were likely to possess 5,000 or more people each".[10] Besides Mohenjo-daro and Harappa, a considerable number of other settlements answers to this. As Chakrabarti observes, "One can possibly say with some emphasis that there were at least 15 Harappan settlements with more than 5,000 people each. The number is likely to be somewhat more when the full data are available for most of the sites".[11]

3. AGRICULTURAL SURPLUS

This naturally raises the question of the basic substance of the city dwellers. A fraction of the population of the comparatively larger settlements could have been direct producers of food—perhaps cultivating the tract of land which every Harappan settlement is sometimes presumed to have around it.[12] But it is impossible to imagine that the city dwellers themselves produced all the food they required, for there must have been fulltime specialists among them engaged in other occupations. According to Fairservis,[13] the composition of the non-agricultural population of Mohenjo-daro was as follows : A) *Administrative*—(1) Priests, (2) Scribes and Seal cutters, (3) Musicians and dancers, (4) Engineers ; (B) *Productive*—(1) Potters, (2)

10. D.K. Chakrabarti in EIP 206.
11. *Ibid* 207.
12. *Ibid* 209.
13. Fairservis in Possehl's ACI 84. Fairservis also gives his idea of the estimated population of the Indus cities : *Ibid* 83.

Weavers, (3) Brick-makers, (4) Masons, (5) Carpenters, (6) Metallurgists and (7) Traders.

This might or might not have been true of all the Harappan settlements, though in this connection it may be worthwhile to remember the following :[14]

> What also seems striking is an almost obvious fact that the basic elements of the Harappan planning seem to be present in all settlements, big or small. At one end of the scale there are sites like Mohenjo-daro and Harappa with their large population and the classic elements of the Harappan civilization while at the other end there are sites like Sutkagendor and Kot Diji which are among the small Harappan settlements with hardly more than a thousand inhabitants and yet possessing some of those distinctive traits of planning which are associated with the largest settlements ... Surkotada was hardly larger than any of these. It contained not merely evidence of systematic planning but also two inscribed seals ... Rao's claim for a population of 15,000 notwithstanding, Lothal was unlikely to have contained more than 2000-2500 people and yet this stands out as a magnificent example of Harappan urban planning and cultural traits. ...Chanhu-daro was not much larger and surely did not contain more than 5000 people on the existing evidence but all the basic traits of a Harappan settlement were here.

Admitting this, are we not permitted also to presume basically the same composition of non-agricultural population in most of the major settlements in Harappan culture? In any case, there seems hardly any scope to doubt that the vast number of city dwellers in the major Harappan settlements—particularly the resident specialists who, in order to specialise, had themselves to be relieved of the direct responsibility of food production—must have subsisted on the surplus produced by the villagers and channelised to the cities. In short, without the assumption of an accumulation of enormous amount of agricultural surplus, it is impossible for us to understand the ancient cities of the Indus.

This brings us to the point which, on Gordon Childe's analysis, forms the first or the most essential precondition for understanding the urban revolution. In recent years, however, there is the feeling among a section of sociologists that Gordon Childe's view must be resisted all along the line. The feeling seems to be prompted by factors more than barely academic.

14. D.K. Chakrabarti, in EIP 207-208.

Gordon Childe's view has its immediate political implications. The production of agricultural surplus and its extraction from the direct producers is inextricably related to the phenomenon of class formation and of the origin of state machinery in service of the privileged classes—an implication politically distasteful to many. Hence is the need felt to reject Gordon Childe's view and offer substitutes to it. Unfortunately, these sociologists, in spite of being unanimous in opposing Gordon Childe, have failed to find an agreed alternative to it. Fortunately, however, it is not necessary for our present purpose to digress into the details of this recent controversy. All that needs basically to be said about it, is already very lucidly said by R. S. Sharma.[15] As he observes :

> The effort to eliminate class and surplus has introduced 'elite', 'status', 'hierarchy', 'decision-making', etc. in their place. The theory of surplus is rejected on the ground that people do not produce more on their own but are compelled to put in more work or more people are mobilized for work. Whatever motives be assigned for producing more—and this will differ from society to society—almost all types of serious investigators admit that only extra produce can support wholetime administrators, professional soldiers, full-time priests, craftsmen and other similar specialists who do not produce their food themselves. The argument that people were compelled to produce more would imply the existence of an organized coercive authority such as the state or at least a protostate represented by a strong chief, but it would not negate the idea of surplus. With increase in production, voluntary or reciprocal gifts made by kinsmen in a tribal set-up marked by low productivity are perverted and converted into compulsory or unilateral payments, for producers are forced to part with a portion of their produce. Whatever be the methods of making people pay, it is clear that these can succeed only when the capacity to pay is created. Surplus plays a key role in the formation of class and leads to the erection of an entirely new type of power structure called the state.

4. POSTULATE OF CENTRALISED POLITICAL POWER

All this leads us to the question of some centralised political power in the vast Indus Valley Civilization.

The imposing granaries unearthed at Mohenjo-daro and subsequently also at Harappa—with a total floor space of over

15. R.S. Sharma, MCSFAI. Intro. p. xv.

800 square metres at both the sites[16]—give us some idea of where the surplus on which the citizens subsisted was stored. However, in default of any telltale evidence for the nature of the machinery used for extracting the surplus from the direct producers, there are controversies about it. Without trying to enter here into these controversies, it is perhaps permissible to observe that whatever might have been the nature of this, the need for postulating some central political power or state structure cannot be evaded. At the presest stage of our discussion, we only quote some of the views expressed on the general need felt for assuming some centralised political power without trying to specify nature.

In 1950, Piggott[17] observes,

> Within the area already described, the uniform products of the Harappa civilization can be traced with the monotonous regularity of a highly-organized community under some strong system of centralized government, controlling production and distribution and no doubt levying a system of tolls and customs throughout the territory under its rule. As we shall see, there is no evidence to imply that the cities of Harappa and Mohenjo-daro were not contemporary : laid out to a common ground-plan, each with its defenced citadel towering above the rest of the town, they seem to have been twin capitals, a northern and a southern, of one united kingdom. One is reminded of historical parallels in North-West India when Sakas and Kushanas ruled from Taxila or Peshawar in the north, and Muttra in the south, over a single state.

In 1968, the Allchins, though without discarding the expression "twin capitals of this extensive state", propose to add some caution in the footnote[18] :

> It is perhaps hardly necessary to mention that this glib sentence conceals the cold archaeological truth, that up to today thre is no positive evidence that the cities were capitals, either of separate states or of a unified 'empire' . . . Generations of archaeologists have felt that some such interpretation better fits the Harappan evidence than any other, but necessarily it remains hypothetical. The reader must therefore draw his own conclusions from the available data : the apparent uniformity of weights and measures, the common script, the uniformity—almost common currency—of the seals, the evidence of extensive trade in almost every class of commodity throughout the

16. B. & R. Allchin, RCIP 182.
17. Piggott PI 136.
18. B. & R. Allchin, BIC 129n.

whole Harappan culture zone, the common elements in architecture and town-planning, the common elements of art and religion. Even if the political and economic unity is admitted, there remain the profound and tantalizing problems of how it came about and how it was maintained. These have yet to be tackled satisfactorily.

But in 1983, the same archaeologists appear to be more categorical and observe :[19]

The cultural uniformity of the settlements over such a wide area leaves no doubt that the relationship between the city-centred communities of agriculturalists and craftsmen, and those who provided the means of transport and communication, must have been a stable one. This in turn indicates a strong and firmly based system which held them together and maintained their relations. Precisely what this system was and whence it drew its authority is not yet clear, but of its existence there can be no doubt, nor that it represented a special achievement in the world of the third millennium B.C.

Finally, we must remark that the indications of the superimposition of a uniform language and script (which seems to be the inescapable inference we may draw from the distribution of the sales and inscriptions), and of a uniform mythology and iconography, over so vast an area, are still and must remain sources of real wonder. They remind us of the similar indications of the rule of Asoka Maurya in the third century B.C., or of the Mughal empire at its height, although neither survived for so long as did their first model. But they are above all the indications of the first great promulgation of an interprovincial 'Indian' style, and as such they carry profound implications for the future of Indian thought and culture.

However the system came into being, it must have been built upon remarkably sound foundations, since the 'Indus Empire' appears to have lasted for around five centuries as a major cultural entity, including a number of major cities and regions, at a time when in other parts of the world the largest effective unit was little more than the city state.

In 1972, Mughal[20] observes :

The picture of the Indus Civilization, as presented to us through many years of excavation, is that of a highly disciplined society, possessing sufficient economic wealth to mobilize labour and to support full-time craftsmen. It also possessed resources to engage in long-distance trade or exchange of products. The existence of interrelated but highly developed socio-political and religious institutions, as reflected through their well-planned cities, public buildings, large fortifications,

19. B. & R. Allchin, RCIP 223-24.
20. M.R. Mughal, in Possehl's ACI 92.

granaries and standardization of material equipment through mass production, is evident. Further excavations at the same sites and new investigations at other sites are adding more detail to the picture.

Further quotations are not necessary. For at least a considerable number of serious archaeologists the impression seems to be unavoidable that there must have been a central political power enforcing its authority over the vast extent of the Indus Civilization. We shall return later to the question of the possible nature of this political power or central authority, which should be a pointer to the general theoretical climate of the Harappan culture, and therefore also to the place of science in it. For the present the point is that the channelization of the surplus from the direct producers to the city granaries, without which the Indus cities remain unexplained, was in all presumption the result of the enforcement of some central state power.

5. PROBLEM OF ORIGIN

Even in 1958, Subbarao commented : "Let it be stated straight away that in the present stage of our studies the formative stages, if any, of this great urban civilization have yet to be explored .... At present it appears to us like Minerva born in a panoply".[21] Such an observation seems to favour the view that the civilization could be an imported phenomenon. Though still endorsed by a section of the archaeologists, it does not appear to be tenable, specially in view of the work done by A. Ghose, Fairservis, Mughal and others, though the unfortunate fact remains that the archaeologists are yet to be agreed on the question of the exact account of the origin and antecedence of Harappan urbanisation. D. K. Chakrabarti in his article on *Origin of the Indus Civilization : Theories and Problems*,[22] gives us an overview of the subject, which we follow here.

In 1924, while first announcing the discovery of the Indus Civilization, Marshall observed : "there is no reason to assume that the culture of this region was imported from other lands

21. B. Subbarao, PI 95. Contrast F.R. Allchins AIC.
22. D.K. Chakrabarti, in FIC 43-50.

or that its character was primarily modified by outside influences".[21] In 1950, Piggott, substantially agreeing with this, asserted : "an origin outside India is inherently improbable, but where and in what form this origin was is quite unknown".[24]

However, Wheeler has been consistently arguing against this view. Though admitting that the Harappan culture is "too individual to be regarded merely as a Mesopotamian colony", he argued that "the idea of civilization came to the Indus from the Euphrates and the Tigris".[25]

Already in 1947-48 he[26] observed :

> And yet the idea of city-life on the developed scale of Mohenjo-daro and Harappa at a time when civic models were few and far between, combined with the certainty that the development in India was considerably later than its equivalent development in Mesopotamia and South-Western Iran, seems to impel the inference that there was some sort of causal relationship between the two. Furthermore, there is at Mohenjo-daro, in contrast for example to Ur, an indication of sudden maturity which suggests the intrusion of a perfected civic scheme... Always with the reservation that our knowledge is incomplete, we seem to have in Mesopotamia the early evolution of an idea, in India the later imposition of the idea perfected. If this inference is correct we are almost driven to suppose that the *civic idea* came to India in some fashion from Mesopotamia or South-Western Iran, but that in India it was re-created by an essentially alien, essentially Indian, cultural environment.

It is difficult indeed to follow an archaeologist when he speaks of the migration of "ideas" ignoring palpable mateiral evidences like those of script, 'seals', brick technology and many other peculiarities, from the point of view of which there is no Mesopotamian influence on the Indus civilization. But Wheeler sticks to his hypothesis and even suggests the possibility of political domination of the Harappan cities by the immigrant Mesopotamians :[27]

> There is the suspicion that the citadel-builders of Mohenjo-daro and of Harappa were innovators arriving with architectural traditions founded elsewhere upon the manipulation of mud-brick and timber, and

23. Marshall in ILN 1924 sept. 20.
24. Piggott 140.
25. Wheeler IC 93-94.
26. Wheeler in AI No. 4 91-92.
27. Wheeler IC 93-94.

imposing themselves upon a pre-existing urban population. The high-built citadels seem indeed to be forwning upon their cities with a hint of alien domination. If so, at Mohenjo-daro that domination must have been dynastic rather than cultural, for the excavations of 1950 hinted at a substantial coutinuity of culture from the pre-citadel into the early citadel phase. These and other possitibilites must be given provisional weight without any undue emphasis.

Whatever may be the rhetorical effect of the expression "frowning upon the cities", it can hardly claim even any "provisional weight" from the archaeological viewpoint. Even the *prima facie* difficulties about the conjecture appear to reduce it into a bundle of contradictions. As Chakrabarti rightly observes, it "does not explain why the dynastic domination failed to have a cultural expression. It also does not explain the genesis of his suggested 'pre-existing urban population'. Interestingly enough, Wheeler argues, on the one hand, that the Indus Civilization was too distinctive to be a colony of any alien group and suggests, on the other, that there was an alien, purely political, domination in the Harappan cities".[28]

In spite of such obvious difficulties about Wheeler's conjecture, it is somewhat strange to note that it found favour with writers like Gordon and Heine-Geldern,[29] though rejected straight-way by Lamberg-Karlovsky, who observes : "The origin and formation of the Indus Civilization have been the source of great speculation but limited evidence. For decades it was commonplace to maintain that the Indus Civilization appeared suddenly, in a mature form, around 2400 B.C., the result of diffusion from Mesopotamia. This view can no longer be maintained . . . It is clear that we are just beginning to understand the prehistoric background to the formation of the Indus Civilization. There can be little doubt that when sufficient excavation is undertaken we will comprehend more fully the independent genesis of the Harappan civilization—as independent a creation as that of Egypt and Mesopotamia. Though all three civilizations were contemporary they were entirely distinctive in their form".[30] To this may be added Chakrabarti's comment : "One may only note that invoking external stimulus

28. D.K. Chakrabarti, in FIC 44.
29. *Ibid.*
30. C.C. Lamberg-Karlovsky, and J.A. Sabloff, AC : NEM 189-192.

in a vague, undefined way absolves the archaeologist of any responsibility to explain or look for culture-change in internal terms in any situation where the data are inadequate".[31]

Fortunately responsible archaeologists these days are not going in for such an easy solution of the problem of the origin of the Indus Civilization. We shall mention here only two examples.

In 1972, Mughal very successfully opens the question of the formative period of the Indus Civilization, drawing our attention not only to the significance of the earlier work of N. G. Majumdar in Sind but also to many recent excavations, like those in Kot Diji, Kalibangan, Mitathal and Siswal, Sarai Khola, Gumla, Jalilpur, Surkotada, Amri, etc. On the basis of these, he observes : "Thus, before the rise of large cities of the Indus Civilization, a widespread cultural phenomenon, constituting early, formative phase of the Harappan culture, had already set a permanent and uniform pattern of essential elements. It would seem that the processes leading to urbanization had already begun during the early third millennium B.C., but it is not possible to reconstruct these fully in the present stage of our knowledge".[32]

In 1983, Bridget and Raymond Allchin[33] have made a systematic review of "the developments of those regions which may be seen to contribute directly or indirectly to the emergence of the Indus civilization" and give us an account of the "formative stage underlying the Mature Indus civilization" which they call "the Early Indus Period." We do not have the scope to go here into the details of the archaeological data reviewed by them mainly from such sites as Amri, Kot Diji, Kalibangan, Jalilpur, Mitathal, Rahman Dheri and Tarakai Qila, and others. Nor is this necessary for our present discussion. Nevertheless, one point needs to be noted here. With all that is known so far about the formative stages of the Indus Civilization, we have still the impression of an extra-ordinarily rapid progress—

31. D.K. Chakrabarti, in FIC 50.
32. M.R. Mughal, in Possehl's ACI 94. See also, Jarrige, Jean-Francois and Richard H. Meadow, "The Antecedents of Civilization in the Indus Valley" in SA Aug. 1980. Vol. 243., No. 2, 122-133.
33. B. & R. Allchin, **RCIP 131ff.**

a qualitative leap as it were—from what the Allchins call "the Early or Incipient urban phase" to the "Mature Indus style". This is evidenced by "such important innovations as writing and all the implicit concomitants of political, administrative and social organization"—changes that were an "intrinsic part of the actual emergence of the cities themselves."[34] "This", observe the Allchins, "probably resulted from the successful control and exploitation of the tremendously productive agricultural potentialities of the Indus plain."[35]

6. AGRICULTURE AND AGRICULTURAL SURPLUS

That the "entire civilization flowered forth as a result of surplus agricultural economy"[36] cannot be doubted. At the same time nothing very spectacular either about the agricultural tools or about the agricultural skill is suggested by the archaeological findings of the Mature Harappan period. As Deshpande observes, "The wherewithal of the community in terms of agricultural equipment that we get at Mohenjo-daro and Harappa consisted of a solitary large hoe (Mackay, *Mohenjo-daro* pl. cvi. 56) made of flint, parallel-sided blades used for cutting corn and sickles".[37] The Allchins observe, "There is as yet comparatively little evidence for the actual tools employed for agriculture".[38]

But the most interesting evidence for agricultural technique comes from the pre-Indus period at Kalibangan, "located on the southern bank of the Ghaggar which is now dry, but must have anciently been a substantial river, as indicated by its span". In the occupation named Period I and identified as pre-Harappan is found an extensive patch of furrow-marks. B. B. Lal[39] describes it thus :

> About a hundred metres to the south of the settlement were identified the remains of an agricultural field, with some of the ploughed furrow-marks still in fact. And no less interesting is the fact that the

34. *Ibid* 167.
35. *Ibid* 166.
36. M.N. Deshpande, in IJHS V. 6, No. 1 1971. 6.
37. *Ibid* 7.
38. B. & R. Allchin, RCIP 192.
39. B.B. Lal, in EIP 69.

pattern of ploughing the field continues to be the same even now in that region. The excavated furrows formed a grid on plan. Thus, one set of the furrows was oriented east-west, while the other ran north-south. The individual furrows in the former were interspaced at a distance of about 30 cm. and those in the latter, at 1.9 m. In the crop-pattern in vogue now, horsegram is grown in the short-distanced furrows and mustard in the long-distanced ones. The choice evidently seems to depend on the size and lateral spread of the respective plants, the latter being bigger than the former.

"This finding", comment the Allchins,[40] "therefore provides a dramatic suggestion that an agricultural practice was already in use during early Indus times which has survived locally till today". "The existence of the furrows", adds Sankalia,[41] "implies that there must have been a plough. Very probably it might have been of wood". Though we do not know whether such wooden ploughs had any metal tip of bronze or copper, the furrows indisputably indicate the use of plough and thus imply that already about a couple of centuries before the Mature Harappan culture, plough for field agriculture was introduced by the pre-Harappans, because the end of the pre-Harappan Period I at Kalibangan is dated as 2700 B.C. while the beginning of Period II of Harappan culture is dated at the same site as 2500 B.C.[42]

We have mentioned all these to make only one point. The vast agricultural surplus that made possible the transition from the incipient urban phase to the urbanization of the Mature Harappan period could not be due to any sudden innovation in the agricultural tool or agricultural technique in its restricted sense. As a matter of fact, the agricultural products that filled the granaries of the Harappan cities did not presuppose a great deal of skill or improved implements. These presupposed, on the contrary, an understanding and control of annual inundation of the rivers.

40. B. & R. Allchin, RCIP 192. The discovery scraps the conjecture of D.D. Kosambi SIH 64, that "The Indus people did not have the plough (which is depicted on Mesopotamian seals) but only a toothed harrow which may be recognised as one of their Indus script ideograms".
41. H.D. Sankalia, SAPTI 63.
42. B.B. Lal, in EIP 94.

The Allchins propose to follow Lambrick, who "from his intimate personal knowledge of Sind, has been able to suggest the way in which the various crops would have been grown, and how they exploited the flooding of the Indus." Wheat and barley, the principal food grains, would have grown as spring (*rabi*)—"sown at the end of inundation upon land which had been submerged by spill from the river or one of its natural flood channels, and reaped in March or April. In modern practice such land is neither ploughed nor manured, nor does it require additional water. Lambrick remarks that 'the whole operation involves an absolute minimum of skill, labour and aid of implements'. It was the first development of this which made possible the development of Indus urbanism. Cotton and sessamum would be sown as autumnal (*kharif*) crops ; they would be sown at the beginning of the inundation and harvested at its close, in the autumn. For these fields surrounded by earth embankments would be required, most probably along the banks of natural flood channels..... Both systems are still in use, and they provided a very convincing explanation of the means by which the Harappans filled their vast granaries".[43] Yet the presumption is that since the accumulation of vast agricultural surplus must have been a precondition of the first urbanization, the control of flood in its turn was a precondition for the agricultural output. [See, however, the brief appendix to this chapter contributed by D. K. Chakrabarty].

7. BRICK TECHNOLOGY AND HARAPPAN CULTURE

This leads the Allchins to emphasise the importance of brick technology—or, more specifically the technology of making and using burnt bricks—for the Harappan urbanization. As they[44] put it :

> We have already mentioned the environment of the Indus Valley and the opportunities it offered once the annual innundation had been understood... A vital necessity of settlement in the Indus plain itself would have been flood defence, and here it seems that burnt-brick must have played an important role. For, in these areas where stone was not readily available (and this includes the majority of the Harappan

43. B. & R. Allchin, RCIP 192
44. *Ibid* 167.

sites) mud-brick would have been rapidly destroyed by rain or flood water. Thus the discovery and utilization of burnt-brick was one factor. It has sometimes been suggested that the Indus Valley could not have produced sufficient timber for this operation... However, Lambrick, writing with many years of administrative experience of Sind, has shown that timber growing along the riverine tracts today is sufficient for all the burnt-bricks made in the province, and anciently cannot have been less abundant.

We may not yet have a very precise knowledge of where and when in the Harappan cultural area the technique of making and using burnt bricks took definite shape and eventual sophistication. Reviewing the developments of those regions which may be seen to contribute directly or indirectly to the emergence of the Indus civilization, the Allchins note that such sites as Mundigak IV,[45] Damb Sadaat,[46] Amri,[47] Kot Diji,[48] Gumla[49] Rahman Dheri[50] and Kalibangan[51] are, generally speaking, indi-

45. B. & R. Allchin, RCIP 133-34 ; J.M. Casal, *Fouilles de Mundigka,* Paris : 1961 ; H.D. Sankalia, in PPIP 314-15 ; W.A.Jr., Fairservis, RAI 122-34, Mughal EHPGIVNB 293ff.
46. B. & R. Allchin, RCIP 134-35 ; Fairservis RAI 319ff ; Mughal EHPGIVNB 225-259.
47. B. & R. Allchin, RCIP 141 ; J. M. Casal, *Fouilles d'Amri* Vol. I, Text. Paris : 1964 ; Sankalia PPIP 331-37 ; Fairservis RAI 175-79 ; M.R. Mughal EHPGIVNB 84-87.
48. B. & R. Allchin, RCIP 145 ; F.A. Khan, "Excavations at Kot-Diji" in *Pakistan Archaeology,* 1965, 11-85 ; H.D. Sankalia, PPIP, 338-42 ; Fairservis RAI 179-84 ; Mughal EHPGIVNB 50ff ; Sankalia in *Antiquity* 1969, 142-44.
49. B. & R. Allchin, RCIP 150 ; A.H. Dani, "Excavations in the Gomal Valley" in *Ancient Pakistan* 1970-71, 1-177 ; Sankalia PPIP 329-31 ; Jim G. Shaffer, "The Indus Civilization : New evidence from Pakistan" in EIP 18-20.
50. B. & R. Allchin, RCIP 150.
51. B. & R. Allchin, RCIP 157 ; Aurel Stein, "A Survey of Ancient Sites along the 'lost' Sarasvati River" in *The Geographical Journal* vol. 99. 1942, 173-182 ; A. Ghosh "The Rajputana Desert. Its Archaeological aspects" in *Bulletin of the National Institute of Sciences of India,* No. 1. 1952, 37-42 ; H.D. Sankalia, PPIP 342ff ; B.K. Thapar, "Excavations at Kalibangan" in *Cultural Forum,* July 1967 ; B. K. Thapar, "Kalibangan : A Harappan Metropolis beyond the Indus Valley" in Possehl's ACI 196-202 ; B.B. Lal, "Kalibangan and Indus Civilization" in EIP 65-97 ; M.R. Mughal, EHPGIVNB 81.

cative of the practice of using sun-dried mud-bricks. But there are also indications of steps being taken towards the making of burnt bricks already in the formative stages of the Harappan culture. We mention here two evidences, namely those of Kalibangan and Banwali.

To Aurel Stein, 'Kalibangan seemed to '"offer interest as marking an extensive site used mainly for burning bricks". He had observed that the two "mounds are composed almost entirely of kiln remains and the painted sherds found among them leave no doubt about the Klin having been worked down to the Kushan period." However, as B. K. Thapar observes, "Stein had obviously failed to identify the Harappan remains including pottery. The scatter of brickbats and potsherds noticed by him on the mound represented in facts the telltale evidence of brick-robbing resorted to at the site for laying the Hanumangarh-Suratgarh section of the rail-track; coincidentally a parallel situation to that of Harappa".[52]

To the story of Harappa we shall presently come. For our immediate discussion the point to be noted is that in Kalibangan we come across the technique of making and using burnt bricks already at the pre-Harappan stage. The data so far published about Kot Diji, as the Allchins[53] sum up, "give us little information regarding house plans. However, throughout these five periods at Kalibangan, both dried brick and stone were used, domestically and for town walls. At this same site there appears already to have been a standardization of brick sizes and although the ratio (3 : 2 : 1) differs from that of the Mature Harappan period, it clearly suggests the way in which the Early Indus anticipates what is to follow". To this needs to be added the following. At Kalibangan, in the pre-Harappan level baked bricks were used as is evidenced by a drain of this level which was made of burnt brick 30 X 20 X 10 cm.[54]

The anticipation of burnt brick in the pre-Harappan stage is also evidenced by Banwali in the Hissa district, Haryana. The culture assemblege gathered during excavations here are assigned to three periods, of which Period I represents Pre-Indus or Kali-

52. B. K. Thapar, in FIC 5. The distance between Hanumangarh and Suratgarh is about 30 miles.
53. B. & R. Allchin, RCIP 157.
54. IAR 1967-68, 45. See also, IAR 1962-3, 20.

bangan culture. This has revealed "bricks, both fired and sun-dried", which were moreover "meticulously standardized, conforming to the ratio 1 : 2 : 3. This ratio was observed earlier at Kalibangan. But the dimensions at Banwali are 12 by 24 by 36 centimetres and 13 by 26 by 39 centimetres as compared to 10 by 20 by 30 centimeters at Kalibagan. In addition there is one aberrant size 24 by 24 centimetres, the thickness remaining still indeterminate".[55]

But all this is yet far from what we come across in Mature Harappan culture, mud-brick being generally used in the pre-Harappan period, inclusive of Kalibangan. "The pre-Harappans lived in well laid-out houses... These were made of mud-bricks whose size was the same as in the case of the fortifications-walls. The style of the masonry was that of placing headers and stretchers in alternate courses, which is commonly known as the English bond. Evidence of plastering the walls with mud and chaff was also available".[56]

In the Mature Harappan Culture, however, we are confronted with an "explosion" as it were in the technique of making and using burnt bricks. The scale which the manufacture of burnt bricks assumed appears literally staggering. Piggott tell us the story of how William Brunton in the mid-nineteenth century plundered the ruins of ancient Harappa for obtaining ballast for the railway line being constructed under his guidance from Lahore to Multan, and, as a result of this, "today the trains rumble over a hundred miles of line laid on a secure foundation of third-millennium brick-bats".[57] What survives this massive brick-robbing in Harappa alone is, to say the least, most imposing. To this is to be added the evidences of brick-robbing for laying the railway line from Hanumangarh to Suratgarh just quoted from Thapar. Besides, what remains at such sites as Harappa, Mohenjo-daro, Kalibangan, Lothal, etc., is imposing, though in Lothal the use of burnt bricks was restricted mainly to the "building of baths, drains and dock which had to be impervious to water".[58]

55. R.S. Bisht, in Possehl's HC 115.
56. B.B. Lal, in EIP 68.
57. Piggott 14.
58. S.R. Rao, *Lothal* (MASI, No. 78) 71.

On sheer quantitative considerations, the manufacture of burnt bricks in Harappa culture is most imposing.

8. BRICKS IN THE FIRST URBANIZATION

We shall try to go here into some detail of brick technology and its evolution in First Urbanization, because, as we shall see, it has very important bearing on our understanding of the making of mathematics in ancient India.

The quantitative considerations apart, what has amazed the archaeologists about brick technology of the Harappan culture is the quality of the bricks used in various types of constructions. Mackay, to whom we owe the first descriptive account of the bricks in Mohenjo-daro, observes :[59] "Well-burnt bricks, and those of Mohenjo-daro are of excellent quality, are practically indestructible and can be used over and over again, provided that a moderate amount of care is taken in removing them from the old walls."

The bricks of Mohenjo-daro are all exceptionally well made, yet have no straw or other binding material. They are always rectangular in shape with the exception of those that were made for special purposes, such as the wedge-shaped bricks almost invariably employed in the construction of wells. The bricks were made in an open mould and struck along the top with a piece of wood, as proved by their streated upper surfaces... The clay that was used seems to have been ordinary alluvial soil like that which is found in the vicinity of Mohenjo-daro today, and is used by the modern brick maker and potter. It is only necessary to dig down a few feet below the surface soil, which is impregnated with salt, to obtain clay of the right consistency... The bricks are exceptionally well baked and range from straw colour to bright red... We do not yet know the type of kiln in which these bricks were baked, but, then as now, there was evidently no difficulty about fuel. Wood must always have been far more plentiful than in Babylonia, where reeds were the only available fuel... As in all brickyards of the modern East, the bricks were laid over large areas for the preliminary drying. The result is that we have found the foot-prints upon them of cattle, crows, dogs... The largest brick yet found at Mohenjo-daro measures 20.25 by 10.5 by 3.5 inches... It was probably made originally to cover a drain... Another brick of large size measures 14.5 by 7.25 by 4 inches. This size of brick is fairly common... The smallest size brick measures 9.5 by 4.35 by 2 inches... In the construction of the bathrooms,

59. Mackay in Marshall's MIC I. 262.

sawn bricks were almost invariably used to ensure the evenness of floor which was considered essential.[60]

Mackay[61] gives the following list of the various sizes of bricks used :

	Size in inches	
1.	9. 5 × 4.35 × 2	Sawn
2.	10. 0 × 5.00 × 2.25	,,
3.	10. 0 × 5.00 × 2.25	,,
4.	10.25 × 5.50 × 2.25	,,
5.	10.25 × 5.00 × 2.25	Moulded
6.	10.35 × 2.50 × 2.00	Sawn
7.	11.00 × 5.25 × 2.35	Moulded
8.	11.00 × 5.50 × 2.25	,,
9.	11. 4 × 5.75 × 2.5	,,
10.	12.00 × 6.00 × 2.25	,,
11.	13.50 × 6.25 × 3.75	,,
12.	14.00 × 6.75 × 3.75	,,
13.	14.00 × 7.00 × 3.25	,,
14.	14.50 × 7.25 × 4.00	,,
15.	20.25 × 10.50 × 3.50	,,

"By far the commonest size of moulded brick is 11 by 5.5 by 2.25 inches ; it occurs at all levels. The moulded brick of small size (no. 5) is very unusual ; possibly it was made for a special purpose... We have not yet succeeded in finding any brick-kilns, but these were probably situated well outside the city area and near an ample supply of wood".[62]

Let us note here specially one point. By far the most commonly used brick in Mohenjo-dado, according to Mackay's mesurement, is 11×5.5×2.25 inches. Basically speaking, this gives us a proportion of 4 : 2 : 1, though with slight discrepancy about the thickness of the brick (which, according to this proportion, should have accurately been 2.75 inches).

Later archaeological work wants us to look at this proportion of 4 : 2 : 1 as having been on the whole characteristic of the standard bricks of Mature Harappan culture. But there are some points to be noted in this connection.

60. *Ibid* 266-67.
61. *Ibid* 267.
62. *Ibid* 268.

The two sites, namely Kalibangan and Banwali, as we have noted, indicate for us the making of burnt bricks already in the pre-Harappan stage, though its use was restricted. The restriction to the use of burnt bricks continued in the Mature Harappan period at both the sites, mud bricks being more generally used also in the Mature Harappan stage. It seems, however, that there is a noticeable change from the pre-Harappan to Harappan stages in the size (and proportion of the different dimensions) of both mud bricks and burnt bricks most commonly used at both the sites. In the pre-Harappan stage at both Banwali and Kalibangan, the proportion of the dimensions of the standard bricks seems to have been 3 : 2 : 1, whereas in the Harappan period it assumes the proportion of 4 : 2 : 1.

In Kalibangan Period I (pre-Harappan), as B.K. Thapar says, "the fortification wall was made of mud bricks (30×20×10 cm ; proportion, 3 : 2 : 1) . . . The use of baked brick was attested by a drain, the size of the brick being the same as that of mud bricks".[63] Discussing the citadel in Kalibangan Period II (Harappan), Thapar continues : "The fortifications were built throughout of mud bricks; two sizes of bricks, 40×20×10 cm. and 30×15×7.5 cm. (ratio 4 : 2 : 1) representing two principal structural phases, were used in the construction, the larger bricks in the earlier phase and the smaller in the later . . . The enclosed area contained some five to six massive platforms of mud bricks (40×20×10 cm. for the earlier phase and 30×15×7.5 cm. for the later)".[64] In the entrance complex of the citadel, again, "two structural phases were recognized, of which the earlier consisted of steps built of mur bricks (40×20×10 cm.) with a riser of 10 cm. and a tread of 40, and the later, perhaps of a ramp, screened by a 1.5 m.-wide wall (30×15×7.5 cm.). Lest the unwary reader carry the impression that in Kalibangan during Period II (Harappan), the small-sized bricks, viz. 30×15×7.5 cm. were introduced only in the later phases, it may be said that bricks of this size have been used right from the beginning of the

63. B.K. Thapar, in Possehl's ACI 197.
64. *Ibid* 199.

occupation for domestic structures. It was only in the fortifications and massive platforms that the large-sized bricks were employed in the earlier phases and smaller ones in the later. It is significant, however, that the ratio of dimension of both sets remains 4 : 2 : 1".[65] The baked bricks, the use of which was confined mostly to drains, had also the same size : 30×15×7.5 cm. i.e. the proportion 4 : 2 : 1.[66] To sum up : In Kalibangan, the transition from the pre-Harappan to Harappan stage indicates also some change in the technique of brick-making, the most conspicuous of which was the change in the proportion of the bricks from 3 : 2 : 1 to 4 : 2 :1.

The same seems to be indicated at Banwali. Referring to Period I (called pre-Harappan or Kalibangan Culture), R.S. Bisht observes, "The bricks, both fired and sun-dried, were meticulously standardized, conforming to the ratio 1 : 2 : 3. This ratio was observed earlier at Kalibangan. But the dimensions at Banwali are 12 by 24 by 36 centimetres and 13 by 26 by 39 centimeters, as compared to 10 by 20 by 30 centimeters at Kalibangan".[67] In Period II, representing Mature Harappan Culture, there is a noticeable change in the proportion of the dimensions of the bricks. "Bricks, both baked and sun-dried, were carefully moulded into various sizes which, except the wedge shaped examples, form two broad groups and always give the ratio of 1 : 2 :4 as regards thickness, width and length. The smaller bricks which variously measure 6 by 12 by 24 centimetres ; 6.5 by 13 by 26 ; 7 by 14 by 28 ; 7.5 by 15 by 30 or 8 by 16 by 32, were used in constructing residential houses. The larger ones, 10 by 20 by 40 centimetres ; 11 by 22 by 44 ; 12 by 24 by 48 and 12.5 by 25 by 50 were used in massive structures such as defences. Fired bricks are normally used in drains, wells and bathing platforms : places with a constant use of water."[68]

In Lothal, the basic proportion of "1 length : 2 breadths" appears to have been generally maintained for the bricks, though there were slight variations in the proportion of thick-

65. *Ibid* 200.
66. Ibid 201.
67. R.S. Bisht, in Possehl's HC 115.
68. *Ibid* 118.

ness. Here is how S.R. Rao[69] sums up the evidences for Lothal bricks :

The Lothal folk did not use kiln-fired bricks for building houses and platforms. They used mostly sun-dried bricks for this purpose... As the climate has not changed much since three thousand years it may be presumed that the rainfall was not very heavy and there was no need of kiln-fired bricks except for building baths, drains and dock which had to be impervious to water. Two types of mud-bricks are easily distinguishable at Lothal... The most common size is $11\times5\frac{1}{2}\times2\frac{1}{2}$ in., but some vary from $12\times6\times3$ in. and $14\times7\times3$ in. to $15\times8\times3\frac{1}{2}$ in. Besides the kiln-fired bricks were well moulded and fired : common size $11\times5\frac{1}{2}\times2\frac{1}{2}$ in. a smaller size measures $9\frac{3}{4}\times5\times2\frac{1}{4}$ in. Radial bricks used in the construction of the wells and the curves of drains are $9\frac{1}{2}$ in. long, $3\frac{1}{4}$ in. thick, and $4\frac{1}{2}$ in. wide at one end and $3\frac{1}{2}$ in. at another. Small bricks measuring $3\times2\times1$ in. and T-shaped bricks were also used in the openings of the drains. A remarkable feature of the masonry of the Harappan sites is the standardization of the size of bricks namely $11\times5\frac{1}{2}\times2\frac{1}{2}$ in. for most of the constructions throughout the vast area covered by the Harappan Civilization. Another noteworthy feature of standardization is the maintenance of a suitable proportion between the length and breadth of the bricks. Whether sun-dried or kiln-fired, the bricks measure 1 length : 2 breadths, e.g. $11\times5\frac{1}{2}$ in. in the case of baked bricks. Apparently, the Harappans must have realized the necessity of maintaining the said proportions with a view to use complete bricks as headers or stretchers in achieving the required thickness of the walls. If the most common size of the mud-brick is compared with the most common size of the baked-brick the difference is within the limits of the permissible fire-shrinkage, namely 8 to 15 percent. This difference further suggests that the moulds were of common size for mud-bricks as well as the burnt ones. Thus it is obvious that the bricks were burnt not only to modern standards but also on modern principles. They have sharp right-angled edges and the sides are parallel.

With this data in mind we may now quote the general observations of K.N. Dikshit[70] on brick technology in the Indus Civilization and after :

The bricks used for the building of houses in Mohenjo-daro and Harappa are well burnt and of excellent proportions, which have excited the admiration of modern engineers in Sind. The most usual size of burnt bricks is 11 by $5\frac{1}{4}''$ or $5\frac{1}{2}''$ with a thickness of $2\frac{1}{4}''$ to

69. S.R. Rao. *Lothal* (MASI, No. 78) 71-72.
70. K.N. Dikshit, 15-16.

$2\frac{3}{4}''$. At no other period has the Indian builder ever struck upon this most business-like size of bricks, and it is remarkable that the evolution of bricks in the historic period from Asoka commences with bricks of about double the length and breadth of the Indus Valley brick. It gradually diminishes in the Kushana, Gupta and mediaeval periods, but never attains the true proportion of length, breadth and thickness as $1 : \frac{1}{2} : \frac{1}{4}$, which makes for an excellent bond. That this ideal proportion was not entirely forgotton is shown by the fact that a later text (*Kasyapa Samhita*) prescribes a proportion of 10 fingers of length to 5 fingers of width and half of the latter for thickness; but it is doubtful whether in actual practice the masons ever followed this in the historic period. Any way, it is clear that the burnt brick of the Indus Civilization has been unexcelled in India and is not comparable with any attempts made in ancient Sumer, Egypt and other countries, till we come down to the Roman times.

Two points about this observation may be specially noted.

First, the standardization of the brick size in the Harappan culture—the basic proportion of 4 : 2 : 1 of the dimensions of the bricks—is indicative of a significant advance, inasmuch as it creates scope for very efficient bonding and therefore also improved masonry and architecture.

Secondly, after noting the excellence of brick technology of the Indus civilization, Dikshit skips over a very long period and takes up the story of the bricks from the time of Asoka. Evidently, from the Maurya period onward, brick technology in India had to reacquire progressive improvement in the still later periods of the Kushanas, Guptas and the medieval times. In any case, Dikshit is absolutely silent about any brick technology in the period intervening between the Indus civilization and the time of Asoka. Why is he thus silent? The answer is quite on the surface. There is no evidence for brick technology in this intervening period, which, chronologically speaking, covered more than ten centuries. In other words, brick technology is conspicuously absent during the "Dark Age" or "Dark Period" intervening the two Urbanizations, notwithstanding an insignificant number of stray evidences like those in Bhagwanpura, Dadheri, etc.,[71] about the exact implications of which the archaeologists are yet to be clear and, in any case, the few burnt bricks found in such sites are not indicative of any brick technology worth taking serious notice of.

71. T.N. Roy, GC 130f.

To sum up this discussion: According to the periodization we have proposed—

(a) the First Urbanization is unmistakably indicative of the tradition of highly sophisticated brick technology, or of the technology of making and using burnt bricks,

(b) the intermediate Dark Period is unaware of it,

(c) the Second Urbanization is indicative of the re-introduction of this technology, though beginning again in a rather humble form.

With these points in mind we proceed to the question of the making of mathematics in ancient India, which, as we are going to see, is inextricably connected with brick technology.

BRIEF APPENDIX
ON AGRICULTURAL TECHNOLOGY AND HARAPPAN CULTURE

by

D. K. Chakrabarti

[When I met D. K. Chakrabarti in Delhi sometimes back, he told me that he was then working on a paper on the agricultural technology in Harappan Culture. Having as I do a good deal of admiration for his work, I immediately requested him to allow me to use his essay as an appendix to my present work. He was kind enough to agree to this. However, it was then almost on the eve of the present manuscript to go to the press and Chakrabarti was evidently in need of some reasonable time to complete the essay in full form. In view of this, he suggested that he would sum up his main points cryptically, so that I could use these as a brief appendix in my work. I am extremely grateful to him for the following note he has sent in the form of personal communication, so that his main ideas could at any rate be communicated to the readers, awaiting of course the publication of his full article: Debiprasad Chattopadhyaya].

"I only wanted to put emphasis on the following points regarding Harappan agriculture :

1. The agricultural ancestry of the Harappans goes back to the 7th millennium B.C. at Mehrgarh in the Bolan valley of the Kachchi plain, and this simple fact suggests that by the time the Harappan civilization came into existence (*c* 2900-2800 B.C.), its agricultural pattern must have been laid out in considerable detail in the earlier—"Early Harappan"—period. The data on the early Harappan agriculture are still meagre, but one remembers the large quantity of wheat at the early Harappan site of Tarakai Qila in the NWEP area.

2. It may be wrong to imagine the Harappan agricultural system within a uniform framework. The agricultural pattern pursued in different areas of its distribution must have been to a great extent different, although our specific ideas are not clear at present. The Nile-like seasonal overflow and the consequent easy cultivation system which H.T. Lambrick postulates for Sind could not very well be true of Punjab, or for that matter, of Gujrat. Double-cropping was possible in all the suitable areas.

3. The Harappan agricultural system seems to show an evolutionary pattern of its own. *Bajra* seems to be peculiar only to the late Harappans in Gujarat. Legumes are specific to the late Harappans in Western U.P. (cf. Hulas). This suggests interaction between the late Harappan agriculture and the neolithic-chalcolithic agriculture of inner India. This is a point which has become sharp in recent years.

4. The terracotta model of a plough at Banawali sets at rest all the hypotheses about the use or non-use of plough in the Harappan civilization.

5. The existence of irrigation canals in the Harappan context has been argued by the French team working on the problem (*Man and Environment,* 1986). This opens a new vista of research.—*D. K. Chakrabarti.*

CHAPTER 5

MATHEMATICS IN ITS MAKING

1. PRELIMINARY REMARKS

With this brief idea of brick technology in Indian history, we now pass on to the problem of the making of mathematics in ancient India. The main points we are going to argue are as follows :

1. A class of texts come down to us with the general title *Sulva-sutra*-s (spelled also as *Sulba-sutra*-s) which are for us the earliest codified documents for the making of mathematics, specially goemetry, in India.

2. The actual mathematical knowledge contained in these, judged specially in the ancient context, cannot but be considered as remarkable.

3. Notwithstanding the usual assumption that this mathematics was created by the Vedic priests, the internal evidences of the texts indicate that it was the direct outcome of the theoretical requirements mainly of the brick-makers and brick-layers, who were using burnt bricks and whose status within the general framework of the social norm of the Vedic priests is at best questionable.

4. With all the uncertainties and controversies about the actual date of these texts, there is no possibility of placing these outside the period intervening the two urbanizations—the "Dark Period" or "Dark Age" of our archaeologists, one of the conspicuous features of which is the loss of the sophisticated tradition of brick technology of the First urbanization.

5. Archaeologically, therefore, we are confronted here with an apparent anomaly. Mathematics came into being in ancient India to meet the theoretical requirements primarily of brick technology in a period in which there is no brick technology—not at least in any notable scale, specially in the sense of the technology of making and using burnt bricks which is presupposed by the *Sulva-sutra*-s.

6. This anomaly can perhaps be resolved by the presumption that the mathematics—though codified much later in the form of the *Sulva-sutra*-s—was actually the creation of the

First Urbanization, in spite of the uncertainty or gap at the present stage of our knowledge of the mode of its transmission to the later period.

7. Though we do not have any direct document embodying mathematical knowledge in the First Urbanization, there are strong circumstantial evidences in favour of the assumption just mentioned. In view of this, it is not easy to dismiss Gordon Childe's postulate of the unknown perishable materials in which this mathematics could have been codified in the Indus Valley Civilization—a postulate on the basis of which he wanted to trace the roots of classical Indian Science to the achievements of the Harappan culture.

The starting point of these arguments, however, is a widely current misconception about the making of mathematics in general.

2. ORIGIN OF GEOMETRY : HERODOTUS AND RECENT CORRECTIONS OF HIS VIEW

According to Herodotus, the Greeks first received the knowledge of geometry from the Egyptians who, in their turn, developed it from the practice of land-measurement required for administrative purposes. Here is the observation of Herodotus,[1] which has found place in most of the histories of science :

> Sesostris also, they declared, made a division of the soil of Egypt among the inhabitants, assigning square plots of ground of equal size to all, and obtaining his chief revenue from the rent which the holders were required to pay him every year. If the river carried away any portion of a man's lot, he appeared before the king, and related what had happened ; upon which the king sent persons to examine, and determine by measurement the exact extent of the loss ; and thenceforth only such a rent was demanded of him as was proportionate to the reduced size of his land. From this practice, I think, geometry first came to be known in Egypt, whence it passed into Greece. The sundial, however, and the gnomon, with the division of the day into twelve parts, were received by the Greeks from the Babylonians.

Depending on this, it is often assumed that the science of geometry developed from the practical requirements of land measurement. As C. Singer puts it, "The development of rights in land demanded some sort of surveying. Greek tradition has

1. Quoted by Smith I.51-52.

it that the inundation of the Nile, by obliterating all landmarks, forced on the Egyptians an annual re-measurement of their fields. Thus *geo-metry* (literally 'earth-measurement') was born".[2]

But such a view is not fully endorsed by archaeology. It is true that the economic and administrative requirements of the urban revolution in Egypt—and also in Mesopotamia—did call forth some geometrical knowledge connected with land measurement. But it was enough for the purpose to have on the whole an approximate knowledge, and not geometry in the sense of an exact science. As Gordon Childe[3] puts it :

The conditions of urban economy .. required some knowledge of geometrical relations. The areas of fields must be determined for estimations of the seed required for sowing them and the rent or tax that might be exacted in respect of them. But for such estimates and assessments absolute accuracy was unnecessary : the bailiff only wanted to know roughly how much grain to allow for each field ; the tax-collector needed a general idea of the yield to be expected. We have seen that even before 3000 B.C. the Sumerians were calculating the areas of fields as the product of length by breadth ; they were, that is, applying the correct geometrical formula for the area of a rectangle. In later documents the areas of irregular quadrangles are calculated by various approximations, usually the mean of the products of the two pairs of adjacent sides. Polygonal fields were divided up into quadrangles and triangles, the areas of which were similarly calculated. In Egypt, even in New Kingdom contracts, the area of a four-sided field is taken as half the sum of two adjacent sides multiplied by half the sum of the remaining sides. In the case of a triangular field, the length of two sides were added together and halved, and then multiplied by half the length of the third side. The documents just examined generally contain plans of the fields in question. The lengths are written in along the sides, but the plans are not drawn accurately or to scale. *The theory that exact geometry arose out of land-hurveying in Egypt or Babylonia is not supported by the evidence at our disposal.*

Where, then, are we to look for the making of geometry as exact science in ancient Egypt and Mesopotamia? Gordon Childe wants us to look for this in the techniques of the masons, architects and engineers. As he puts it, "On the other hand [i.e. as contrasted with the land-serveying administrators], architects and engineers often required more exact calculations to

2. Singer 4.
3. Childe MMH 205-7.

fulfil the tasks imposed on them. The accuracy of a pyramid was a matter of ritual significance. To secure it, the sizes of the blocks facing it must be accurately calculated".[4]

The pyramids were made, of course, of blocks of stone rather than bricks. But J.D. Bernal[5] very explicitly discusses the question of the making of mathematics as connected with brick technology. We quote him at some length.

The operation of building itself also contributed, probably even before land survey, to the foundation of geometry. Originally, town buildings were simply village huts made of wood or reeds. In cities, with a restricted space and danger of fire houses of pise or rammed mud were a great improvement. The next step was to have even greater consequences : the invention of the standard moulded block of dried mud — the brick. The brick may not be an original invention, but a copy, in the only màterial available in the valley country, of the stone slabs that came naturally to hand for dry walling in the hills. Bricks are difficult to fit together unless they are rectangular, and their use led necessarily to the idea of the right angle and the use of the straight line—originally the stretched line of the cordmaker or weaver.

The practice of building in brick, particularly of large religious buildings of pyramid form, gave rise not only to geometry, but also to the conception of areas and volumes of figures and solids reckonable in terms of the lengths of their sides. At first only the volume of rectangular blocks could be estimated, but the structural need for tapering or buttering a wall led to more complicated shapes like that of the pyramid. The calculation of the volume of a pyramid was the highest flight of Egyptian mathematics and foreshadowed the methods of the integral calculus.

Also from building came the practice of the plan to scale. Such a plan for a town together with the architect's rule is for instance shown in the status of Gudea of Lagash in *c.* 2250 B.C. With these mathematical methods an administrator was able to plan the whole operation of brick or stone building in advance. He could estimate accurately the number of labourers wanted, the amount of materials and food they would need, and the time the job would take. The techniques were readily extendable from the city to the country in the lay-out of fields, the calculation of their areas, and the estimate of their yields for revenue purposes. This is the origin of mapping and surveying. It was this practical use that later gave rise to the term of geometry—land measurement. Mathematics, indeed, arose in the first place as an auxiliary method of production made necessary and possible by city life.

4. *Ibid* 207.
5. Bernal 121.

3. R S. SHARMA AND SULVA GEOMETRY

Notwithstanding etymology (*geo-metry = earth-measurement*), the theory of geometry, i.e. in the sense of exact science, emerging from the practice of land-measurement for administrative purposes is thus no longer tenable. With this point in mind, let us return to the Indian evidences.

We have no direct evidence, of course, of any practice of land-measurement in the Indus Civilization. However, judging from the circumstance that this civilization thrived on agricultural surplus collected from a vast area, one is inclined to presume that some such practice was prevalent also in the Indus Civilization. The presumption seems to be supported by the analogy of the two other primary centres of urban revolution, namely Egypt and Mesopotamia. The analogy of Egypt and Mesopotamia, again, would lead us to think that the administrative requirements in the Indus Civilization could have been satisfied only with some approximate knowledge of the area of the land surveyed rather than geometry as an exact science.

Some practice of measuring land with rods or poles is perhaps suggested—though desultorily—by a few verses of the *Rgveda*.[6] But it is difficult to be exact about what is really implied by such archaic references, specially in view of the fact that in the Rgvedic period wealth was basically conceived in terms of cattle rather than land or agricultural products. In any case, there is no hint whatsoever either in the Rgvedic verses or even in Sayana's commentary on these of anything that may be taken as directly or indirectly indicating any geometrical interest of the early Vedic poets.

The first references to land-measurement for administrative purposes are to be found in the Pali sources, which belong to the earlier period of Second Urbanization. Since the rope or *rajju* was used for these measuring purpose, the officer entrusted with the work was called *rajju-gahaka-amacca*, or per-

6. *RV* i.100.18 ; i.110.5 and iii.38.3. See Srinivasan 7. It is usually assumed that the early Rgvedic people were pastoral nomads, which if true would hardly require of them the technique of land measurement. The earliest reference in Vedic literature to what could perhaps been mathematics is to be found in *Ch. Up* vii. 2, where the word *rasi* is interpreted by Sankara as mathematics.

haps simply as *rajjuka,* literally "the rope-holder". As B. B. Datta observes, "In the Pali literature, we find the terms *rajjuka* or *rajju-gahaka* (rope-holder) for the king's land-surveyor. The first of these terms appears copiously... in the inscriptions of Emperor Asoka".[7]

R.S. Sharma, while discussing the tithe system of the period, naturally goes into some detail of the evidence[8]:

> The two references (in the *Jatakas*) which relate to the measuring of field by royal officers, are capable of being interpreted in a way which may suggest some sort of 'ground rent'. Buhler compares the *rajjugahaka-amacca* with the Land Revenue Settlement Officer of British India and suggests that measurement was used for assessing 'ground rent'. But Fick surmises that land was measured either to form an approximate idea of the amount of rent payable by the subjects to the king or to determine the average produce to be brought to the king's store-room. Nevertheless, the fact that in measuring the field the *rajjugahaka-amacca* was conscious of doing nothing which might cause loss either to the *raja* or to the *khettasamika* ('the owner of the field') or *kutumba* ('the royal property') lends strength to Buhler's hypothesis that the land was measured for the purpose of levying rent on it.

Because it is outside the scope of R.S. Sharma's present theme of discussion, he does not raise in this connection the question of the making of geometry from this practice of land-measurement. Nevertheless, it is relevant for our purpose to note in this connection that the two references to the *rajju-gahaka-amacca* or 'rope-holding officer' found in the Pali *Jataka*-s do not mention geometry. As a matter of fact, in the entire voluminous literature called the *Jataka*-s we have no hint of geometry anywhere. The repetition of the stereo-typed phrase of "accomplishment in the eighteen branches of learning", which we come across in this literature,[9] is of course, not explained in the *Jataka*-s, though from other Pali sources we are left to presume that these refer to divination, auguries, interpretation of prognostics, etc.[10] which had some sort of prestige for the popular minds in those days, though, the Buddha himself is supposed to have disapproved of these as undesirable

7. B. B. Datta SS 9.
8. R. S. Sharma in SSP I.i. 7-8.
9. Cowell *Jatakas* i. 126, 203, 285 ; ii. 60, 168, 287 ; iv. 104, 105.
10. See, e.g., The *Mahasilam,* Tr., *SBE.* XI, 196 ff.

branches of learning. In any case, in spite of the mention of the practice of land-measurement in the period, there is no evidence to show that geometry as an exact science emerged from this practice.

At the same time, there seems to be a rather tricky point in this connection. The earliest Indian documents embodying geometry as exact science, as we have already said, are called the *Sulva-sutra*-s and the word *sulva* (or *sulba*), to which this body of literature owes its name, does mean the "rope". It is in fact an elitist-esoteric equivalent for the word *rajju*. Since the word *rajju* is used in connection with land-measurement—and perhaps largely under the older hypothesis that geometry originated in Egypt from the practice of land-measurement—R.S. Sharma in his *Material Culture and Social Formations in Ancient India* ventures the hypothesis that this practice of land-measurement by rope could have originally been at the basis of the Sulva geometry. As he puts it, "For fixing individual possession of fields and assessing taxes knowledge of measurement was necessary. Methods of calculating the areas of the circle, rectangle, etc., or the method of converting circles into squares, though prescribed in the religious context in the *Sulvasutra*-s, may have arisen in response to the needs of field agriculture".[11]

The hypothesis seems to be a hasty one. What is embodied in the *Sulva-sutra*-s, as we are going to see, is geometry as an exact science—much more than the approximate calculations of areas etc. which, as Gordon Childe has shown, can be the outcome of the early administrative requirements of field-measurement. Secondly, though the texts derive their name from the rope or string—*sulva*—and though use of the string or the rope is of much importance for the making of geometry in these texts, the use of the rope in the *Sulva-sutra*-s has little to do with field-measurement, beyond perhaps the measurement of the ground-plan of certain brick structures. The construction of these bricks structures form, in a sense, the be all and end all of the *Sulva-sutra*-s and the use of the string or rope is concerned above all with the techniques of making or laying these bricks. The technological context of the use of "rope", in short, is different altogether. It is the context in

11. R. S. Sharma MCSFAI 108.

which the string or rope is still found to be used by the masons and architects. In fact, if it is at all conceivable to scrap the brick technology from the Sulva texts—which, as we shall see, is really not conceivable—practically nothing that can be called geometry—or more broadly mathematics—is at all left in the *Sulva-sutra*-s. This consideration, by itself, seems to make R.S. Sharma's hypothesis unacceptable. Besides the general consideration remains that nowhere in Indian literature we come across even any remote suggestion connecting the administrative requirements of field measurement with the making of geometry even as an approximate knowledge, not to speak of geometry as an accurate science, which we have in the Sulva texts.

At the same time, all this leads us to a situation which, from the archaeological viewpoint at any rate, is apparently anomalous. The Sulva texts belong to a period which had no brick technology. Before passing on to discuss this, let us have a preliminary idea of how the Sulva geometry is inconceivable without brick technology.

4. BRICK TECHNOLOGY AND MAGICO-RELIGIOUS BELIEFS

The main theme of the *Sulva-sutra*-s is the construction of certain brick-structures. The structures range from comparatively simpler to highly complicated ones and the texts concentrate mainly on the latter. Without developing a body of mathemetical knowledge it was not possible to be accurate about the constructions. From this point of view, it was necessary to develop the mathematical—particularly geometrical—knowledge, and it is this that we read in the Sulva texts.

What complicates matters, however, is a different point. The brick-structures were supposed to be not just brick structures constructed for secular purposes. These were called *citi*-s or *agni*-s—meaning altars required for the performance of Vedic sacrifices and the sacrifices were intended to have magico-religious efficacy. Hence, apparently at any rate, the question of the making of mathematics remains in the Sulva texts as somehow interlinked with a body of magico-religious system.

In such a circumstance, it is first of all necessary to be clear about one point. How far is the body of the magico-religious beliefs *internally* connected with the technological questions concerning the brick-constructions, which are directly or overtly related to the mathematical theme of the Sulva texts?

Our point is: *While there is a necessary connection between the technological problems and the mathematics in the Sulva texts, there is no such connection between this mathematics and the system of magico-religious beliefs.* As a matter of fact, the body of magico-religious beliefs is totally extrinsic to the mathematics of the Sulva texts. This is clear from a few obvious considerations.

First, the same mathematical problems would have remained in tact had the same brick-structures been required to serve some purpose other than the magico-religious ones. Thus, for example, a very important form of these brick-structures is required to have the shape of a falcon with specified size and made of a specific number of bricks arranged in a specified number of layers. According to the system of magico-religious beliefs, the use of such a structure in the sacrificial ritual ensures for the *yajamana* or rich patron financing the sacrifice the quick attainment of heaven. However, if we think of a rich patron wanting to have the same bird-like brick structure for some other purpose—say, as a decoration for his pleasure garden or for the play ground of his children—the technological requirements along with the collateral mathemetical problems would have remained identical, though without being associated with the body of magico-religious beliefs.

Secondly, with all the importance seemingly attached to the brick-structures for magically ensuring the desired results, the priest-class recommending the construction of these tacitly admit that the same results could be magically ensured for the *yajamana* without the physical construction of the brick-structures. Thus, as we shall later see in some detail, the priests prescribing the rituals themselves claim that the results may as well be obtained with altars made only of spells (*chandas-citi*) or mind-made altars (*manomaya citi*) which can only mean imaginary altars as substitutes for physically constructed ones. Thus the actual or physical construction of the structures are really not so essential for the magico-religious belief-system as these are often thought to be. So also is the mathematical knowledge, which, however, is essential for the physical construction of the structures, irrespective of the circumstance of their serving magico-religious or any other secular purpose.

Pending a fuller discussion of the brick technology discussed in the *Sulva-sutra*-s, we quote here a few observations of Thibaut to have some idea of why a good deal of mathematical knowledge was essential for the physical construction of the structures, which, in priestly terminology, were called the altars —*citi*-s or *agni*-s. Discussing the more elaborate ones, he[12] observes :

> Every one of these altars had to be constructed out of five layers of bricks, which reached together the height of the knee ; for some cases ten or fifteen layers and a correspondingly increased height of the altar were prescribed. Every layer in its turn was to consist of two hundred bricks so that the whole *agni* contained a thousand ; the first, third, and fifth layers were divided into two hundred parts in exactly the same manner ; a different division was adopted for the second and the fourth so that one brick was never lying upon another brick of the same size and form.

Secondly, the *Sulva-sutra*-s discuss not merely one type of brick-structure. The elaborate brick-structures discussed in these are supposed to have various shapes—those of the falcon with stretched wings, of the chariot wheel, of the triangular forepart of a cart (called *prauga*), of the tortoise, etc. Of these altars with various shapes, as Thibaut observes, the falcon-shaped one was the "most primitive".[13] For this was specified an area of seven and half square *purusa,* i.e. "the height of a man with uplifted arm. This rule was valid at least for the case of the *agni* being constructed for the first time ; on each subsequent occasion, the area had to be increased by one square *purusa*".[14] But this was not all. As Thibaut[15] adds,

> Now when for the attainment of some special purpose, one of the variations enumerated above was adopted instead of the primitive shape of the *agni,* the rules regulating the size of the altar did not cease to be valid, but the area of every *citi,* whatever its shape might be—falcon with curved wings, wheel, *prauga,* tortoise, etc.—had to be equal to 7½ square *purusa*-s. On the other hand, when at the second construction of the altar one square *purusa* had to be

12. G. Thibaut, *On the Sulva Sutras.* Reprinted in *SHSI* II. 419.
13. *Ibid* II. 420.
14. *Ibid.* II. 420.
15. *Ibid.* II. 420-21.

added to the $7\frac{1}{2}$ constituting the first *citi,* and when for the third construction two square *purusa*-s more were required for the shape of the whole, the relative proportions of the single parts had to remain unchanged. A look at the outline of the different *citi*-s is sufficient to show that all this could not be accomplished without a certain amount of geometrical knowledge. Squares had to be found which would be equal to two or more given squares, or equal to the difference of two given squares; oblongs had to be turned into squares and squares into oblongs; triangles had to be constructed equal to given squares or oblongs, and so on. The last task and not the least was that of finding a circle, the area of which might equal as closely as possible that of a given square.

To all this is to be added another point. Obviously enough, only one standardised brick-type with fixed shape and size could not meet the requirements of the construction of varied type of structures. In the *Sulva-sutra*-s, therefore, are specified how many varieties of bricks were required for the detailed demands for the construction of each type of the structure and we shall see into what meticulous details of calculations the texts had to go in order to be precise about the shape and size of each brick-type and about the mode of arranging these in different layers.

It is for meeting these essentially technological questions connected with brick-making and brick-laying that the Sulva mathematics came into being, notwithstanding the circumstance that in the Vedic text the brick-structures are said to be connected with a body of magico-religious beliefs—beliefs that are totally extrinsic to the Sulva mathematics. That is why we have observed that if we ignore or overlook the technology of brick-making and brick-laying of the Sulva texts, nothing substantial will be left in these having any mathematical interest, while it is possible to scrap the entire body of magico-religious beliefs from the texts without in the least affecting their mathematical contents.

Before we pass on to the nature of the mathematical contents of the Sulva texts however, we are confronted with a serious problem. If the Sulva mathematics is inconceivable without sophisticated brick technology, sophisticated brick technology is not easily conceivable in the period of the Sulva texts. In other words, from the archaeological point of view, we have here some kind of an anomaly: mathematics thriving essentially on sophisticated brick technology is found to be embodied in texts of a period where there is no brick technology—not to speak of any sophisticated form of it.

5. AN APPARENT ARCHAEOLOGICAL ANOMALY

It is impossible, of course, to be exact about the date of *Sulva-sutra*-s. Depending on various circumstantial evidences, the modern scholars have proposed various possible dates for these works—or, more properly, the dates of the authorities with whose names the works are connected—Baudhayana, Apastamba, Katyayana, etc. These suggested dates range from 800 B.C. (Kane) to 250 B.C. (Keith), though of these 800 B.C. seems to be an exaggeration of the antiquity of the works just as 250 B.C. seems to be underestimating it. Without trying to enter into this chronological controversy, however, it may be permissible to assert that the actual date of the Sulva texts cannot fall outside the period intervening the two urbanizations, i.e. the "Dark Age" or "Dark Period" of the archaeologists.

This point is of material importance for our discussion, because an important feature of the period intervening the two urbanizations, as we have already seen, is the loss of the brick technology of the first urbanization. The re-introduction of brick technology—or, to be more specific, the technology of making and using burnt bricks—is generally admitted to begin again with what the archaeologists call the NBP period, i.e. the period the main index to which is a pottery-type called the Northern Black Polished Ware. This period, again, is roughly divided into two phases—the first dated 600-300 B.C. and the second 300-100 B.C. It is only in the second of these phases that we come across the re-introduction of brick technology in any noticeable scale. As R.S. Sharma summarily puts it, "while the second phase of the NBP is marked by.... burnt brick structures, occasional tiles and ringwells, the first phase is marked by the absence of burnt brick structures and ringwells".[16]

The introduction of NBP Ware is generally viewed as foreshadowing the Second Urbanization or the beginnings of early historical cities, a number of which are mentioned specially in early Pali literature. Here is how, in 1973, A. Ghosh[17] sums up the archaeological evidences for the use of burnt bricks in

16. R. S. Sharma MCSFAI 91.
17. A. Ghosh CEHI 68-70.

connection with the formation of the cities in early historic India :

..of the use of burnt brick there is not much evidence in the earlier life of the cities. At most sites where evidence exists, burnt brick came into vogue either in the late phase of the Northern Black Polished Ware or, more commonly, in a still later period. Taking into consideration the metropolitan cities first, we find brick structures in the Northern Black Polished Ware levels, but it is not known from the published notices whether they were from the early or late levels : at Pataliputra, the origin itself of which being later than the advent of the Ware, it might be presumed that the structures belonged to a late phase ; Vaisali, where there was a single brick wall in pre-Sunga levels but in the Sunga level and onwards there was a net-work of brick structures : Ujjain, where there were mud, mud-brick and brick walls ; Besnagar (Vidisa) ; and Ahicchatra, where the use of burnt brick in the earlier period is attested, but where there was a free use of the material in the pre-Kushan, Kushan, and later levels.

Elsewhere we have explicit knowledge that burnt brick appeared only in the late phase of the Ware or even later. At Hastinapura, in Sub-period I of Period III, with that Ware, there were only two drains and a small wall, and in Sub-period III a long wall, followed by a large number of walls in Period IV, when the Ware had disappeared. At Rajghat, there were brick structures only in the late phase of the Ware, but in the next epoch there were a large number of structures. Mathura, with scanty burnt-brick remains in Sub-period I, had a vigorous building-activity in Sub-period III of Period II, both with the Ware. At Charsada, many of the early layers were associated with mud-brick and only the later ones with burnt brick. Comparable evidence is available at Tilaurakot, Atranjikhera, Sonpur and Chirand and other sites. The evidence of Kausambi is no less significant : here too burnt-brick structures appear well after the introduction of the Northern Black Polished Ware.

Outside north India, at Navdatoli the first burnt-brick structure appears after 400 B.C. At Nasik, Nevasa and Tripuri the use of brick is post-Mauryan.

Evidence is thus complete that burnt brick became popular very well after the appearance of the Northern Black Polished Ware ; it became common only in the second century B.C. and abundant even later on. The early cities were contented with mud and mud-brick structures where stone was not available, with the possibility of wooden structures, the remains of which have not survived. It has been said : 'In India, till recently the existence of kiln-burnt brick houses distinguished the town from the village, and this could serve as a yardstick even in classifying older habitations'. [Y. D. Sharma] An application of this criterion would deny a civic status even to those places which were renowned cities at the time of Buddha.

More evidences are perhaps not necessary to show the absence of the technology of making and using burnt bricks—not at least in any significant form—after the decline or final disruption of the First Urbanization in c. 1750 B.C. and before considerable progress towards the Second Urbanization, in the later phase of the NBP Ware period in c. 300 B.C. But it may be useful to have here some clarifications.

Apart from a stray baked brick found in Sanghol[18] and a fragmentary one found in Hastinapur[19] or a broken one found in Ahichchhatra[20]—which are somehow associated with PGW culture but the evidences of which really prove nothing—G.R. Sharma, the excavator of Kausambi, claims to date the use of burnt brick in this site in 1025 B.C.,[21] connecting it with the Harappans rather than the PGW people. As he[22] puts his claim :

> The early defences at Kausambi closely recall the Harappan citadel. The mud-packed rampart revetted externally with baked bricks in the so-called English bond in alternate courses of headers and stretchers, battered back to angles of 20° to 40°, bastions at intervals, rectangular towers and underground passage built on corbelled arch, are significant features of architecture at Kausambi with prototypes for each one of them in Harappan architecture. The very idea of town life was so far unknown in the Gangetic Valley. The defences show that in the first centuries of the first millennium B.C. Kausambi developed as a town fully equipped for its protection by the magnificent defences built on the Harappan pattern. Evidently, this was not an achievement of the P. G. Ware culture which shows a distinct aversion to the very concept of urban life in its earlier settlements in the Ghaggar Valley, the Punjab and Western U.P. Nor can it be associated with the Red Ochre-washed Ware. It is equally significant that P. G. Ware occurs at Kausambi two structural periods after the origi-

18. IAR, 1977-78. 43.
19. AI, No. 9-10, 17.
20. IAR, 1963-64, 43.
21. G. R. Sharma EK 22.
22. *Ibid*. 6. B. B. Lal (in HC, ed. Possehl, p. 336) comments : "Sharma's dating of the Kausambi's fortifications has been challenged by K. K. Sinha (1973) and A. Ghosh (1973). The grounds they have adduced against such an early chronology are quite valid and one would have expected the excavator of Kausambi to rethink the matter. Instead, he has come out with a renewed vigor about the Harappan influence on Kausambi."

nal construction of the defences. The recent discovery at Alamgirpura (District Meerut, U.P.) has established definite evidence of the penetration of the Harappan culture in the Ganga-Yamuna Doab. If the Harappans could reach the banks of the Hindon, a tributary of the Yamuna, the percolation and the survival of the Harappan influences at Kausambi only 300 miles down the Yamuna, is more than likely.

However, generally speaking, the view has not found favour with other serious archaeologists. The dating of Kausambi fortification by G.R. Sharma is rejected by K.K. Sinha[23], A. Ghosh[24] and others as too early. On the authority of Wheeler[25], R.S. Sharma argues that this date "cannot be pushed beyond 550 B.C. In fact the discovery of a cast copper coin may bring its date down to around 300 B.C."[26] B.B. Lal—besides rejecting G.R. Sharma's dating of the burnt bricks at Kausambi—vigorously argues against the possibility of any Harappan influence on the site.[27] In any case, the Kausambi excavation does not prove the technology of burnt bricks in the "Dark Age" intervening the two urbanizations. We shall later come to the question of G.R. Sharma's claim to unearth the ruins of an actual Vedic fire altar (*syenaciti*) at Kausambi.

But we cannot ignore or overlook in this connection the evidences unearthed by J.P. Joshi's explorations and excavations during the field seasons of 1975-76 and 1976-77 at Bhagwanpura (District Kuruksetra) and Dadheri (District Ludhiyana), revealing the use of burnt bricks. The excavations at Bhagwanpura, according to Joshi, "revealed a two-fold sequence of cultures designated as sub-Period IA and IB within a deposit of 2.70 m. showing for the first time that the Late Harappan Culture was interlocked with Painted Grey Ware Culture".[28] Describing the structures of Sub-Period IB, Joshi observes, "At first the people were living in round or semi-circular huts... In the next stage, the houses were built of mud walls... The third structural phase was associated with houses built of baked bricks of different sizes. Due to ploughing operation, all the

23. K. K. Sinha, in RIA, 1973. 231-38.
24. A. Ghosh, CEHI 81.
25. Wheeler EIP 130.
26. R. S. Sharma MCSFAI 59.
27. B. B. Lal, in Possehl (ed) HC, 336.
28. J. P. Joshi, in ME 1978, 98.

structures have been damaged. Whatever bricks were found *in situ* conformed to the following sizes: (1) 20×12×8 cms. (ii) 12×12×8 cms., (iii) 29×22×12½ cms. (wedge shape), (iv) 20×30×8 cms., (v) 16×12×4 cms. Some of the bricks have deep finger marks".[29]

Further :[30]

At Dadheri, in a 6 m. cultural deposit a three-fold sequence of cultures was identified. Of these, the lowest, Sub-period IA, is represented by pure Late Harappan Culture, closely followed by Sub-period IB wherein Painted Grey Ware and Late Harappan pottery are found together.. In Sub-period IA evidence of mud-walled houses and huts is available. Other important finds include a huge storage jar (Pl. VIII) with late Harappan Painted and incised wavy lines of pre-Harappan tradition and late Harappan pottery of the usual type, copper objects, terracotta beads, wheels and round cakes, faience bangles and a terracotta painted bull.

In Sub-period, IB, Painted Grey Ware, black ware, grey ware, red ware and typical late Harappan pottery is available. In this Sub-period, three structural phases have been recognized. At first the people were living in semi-circular huts as attested to by the discovery of post holes. Three oval structures of burnt earth probably of religious character came from this phase. In the next stage, the houses were built of mud walls. One such room measuring 1.10× 2.50 of a house complex has been noticed. The last phase is represented by a wall made of bricks, brickbats and brick jelly. Two sizes of burnt bricks, (12 × 12 × 7 cms., 25×20×5 cms.) have been found. Other finds from this Sub-period include terracotta beads, copper ring, terracotta wheels and faience bangles. No Iron has been found.

The last phase of occupation of the site belongs to the medieval times. The finds from this Period II include remains of a mud wall, typical medieval plain and painted pottery and terracotta figurines and games-men.

Much, of course, remains to be clarified about the above. The dates assigned to the Bhagwanpura finds by Joshi range from 1500 to 1000 B.C.[31]; but we are not told about the dates assigned to the different Sub-Periods at the site. As R.S. Sharma rightly observes, "Only the publication of the full report can throw light on the stratigraphical position of these bricks which appear to be rather unusual if the PGW-

29. *Ibid.* 98-99.
30. *Ibid.* 99-100.
31. R. S. Sharma, MCSFAI, 23 ; see also p. 33 note 6.

iron phase is placed roughly in 1000—500 B.C. So far no Carbon-14 dates have been made available".[32] Besides, B.B. Lal expresses strong doubts about the use of the expression of "Late Harappan" in the context of this site and comments: "However, the point to be emphasised is that the Bhagwanpura culture complex, composed of what can be termed an amalgam of the *n*-th generation of Harappans; the *n+x*-th generation of pre-Harappans and the *y*-th generation of Harappan cousins, *was in no way urban.* It was in this essentially rural setting that the meeting with the PGW Culture took place, the period of overlap being termed IB at Bhagwanpura, and likewise at Dadheri".[33]

In any case, from the point of view of what we are now discussing, namely brick technology, the discovery of some burnt bricks at Bhagwanpura and Dadheri, if assumed to belong to *c.* 1000 B.C., cannot but appear to be unexplained and an extremely odd phenomenon—as odd indeed as their sizes and proportions of their sides which answer neither to those of First Urbanization nor to any of the large variety of bricks we read in the Sulva texts. However, in default of an exact dating of the Bhagwanpura and Dadheri bricks and in the context of what is overwhelmingly obvious about archaeological data of the PGW sites in general, it may be permissible to work on the assumption that the technology of burnt bricks in any significant sense is absent throughout the period intervening the two urbanizations.

6. R. S. SHARMA'S THEORY OF MUD-BRICKS

We are thus confronted here with a serious problem. If—as we have already said and as we are later going to show in some detail—the mathematics embodied in the Sulva texts is inconceivable without the assumption of very sophisticated brick-technology, how are we to understand the fact that the Sulva texts themselves belong to a period in which, archaeologically speaking, the technology of making and using burnt bricks is conspicuous by its absence?

32. *Ibid,* 66 note 39.
33. B. B. Lal, in Possehl (ed.) HC 338.

R.S. Sharma seems to suggest a somewhat easy way out of the difficulty. His point is that the bricks mentioned in the Vedic texts—inclusive of the *Sulva-sutra*-s—need not be conceived as baked or burnt bricks at all. These were unbaked mud-bricks instead, which are easily conceivable in the PGW sites that are usually viewed as Vedic settlements, and hence the making of these is easily conceivable as forming part of the technology known to the Vedic people. As he[34] observed :

> The PGW mud-brick walls found at Hastinapur remind us of later Vedic references to bricks in connection with the construction of altars ; seven brick names are found in the *Taittiriya Samhita,* nine in the *Kathaka Samhita,* and eleven in the *Maitrayani Samhita.* In the *agnicayana,* the stacking of the bricks for the fire altars which is made obligatory in the *mahavrata* and optional in other *soma* sacrifices, the building of the *uttaravedi* involves five courses of bricks, making 10,800 bricks in all, in prescribed patterns often in the form of a bird with outstretched wings. But generally the PGW sites, except at Bhagwanpura and a few other places where the fire burnt bricks have been reported but not accounted for, do not yield fire-baked bricks ; similarly the later Vedic texts do not know of these. Of course a battered facing of brick on the mud ramparts of Kausambi has been discovered, but it cannot be pushed beyond 550 B.C. In fact the discovery of a cast copper coin may bring its date down to around 300 B.C. Therefore, the bricks mentioned in the Vedic texts were not generally baked in fire. A potter's kiln of the PGW level has been discovered in Atranjikhera. Such a kiln is known by *apaka* (Hindi *ava*) in the Vedic texts, but no term for brick-kiln is found in Vedic sources. *The old Vedic practice of using unbaked bricks for religious purposes continues* in Maharashtra and possibly in the other parts of the country. The total picture of PGW settlements does not warrant their characterization as urban, as has been done by Wheeler ; at best they can be called proto-urban towards the end of the PGW period. The later Vedic texts do not know of urban life. Kampila, the capital of Pancala, may have been an administrative settlement. The term *nagara* occurs in an Aranyaka and *nagarin* in two Brahmanas which are not earlier than 600 B.C.

R.S. Sharma does not mention here the *Sulva-sutra*-s or the mathematics embodied in these. This is perhaps because of his view already mentioned that the mathematics embodied in the Sulva texts arose our of the administrative requirements of land-measurement—a view evidently originating from the observation of Herodotus but rejected by archaeological-technological

34. R. S. Sharma, MCSFAI 58-59. Emphasis added.

considerations mentioned by Gordon Childe and J.D. Bernal. However, the point is that Sharma discusses here the question of *agnicayana* or the construction of the fire altars with bricks and though the Yajurvedic texts like *Taittiriya-samhita* and even the *Brahmana*-s enable us to understand the magico-religious beliefs imputed to these brick-structures, the *Sulva-sutra*-s are for us the most essential texts for understanding the technique for the physical construction of such structures. Hence is the obvious difference in the view of the bricks taken in the *Taittiriya-samhita* etc. and the Sulva texts—a difference which we shall later discuss. For the present let us concentrate on the other point stressed by R.S. Sharma in his observation just quoted, namely that the *Vedic texts refer to only mud-bricks and not burnt-bricks* because from the archaeological view-point only the former is to be expected in the PGW sites usually associated with the Vedic peoples.

But is it a fact that the Vedic texts speak only of mud-bricks and are unaware of burnt bricks? The answer is evidently in the negative.

Already in the *Satapatha Brahmana*[35] we read:

> Having gathered both that clay and water, he made a brick: hence a brick consists of these two, clay and water. He considered, 'Surely if I fit this (matter) such as it is unto my own self, I shall become a mortal carcase, not freed from evil: well then, I will bake it by means of fire'. So saying, he baked it by means of the fire, and thereby made it immortal.. Hence they bake the bricks with fire; they thereby make them immortal.

The *Baudhayana Sulva-sutra*, too, while discouraging the use of over-burnt bricks (ii.55) and also by advising how to make up for "that which is lost by the heat and the burning from the right size of the bricks" (ii.60), is indicative of burnt rather than simple mud-bricks. Here again the word used is *paka* which means firing. For unbaked brick the word would have been *ama*.

7. BURROW ON 'ARMA' AND 'ARMAKA'

So the anomaly remains. We do come across references to burnt bricks in the literature of a period in which—archaeo-

35. *Satapatha Brahmana,* vi. 2-1.8-9. The text, by clearly using the expressions *agnina pacani* and *aganina apacat* leaves absolutely no scope for doubt that the bricks spoken of were burnt in fire.

logically speaking—the technology of making and using these cannot be admitted. How are we to account for it?

At the present stage of research, it is evidently premature to expect a full answer to this question. What is possible, nevertheless, is to raise some counter-questions that may perhaps be pointers to further research.

The first question that occurs to us in this connection is: Could it be that the Vedic peoples, though without the know-how of making and using burnt bricks, were acquainted with ready-made bricks, i.e. with bricks made and used by others centuries before, or to be more specific, during the period of the First Urbanization, when highly sophisticated brick-technology was an accomplished fact, and when, therefore, the possibility of the emergence of mathematics from this technology cannot be *prima facie* impossible? If there be anything in such a possibility, the presumption would be that the roots of the mathematics of the *Sulva-sutra*-s are to be traced back to the Harappan culture, though we are yet to know how it was trtnsmitted and codified in the Sulva texts many centuries later.

The primary evidences for the making of mathematics in Harappan culture are no doubt to be searched from the archaeological data. Before passing on to these, however, we may try to review some circumstantial evidences of the later period indicating the general possibility of the Vedic people being acquainted with ready-made bricks of the Harappan period.

We begin with the brief but exceedingly interesting article by T. Burrow *On the Significance of the Term arma- armaka- in Early Sanskrit literature.*[36] The article needs to be read in full, though we have the scope here to mention only some of its salient points.

Though the word *arma* fell into disuse in classical Sanskrit literature, Burrow draws our attention to its use in Panini's grammar and the *Kasika* commentary on it as illustrating rules "concerning the accentuation of certain compounds having this word as last member". This list from the grammatical literature given by Burrow includes Bhutarma, Adhikarma, Sanjivarma, Madrarma, etc. On the authority of V.S. Agarwala[37], Burrow observes, "All these are place names and the

36. *Burrow in JIH,* Vol. 41. 1963, 159-166.
37. V. S. Agrawala, *India as known to Panini,* 66-67.

element arma- at the end means a ruined site or settlement".[38] The meaning 'ruined site or settlement' is taken from the commentators who use the expression *vinastagrama*—, which in later times literally means 'ruined village'—*grama* meaning village, while the word for city is *nagara*. But on the authority of Agrawala, again, Burrow shows that, in the context of its use in *Panini*, it should rather mean ruined cities[39] :

As V. S. Agrawala points out, the two terms *grama* and *nagara* were used indiscriminately in the Vahika country (i.e. the Punjab), although distinguished in Eastern India as 'village' and 'town'. In the case of these place names one would expect that the meaning required would be 'town' or 'city' rather than 'village', since any material remains of mere villages would be insignificant and would not qualify for a name having general currency and for that reason being worthy of mention in the grammatical work of Panini. To account for this we must assume that they were substantial ruins, and that the word *grama-* in this context is to be taken in the sense in which it was used in the Vahika country. From this one might be tempted to think that the situation of these ruined sites was in fact in the Vahika country. Certainly the only one that can be localised from the name—Madrarma—does not (*sic*) belong to this region, and it would not be surprising if the same applied to the other names formed on this pattern, particularly when we bear in mind the fact that in Panini's geography there is a distinct bias towards the North-west.

Where, then, were these ruined cities ? Could these be earlier Aryan settlements or the ruined cities of the ancient Indus civilization, which, in the Vedic period, were presumably quite abundant in the North-west ? Burrow[40] answers :

It is to be remembered that in Panini's time (say fourth century B.C.) the period of urban civilization for Aryan India was of comparatively recent origin so that although one might expect a few such deserted or destroyed cities (and presumably Navarma- was one of these), one would not expect them to be so thick on the ground as this large collection of names would indicate. On the other hand the number of Indus sites of the required magnitude already identified in this region is sufficient to explain this long string of names (which is to be taken as typical and not exhaustive), and it is to be remembered that during the early Aryan period the ruins of many Indus cities must have formed a conspicuous feature of the countryside.

38. Burrow, *op. cit.* p. 159.
39. *Ibid.* 159-60.
40. *Ibid.* 160.

This leads Burrow to search for evidences of the use of the word *arma* (or its derivative *armaka*) in literature earlier than Panini. And he comes across a number of these in the Vedic literature, some of which appear to have much significance. Here are a few of these.

The *Latyayana Srautasutra* X. 18.3 says : "On the Sarasvati. there are ruined sites called Naitandhava ; Vyarna is one of these". Other Vedic texts like *Apastamba Srautasutra* (xxiii. 13.12), *Sankhyayana Srauta-sutra* (xii. 29.28), *Pancavimsa Brahmana* (xxv. 13.1) also mention Naitandhava and Vyarna, though the one first quoted is very significant. As Burrow observes, "The mention of these ruined sites with the precise information about their location, informing us that they were situated along the Sarasvati, is exceedingly valuable information, since it is now well established that the Indus sites are a feature of this region. A recent excavation of one of these sites, at Kalibangan on the south side of the Ghaggar (ancient Sarasvati) has demonstrated the importance of this region as a centre of the Indus civilization".[41] The *Latyayana Srautrasutra* (x.19.9) speaks also of ruined site (*armaka*) along the right bank of the Drsadvati—a location of Harappan ruins again. Other references to *arma* and *armaka* in the Vedic literature, according to Burrow's argument, indicate the destruction or devastation of the cities of the Indus civilization by the invading Vedic Aryans—a view about which there is much debate in recent times and to which we shall later return. For the present our point is that Burrow's interpretation of *arma* (and *armaka*) as ruined Indus sites seems to be of much importance for what we are going to argue. Let us see why it is so.

Burrow[42] observes,

A compound *arma-kapala* meaning 'a tile from a ruined site' occurs not infrequently in the *Srautasutras* (e.g. *Baudhayana* ix.1.3, etc.) where it appears among a list of paraphernalia for a sacrifice. In this connection the *Vadhulasutra* glosses : *atha yad armakapalani bhavanti armad evainam tat prthivyah sambharati*—'since there are tiles from a ruined site, in this respect he assembles it (the fireplace) from a ruined site of the earth.' From these *sutra* references we

41. *Ibid*. 162.
42. *Ibid*. 161.

gather that *arma*-s or ruined sites were a commonplace thing in the Vedic period, since these *arma-kapalani* prescribed in the ritual appear to have been readily available. This is in agreement with the fact that material remains of the Indus Civilization have been located in abundance throughout the territory occupied by the Vedic Indians subsequent to its downfall.

The 'paraphernalla for a sacrifice' referred to above are the special type of earthen vessel prescribed for the ritual use, a fuller discussion of which we have in the article *Vedic Literature on Pottery* by Shivaji Singh.[43] In this article Singh mentions the priestly instructions concerning the preparation of clay for the making of the earthen vessels to be used in the Vedic sacrifice. According to the *Taittiriya-samhita,* as Singh observes, "potsherds collected from ancient deserted sites (*arma-kapala*), sand (*sarkara*) and hairs (*ajaloma* and *krsnajinaloma*) were to be mixed with clay".[44] In the notes, Singh adds that Sayana, commenting on the Brahmana portion of *Taittiriya-samhita* iv.1 explains *arma-kapala* as (potsherds from) *cirakala-sunya-grame bhumau avasthitani puratanani*, i.e. ancient (potsherds) existing in the eternally deserted cities. (We have already noted Burrow's argument why in this context the word *grama* should preferably be taken to mean 'city' rather than 'village'). Since, obviously enough, no city can be "eternally deserted", we have to take Sayana here as referring to cities that remained deserted from a very ancient period. As far as our present knowledge goes, only the ancient Harappan ruins can answer to what is meant by Sayana.

To sum up the discussion so far : From the ancient Vedic literature to Panini we come across references to ruined cities as *arma,* which, moreover, were presumably quite commonplace in north-west India. Secondly, according to the ritual instructions, potsherds from these ruined cities had to be collected for the preparation of clay, from which to fashion earthen vessels for Vedic sacrifices.

To these we have to add here only one point. Archaeologists have no doubt found plenty of potsherds from the ruined Harappan sites. However, if anything is much more conspicuous about the ruined cities like Harappa, Mohenjo-daro and Kalibangan it is the heap of burnt bricks. Even after

43. See B. P. Sinha (ed) PAI 301-13.
44. *Ibid*. 307.

the brick-robbery in massive scale and using these as ballasts for the railway lines from Lahore to Multan and from Hanumangarh to Suratgarh, the burnt bricks still surviving in the ruined sites of the Indus Valley civilization are most imposing. If, therefore, the Vedic priests were actually collecting potsherds from these ruined sites, it is not difficult to conceive how they could speak of burnt bricks without acquiring the technology of making and using these.

Do we have here a clue to the apparent anomaly of the references to burnt bricks in the literature of a period which, archaeologically speaking, is unaware of the know-how of making and using burnt bricks?

Not that such a supposition is free from difficulties. We do not have in the Vedic literature any reference to the collection of *arma-istaka* or bricks from ruined sites as we have to that of *arma-kapala*-s. Nor have we any direct archaeological evidence of the re-use of ready-made bricks for the construction of the Vedic sacrificial altars. The earliest reference to the fire altars in Vedic literature are to be found in the *Yajurveda*—particularly the *Taittiriya-samhita*. However, as we shall presently see, the text is much too interested in describing the mysterious magical efficacy of the bricks to give us any physical description of these—descriptions which could have enabled us to compare these bricks with the Harappan ones. Nor is it possible for us to make much of the quaint brick-names of the *Taittiriya-samhita*, some idea of which we shall presently have. On the whole, it may not be an error to think that the *Taittiriya-samhita*, viewing as it does the bricks as highly mysterious entities with wonderful magical potency seems to suggest that the priests in the *Taittiriya-samhita* knew of bricks without knowing what these really were.

Things would have been helpful for us had the archaeologists been able to discover any actual ruin of a sacrificial altar belonging to the Yajurvedic period. But the fact is that nothing like that is so far discovered. The earliest archaeological evidence of what is claimed to have been a Vedic fire-altar—a *Syenaciti*—comes from the excavation of Kausambi. According to G.R. Sharma, the excavator of Kausambi, the site has revealed not only the remains of a brick built *Syenaciti* but also other relics like animal and human bones reminiscent of the ritual prescriptions of the Vedic priests.[45] But he has not tried to correlate the brick-sizes in the ruins of this "altar" with what are mentioned in the Sulva texts regarding the construction of the *Syenaciti*. The average brick-sizes mentioned by G.R. Sharma as having been generally used in Kausambi, namely 19.5×13×2.75 inches[46] answer neither to the known brick-sizes of the Harappan ruins nor to those prescribed in the Sulva texts. In any case, even admitting his claim that the structure unearthed represents the remains of a *Syenaciti*[47], its date cannot be pushed back to the Yajurvedic period. As Y.D. Sharma observes, "The sacrifice is believed to have been performed by the founder of the Mitra dynasty whose coins have been recovered in abundance from corresponding levels".[48] The coins of Mitras could not be earlier than the second century B.C.[49]. Other archaeological evidences suggesting Vedic fire-altar are much later. One of these, e.g., come from the excavation of Jagatram, about 30 miles to the North-west of Dhera-dun, and is dated about the third century A.D.[50]

Such, then, is our present knowledge of the archaeological evidences about the Vedic fire altars and it is no use speculating on what might have eluded the archaeologists' spade so far. Nevertheless, the fact remains that we have in the late Vedic literature unmistakable evidence of mathematics emerging from the requirements of brick technology while the authors of this literature—without any knowledge of the technology in their times—could conceivably have any acquaintance with it only in the ruined Harappan sites. Could this be a pointer to the possibility of the whole thing—the brick technology as well as the mathematics emerging to meet its requirements—

45. G. R. Sharma, EK, pp. 87-206.
46. .*Ibid*. 27.
47. Lal B.B, personal correspondence dated 17th July, 1985, observes : "any way, my considered view is that the brick assemblage concerned is not at all a Syenaciti. It represents the collapse of an adjacent brick-wall pertaining to the fortifications." We are expecting the publication of his detailed discussion of the point in the *Puratattva*.
48. Y.D. Sharma, ARMM. I. 55.
49. Bela Lahiri, 90.
50. Excavated by T.N. Ramachandran, I.A. 1953-54, 10-11, See also B.B. Lal in CF Dec, 1961, 36.

actually developed in Harappan culture and somehow transmitted to the Vedic peoples of later times? It is premature at the present stage of our knowledge to try to arrive at any categorical answer to this question. What is possible—and perhaps also necessary—is to note some relevant points which may stimulate further research.

8. EVIDENCE OF THE SATAPATHA BRAHMANA

After the *Taittiriya Samhita* the question of Agnicayana or that of the ritual building of the fire-altar is elaborately discussed in the *Satapatha Brahmana.* As contrasted with the *Taittiriya Samhita,* in which the bricks are viewed as some mysterious entities with quaint names and quaint magical potency, the *Satapatha Brahmana* takes on the whole a comparatively realistic view of these, though of course without scrapping the associated ancient magico-religious beliefs. It is only in the Sulva texts that we come across a well-defined technological view of the bricks, specifying their exact shape, size, etc. We shall later return to some details of these. For the present, we have another interesting point to note about the *Satapatha Brahmana.*

Like the other *Brahmana-s,* the *Satapatha,* too, comes down to us as appended to one recension of the *Yajurveda.* In the Yajurvedic text there is constant mingling of magical formulas with explanatory portions of which only the latter is strictly called the Brahmana-s. The class of priests called the Adhvaryus wanted entirely to separate exegetic matter from the magical formulas or spells proper. The name given to the school of Adhvaryus responsible for the preparation of the *Satapatha Brahmana* is Vajasaneyins, its origin being ascribed to one Yajnavalkya Vajasaneya. "The *Brahmana* of the Vajasaneyins bears the name *Satapatha* i.e. the *Brahmana* of 'a hundred paths', because it consists of a hundred lectures".[51]

We have noted it mainly to emphasise one point. Yajnavalkya is expected to be the authority for the theological discussions in the text. But the text in the form in which it reaches us does not satisfy the expectation. We quote Eggeling[52] at some length who draws our attention to this:

51. Eggeling in SBE vol. xii. Intro. pp. xxvii-xxviii.
52. *Ibid Intro.* xxxi.

As regards the earlier portion of the work, however, it is a remarkable fact that, while in the first five books Yajnavalkya's opinion is frequently recorded as authoritative, he is not once mentioned in the four succeeding *kandas*. The teacher whose opinion is most frequently referred to in these books is Sandilya. This disagreement in respect of doctrinal authorities, coupled with unmistakable differences, stylistic as well as geographical and mythological, can scarcely be accomunted for otherwise than by the assumption of a difference of authorship or original redaction. Now the subject with which these four *kandas* are chiefly concerned, is the *agnicayana* or construction of the sacred fire-altar. For reasons urged by Professor Weber, it would appear not improbable that this part of the cermonial was specially cultivated in the north-western districts; and since the geographical allusions in these four *kandas* chiefly point to that part of India, while those of the other books refer almost exclusively to the regions along the Ganges and Jumna, we may infer from this that the fire-ritual adopted by the Vajasaneyins at the time of the first redaction of their texts—that is, of the first nine *kandas,* as far as the Brahmana is concerned—had been settled in the north-west of India.

Here, however, we meet with another difficulty. The tenth book, or Agnirahasya, deals with the same subject as the preceding four *kandas*; and here also Sandilya figures as the chief authority, while no mention is made of Yajnavalkya.

What concerns our present discussion is not the question of the actual redaction of the *Satapatha Brahmana* in its present form. But the pointer to north-west India as the region from which it appears to incorporate within itself the matters concerning the Agnicayana or the ritual building of the fire-altar—and therefore by implication also of the brick-technology required for the altar construction—can by no means be ignored or overlooked. In fact, it acquires much more importance for us today than it could possibly have during the time of Weber (1805-1901) on whose discussion[53] of the geographical allusions in those portions of the *Satapatha Brahmana* Eggeling depends. The same is true of Eggeling's comment just quoted which was first published in 1882. Nothing whatsoever was known in their times of the imposing civilization in north-west India with its remarkably developed brick technology. With the discovery of the Indus Valley civilization, however, the peculiarities already noted by Weber and Eggeling cannot but lead us to a new question. Could it be that the brick-structures

53. A Weber, Ind. Stud. I., 187 seq. and XIII. 266ff.

spoken of in the *Satapatha Brahmana* are indicative of the heritage of the Harappan culture? The recent discovery of some structures generally considered as "fire altars" specially at Kalibangan[54] and Lothal[55] apparently strengthens the possibility, though it remains for us to see the hazards about any hasty conclusions about these.

9. THE QUESTION REOPENED

With the dramatic expansion in our understanding of the ancient Indian history by the recent archaeological work, the entire question of the Agnicayana is reopened by a section of the modern scholars. Some of them are trying to argue that the ritual was borrowed by the Vedic people from the Harappans. The question of the ritual as such is, of course, outside the scope of our own discussion. Nevertheless, some of the recent views expressed do interest us, because the question of the Agnicayana is inextricably connected with brick technology. As a matter of fact the vital dependence of this ritual on brick technology is used by some scholars as an evidence of the ritual itself having been a Harappan survival.

As Hyla Stuntz Converse[56], in her article *The Agnicayana Rite: Indigenous Origin?* very strongly argues:

> The question of brick is of major importance. The Harappa civilization, whose last, flood-damaged strongholds in the north were overthrown by the invading Aryans in battles commemorated in the *Rgveda,* was a brick-using culture. The Harappans used millions of kiln-fired bricks as well as countless sun-baked ones.. The bricks of the Harappa civilization in its mature phase were beautifully made, well fired, and standardized in size..
>
> Now, in the whole of the *Rgveda* there is no word for brick, nor any descriptive phrase for bricks. So far no ruins of brick dwelling have been found that can be attributed to the Aryans in the early Rg-Vedic period.. There are also no references to bricks in the *Rg-Veda Brahmanas* and outside of the Agnicayana sections of the Samhitas and Brahmanas of the Yajurveda tradition, no significant reference to bricks occur in these or in the *Samaveda Brahmanas.* Thus, in the *Brahmanas,* when references to brick begin to appear, their use is confined to one specialized rite, and the rite itself is found only

54. Allchins RCIP 183 and 303.
55. S.R. Rao, LIC 139-40.
56. H.S. Converse, in HR Vol. xiv, No. 2, 83-84.

in the Yajurveda tradition. The fire altars in the other rites were made of packed earth, not bricks.

The size of the bricks to be used in the rite was one foot square, and half-bricks were also to be used (SB vii. 5.3.2 ; viii. 7.2.17). This size and shape corresponds very closely to that of the Harappa bricks described above. The lack of any bricks in the early Vedic tradition and the presence of bricks in large numbers and of the same size in the adjacent indigenous Black-and-Red Ware territory suggest that the Black-and-Red Ware culture is the source of the Agnicayana brick-making skills.

It is not necessary for our present purpose to enter into the technicalities of the Black-and-Red Ware pottery, which Converse readily takes as "the distinctive trait of the indigenous non-Vedic culture". Other archaeologists do not agree to this. A. Ghosh, for example, wants to see the Black-and-Red Ware in the expression *nilalohita* of the *Atharvaveda.*[57] Though Converse herself for her basic argment wants to depend much on the assumption of the Black-and-Red Ware being non-Vedic or pre-Vedic, the main drive of her argument is not basically affected by it, because she offers other evidences for the possible Harappan origin of the Agnicayana, the main point of which remains unaffected by what the archaeologists may finally conclude about the Black-and-Red Ware.[58] The most important of these arguments seems to be that Agnicayana is not conceivable without brick technology, "although the Vedic Aryans were not a brick-making people."

Frits Staal, in his *The Ignorant Brahmin of the Agnicayana,*—in which he argues that the apparently quaint concept of the "ignorant" priest occurring in the Vedic literature makes "sense only if he is a representative of a pre-Vedic fire cult"—finds it necessary to come back to the evidence of the brick technology. After reviewing the literary evidences which he considers important for his main arguments, he[59] adds :

If we wish to understand more, we have to go beyond the texts and place the Agnicayana in a wider historical perspective. The techniques for firing bricks, which we meet for the first time in Yajurvedic texts dealing with the Agnicayana, could not have been im-

57. A. Ghosh, CEHI 6.
58. H.N. Sing, "Black-and-Red Ware" in EIP 267-281.
59. F. Staal, in ABORI 1978, 345.

ported by the Vedic nomads who had earlier entered the subcontinent from the northwest. Nomads have no need for bricks. The bricks, in fact, occupy an exceptional position in the Vedic ritual, when all implements are made of perishable materials, and are taken away, destroyed, burnt or immersed in water after a ritual performance has been completed..

For the firing of bricks we have to look in a non-nomadic direction—i.e., for a sedentary civilization. The fact which springs to mind is that in the Harappa civilization the use of fired bricks was normal and widespread. We need not jump to the conclusion that the Agnicayana was a Harappa ritual, though Converse (1974) has certainly forwarded weighty arguments (together with some incorrect ones) in support of such a hypothesis. Whatever else may be true, it is certainly reasonable to suppose that knowledge of the techniques for firing bricks was preserved among the inhabitants of the subcontinent even after the Harappa civilization had disappeared. The Vedic nomads, who by the time of the *Yajurveda* were in close interaction with the indigenous population and had intermarried and accepted many of them within their fold, adopted these techniques in the construction of the Agnicayana altar.

Whether the "technique for firing bricks" is already to be found in the *Yajurveda* is, of course, a very doubtful point. Neither the internal evidences of the *Yajurveda* nor the archaeological data we have of the period of the formation of the text point to such a possibility though the *Satapatha Brahmana* clearly speaks of burning he bricks. We have already seen, how—without this technology—the Yajurvedic priests could possibly speak of the *istaka*-s. On this point, therefore, Staal's observations may be in need of some modification.

C. G. Kashikar critically reviews the opinions of both Converse and Staal. Mainly on theological grounds he argues that the Agnicayana could as well be "an extension of Vedic Aryan rituals". Also the apparently quaint expression "ignorant Brahmin," Kashikar argues, can have an alternative explanation, mainly from the theological viewpoint again. What concerns our own discussion, however, is not theology but brick technology. On this point Kashikar[60] observes :

It is true the *Rgveda* does not contain any reference to bricks, even though the existence of some kind of pottery can be thought of in the Rgvedic period. The *Brahmanas* of the *Rgveda* and *Samaveda* cannot be expected to contain any reference to bricks because the

60. C.G. Kashikar, in ABORI, 1981. 125.

executive function of the piling up of the altar was outside their scope. Even in the *mantra* and *brahmana* portions of the *Yajurveda* one cannot expect any reference to brick outside the piling up of the fire-altar.. It is obvious that the Agnicayana rite was introduced in the Vedic rituals after the Vedic people became conversant with the use of bricks. As already noted, Vedic Aryans came into close contact with the Dravidians and other people in the north-west region of India even in the days of the *Rgveda,* and the process of cultural give-and-take had already started. The racial admixture, at least to a certain extent, was a natural consequence. The Vedic people, characterised as a racial complex, might naturally be expected to have adopted a number of worldly things from the indigenous people, among which the use of fired bricks for housing purposes must have been included.

But the only reference Kashikar gives from the Vedic ritual literature for the use of bricks for housing purposes is from Kesava's *Paddhati,* a 13th century commentary on the *Srauta-sutra*-s.[61] The text is obviously too late to prove his thesis. It seems, thus, that after accepting bricks originally for ritual purposes, the Vedic people had to wait long for the technology to be used for house construction.

More interesting, however, is Kashikar's argument against Converse, who tries to correlate the brick-sizes of the Harappan cities with these mentioned in the *Satapatha Brahmana.* Argues Kashikar[62] :

> The comparison would hardly serve any purpose because the *citi,* even following the *Satapatha Brahmana,* would require bricks of various sizes and patterns. Even though the *Satapatha Brahmana* directly mentions only a few of them, the other sizes mentioned in the *Katyayana Srautasutra* and the *Istakapurana Parisista* need to be presumed even in the case of *Satapatha Brahmana.* Thus square bricks with each side measuring 24, 18, 12 or 6 *angulas* are mentioned. Besides these, oblong bricks of various sizes were also required. There are other varieties of the altar which need bricks of acute and obtuse angles and also curved bricks.

But all this really brings us back to the point we have been trying to drive at. Already the *Satapatha Brahmana* presupposes a long and sophisticated tradition of brick technology, while—archaeologically speaking—there is no such tradition in the immediate context of the text. On the contrary, there is only

61. *Ibid* 126, note 13.
62 Ibid 124-125.

one tradition from which the ritual texts like the *Satapatha Brahmana* could possibly borrow or assimilate such sophisticated brick technology. And that is the tradition of the Mature Harappan Culture.

10. PREHISTORY OF SULVA GEOMETRY

But let us return to the question of geometry and the Sulva texts.

The *Sulva-sutra*-s introduce us to a different technological climate. The bricks are no longer just mysterious entities with quaint magical potency, as these basically appear to be in the *Yajurveda*. Though without questioning the Yajurvedic priests and even accepting their dictations, the *Sulva-sutra*-s in fact take a different view of these. The bricks are deliberately made according to certain specific shapes and sizes ; these are deliberately dried and burnt. Attempt is made to determine exactly how much in size and area these lose as a result of drying and firing,[63] so that provision may be kept in the original mud brick for this shrinkage and eventually burnt brick is obtained according to their exact area required for the altar-making. Care is taken to see that the bricks are neither overburnt nor underburnt[64] and thereby become useless for construction purposes. In short, brick technology is discussed in the *Sulva-sutra*-s not from the standpoint of any outsider marvelling at the bricks but from that of the technologists themselves. More than this. The *Sulva-sutra*-s do not merely discuss the technology of *making* bricks ; the texts are as a matter of fact much more interested in the technique of *using* these bricks for the construction of certain pre-conceived structures, however complicated these might have been. However, what is most remarkable about the *Sulva-sutra*-s is that these are not mere manuals or handbooks for the craftsmen and the technicians ; these indicate also the awareness of developing the exact theoretical propositions which, as actually developed in the *Sulva-sutra*-s, come down to us as the earliest documents of geometry in ancient India.

The standardization of the theoretical knowledge in the form of geometrical propositions, along with the development of the terminologies required for these, evidently presuppose a pro-

63. *Baudh. Ss* ii 60.
64. *Baudh. Ss.* ii. 55.

longed period. Since all this is the outcome of the techniques of brick making and brick laying, the formation of the Sulva geometry presupposes a long and sophisticated tradition of brick technology. However, if we are to place the *Sulva-sutra*-s roughly in the fifth or fourth century B.C.—as the modern scholars propose to do—we find practically nothing in the immediate historical context of these texts to suggest any long and sophisticated tradition in brick technology. How, then, are we to understand the formation of the Sulva geometry?

B. B. Datta concedes to the possibility of the Sulva geometry having some anterior stage of development. He speaks of "the growth and development of the Hindu Geometry from its earliest state down to the one in which we find it now in the *Sulba*".[65] Thibaut also suggests that the Sulva geometry necessarily presupposes a long time of development. As he puts it, "Regarding the time in which the *Sulva-sutra*-s may have been composed, it is impossible to give more accurate information than we are able to give about the date of the *Kalpa-sutra*-s. But whatever the period may have been during which the *Kalpa-sutra*-s and *Sulva-sutra*-s were composed in the form we have now before us, we must keep in view that they only give a systematically arranged description of sacrificial rites which had been practised during long preceding ages".[66] But where are we to look for the "long preceding ages", the formative period of the Sulva geometry? Working as he did in the last quarter of the nineteenth century, Thibaut himself had nowhere else to look for it than in the earlier strata of the Vedic literature itself. And he fails to find there anything actually foreshadowing the Sulva geometry. The fact is that as we move backward from the *Sulva-sutra*-s to the earlier strata of the Vedic literature, we are confronted more and more with the mystery-mongering about the bricks and fire altars rather than either with the techonology of actually building these or with the geometrical calculations required by the technology. Thibaut, therefore, wanted somehow to imagine that the formation must have had a pre-history among the Vedic priests themselves—the Adhvaryus—though without knowing what it actually had been. As he puts it, "The rules for the size of

65. B.B. Datta SS 20.
66. Thibaut, in SHSI, ii. 472.

the various Vedi-s, for the primitive shape and the variations of the *agni* etc. are given by the *Brahamana*-s, although we cannot expect from this class of writings explanations of the manner in which the manifold measurements and transformations had to be managed. Many of the rules which we find now in *Baudhayana*, *Apastamba* and *Katyayana*, expressed in the same or almost the same words, must have formed the common property of all Adhvaryu-s *long before they were embodied in the Kalpa-sutra*-s which have come down to us".[67]

Writing on the *Sulva-sutra* in 1875, as Thibaut did, what else could he do than try to convince himself that the Sulva geometry must have had its origin somewhere in the earlier stage of the tradition of the Vedic priests themselves, though without at all knowing where it could be.

67. *Ibid.*

CHAPTER 6

TECHNICIANS AND THE VEDIC PRIESTS

At the very outset of the present discussion, it may be useful to try to be clear about what we are going to argue. On circumstantial evidences—specially from the evidence of the *Sulva* geometry being indisputably the outcome of sophisticated brick technology, which was a prominent feature of the Harappan culture but which was totally absent in what is usually called the Vedic period—we are led to the view that this geometry was in all presumption developed in the Harappan culture. In the Harappan culture, however, the priests or priestly corporations were very likely to have a dual role. On the one hand, they were the 'organisers of production', while on the other, they were the 'administrators of superstition'—a point which we hope to discuss in some detail in our Chapter X. From this viewpoint, there is nothing *prima facie* impossible for them to have been the makers of the geometry under consideration, because as 'organisers of production' they were in need of having geometry as a conscious science. Therefore, our present argument is not that the *priest class as such* could under no circumstance be the makers of the geometry that comes down to us in the *Sulva* texts. What we are going to argue here is the *prima facie* difficulty of the *Vedic priests*, having been the makers of this geometry, whose social function and social position must have been basically different from those of the Harappan priests. From the Vedic literature itself, we have nowhere the impression that these priests were any longer the 'organisers of production'. Cut off from this social function, they became some kind of social parasites, subsisting mainly on the *daksina* or sacrificial gifts received from the patrons financing the sacrifices. In short, they were left only with the other priestly function, namely that of being the 'administrators of superstition'. What we are going to discuss in the present chapter is thus not the general question of the priest class having anything to do with the making of mathematics but specifically the possibility of the Vedic priests having the role.

1. PRELIMINARY REMARKS

Our discussion so far has hinged on one point. The mathematics in the Sulva texts was the outcome of the theoretical requirements of the technology of making and using burnt bricks, notwithstanding a body of magico-religious beliefs somehow sought to be associated with the brick-structures.

But this strongly goes against the usual assumption about the making of mathematics in ancient India, according to which it was created by the Vedic priests to meet the requirements of their sacrificial rituals.

It was A.C. Burnell (1840-1882), who, in his catalogue of the *Collection of Sanskrit Manuscripts* (1869), first drew our attention to Sulva texts for understanding "the earliest beginnings of geometry among the Brahmanas".[1] Since then the texts are extensively discussed by a number of outstanding modern scholars like A. Burk, G. Thibaut, B.B. Datta and many others, to whom we are basically indebted for explicating the mathematical contents of these texts. In our own discussion, we shall have to draw freely from their writings. In the present chapter, however, we shall try to explain why we feel obliged to differ from them on a point that appears to us to be historically very significant. All of them have on the whole assumed this mathematics was created by the Vedic priests. Let us first have some idea of this assumption and the *prima facie* grounds for it.

2. ORIGIN OF GEOMETRY : HERODOTUS AND RECENT VIEW

The usual assumption that Sulva mathematics arose to meet the requirements of the Vedic sacrifices leads Thibaut to the extent of arguing that here at last we have an unquestionable evidence for the genuinely Indian contribution to science. As he[2] puts it :

> We have been long acquainted with the progress which the Indians made in later times in arithmetic, algebra, and geometry ; but as the influence of Greek science is clearly traceable in the development of their astronomy, and as their treatises on algebra, etc. form but parts of astronomical text books, it is possible that the Indians may have

1. G. Thibaut in *SHSI* Vol. II, 417.
2. *Ibid.* 415-16.

received from the Greeks also communications regarding the methods of calculation. I merely say possible, because no direct evidence of such influence has been brought forward as yet and because the general impression we receive from a comparison of the methods employed by Greeks and Indians respectively seems rather to point to an entirely independent growth of this branch of Indian science. The whole question is still unsettled, and new researches are required before we can arrive at a final decision.

While therefore unable positively to assert that the treasure of mathematical knowledge contained in the *Lilavati*, the *Vijaganita*, and similar treatises, has been accumulated by the Indians without the aid of foreign nations, we must search whether there are not any traces left pointing to a purely Indian origin of these sciences.

Thibaut sees these in the *Sulva-sutra*-s the geometry embodied in which arose out of the requirements of the Vedic sacrifices. What, then, were the requirements of the Vedic sacrifices that is supposed to bring geometry into being? B.B. Datta explains, "According to the strict injunctions of the Hindu Sastra (or 'Holy Scriptures'), each sacrifice must be made in an altar of prescribed shape and size. It is stated that even a slight irregularity and variation in the form or size of the altar will nullify the object of the whole ritual and may even lead to an adverse effect. So the greatest care has to be taken to have the right shape and size of the altar".[3] Such a demand for precision could not be met without developing a sufficient amount of geometrical knowledge, and it is this that we find in the Sulva texts. Geometry thus came into being to meet the requirements of Vedic sacrifices.

This view of the origin of geometry seems to be endorsed by the immediate context as belonging to which the *Sulva* texts come down to us. It is the context of the sacrificial literature of the Vedic priests. We may thus have here a few words on this literature, specially those parts of it to which we shall have to refer repeatedly in the course of our own discussion.

The first fullfledged literature with the special theme of the sacrifices is called the *Yajurveda*. Tradition wants us to believe that there was once a very large number of schools of the Vedic priests, each with its own version (or recension) of the *Yajurveda*. A number of these actually survive for us,

3. B.B. Datta SS. 20.

of which we shall have to refer specially to the *Taittiriya Samhita* in which the question of Vedic fire altars is elaborately mentioned.

It is, of course, impossible to be exact about the date of this text, though on various considerations serious Vedic scholars think that the end of the period of the *Yajurveda* could not be later than 800 B.C.

Next in importance from the priestly point of view is a class of texts known as the *Brahmana*-s. These are fabulous in bulk, discussing all sorts of details of the sacrifices and often expressing sharp differences of opinion on ritual trivialities and "theological twaddles". Of these texts, the *Satapatha Brahmana* contains a long portion on the question of the fire altars and hence has special significance for purposes of our own discussion. It is impossible, again, to be exact about the date of the text, though the modern scholars are generally inclined to think that the end of the age of the *Brahmana*-s could not be later than 600 B.C.

Though exclusively concerned with the Vedic sacrifices, the *Brahmana* texts are really too vast, too complicated and often also too obscure and controversial to be of actual use for the practising priests. The need was thus felt for some kind of handbooks or manuals for their use. To meet this requirement, there grew in course of time another class of literature called the *Kalpa-sutra*-s. "They arose out of the need for compiling the rules for sacrificial ritual in a shorter, more manageable and connected form for the practical purposes of the priests. *Kalpa-sutra*-s dealing with the Srauta sacrifices taught in the *Brahmana*-s are called *Srauta-sutra*-s, and those dealing with domestic ceremonies and sacrifices of daily life, the Grhya-rites, are called *Grhya-sutra*-s".[4]

Directly attached to the *Srauta-sutra*-s are the *Sulva-sutra*-s: though separately classified under the *Kalpa-sutra*-s, the *Sulva-sutra*-s come down to us as forming parts of the *Srauta-sutra*-s. But there is another class of literature belonging to the *Kalpa-sutra*-s. These are called the *Dharma-sutra*-s and deal with religious as well as secular laws. Through these works, observes Winternitz, the Brahmins "succeeded in transforming the law of ancient India to their own advantage, and in making

4. Winternitz, I, 272.

their influence felt in all directions".[5] We shall later see the importance of these legal texts for understanding the generai theoretical temper and sociological ideas of the Vedic priests, which, in their turn, are of immediate relevance for the genertl problem of the formation of exact science in ancient India. For the present the point is that since the *Srauta-sutra*-s and the *Dharma-sutra*-s have the same origin, it will be a methodological error to try to understand the former without relating it to the latter. However, let us try first to be clear about the *Sulva-sutra*-s.

The Vedic priests belonged to various schools like those of Baudhayana, Apastamba, Katyayana, etc., each having *Kalpa-sutra* of its own. The *Sulva-sutra*-s, as appended to the *Kalpa-sutra*-s, also come down to us in somewhat different versions, —though on the whole the differences are minor. Of these, the *Sulva-sutra* of Baudhayana is viewed as the oldest and most important one for understanding the making of mathematics in ancient India. Next in importance is the *Sulva* text associated with the name of Apastamba. Five other *Sulva-sutra*-s are known to the modern scholars, as belonging to the schools of Katyayana, Manava, Maitrayana, Varaha and Vadhula. But these are of comparatively lesser significance.

It is impossible, of course, to be exact about the date of the *Sulva-sutra*-s. On various circumstantial evidences, however, it is believed that the dates of the Sulva texts associated with the names of Baudhayana and Apastamba are not likely to have been later than the third or second century B.C., though this could be somewhat earlier—but not earlier than the fifth century B.C. But these are to be understood as possible dates of the codification of the Sulva texts. The actual mathematical knowledge being codified in these could have come down from a distant past.

3. "WHAT IS TO BE DONE" & "HOW IS IT TO BE DONE"

Judging from the fact that the *Sulva-sutra*-s come down to us as forming parts of the priestly handbooks, nothing seems to be *prima facie* more natural than the usual assumption that the knowledge embodied in these developed in the priestly

5. *Ibid*. 275.

circles to meet certain requirements of the priest-craft. But the actual question of the making of mathematics in ancient India is much more complicated than can be answered merely by judging the context of the Sulva texts.

For this purpose, it is essential first of all to differentiate between two questions concerning the brick-structures, called the *citi*-s or *agni*-s. These questions are : First, "What is to be done ?" Secondly, "How is this to be done ?" We are going to see that the Vedic priests, *in the capacity of priests in the strict sense*, could be concerned only with the first of these two questions. However, the second question is clearly outside the scope of priestly specialization and could be faced only by technicians—the highly skilled brick-makers, masons, architects or engineers. Therefore, even admitting that a section of the Vedic priests could have somehow taken an absorbing interest in the second question, we have to admit further that these priests were taking the interest not in their priestly capacity but in the extra-priestly capacity of technicians and craftsmen.

However, it is difficult to imagine even this, because—as we shall later see—such a technological interest could be taken only by flouting not only the Dharmasastra norm but also a fundamental requirement of the priestly ideology favouring deliberate mystification, and hence rejecting the demand for precision without which the craftsmen and technicians could not operate.

To this needs to be added only one point. The mathematical content of the *Sulva* texts developed not from the attempt to answer the first question, viz. "What is to be done ?", but from the attempt to answer the second one, viz., "How is it to be done ?" Thus the possibility of the priests, in the restricted capacity of priests proper, creating this mathematics is a remote one. The presumption, on the contrary, is that those who created this mathematics did it in the capacity of the technicians, notwithstanding the fact that the mathematical knowledge—though developed to meet the theoretical requirements of the technologists—was somehow codified in texts that were appended to the priestly handbooks.

Let us see some of the main points for taking such a view.

"The Vedic sacrifices", observes Datta, "are mainly of two classes : *Nitya* (or 'indispensable', 'obligatory') and *Kamya* ('optional', 'intentional'). The performance of the sacrifices of the former class is obligatory upon every Vedic Hindu... But it is not so with the sacrifices of the second kind. For they are to be performed each with the sole motive of achieving a special object".[6]

It may be convenient for our discussion to begin with the second kind, the fire-altars recommended for which are quite elaborate and hence their physical construction required a good deal of technical skill.

The first question obviously is : Why did the Vedic priests recommend such elaborate fire-altars with definite shapes and sizes? The standard answer to this is to be found in the *Taittiriya Samhita* (v.4.11) which, though often quoted, may be quoted over again in rough English rendering :

> He who desires cattle should pile a piling with the metres ; the metres are cattle ; verily he becomes rich in cattle. He should pile in hawk shape who desires the sky(heaven) ; the hawk is the best flier among birds; verily becoming a hawk he flies to the world of heaven. He should pile in heron form who desires, 'May I be possessed of a head in yonder world' ; verily he becomes possessed of a head in yonder world. He should pile in the form of an Alaja bird, with four furrows, who desires support ; there are four quarters ; verily he finds support in the quarters. He should pile in the form of a triangle, (*prauga,* literally the fore-part of a cart), who has foes ; verily he repels his foes. He should pile in triangle form on both sides, (*ubhayata-prauga* = rhombus), who desires, 'May I repel the foes I have and those I shall have' ; verily he repels the foes he has and those he will have. He should pile in the form of a chariot-wheel, who has foes ; the chariot is a thunderbolt ; verily he hurls the thunderbolt at his foes. He should pile in the form of a wooden trough who desires food ; in a wooden trough food is kept ; verily he wins food together with its place of birth. He should pile 'one that has to be collected together' (? *Paricayyacit*), who desires cattle ; verily he becomes rich in cattle. He should pile one in a circle, who desires a village ; verily he becomes possessed of a village. He should pile in the form of a cemetery, who desires, 'May I be successful in the world of the fathers' ; verily he is successful in the world of the fathers.

Let us see some of the points mentioned here in simpler language. For ensuring the acquirement of cattle for the *Yajamana* or patron financing for the sacrifice, the altar is to be made

6. B.B. Datta SS. 20.

of 'metres' (*chanda*). Hence such an altar is called *Chandasciti.* For ensuring heaven for the said patron, the altar is to be built in the shape of a falcon (*Syenaciti*), or, perhaps alternatively, also in the form of a heron (*Kanka-citi*). For ensuring "support in the four quarters" for the sacrificer, the altar is to be built in the shape of a certain species of bird called *alaja* (*Alajaciti*). For the annihilation of the present rivals, the altar is to be built in the shape of the "forepart of the poles of the chariot" or *prauga,* i.e. in the shape of an equilateral acute-angled triangle (*Praugaciti*). For the annihilation of the rivals both present and future, the altar is to be built in the shape of two such triangles joined together at the base (*Ubhayata-praugaciti*), though also an altar made in the shape of the chariot-wheel (*Rathacakraciti*) seems to have been recommended for the same purpose. An altar built in the shape of the funeral pyre (*Smasanaciti*) is recommended to ensure for the sacrificer the attainment of "the world of the fathers". And so on.

In the passage of the *Taittiriya Samhita* just quoted, there seems to be a thin attempt no doubt of connecting the different altar-shapes with the different desires these are supposed to fulfil. But the logic connecting the two is clearly arbitrary, or based at best on magical imagination, more than which is evidently not needed for the priestly purposes. In short what all this means is that the priests wanted the *Yajamana*-s or rich patrons financing for the sacrifices to believe that the sacrifices performed with fire-altars of specific designs ensured the fulfilment of the desires the rich patrons then cared for—desires for obtaining cattle, food, annihilation of rivals, attainment of heaven, and so on.

For our present discussion we have to note here specially one point. So far as the magical beliefs are concerned, there is evidently no need for any mathematics. But it is different when it comes to the question of the physical construction of the fire-altars. The *Yajurveda* text (*Taittiriya Samhita*) just quoted seems on the whole satisfied with the general prescription that certain shapes of the fire-altars magically ensured the fulfilment of certain desires. But such formulations in general terms evidently failed to satisfy the priests or priestly corporations perhaps of the later times. There grew among them the further demand that the fire-altars had to satisfy a lot of further details. Thus, for example, each altar had to be constructed with a specific number of bricks arranged in a specific manner and answering to very precise sizes. At the present stage of research, it may be premature to venture any hypothesis as to how, where and why such detailed demands grew among the priests. But judging from the circumstances that in the *Sulva-sutra*-s the authority of the *Brahmana* texts is usually mentioned in justification of the demands for details, it may not be wrong to presume that the demands grew during the period of the *Brahmana*-s.

We can perhaps have from Eggeling's analysis of the general trend of the *Brahmana*-s some clue to the tendency of this. The addition of all sorts of complicated details to the sacrificial performances could make these extremely awe-inspiring, and hence conducive to the hierarchical aspirations. As he[7] puts it,

> The *Brahmanas*, it is well known, form our chief, if not our only, source of information regarding one of the most important periods in the social and mental development of India. They represent the intellectual activity of a sacerdotal caste which, by turning to account the religious instincts of a gifted and naturally devout race, had succeeded in transforming a primitive worship of the powers of nature into a highly artificial system of sacrificial ceremonies, and was ever intent on deepening and extending its hold on the minds of the people, by surrounding its own vocation with the halo of sanctity and divine inspiration. A complicated ceremonial, requiring for its proper observance and consequent efficacy the ministrations of a highly trained priestly class, has ever been one of the most effective means of promoting hierarchical aspirations. Even practical Rome did not entirely succeed in steering clear of the rock of priestly ascendancy attained by suchlike means.

We have indeed some idea in this of the possible motivation behind the priestly tendency to make the sacrifices—inclusive of the construction for their altars—highly complicated. But the motivation itself is of no relevance for our understanding of the mathematical contents of the *Sulva-sutra*-s, for the simple reason that this mathematics does not follow from it.

What we are trying to argue and reemphasise is that the discussion of the fire-altars needs to be understood from two different standpoints. The first is the standpoint of the magico-

7. Eggeling SBE, Vol. XII. intro. pp. ix-x.

religious beliefs associated with these. The second is the standpoint of technology. For the second, the crucial question is : How exactly to construct the altars of specific shape, specified size and by using a definite number of bricks and how to vary the shape of the altars without changing their area, and so on.

The technicians—and the technicians alone—were competent to answer this question, and while answering it, they had to develop a significant amount of mathematical knowledge, or at last depend upon it.

It is interesting to note that the internal evidences of the Vedic tradition itself virtually admit this. The evidences are two-fold—positive as well as negative. Positively speaking, there are admissions within the Vedic tradition (or the tradition fully endorsed by Vedic orthodoxy) that though the Vedic priests, in their priestly capacity, wanted to have various complicated brick-structures constructed as fire-altars, they were also aware of the simple fact that for the physical construction of these structures it was necessary to depend on the know-how of the manual workers like the brick-makers, masons and architects. Negatively, the Vedic priests also suggested ways of ensuring the fulfilment of the same magico-religious designs while at the same time completely by-passing the problem of the physical construction of the brick-structures, apparently because they were also aware of the fact that skilled technicians for the purpose were not readily available.

4. "THUS WE ARE TOLD"

Reviewing the *Sulva-sutra*-s Thibaut observes, "But the chief interest of the matter does not lie in the superstitious fancies in which the wish of varying the shape of the altars may have originated, but in the geometrical operations without which these variations could not be accomplished."[8]

Surprisingly enough, in a sense this seems to be true also of the actual builders of the brick-structures whom we meet in the *Sulva* texts. Not that they expressed any doubt concerning the magico-religious efficacies imputed to the different altar

8. G. Thibaut in SHSI, Vol. II, 419-20

designs by the Vedic priests. It is inconceivable that they could have used the expression "superstitious fancies" for such beliefs, as Thibaut does. Nevertheless, for all that we can infer from the internal evidences of the *Sulva-sutra*-s, these technicians looked at such beliefs as belonging to the province of the "others", not their own. As for themselves, they took an absorbing interest only in the technological problems involved in physically constructing the brick-structures, in trying to solve which they developed their geometry and this for the simple reason that without geometry these technological problems could not be solved.

In the texts, therefore, we frequently come across expressions implying "Thus we are told" or "Such is the authoritative instruction", very clearly indicating that the instructions themselves were outside the main scope or the main theme of the *Sulva* texts. These are simply taken for granted and the main purpose of the texts is to explore ways for successfully executing these instructions.

This is virtually admitted by B. B. Datta, who otherwise subscribes fully and strongly to the most orthodox Vedic beliefs. After giving "in brief, a resume of the more salient points in the elaborate and minute in details specifications of the shape and size of the principal sacrificial altars and of the geometrical knowledge presupposed in their construction", he finds it necessary immediately to add :[9]

> What should be particularly emphasized now is the fact that those specifications are not due to the authors of the *Sulba* themselves. They do not even pretend to make any such claim. On the other hand, they have often and then expressly admitted to have taken them from earlier works. We, in fact, find that numerous passages of *Baudhayana* and *Apastamba Sulba* dealing with the spatial magnitudes of sacrificial altars as well as with the methods of their construction, end with the remark *iti vijnayate* [or 'it is known', 'it is recognised or prescribed (by authorities)]. Sometimes *iti abhyupadisanti* ('thus they teach') or *iti uktam* ('it has been said'), is used in the same sense. It has been rightly pointed out before by Garbe that all those passages of *Apastamba* are literal quotations from the *Taittiriya Brahmana* or from the *Brahmana*-like portions of the *Taittiriya Samhita* or *Aranyaka*. That is exactly true also of the similar passages of Baudhayana. This writer is occasionally more explicit about his sources.

9. B.B. Datta SS. 25.

Expressions like *iti vijnayate, iti abhyupadisanti, iti uktam,* etc. which frequently occur in the *Sulva* texts, are exceedingly interesting. These clearly imply that "what is to be done" is prescribed by "others", rather than the technicians and mathematicians whom we meet in the texts themselves. Thanks to the highly competent textual exploration by scholars like Garbe, we have also a clear idea of who these "others" were : we meet them in the priestly manuals *par excellence* like the recensions of the *Yajurveda* and more particularly in the *Brahmana*-s. It needs at the same time to be noted that the authorities in such priestly manuals are quoted in the *Sulva-sutra*-s *not in the context of the technological and mathematical questions discussed in the texts but only in the context of what is being dictated to them or of what they are asked to execute by the priests, leaving the technological as well as the mathematical problems of the actual execution of the work to be solved by them.*

5. AN EXAMPLE : BAUDHAYANA'S PROCEDURE

We shall mention here only one example of this, though adding that this is fairly typical of the major *Sulva* texts.

The oldest and by far the most important of these texts for the purpose of understanding the making of mathematics in ancient India is the *Sulva-sutra* associated with the name of Baudhayana. In this, the expression *vijnayate* signifying what is prescribed by the priestly sources first occurs as the sixtyfifth *sutra* or aphorism of the first chapter. It is, therefore, relevant to have a brief idea of the following points :

(i) What is discussed in the preceding sixtyfour *sutra*-s or aphorisms which refer to no authoritative instruction whatsoever may be assumed by us as characteristic contribution of the maker or makers of the mathematical contents we have in the text itself.

(ii) What actually is the nature of the authoritative instruction coming down to Sulva text and what moreover is the real source of this authoritative instruction.

(iii) What is discussed in the subsequent *sutra*-s intended to solve the problems of the actual execution of the authoritative instruction.

After stating in the first *sutra* that the purpose of the work is to discuss the questions concerning the construction of the fire-altars, the *Baudhayana Sulva-sutra* asserts in the second *sutra* that for this purpose it is necessary at once to be clear about the units of measurement. Thus it opens with an account of linear measurement on which all the calculations in the text are based. The basic unit for this is called an *angula,* literally "the finger", which is specified by the *Arthasastra*[10] as "the maximum width of the middle part of the middle finger of an average person". J.F. Fleet[11] gives its modern equivalent as 0.75 inch (= 19.049 mm.), though V.B. Mainkar[12] calculates it to be 17.78 mm.

However, evidently for the purpose of finer calculations, our text adds that an *angula* is divisible into 14 *anu*-s or 34 *tila*-s. The mathematical utility for such sub-division is discussed by Thibaut.[13] For our present purpose let us have the table of linear measurement of the *Baudhayana Sulva-sutra,* which we have up to 21st *sutra* of the text and to which we shall have to refer repeatedly for understanding the mathematical calculations in it.

Here is Baudhayana's table of linear measures :

1 *angula*	=	14 *anu*-s or 34 *tila*-s
1 "Small" *pada* (*ksudrapada*)	=	10 *angula*-s
1 *pradesa*	=	12 *angula*-s
1 *prtha* or 1 *uttarayuga*	=	13 *angula*-s
1 *pada*	=	15 *angula*-s
1 *isa*	=	188 *angula*-s
1 *aksa*	=	104 *angula*-s
1 *yuga*	=	86 *angula*-s
1 *janu*	=	32 *angula*-s
1 *samya* or 1 *bahu*	=	36 *angula*-s
1 *prakrama*	=	2 *pada*-s (i.e. 30 *angula*-s)
1 *aratni*	=	2 *pradesa*-s (i.e. 24 *angula*-s)

10. *Arthasastra,* ii. 20. 7.
11. J.F. Fleet in JRAS, 1912, 229-239.
12. V.B. Mainkar in FIC, 147.
13. Thibaut in SHSI, II, 433. Taking *tila* as "Sesame grains", 34 of this "put together with their broad sides" makes a much longer length. Hence it is better to accept Thibaut's interpretation that *tila* here is taken in some other technical sense required for finer calculations.

1 *purusa*	=	5 *aratni*-s (i.e. 120 *angula*-s)
1 *vyama*	=	1 *purusa* (i.e. 120 *angula*-s)
1 *vyayama*	=	4 *aratni*-s (i.e. 96 *angula*-s)

After enumerating these units of linear measures, our text passes straightway to formulate a number of purely geometrical problems and offers their solutions from an essentially practical point of view—i.e. the point of view from which a craftsman or technician tries to solve these, having at his disposal nothing more than such simple equipments like the cord and the pole. To put in current terminologies these problems (and propositions) are :

1. How to construct (or draw) a square[14] the length of its side being given? The text gives two methods for the purpose, adding subsequently also a third one.
2. How to construct an oblong or a rectangle (*dirgha-caturasra*), its length and breadth being given.
3. The proposition that the square on the diagonal (*aksnaya*) of a given square is twice as large as that of the given square.
4. To construct a square whose area is three times the area of a given square. The square on the diagonal of an oblong (rectangle) is equal to the sum of the two squares on the two sides. This is shown in the cases of the oblongs the two sides of which are (a) 3 and 4; (b) 12 and 5; (c) 15 and 8; (d) 7 and 24; (e) 12 and 35; (f) 15 and 36. This, it may be noted, is in essence the proposition usually associated with the name of Pythagoras, though in our text we have hint also of the formulation of the proposition in general terms.
5. The way of making a square equal in area to the combined areas of two other squares of different sizes.
6. The way of making a square having an area equal to the difference of two given squares.
7. To construct a rectangle (or an oblong) whose area is equal to the area of a given square.

14. The more exact term for square is *samacaturasra,* which occurs in i.52 of the text.

8. To construct a square whose area is equal to that of a given rectangle (or an oblong).
9. The way of transforming a square into an isosceles trapezium, whose shorter side is given as lesser than the side of the square.
10. To construct a triangle equal in area of a given square.
11. To construct a rhombus equal in area of a given square.
12. The way of turning a square into a circle.
13. The way of turning a circle into a square.

Modern scholars like Burk, Thibaut, Datta and others—and following them recently also S.N. Sen and A.K. Bag—have extensively discussed the mathematical knowledge required to solve these propositions—knowledge which is already contained in the Sulva texts, though as to the texts themselves, composed as these were in cryptic aphoristic style, we have often to depend on their commentaries for a fuller appreciation of this mathematics. However, it would require long digression from our main argument to go here into the details of this mathematics. For purposes of procedural advantage, therefore, we have given it in the form of a separate appendix. The readers are thus referred to Appendix I of the present work, which is prepared by Professor Subinoy Ray of the Department of Mathematics, St. Paul's College, Calcutta.

For what we have been discussing here, one point needs immediately to be noted and re-emphasised. Beginning with the enumeration of the units of linear measures and upto the geometrical theme we have just quoted, our text finds no need whatsoever to mention or quote any authority or use the typical expression like *vijnayata* etc., implying 'thus is being told' or 'such is the authoritative instruction' etc. What is discussed upto this point is mathematics and mathematics alone.

Besides, it refuses to allow this mathematics to be in any way mixed up with the priestly view of the efficacy of the different brick-structures, which in our eyes appears to be totally arbitrary or fanciful, though it might have formed part of the priestly beliefs. One obvious reason for this strict separation of mathematics with the priestly ideology is that the allegedly mystical or mysterious imports that constitute the essence of their dictations do not and cannot agree with the strict rigour and accuracy required by the mathematical calculations. So the two are best kept separated. At the same time, whatever have been the motivations of the priests to get altars constructed with peculiar shapes and sizes, those to whom come down the dictations of physically constructing these, were well-aware of the mathematical equipment needed for the purpose. Hence our text begins with this. In any case, it is a remarkable feature of the *Baudhayana Sulva-sutra* that so far as it discusses mathematics, it restricts itself to mathematics alone, and therefore remains aloof from the magico-religious beliefs.

6. MATHEMATICS TO MEET THEOLOGICAL TWADDLE

Only after the mathematical discussions essential for technological requirements are over, the text mentions something that really belongs to the priestly province, i.e. theme of the sacrificial ritual. Significantly, the text also uses for the first time the expression *vijnayate* or 'thus we are being told', in connection with this theme. Here is how the text reads :

"The place of the *ahavaniya* fire is to be found by starting from the *garhapatya* fire (and measuring towards the east).

"In this matter, following is the instruction (*vijnayate*).

"The Brahmin is to construct the (*ahavaniya*) fire at a distance of eight *prakrama*-s (i.e. 8×30=240 *angula*-s) (to the east from the *garhapatya*), the Rajanya (Ksatriya) at the distance of eleven, the Vaisya at the distance of twelve.

"With the third part of the length (of the distance between *ahavaniya* and *garhapatya*) he is to make three squares following upon each other (touching each other) ; the place of the *garhapatya* is in the north-west corner of the western square, the place of the *daksina agni* (=*anvaharya-pacana*) in the south-east corner of the same square ; the place of the *ahavaniya* in the north-east corner of the eastern square."[15]

What are the fire altars for which the names chosen are *ahavaniya*, *garhapatya* and *daksina*? Why are the distances and relative directions recommended for them? These are questions completely outside the scope of our Sulva texts and hence the *Baudhayana Sulva-sutra* shows no interest whatsoever in

15. *Baudhayana Sulva-sutra* 1.64-67. The references to the Sulva text are based on Thibaut's edition of it originally published in the *Pandit* 1874/5-1877.

raising these. We are simply told that such and such are the authoritative instructions and the text is concerned only with the technique of carrying out the instructions. The problems of "why" and "how" are different altogether and the *Sulva* text is concerned with the second and the second alone.

Impelled by curiosity, however, one may be inclined to ask : wherefrom do the instructions actually come ? As is only to be expected, the source of the instructions is the priestly literature. In the case under consideration, it is the *Satapatha Brahmana.* Impelled by natural curiosity, again, one may be inclined to go back to this source to see how exactly are the instructions expressed there. When one does this, one cannot but have the feeling of entering into a different thought world altogether—a realm of capricious assumptions strung together at best by a thin string of pure magical belief. The *Satapatha Brahmana* simply declares :

"He may lay it (the Ahavaniya) down at the distance of eight steps (from the Garhapatya) ; for of eight syllables, doubtless, consists the *gayatri* : hence he thereby ascends to heaven by means of the *gayatri*.

"Or he may lay it down at the distance of eleven steps ; for of eleven syllables, indeed, consists the *tristubh* : hence he thereby ascends to heaven by means of the *tristubh*.

"Or he may lay it down at the distance of twelve steps ; for of twelve syllables, indeed consists the *jagati* : hence he thereby ascends to heaven by means of the *jagati*. Here, however, there is no (fixed) measure : let him, therefore, lay it down where in his own mind he may think proper. If he takes it ever so little east (of the Garhapatya), he ascends to heaven by it."[16]

There is really nothing in it than what Max Muller aptly describes as a 'theological twaddle'—twaddles that crowd the *Brahmana* literature. It requires an extra-ordinarily inflated veneration for the Vedas to see in this anything having even a semblance for any enthusiasm for science. What is palpable instead is the anxiety to tempt the sacrificer to move rather easily to the heavenly region : he may do it only by moving either eight or eleven or twelve steps. Remarkably enough,

16. *Sat. Br.* i. vii. 3. 23-25.

while prescribing these the priests apparently forget—or perhaps conveniently overlook—what they themselves develop as 'metrical science' or *chanda*, evidently required by the oral tradition of the pre-literate poets whom we meet in the *Rgveda*. In their own understanding of the metres, the Gayatri does not actually consist of only eight syllables ; the number of syllables in this metre being twentyfour. So also are the cases of the metres called Tristubh and Jagati which, instead of consisting of eleven and twelve syllables, actually consist of fortyfour and fortyeight syllables. The priests in the *Satapatha Brahmana* must have been well-aware of this. Could it, then, be that in prescribing the instructions quoted, they were deliberately shortening the standard number of the syllables in the metres (by counting only the number of the syllables contained in the first line of each metre) to spare the sacrificer the trouble of walking longer to reach the heaven?

In any case, there also remains an obvious arbitrariness in the instruction as mentioned in the *Satapatha Brahmana*. If it is possible to reach the same results by allowing only 'eight steps' as the distance between the two fire altars, why do the priests bother to mention alternative prescriptions of eleven and twelve steps ?

It is for the specialists in the Vedic ritual literature to find out if there is any authority intervening between the *Satapatha Brahmana* and the *Baudhayana Sulva-sutra*, who notices such an arbitrariness and wants to remove it by suggesting that the three alternatives proposed are really intended for sacrificers belonging to the three upper castes—Brahmins, Rajanyas and Vaisyas—i.e. giving the instructions the form in which these reach the Sulva text ?

But whatever the case may be, the point to be noted is that such prescriptions, *as ritual prescriptions,* have nothing to do with the actual problems discussed in the *Sulva* texts, beyond the bare admission, of course, of mentioning *what is required to be done*. The distinctive theme of the *Sulva* texts is different altogether, namely *how is this to be done*? This difference is crucial for our understanding of the history of science in ancient India. Whatever might have been the motivation of the priests dictating certain orders to be carried out in the interest of their sacrificial ritual, that has no bearing at all for the making of science in ancient India as found in the *Sulva* texts. These texts take an exclusive interst in the technological-mathematical problems without solving which the dictations that came down to the technicians in the texts could not be executed in actual practice, with as much precision as was perhaps feasible in the ancient context.

7. B.B. DATTA'S ANALYSIS OF A PROBLEM

Compared to many other technological and mathematical problems sought to be solved in the *Sulva* texts, the problem that we have just mentioned, *viz.* that of ensuring the precise distance and relative directions of the three fire altars, may appear to be somewhat simple. Really speaking, however, it may be an error to take an over-simplified view of the whole thing, for the *Sulva* text actually suggests three different methods of solving the problems, from the analysis of which B.B. Datta is inclined to see the prodigious first step—though understandably only with approximate success according to our modern standards—taken to work out the value of $\sqrt{5}$ and $\sqrt{2}$. We quote him[17] at some length :

There seems to have been a serious attempt, though without much success, to find an approximation to the value of the surd $\sqrt{5}$. The occasion was to define clearly the relative positions of the three principal and oldest known fire-altars, viz., the Garhapatya, Ahavaniya and Daksina. Baudhayana's rules to determine their positions are these :

'With the third part of the length (i.e., the distance between the Garhapatya and Ahavaniya) describe three squares closely following one another (from the west towards the east) ; the place of the Garhapatya is at the north-western corner of the western square ; that of the Daksinagni is at its south-eastern corner ; and the place of the Ahavaniya is at the north-eastern corner of the eastern square.' (BSS. i. 67).

'Or else divide the distance between the Garhapatya and Ahavaniya into five or six (equal) parts ; add (to it) a sixth or seventh part ; then divide (a cord as long as) the whole increased length into three parts and make a mark at the end of two parts from the eastern end (of the cord). Having fastened the two ends of the cord (to the two) poles at the extremities of the distance between the Garhapatya and Ahavaniya, stretch it toward the south, having taken it by the mark and fix a pole at the point reached. This is the place of the Daksinagni'. (*BSS*. i .68).

17. B.B. Datta SS. 203-5.

'Or else increase the measure (between the Garhapatya and Ahavaniya) by its fifth part ; divide (a cord as long as) the whole into five parts and make a mark at the end of two parts from the western extremity (of the cord). Having fastened the two ties at the ends of the east-west line stretch the cord towards the south having taken it by the mark and fix a pole at the point reached. This is the place of the Daksinagni' (*BSS.* i.69).

The second is also given by Apastamba. A rule leading to the same result as the first one above, though defined differently, is stated by Katyayana and Manu. Katyayana has specified the relative positions of the three fire-altars also in a new way :

'Divide the distance between the Garhapatya and Ahavaniya into six or seven parts ; add a part ; then divide (a cord) equal to the total increased length into three parts, etc.' (*KSS* i. 27).

The rest of this rule is the same as the latter portion of the second rule above and hence need not be mentioned.

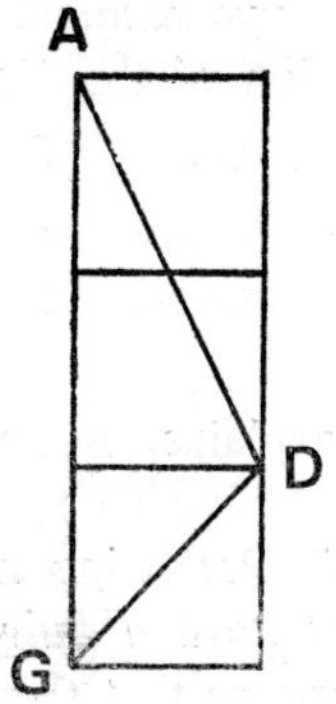

Let b denote the distance beetween the Garhapatya and Ahavaniya, that is AG. Then from the different specifications given above we obtain the following values for AD and GD :

$$\text{AD} = \tfrac{b}{3}\sqrt{5},\ \tfrac{4}{5}b,\ \tfrac{7}{9}b,\ \tfrac{16}{21}b,\ \tfrac{18}{25}b$$

$$\text{GD} = \tfrac{b}{3}\sqrt{2},\ \tfrac{2}{5}b,\ \tfrac{7}{18}b,\ \tfrac{8}{21}b,\ \tfrac{12}{25}b$$

If it be assumed that the relative positions of the three fire-altars were meant to be the one and the same, in all cases though expressed differently, then we shall have the following approximations to the values of $\sqrt{5}$ and $\sqrt{2}$:

$$\sqrt{5} = 2\tfrac{2}{5},\ 2\tfrac{1}{3},\ 2\tfrac{2}{7},\ 2\tfrac{4}{25}$$
$$= 2.4,\ 2.333\ \ldots,\ 2.285\ \ldots,\ 2.16.$$

$$\sqrt{2} = 1\tfrac{1}{5},\ 1\tfrac{1}{6},\ 1\tfrac{1}{7},\ 1\tfrac{11}{25}$$
$$= 1.2,\ 1.166\ \ldots,\ 1.142\ \ldots,\ 1.44.$$

Since according to modern calculation $\sqrt{2} = 1.414213 \ldots$ and $\sqrt{5} = 2.23607 \ldots$ none of the above values can be said to be a fair approximation, perhaps except the values $\sqrt{5} = 2^{2}/_{7}$ and $\sqrt{2} = 1^{11}/_{25}$ which are correct up to the first place of decimals.

It is for the mathematicians to judge if there is any tendency to over-rate the mathematical potentials in the *Sulva* texts in the analysis of Datta just quoted. For us, however, there are other interesting points to note regarding what is dictated by the priests about the construction of the three relatively simple fire altars called Garhapatya, Ahavaniya and Daksina.

Serious problems are created by what the priests further dictate about the construction of the Garhapatya fire altar. Thus, for example, in the *Baudhayana Sulva-sutra* (ii. 61 ff.) we read :

"It is of the measure of *vyayama*. This is the instruction (*vijnayate*) for the construction of the Garhapatya.

"According to one authority, it (Garhapatya) has the form of a square (*caturasra*).

"According to other authorities, it has the form of a circle (*parimandala*)".

All this, as Thibaut explains, means : "The Garhapatya is either a square the side of which is one *vyayama* long or a circle of the same area." Preferring not to ignore any of the two authoritties, the *Baudhayana Sulva-sutra* takes both with equal seriousness. This means that the text is immediately confronted with two problems :

First, how to construct a square the area of which is one *vyayama* X one *vyayama*. It needs to be noted that one *vyayama* = 4 *aratni*-s, each *aratni* = 2 *pradesa*-s and each *pradesa* = 12 *angula*-s. That is, one *vyayama* = 96 *angula*-s. The first problem, thus, is to draw a square with an area of 96 angula-s X 96 angula-s.

Secondly, since the shape of the Garhapatya is required to be either a square or a circle though *without affecting its size,* there is also the problem of converting a square into a circle, or, to be more specific, converting a square of 96 *angula*-s × 96 *angula*-s (i.e. 9216 sq. *angula*-s) into a circle having exactly the same area, and *vice versa.*

From the mathematical viewpoint the problem of circling a square or squaring a circle was not an easy one—in fact it is later found to be an impossible one.[18] But the *Baudhayana Sulva-sutra* cannot evade it, because it reoccurs in the context of the fire-altar called Dhisnya which, according to the priestly dictation, must have the shape of either a square or of a circle evidently of the same size.[19]

How far the mathematicians in our *Sulva* texts successfully solve the problems of circling a square or of squaring a circle is, of course, a different question. Eminent modern scholars like Burk, Thibaut, Datta and others have tried to go into it in some detail, though without necessarily agreeing with each other. Evidently enough, while re-examining this problem in the general background of modern mathematical knowledge, the scholars remain exposed to the dual risks of either over-rating or under-rating the scientific potentials in these ancient texts using the ancient style of cryptic mnemonical aphorisms. Since these aphorisms by themselves are often not explicit in what precisely these are intended to convey, our scholars are understandably often obliged to depend more or less heavily on the later commentators on the *Sulva* texts. Such a procedure, though methodologically somewhat inescapable, makes them exposed to another risk, namely that of historical anachronism, inasmuch as at least some of these major commentators—on their own admission—drew on much later mathematical works, like those of Aryabhata (born A.D. 476) and even Bhaskara II (born A.D. 1114), when mathematics in India takes a sharply new turn with intricate methods of new calculations, notwithstanding—as Thibaut shows—the survival in later Indian mathematics of some ancient nomenclature like *varga, karani,* etc. of the *Sulva* texts. But all this does not and must not mean that the mathematical achievements recorded in the Sulva texts, specially as judged in their historical context, cannot but be considered as highly remarkable, whatever might have been the intrinsic worth of priestly beliefs the demands of which this mathematics was required to meet.

To sum up the discussion so far: The starting point for our understanding of the *Sulva* texts, is to differentiate bet-

18. D.E. Smith HM II, 302f.
19. *Baudh. S. S.*, ii. 73.

ween two questions. First : *What* is to be done? Secondly : *How* is this to be done? The first of these two questions is really extrinsic to the essential theme of the *Sulva* texts which concentrate—and concentrate on the whole exclusively—on the second question. The *Sulva* texts themselves make this abundantly clear by repeating over and over again typical expressions like *vijnayate, iti abhi-upadisanti* etc., implying that what is to be done is dictated by sources external to the texts themselves. Significantly, such expressions invariably occur in the context of the ritual prescriptions connected with the nature of the brick-structures called fire altars. This, in other words, means that the dictations come from the Vedic priests, to whose literature—specially the *Taittiriya Samhita* and the *Satapatha Brahmana*—the priestly dictations are naturally traced by the modern scholars, and sometimes even frakly admitted by the Sulva texts themselves.

8. ROLE OF THE TECHNICIANS

Commenting on the methods suggested by the *Sulva-sutra*-s of squaring the circle and *vice versa,* Thibaut says, "Theirs was not the love of disinterested research which distinguishes true science, nor the inordinate craving of undisciplined minds for the solution of riddles which reason tells us cannot be solved; theirs was simply the earnest desire to render their sacrifice in all its particulars acceptable to the gods, and to deserve the boons which the gods confer in return upon the faithful and conscientious worshipper".[20]

Such an observation seems to be in need of a number of amendments from different standpoints.

A specialist in Vedic studies would perhaps prefer to look back at the Vedic rituals as but magic techniques for obtaining the desired results by the intrinsic efficacy of the ritual acts and spells, leaving practically no scope for any god or gods to come effectively into the picture. The Vedic literature—particularly the earliest one or the *Rgveda*—mentions a whole host of gods and demi-gods no doubt. In the ritualist

20. Thibaut, in SHSI, 445.

view, however, they are denuded of divine functions like conferring the boons etc. As is rather well-known, the Purva-mimamsa phliosophy—which is the direct outcome of the theoretical presuppositions of the standpoint of the Vedic ritualists—finds it necessary not only to deny the existence of god but also of going to the extent of arguing that the names of the famous gods of the ancient Vedic pantheon are mere words after all, having efficacy only as forming parts of the magic spells. As Nilakantha Sastri, after reviewing the Purva-mimamsa view of the Vedic gods, observes: "Is the sound 'Indra' then, all that is left of the great Vedic hero or god? It may be so, Mimamsa is not concerned with that, in effect it does not know."[21] This is true, though not the whole truth. Jaimini in his *Mimamsa-sutra* and also his commentator Sabara elaborately and very strongly argue that what is important above all is the ritual act and the names of the gods have no more function in it than forming parts of the right spells.[22] Thus, on the whole, the Christian or quasi-Christian view of God which Thibaut virtually imputes to the Vedic priests is hardly acceptable from their expressed standpoint.

Secondly—and this is much more serious an objection to the observation of Thibaut—he takes it for granted that genuine science is in fact only a disinterested pursuit of pure knowledge. Though such an elitist understanding of science intended to keep it aloof from any contribution to it of the manual workers—the craftsmen and the technicians—had once been popular with the historians and philosophers of science, it is exploded today by the more objective researches of Joseph Needham, J. D. Bernal, Benjamin Farrington, Gordon Childe and many others. And long before them, Prafulla Chandra Ray had very convincingly shown that by far the most important factor accounting for the decline of scientific spirit in India had been the caste system which prevented the elites to draw upon the experiences of the manual workers relegated to the lower strata of the society, which evidently implies that science draws its real nourishment from manual workers.

21. Nilkantha Sastry in *IA,* vol. 50. 241.
22. *Mimamsa-sutra,* iii. 344; viii. 1.32-34; ix, 1.10 etc., and Sabara thereon.

As a matter of fact, one of the main interests of the *Sulva-sutra*-s in the history of science is that these make it very palpable that the mathematics therein is inconceivable without being connected with the contributions of the manual workers —the craftsmen and the technicians, specially the brick-makers and the brick-layers. When, therefore, Thibaut observes about the makers of mathematics in ancient India that "theirs was not the love of disinterested research", he tells us something of fundamental importance about them, though obviously not in the sense in which he himself means it. Working under the influence of the usual assumption that this mathematics was the creation of the Vedic priests, Thibaut assumes that it was the result of the attempt to please the gods with sacrifices, whereas the fact seems to be that the priests had no more role in the matter than dictating *what is to be done*, whereas the mathematics we have in the texts was the outcome of the theoretical requirements of those who were confronted above all with the question of *how is this to be done*.

9. ADMISSION IN THE VEDIC TRADITION

That for the actual construction of the brick-structures—called *agni* or *citi* in the priestly terminology—technicians outside the circle of the Vedic priests were required is virtually admitted by the Vedic tradition itself. In the appendix to the *Katyayana Srauta-sutra* we read that those who were required to execute 'what is to be done' need to have the following qualifications :

sastrabudhya vibhagajnah parasastra-kutuhalah/
silpibhyah sthapatibhyah ca adadita matih sada//[23]

"One who is engrossed in the subject, knows the technique of division, has the curiosity to know the science of the others and always pays proper attention to the work of the artisans and architects."

The word *vibhagajnah,* literally "one who knows the technique of division", could have been an old way of referring to

23. S.D. Khadilkar's, *Katyayana Sulva Sutra* (Poona : 1974). This occurs as the sixth verse of chapter vii. That seeems to be a case of editorial oversight, for the sloka does not form part of the *Katyayana Sulva-sutra.* See, B.B. Datta SS 4.

those that had knowledge of what we call geometry and allied calculations. But there is no scope at all to speculate on the meanings of *silpi* and *sthapati*, which clearly mean the artisan and architect. More interesting, however, is the expression *parasastra-kutuhalah*—"having curiosity to know the science of others." This seems to speak volumes. It is not easily conceivable that the priests, *as priests*, would admit the importance of any *Sastra* other than their own, i.e. their priest-craft, consisting mainly of ritual details, magical beliefs and theological or quasi-theological disputations. In the expression "having curiosity to know the science of others", it is tacitly admitted, however, that this is not enough for the purpose of the physical construction of the most complicated brick-structures with specific shapes and size. How can the priest, with all his claim to the most profound wisdom but equipped only with the knowledge of the priest-craft, at all confront the problem of the physical construction of the complicated brick-structures which require above all whole-time specialisation of the brick-makers, masons and architects ?

Accordingly, the *Ramayana* also mentions that for the performance of the sacrificial ritual—priests apart—are required the following classes of persons : manual worker (*karmantika*), artisan (*silpakara*), carpenter (*vardhaki*), digger (*khanaka*) and mathematician (*ganaka*).[24] Evidently, it is not conceivable that the *Ramayana* specially in the form in which it comes down to us could flout the priestly norm, which, in other words, means that all this had implicit acceptance of the priest-class.

At the same time, the acceptance could have been rather grudging than enthusiastic, because according to the *Dharma-sastra* norm— which, as already noted, was the creation of the same priestly corporations that produced the *Kalpa-sutra*-s —the social status of such manual workers is supposed to be low—certainly much lower than of the priests themselves.

An enthusiast for Vedic culture, in defence of the view that the Sulva mathematics could not but be the creation of the Vedic priests, may argue that there is nothing to prevent the assumption that a section of the Vedic priests might have developed an all absorbing interest in brick-making and brick

24. *Ramayana.* Balakanda, xii. 6-9 ; xiv. 26-28.

laying essential for the construction of the fire altars, and, drawn by the theoretical requirements of these technologies, could have been the makers of mathematics. Such an assumption, however, would necessitate the further assumption that these priests could do it *not in the capacity of priests in the real sense but in the capacity of craftsmen and technicians*—flouting for the purpose the Dharmasastra norm generally contemptuous of the manual workers.

But an interesting evidence of the *Mahabharata* seems to pour cold water on such an enthusiasm.[25] According to this the king wanting a sacrifice to be performed, got priests (*rtvij*) to measure the space for altar-construction and employed learned Brahmins with specialised knowledge of priest-craft to construct the altar. Eventually, however, it was realised that there was some bungling somewhere about the whole procedure—specially about matters concerning measurements—which would have frustrated the entire sacrificial performance. It is, however, of crucial importance to note that this bungling was detected not by the priests themselves but by a mason (*sthapati*) well-versed in architectural techonology (*vastu-vidya*). Thus the legend admits, though negatively, that something more than mere specialisation in priest-craft was needed for the construction of the structures called sacrificial altars and this was the know-how of the masons and architects. To this needs to be added only one point. The *Mahabharata* as a whole, in the form in which it reaches us, is accepted as something sacred by the orthodox Vedic tradition and every legend in it is fully endorsed by Vedic orthodoxy.

10. EVADING THE ENTIRE PROBLEM OF THE PHYSICAL CONSTRUCTION OF THE BRICK-STRUCTURES

Let us begin by reiterating a simple point. In the Sulva texts, authoritative instructions concerning "what is to be done" indicated by expressions like *vijnayate*, *iti-abhi-upadisanti* etc. invariably come down from the priestly literature like the *Taittiriya-Samhita* and the *Brahmana*-s. But nowhere in these texts such expressions occur in the context of discussing the problem of "how is this to be done", though the discussion

25. *Mahabharata*, Adi-parva. LXI. 9 ff.

of the second constitutes the special peculiarity of the Sulva texts, inclusive of their mathematical contents. This peculiarity of the texts seems to give us the impression that if the Vedic priests are at all to be credited with the making of this mathematics, it could at best be in a very indirect sense, like the sense in which the god-kings of ancient Egypt wanted the pyramids to be constructed and thereby providing the stimulant for the making of mathematics among the actual architects, masons, engineers and others, to whom came down the task of the physical construction of the structures—construction that required the development of mathematical knowledge. But let us leave the problem of the assessment of this analogy to those that specialise in the history of ancient Egypt, for the possibility remains that the Egyptian god-kings were not so fully parasitical after all as were the Vedic priests—the latter subsisting only on the *daksiṇa*-s or fees for performing the sacrifices in the interest of their rich patrons or *yajamana*-s and whose law-codes frowned upon manual labour on the part of the nobles and priests.

But there remains another important point to be noted in this connection. The dictation of "what is to be done" evidently presupposes the feasibility of its execution. To borrow an example often used in Indian philosophical literature, you cannot dictate anybody to get oil by pressing sand, because that is beyond feasibility. The construction of the complicated brick-structures required by the priests, therefore, must have been feasible, in order not to make the dictation meaningless. This means that there must have been technicians capable of physically constructing such brick-structures. There is no difficulty in assuming the existence of such technicians during the period of First Urbanization, or that of the Mature Harappan Culture, when highly sophisticated brick-technology was an accomplished fact. But it is not easy to assume ready availability of such technicians during the period intervening the two urbanizations, i.e. during what is often referred to as the "Dark Age" or "Dark Period" by the archaeologists—a period to which the Sulva texts belong. However, since the consideration of basic feasibility wants us to assume that there were technicians capable of executing what was dictated to them, the availability of at least some such techinicians even in this period cannot be denied outright.

At the present stage of research, it is of course impossible to ascertain their identity. One is tempted to presume that they could be stray stragglers of the Harappan technicians, though how the technique could survive among them in spite of falling into general disuse is not easy to understand. One may perhaps hope that with the increasing interest of the recent archaeologists to explore and excavate the Late Harappan and Post Harappan Sites, some new light may eventually be thrown on the problem. For the present, however, it remains one of the many unresolved problems of ancient Indian history.

What is not obscure, however, is another point. Admitting the existence of some stray technicians capable of actually constructing the complicated brick-structures in the period under consideration, it is not easy to conceive that their services were readily available in the Vedic settlements for making the fire-altars. The Vedic priests must have been aware of this difficulty more than anyobdy else. Hence must have also been the need felt by them for bypassing the whole problem of the physical construction of the brick-structures, without affecting in any way their basic claim to the magico-religious efficacy of the structures themselves. The way in which they actually solved this problem appears to us as amazingly simple. Some imaginary structures could be as good as the physically constructed ones. This leads us to see their ingenuous theory of *chandascit* and *manomaya* or *manascit*.

Already in the *Taittiriya-samhita* it is declared, "He who desires cattle should pile a piling with the metres (*chandascit*) ; the metres are cattle ; verily he becomes rich in cattle."[26] But what is meant by this ? What is meant by *chandascit* ? Datta[27] gives a very lucid answer to this :

> In case of the Chandasciti, the *agnicit,* ('fire-altar builder') draws on the ground the Agni (altar) of the prescribed shape, ordinarily of the primitive shape of the falcon. He then goes through the whole prescribed process of construction imagining all the while as if he is placing every brick in its proper place with the appropriate *mantra*-s (spells). The *mantra*-s are, indeed, muttered but the bricks are not

26. *Taittiriya Samhita,* V. 4.11.1.
27. B.B. Datta SS. p. 3n.

actually laid. Hence the name Chandasciti, i.e. the *citi* or altar made up of *chandas* or Vedic *mantra*-s instead of bricks or loose mud pieces.

So that is not a real structure but the imaginary substitute for a structure, for the construction of which the entire problem of recruiting skilled craftsmen could be conveniently evaded, but, which, from the viewpoint of the priestly claims, could be equally efficacious in the matter of magically ensuring the financiers for the sacrifice a good deal of wealth—cattle having been the measure for counting wealth for the early Vedic peoples.

There are evidences to think that the priestly ingenuity of suggesting imaginary altars as substitutes for the physically constructed ones was not confined to this simple one.

Commenting on the *Brahma-sutra*[28] both Sankara and Ramanuja found it necessary to examine the view of the ritualists concerning what was called *manomaya* or *manascit*, literally "mind-made altars." Apparently, there was talk of a very large number of these in the priestly circles, because Samkara mentioned thirtysix thousand varieties of these! The names of some of these speak for themselves. Thus, we read of *vakcit*. literally 'altar made of the organ of speech' or, more simply, verbally constructed altar; *caksuscit,* literally 'altar made of the organ of vision', or, more simply, visually constructed altar; *srotracit*, literally 'altar made of the auditory organ', or, more simply, sound-made altar, etc.

The claim of the ritualists, intended to be examined by the *Brahma-sutra* was very lucidly put by Ramanuja[29] :

> That the altar built of thought is an optional substitute for the altars built of bricks, and of the nature of an action, appears therefrom also that the clause 'of these each one is as great as that previous one', explicitly transfers to the altars of mind, and so on, the powers of the previous altars made of bricks. All those altars, thus having equal effects there is choice between them. The altars of mind, and so on, therefore are auxiliary members of the sacrificial performance which they help to accomplish, and hence themselves of the nature of (ritual) action.

Not that Samkara and Ramanuja—or, for that matter, presumbaly also the author of the *Brahma-sutra*—subscribed to

28. *Brahma-sutra*, iii. 3.44 ff.
29. Ramanuja on *Brahma-sutra,* iii. 3.45. Tr. Thibaut, *SBE,* XLVIII, 669.

this view. The view is mentioned as a *purvapaksa,* i.e. the opponent's view intended only to be rejected. But the reason for this needs to be clearly noted. Samkara and Ramanuja, like the author of the *Brahma-sutra* itself, were opposed to the ritualistic view as such. In traditional terminology, they represented the *jnana-kanda* or "knowledge branch" of the Vedas, as constrasted with and sharply opposed to what was called the *karma-kanda* or "the ritual branch" of the Vedas. In other words, this was a controversy between the philosophical standpoint claiming to draw its sanction from the Upanisads and priestly standpoint in its restricted sense claiming to draw its sanction above all from the *Yajurveda* and *Brahmana*-literature. The view that the mind-made altars were suitable substitutes for brick-built altars was contested by Samkara and Ramanuja mainly because it formed part of the *karma-kanda* or the ritualistic standpoint of the Vedic priests. But what, indeed, could the priests do than to recommend such imaginary or mind-made altars as substitute for the brick-built ones, when the physical construction of the latter required highly skilled masons and architects not easily available in the Vedic settlements, which are archaeologically connected with the Painted Gray Ware sites unaware of brick technology?

The evidence for *chandascit* and *manascit* are, therefore, crucial. These prove that there was *no necessary connection between the physically constructed brick-structures and the magico-religious beliefs imputed to these by the Vedic priests.* Thus the know-how of the brick-constructions was not obligatory for the priests—whose priest-craft itself was so elaborate that it presumably required whole-time specialisation—notwithstanding the circumstance that the Sulva texts survive for us as appended to the manuals of the practising priests.

11. QUESTION OF VOCABULARY AND TERMINOLIGY

Further, certain internal evidences of the Sulva texts do not easily cohere with the view that these could be the creations of the Vedic priests. One of these is that of vocabulary and terminology. From this point of view where the Sulva texts most sharply turn away from the *Taittiriya-samhita* and even from the *Satapatha Brahmana*—is about the brick-names. Before passing on to discuss it, we may briefly note certain other hints about the vocabulary of the Sulva texts suggesting or hinting more of the association with plebian craftsmen than the priestly elites.

The very word *Sulva* for rope or cord is so elitist-esoteric that in the whole range of Sanskrit literature it is difficult to come across it outside the titles of the texts. Within the texts, however, the word is never used, the cord or rope being referred to as *rajju*. This peculiarity of the texts is already noted by Thibaut who observes : "I may remark at once that the *sutra*-s themselves do not make use of the term *sulva* ; a cord is regularly called by them *rajju*."[30] What needs to be added to it is only a simple point. The ward *rajju* is in fact so plebian that it is the same also in Pali. A craftsman would have been easily familiar with it and use it as part of his own vocabulary rather than the word *sulva,* which was presumably current only among the priestly elites.

Another peculiarity of the texts which cannot be easily overlooked is the use of the word *atman* to mean the body and the body alone. In this matter, there is no scope whatsoever for any speculation to suggest a theological or philosophical alternative. In the Sulva texts the word *atman* is invariably used to mean the body in its crass physical sense. But this is not easily conceivable in the circle of the Vedic priests and their theologico-philosophers specially after the Upanisads, which Vedic orthodoxy wants us to accept as but appendices to the *Brahmana*-texts and in which the word *atman* acquires the sense of Pure Spirit or the Pure Soul with the vigorous emphasis that this soul must not be confused with the body, which is at best a temporary (but on the whole unfortunate) dwelling place for the soul.

One is almost tempted to see in the Sulva vocabulary the much-maligned philosophical view called *dehatmavada,* i.e. the view equating the *deha* or body with the *atman*. Such a view, as is rather well-known, is characteristic of the Lokayatas, a name that literally means "that which is prevalent among the people", or, as Samkara more explicity states, is characteristic of "the common uncultured mob" (*prakrtajanah*).[31] The

30. Thibaut, G. in SHSI II. 416.
31. D. Chattopadhyaya, *Lokayata,* 1-2.

technicians and craftsmen would easily belong to this category ; but it is inconceivable that the Vedic priests or their ideologues could come under it or accept even tacitly a philosophical view usually characteristic of them.

But more of the question of the general theoretical temper later. For the present let us have a brief survey of the view of bricks in the Vedic literature.

12. BRICKS IN THE YAJURVEDA : TAITTIRIYA SAMHITA

The word for brick in the Vedic literature is *istaka.* Though the controversy over the exact origin of the word is still going on, it may be rash for us to ignore outright the view of Przylusky and others[32] who strongly argue that it is Dravidian in origin, though eventually borrowed by the Vedic peoples. In any case, the fact is that the word is absolutely unknown to the *Rgveda* and, in the Vedic tradition, it first occurs in the *Yajurveda.* But the *Yajurveda* appears to be more interested in the mysterious or magical efficacy of the bricks than telling us what these things actually are. Here is a typical example of this from the *Taittiriya-samhita*[33] :

> May these bricks, O Agni, be milch cows for me, one, and a hundred, and a thousand, and ten thousand, and a hundred thousand, and a million, and ten million, and a hundred million, and a thousand million and ten thousand million, and a hundred thousand million, and ten hundred thousand million, and a hundred hundred thousand million ; may these bricks, O Agni, be for me milch cows, sixty, a thousand, ten thousand unperishing ; ye are standing on holy order, increasing holy order, dripping ghee, dripping honey, full of strength, full of power ; may these bricks, O Agni, be for me milkers of desires named the glorious yonder in yon world.

In such gambols of pure phantasy, it is difficult to see the physically made bricks, not to speak of the hint of brick technology in any significant sense. The desultory interest in the bricks as such can easily be judged from the use of the brick-names randomly connected with their magical potency which we read in the text. Here are just a few examples :

32. J. Przyluski in IHQ, Vol. VII, 735 f.
33. *Taittiriya-samhita.,* IV. 4. 11.

apanabhrt	or	'nourisher of *apana* (*Vayu*)' bricks.[34]
pranabhrt	or	'nourisher of *prana* or breath' bricks.[35]
asvini	or	'connected with Asvin (an asterism or the twin-god-Asvins)' bricks.[36]
vayasya	or	'*vayas* (? vigour)' bricks.[37]
brhati	or	'*brhati*-metre' bricks.[38]
valakhilya	or	'valakhilya (derivation doubtful)' bricks.[39]
aksayastomiya	or	'*aksaya* (? circuitous) + *stoma* (collection)' bricks.[40]
samyani	or	'*samyana* (= a mould ; a vehicle ; going together)' bricks.[41]
vyusti	or	'vyusti (= breaking of dawn ; increase ; prosperity)' bricks.[42]
nakasat	or	'nakasat (= sitting or dwelling in the sky)' bricks.[43]
pancacoda	or	'having protuberance' bricks.[44]
vikarni	or	'earless' bricks.[45]
svayamatrnna	or	'full of natural holes' bricks.[46]
sayuj	or	'united or comrade' bricks.[47]
bhuyaskrt	or	'augmenting or increasing' bricks.[48]
vrstisani	or	'rain-acquiring' bricks.[49]
asapatnya	or	'rival-removing' bricks.[50]
viraj	or	'ruling far and wide' bricks.[51]

34. *Ibid* IV. 3.3 see, Keith VBYS. p. 328 note 3.
35. See, Keith *Ibid.*
36. *Tait. Sam.* iv. 3.4, See Keith *Ibid,* 329, note 4.
37. *Ibid,* iv. 3.5, See Keith *Ibid* 330, note 6.
38. *Ibid.* iv. 3.7, see Keith. *Ibid.,* p. 331 note 4.
39. *Ibid.*
40. *Tait. Sam.* iv. 3.8. See, Keith, VBYS 332, note 1.
41. *Ibid,* iv. 3.10. See Keith *Ibid,* 333, note 2.
42. *Ibid,* iv. 3.11 Keith *Ib.* 334, note 1.
43. *Ibid,* iv. 4.2. Keith *Ib.* 340. note 4.
44. *Ibid,* iv. 4.3. Keith *Ibid,* 341, note 3.
45. *Ibid.*
46. *Ibid,* v. 3.7. See, Keith *Ib.* 425, note 1.
47. *Ibid,* iv. 4.5. Keith *Ibid* 346, note 1.
48. *Ibid,* v. 3.11, Keith *Ibid,* 347 note 1.
49. *Ibid,* v. 3.10. Keith *Ibid,* 426 note 3.
50. *Ibid,* v. 3.5. Keith *Ibid,* p. 423. n. 2.
51. *Ibid.*

visvajyotis	or	'representing fire, wind and sun' bricks.[52]
rtavya	or	'related to seasons' bricks.[53]
vihavya	or	'invoked, invited or desired' bricks.[54]
apasya	or	'watering' bricks.[55]
dravinoda	or	'granting wealth' bricks.[56]
ayusya	or	'conducive to long life' bricks.[57]
vamabhrt	or	'supporting the desirable as gold, horses etc.' bricks.[58]
adhipatni	or	'female sovereign' bricks.[59]
ajyani	or	'free from injury' bricks.[60]

metre-bricks, like *gayatri, trsthubh, jagati* etc. bricks.[61]

naksatra-bricks	or	'asterisms like *Krttika*, *Visakha*, etc. bricks.[62]
durva	or	*durva*-grass bricks.[63]
ghee	or	'melted butter' brick.[64]

etc. etc.

Let us, however, try to be clear about one point. Being a ritual text after all, the *Taittiriya-samhita* is interested in naming the bricks with an exclusive interest in their ritual functions rather than their shape, size, construction, etc. What it seems to be referring nevertheless is to real bricks, sometimes specifying their number and even the manner of placing these.

Thus, for example, *ghee* or 'melted butter'-brick is not intended to mean 'brick made of melted butter', no more than

52. *Tait. Sam.* v. 3.9. See Keith. VBYS 426. n. 2.
53. *Ibid*, v. 3.1. Keith *Ibid*. 418 n. 1
54. *Ibid* iv. 7.14. Keith *Ibid*, 386 n. 4. & 438.
55. *Ibid*, v. 2.10. Keith *Ibid*, 414, n. 1 ; 418, n. 1.
56. *Ibid*, v. 3.11. Keith *Ibid*, 427 note 1.
57. *Ibid*.
58. *Tait Sam.* v. 5.3. See Keith. VBYS 442.
59. *Ibid* v. 4.2 Keith *Ibid* 429. note 2.
60. *Ibid*, v. 7.2. Keith *Ibid*, 468, note 2.
61. *Ibid*, iv. 4.4. Keith *Ibid*, 342. n. 8.
62. *Ibid*, iv. 4.10. Keith *Ibid*, 349. note 1 etc.
63. *Ibid*, v. 2. 8. Keith *Ibid*, 412.
64. *Ibid*, v. 3.10. Keith *Ibid*, 426 n. 3.

'water'-brick (v.2.10) is intended to mean 'brick made of water' or '*durva*-grass-bricks to mean bricks made of *durva-grass*. What seems to be referred to are real bricks—perhaps even baked ones. But the text is not interested in such physical aspects of the bricks, which appears to have been taken for granted. As a manual of the Vedic priests, it is simply interested in their magical potency and the apparently quaint names of the bricks hinge on this. Here is a passage from the *Taittiriya Samhita*[65] which, however tiring it may appear to the modern readers, is likely to illustrate the point :

He puts down the metre bricks ; the metres are cattle ; verily he wins cattle ; the good thing of the gods, cattle, are the metres ; verily he wins the good thing, cattle. Yajnasena Caitriyayana taught this layer ; by this he won cattle ; in that he puts it down, he wins cattle. He puts down the Gayatris on the east ; the Gayatri is brilliance ; verily at the beginning he places brilliance ; they contain the word 'head' ; verily he makes him the head of his equals. He puts down the Tristubhs ; the Tristubh is power ; verily he places power in the middle. He puts down the Jagatis ; cattle are connected with the Jagati ; verily he wins cattle. He puts down the Anustubhs ; the Anustubh is breath ; (verily it serves) to let the breaths out. Brhatis, Usnihs, Panktis, Aksarapanktis, these various metres he puts down ; cattle verily he wins various cattle ; variety is seen in his house for whom these are put down, and who knows them thus. He puts down an Atichandas ; all the metres are the Atichandas ; verily he piles it with all the metres. The Atichandas is the highest of the metres ; in that he puts down an Atichandas, he makes him the highest of his eqauls. He puts down two-footed (bricks) ; the sacrificer has two feet ; (verily they serve) for support.

For all the gods is the fire piled up ; if he were not to put (them) down in unison, the gods would divert his fire ; in that he puts (them) down in unison, verily he piles them in unison with himself ; he is not deprived of his fire ; moreover, just as man is held together by his sinews, so is the fire held together by these (bricks). By the fire the gods went to the world of heaven ; they became yonder Krttikas ; he for whom these are put down goes to the world of heaven, attains brilliance, and becomes a resplendent thing. He puts down the circular bricks ; the circular bricks are these worlds ; the citadels of the gods are these worlds ; verily he enters the citadels of the gods ; he is not ruined who has piled up the fire. He puts down the all-light (bricks) ; verily by them he makes these worlds full of light ; verily also they support the breaths of the sacrificer ;

65. *Ibid* v. 3 8-10. See, Keith *Ib*. 426-27.

they are the detities of heaven verily grasping them he goes the world of heaven. He puts down the rain-winning (bricks) ; verily he wins the rain. If he were to put (them) down in one place, it would rain for one season ; he puts down after carrying them round in order ; therefore it rains all the seasons. Thou art the bringer of the east wind, he says ; that is the form of rain ; verily by its form he wins rain. With the Samyanis the gods went (sam ayus) to these worlds ; that is why the Samyanis have their name ; in that he puts down the Samyanis, just as one goes in the waters with a ship so the sacrificer with them goes to these worlds. The Samyanis are the ship of the fire ; in that he puts down the Samyanis, verily he puts down a boat for the fire ; moreover, when these have been put down, if the waters strive to drag away his fire, verily it remains unmoved. He puts down the Aditya bricks ; it is the Adityas who repel from prosperity him who being fit for prosperity does not obtain prosperity ; verily the Adityas make him attain prosperity. It is yonder Aditya who takes away the brilliance of him who having piled up a fire does not display splendour ; in that he puts down the Aditya bricks, yonder sun confers radiance upon him ; just as yonder sun is radiant, so he is radiant among men. He puts down ghee bricks ; the ghee is the home dear to Agni ; verily he unites him with his dear home, and also with brilliance. He places (them) after carrying (them) round ; verily he confers upon him brilliance not to be removed. Prajapati piled up the fire, he lost his glory, he saw these bestowers of glory, he put them down ; verily with them he conferred glory upon himself ; five he puts down ; man is fivefold ; verily he confers glory on the whole extent of man.

From the standpoint of the Sulva texts, this is an example of the priestly dictation—of "what is to be done"—wrapped up in all sorts of magical beliefs. For the *Sulva-sutra*-s, however, the question of overriding importance is "How is this to be done?" For answering this question what is important above all is the physical descriptions of the bricks. It is, therefore, natural for the Sulva texts to relegate magico-religious brick-names to the province of the "others", perhaps with due reverence. But the point is that the whole host of magico-religious brick-names proved absolutely useless for the craftsmen and technicians engaged in the work of physically constructing the brick-structure. In the Sulva-texts, therefore, we come across a new set of terminologies referring to the bricks, notwithstanding the rather desultory remains of a few magico-religious brick-names having no specific function for the real construction purpose. The overriding interest of the brick-names in the *Sulva-sutra*-s is mathematical.

But before we pass on to the bricks in the *Sulva-sutra*-s we may have a brief note on the bricks in the *Satapatha Brahmana* which we are required by the Vedic scholarship to accept as text intervening between the *Yajurveda* and the *Kalpa-sutra*-s.

13. BRICKS IN THE SATAPATHA BRAHMANA

It would be wrong, of course, to overlook or ignore the basic fact that the *Satapatha Brahmana,* too, is essentially a ritual text and is hence obliged to remain within the general framework of the system of magico-religious beliefs of the Vedic priests. It is nevertheless interesting to see how we have in it occasional glimpses of a somewhat different understanding of the bricks—an understanding in which bricks are accepted as genuinely physical entities rather than mere words carrying the baggage of magico-religious beliefs.

In the *Satapatha Brahmana*[66] we come across some attempt to explain the etymology of both *citi* (the fire altar) and *istaka* (brick). Thus :

> Now it was those five bodily parts (*tanu*) of his (Prajapati's) that became relaxed,—hair, skin, flesh, bone, and marrow,—they are those five layers (of the fire-altar) ; and when he builds up the five layers, thereby he builds him up by those bodily parts ; and inasmuch as he builds up (*ci*), therefore they are layers (*citi*).

The whole thing is wrapped up in a myth no doubt. Nevertheless we can have in it the glimpse of a five-layered altar actually or physically constructed and the suggested etymology of *citi* from *ci* or the actual work of building up or physical construction cannot perhaps be easily dismissed. Unfortunately, it is not so in the case of the etymology suggested for *istaka*[67] :

> In the fire the gods healed him (Prajapati) by means of oblations ; and whatever oblation they offered that became a baked brick and passed into him. And beecause they were produced from what was offered (*ista*), therefore they are bricks (*istaka*). And hence they bake the bricks by means of fire, for it is oblations they thus make.
>
> He spoke, 'Even as much as ye offer, even so much is my happiness' ; and inasmuch as for him there was happiness (*ka*) in what was offered (*ista*), therefore also they are bricks.

66. *Sat. Br.* vi. 1.2. 17 See Eggeling SBE XLI. 152.
67. *Ibid* vi. 1.2. 22-33. Eggelling SBE XLI. 153.

This etymology of *istaka* is often considered fanciful by the modern scholars and it is bound to be so if it is non-Vedic in origin. But not so are the bricks and the mention of firing them. What is referred to are physical bricks, physically baked, notwithstanding the general framework of the priestly myth within which these are mentioned.

The text also mentions the real or physical materials which the bricks are made of: "the bricks consist of clay and water",[68] "he makes bricks from clay".[69] Sometimes the size of the bricks and the way of arranging these are also specified, though with a thin line of pseudo-logic intended to rationalise the discourse. Thus[70]:

> Now, then, of the measures of the bricks. In the first and last layers let him lay down (bricks) of a foot (square), for the foot is the support; and the hand is the same as the foot. The largest (bricks) should be of the measure of the thigh-bone, for there is no bone larger than the thigh-bone. Three layers should have (their bricks) marked with three lines, for threefold are the worlds; and two (layers may consist) of (bricks) marked with an indefinite number of lines, for these two layers are the flavour, and the flavour is indefinite; but all (the layers) should rather have bricks marked with three lines, for threefold are all the worlds.

In such passages of the *Satapatha Brahmana* we seem to have a faint glimpse of real technologists though as required by the general nature of the text the techniques are referred to as wrapped up with mysticism and magic. Besides, it must not be forgotten that a very large number of the earlier magico-religious—and, therefore, irrelevant from the technological viewpoint—brick-names survive in the *Satapatha Brahmana,* re-cataloguing of which may prove tedious to the modern readers.

14. BRICKS IN THE SULVA SUTRA-S: GENERAL OBSERVATIONS

With the *Sulva-sutra*-s we seem to enter into a new world altogehter. It is the world of the craftsmen or technologists—mainly brick-makers and brick-layers—keenly conscious of the need of precise calculations. A bare list of the varieties of

68. *Ibid* vi. 1.2. 34; vi. 2.1.8.
69. *Ibid* vi. 5.3.7.
70. *Ibid* viii. 7.2.17.

bricks required to construct structures of various shapes may be enough to see this point. We shall give this list mainly from the *Baudhayana Sulva-sutra*. For the calculations in this list the text follows the table of linear measures with which it opens. We have already quoted this table in Section 5 of the present chapter, to which it will be necessary for us to refer constantly.

But let us begin with a number of preliminary observations on the bricks in the *Baudhayana Sulva-sutra*, which gives us almost an impression of fastidiousness of the technicians we meet in the text[71] :

> No brick, which has been formed by breaking is to be employed for the constructions of the *agni*, (i.e. a triangular brick which has been formed by dividing a square-shaped brick is forbidden).
>
> No brick, which is cleft (for example, square-shaped brick made out of two triangular bricks) is to be employed.
>
> And no brick which is damaged in any way.
>
> And no brick of black colour (which may be produced by the burning, etc.).
>
> And no brick is to be employed which has some mark (an impression of some foreign body, etc.).

With these points in mind, let us next have a few clarifications about the more complicated brick-structures called Kamya-agni-s or fire altars needed for sacrificial rituals supposed to ensure for their financiers the fulfilment of certain specific desires :

1. The total area to be covered by such a brick structure must be 7½ square *purusa*-s. Let us try to see what it means according to our modern system of measurement. One *purusa* = 120 *angula*-s, and, according to V. B. Mainkar's estimate, one *angula* = 17.78 mm. Thus one *purusa* = 120×17.78 mm. = 2133.6 mm. One square *purusa* = 2.1336×2.1336 meters = 4.552 square meters. Seven and half square *purusa* = 4.552×7.5 = 34.14 square meters, or 367.48 square feet. This, in short, means that the total area to be covered by such a brick structure is not negligible even in modern standard.

2. Each structure must be made of five layers of bricks, though the total height of the structure is to be one *janu* or the knee. This gives us some idea of the average height of

71. *Baudh. Sulva-sutra* ii. 52-56. Tr. Thibaut.

the bricks. As the *Baudhayana Sulva-sutra* says, " The fifth part of *janu* is to be taken as the height of the bricks so that the height of the whole *citi* with its five *prastara*-s (layers) is equal to one *janu*".[72] In other words, one *janu* being equal to 32 *angula-s,* the height of each brick is to be $6^2/_5$ or 6.4 *angula* or 113.792 mm. 4.48 inches, or roughly say 4½ inches. But provision had also to be made for making some thinner bricks to meet the requirements of some special type of structure. As the *Baudhayana Sulva-sutra* adds : "The height of the *nakasat*-s and *pancacoda*-s is to be made with half that measure (of height)".[73] Such bricks, thus, are to have the height of $3^1/_5$ *anugla*-s, or, according to our modern measurement, 56.89 mm.

3. The number of bricks to be used for each layer is specified as 200. The total number of bricks for each structure is thus 1000. Lest, however, this number is disturbed, in the special cases in which the thinner bricks are technologically required, it is added "If the intention is to give all the *prastara*-s (or layers) two hundred bricks, then the *pancacoda*-s and *nakasat*-s are to be considered as forming one number together".[74]

4. While arranging the bricks in different layers, the cleft or meeting of the edges of the bricks used in the lower and its immediately upper layer is to be avoided. Such clefts are technically called the *bheda*-s. As Thibaut explains, "Every *agni* consists of five layers (*prastara-citi*) of *bricks* (200 bricks in each layer). Now the bricks in the second layer, for instance, had to be arranged in that way that none of them ever covered exactly a brick of the first layer, but that the line in which two bricks of the first layer were meeting was covered by a brick of the second layer. The same rule had to be observed with regard to the second and third layers, and so on".[75] The importance of this rule from the technological view-point is obvious, for without observing this the structure as a whole would not have any stability and would

72. *Ibid*. ii. 58.
73. *Ibid*. ii. 59.
74. *Ibid*. ii. 28.
75. Thibaut on *Baud Sul. Su.* ii. 23.

crumble down, specially when there was no provision for the use of mortar for joining the bricks. Interestingly, however, the brick-layers in the Sulva texts also knew that to maintain the exact size and shape of the brick-structures they were required to execute, this rule of avoiding the *bheda*-s could not be adhered to too rigidly, or that there were cases when exceptions had to be made to it. As the *Baudhayana Sulva-sutra* puts the point : "These *bheda*-s do not occur (are not to be considered as such) on the outer periphery of those *agni*-s (for there of course the edges of the bricks of all five layers must meet). And secondly not (i.e. the rule is not valid) in the case of the angles and of the sides. (This refers to the *sararatha-cakra-citi,* where unavoidable *bheda*-s occur in the centre of the *nabhi*—the nave of the wheel and on the side of the *ara*-s, the spokes)".[76]

5. In the text itself, the need is naturally felt for giving separate lists of brick-types for the construction of structures with different shapes. Evidently, the kind of bricks required for the construction of a structure having the shape of a falcon with stretched wings are not suitable for the purpose of constructing a structure having the shape of a chariot wheel with or without the spokes. Even for the same structure like the falcon-shaped one, bricks of different sizes and shapes are required. So it is no use to look for any standardised brick-type in the Sulva texts. The technicians in our text had to improvise a large number of brick-types to execute what they were asked to do. What amazes us even today is the mathematical precision they maintained for the purpose and the way in which they mentioned the names of the brick-types almost invariably using mathematical terms to indicate their shapes and sizes.

15. BRICK-TYPES : SOME EXAMPLES

With these general points in mind, let us try to have some idea of the technological-mathematical questions discussed in the Sulva texts concerning the brick structures called *agni* or *citi* in the priestly terminology. It will not be within the scope of our present discussion to review all the brick-types for the

76. *Baudh. Sul. Su.* ii. 24-25.

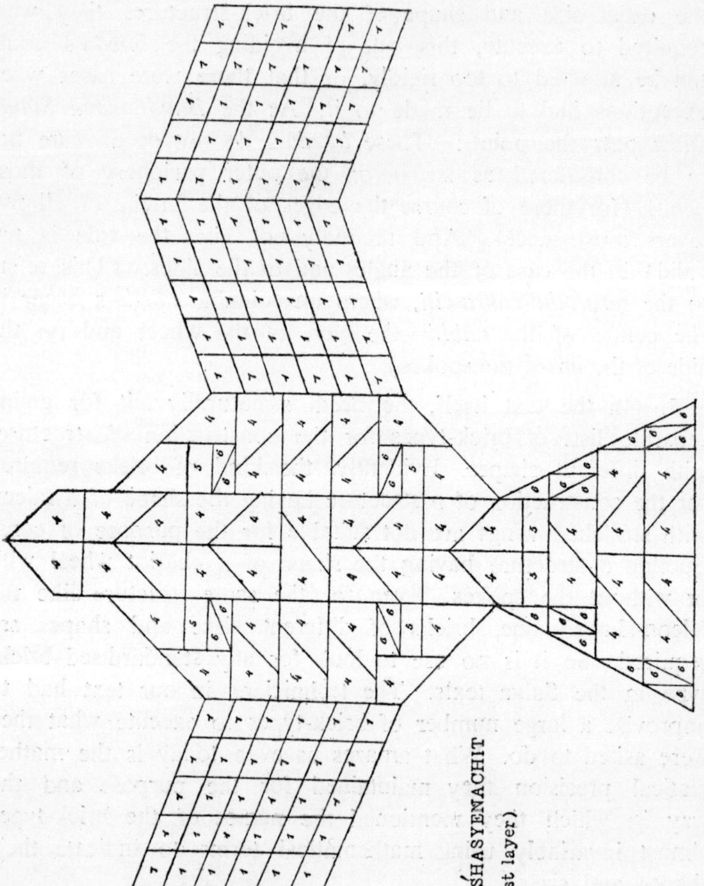

VAKRAPAKSHAŚYENACHIT
(first layer)

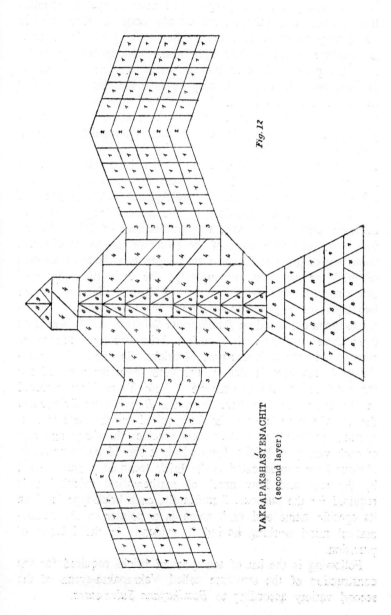

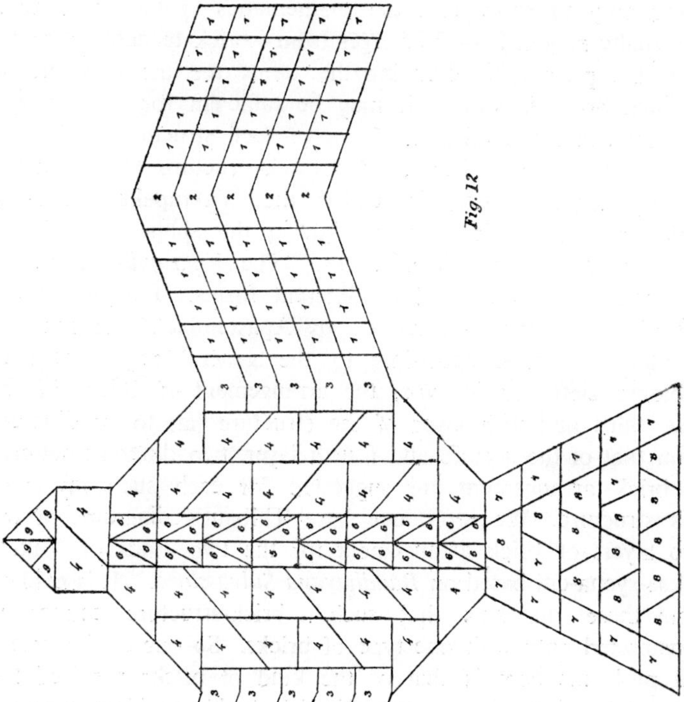

Fig. 12

VAKRAPAKSHAŚYENACHIT
(second layer)

making of the different structures discussed in the texts. Nor is it necessary for our purpose, because competent scholars like Thibaut and others have already done it. Our purpose being only to show that the mathematics in the Sulva texts is actually rooted in—or emerges from—brick technology rather than the priest-craft dictating the terms for the construction of such brick structures, it may be sufficient for our purpose to have some general idea of one of such structures. We propose to concentrate here on the brick-structure which, in the priestly nomenclature, is called the Vakrapaksa-syena-citi, which means a brick-structure having the shape of a falcon with outstretched and bent wings. "The *Vakrapaksa-syena* itself could be constructed in different forms. Two forms are described by Baudhayana, two by Apastamba".[77] It may be recalled here that according to the general requirement of avoiding clefts or *bheda*-s, the arrangement of bricks in the first, third and fifth layer of the structure has to be different from that of the second and fourth layer, two different patterns of brick arrangement are suggested for each such layers of this structure. We reproduce here Thibaut's diagrams of the two layers of brick arrangement for the two kinds of Vakrapaksa-syena-citi based on *Baudhayana Sulva-sutra.* It is evident from these diagrams that such a brick-structure cannot be constructed only with one type of bricks. So one of the essential problems here is that of the kind of bricks required for the construction. As an example of how this problem is solved in the texts, we give here from the *Baudhayana Sulva-sutra* the list of ten varieties of bricks required for the construction. It needs to be noted that the text specifies the shape and size of each variety of bricks, from which it should be immediately obvious how sophisticated is the brick technology presupposed by the text and how much of mathematical calculation is required for the purpose. Significantly, each brick-type is given its specific name and each name is a pointer to the mathematical mind working behind it, evidently in the interest of precision.

Following is the list of ten types of bricks required for the construction of the structure called Vakrapaksa-syena of the second variety according to *Baudhayana Sulva-sutra.*

77. Thibaut in *SHSI,* II 449.

1. *Pancami*, Literally one-fifth. These are square shaped bricks, the side of which is one-fifth of a *purusa,* i.e. 24 *angula*-s. In other words, these square shaped bricks have the measure of 24×24 *angula*-s or 576 square *angula*-s. Lest there be any vagueness about the exact size of such bricks, Thibaut explains that had it been possible to cover the entire area of 7½ square *purusa*-s only with such bricks, the total number of bricks required would have been 187½ for each layer, "One square *purusa* contains 25 such bricks, therefore 7½ square *purusa*-s = 187½ *pancami*-s".[78] The calculation is quite simple: Each brick = 24×24 *angula*-s = 576 square *angula*-s; 25 such bricks, 576×25 = 14,400 square *angula*-s or one square *purusa*; 576×187.5 = 108,000 square *angula*-s = 7½ square *purusa*-s. However, it is specified that each layer is to consist of only 200 bricks, which means that the construction of the altar is more than the question of stacking only a certain number of the *pancami* bricks. Bearing this in mind it is easy to judge that a brick structure as complicated as to have the shape of a falcon with out-stretched and bent wings cannot be constructed with such square bricks alone: various other types of bricks with various shapes and sizes are evidently needed for the purpose, i.e. over and above a certain number of these square bricks. Hence our text passes on to mention bricks of other types required for the purpose.

2. *Adhyardha,* Literally longer by half. Since this is mentioned in the immediate context of the *pancami* bricks, we may—for our purpose—assume that the expression presupposes *pancami* and hence we have to understand these as (*pancami*)-*adhyardha.* The *Baudhayana Sulva-sutra* describes the first two forms of bricks as "Bricks the side of which (i.e. one side of which) is equal to the fifth part of a *purusa* (i.e. half of the side of the *pancami* (24 *angula*-s) and bricks which are longer on one side by a half (i.e. longer by half of the side of the *pancami* (i.e. 24+12=36 *angula*-s)"[79]. In simpler language, it means this second type is oblong-shaped bricks the size of which is 24 *angula*-s×36 *angula*-s or 864 square *angula*-s in area.

78. Thibaut on *Baudh. Sul. Su.* iii. 106.
79. *Baudh. Sul. Su.* iii. 118.

3. *Sapada,* literally with an extra one-fourth. Occurring, again, as it does in the immediate context of the *pancami*, we can perhaps use for it the expression (*pancami*-) *sapada*. This means oblong bricks one side of which is equal to that of the *pancami* (i.e. 24 *angula*-s) and the other one-fourth more than this (i.e. 24 *angula*-s + 6 *angula*-s = 30 *angula*-s). In short, these are rectangular bricks measuring 24 *angula*-s × 30 *angula*-s, or 720 square *angula*-s.

4. *Ardhya,* literally measuring half. In the immediate context of *pancami,* we can perhaps use for it the expression (*pancami*-) *ardhya*. It needs to be remembered here that the *Baudhayana Sulva-sutra* had already instructed that in the context of bricks, the process of having a square or rectangle is to be understood as by drawing a diagonal on it. As the text puts it, "(If there are mentioned half-bricks, quarter-bricks, etc.), the division (of the original brick) is always to be made following the diagonals in case there is no special direction given".[80] Accordingly, half of *pancami* brick mentioned here refers to the isosceles right-angled triangle obtained by drawing a diagonal on the *pancami* brick. *Pancami* brick being 24×24 *angula*-s, the triangular brick referred to here as *ardhya* shall have the three sides as 24 *angula*-s by 24 *angula*-s by 33.94 [i.e. $\sqrt{(576+576)} = \sqrt{1152} = 33.941$ *angula*-s]. Its area = half of 576 = 288 sq. *angula*-s.

5. *Padya,* literally measuring a quarter. In the general context of the *pancami*-s, these bricks are to be understood as (*pancami*-) *padya*-s or quarters of *pancami*-s. Following the general rule already quoted, viz. that expressions like half-bricks and quarter-bricks are to be usually measured by drawing diagonals on the rectangular or square bricks under consideration, by the *padya*-s are to be understood here as triangular bricks the size of which is to be obtained by drawing two diagonals on a *pancami* brick—the size of each *pancami* brick thus giving four equal triangular bricks. A *pancami* brick having the size of 24×24 *angula*-s, the size of each *padya* brick is equal to 24 *angula*-s by 16.97 *angula*-s by 16.97 *angula*-s. Its area being one-fourth of 576 = 144sq. *angula*-s.

80. *Ibid*. iii. 65.

6, 7 and 8. The next three brick sizes are cryptically referred to by our text as *tatha adhyardhayah.* Thibaut translates it as "In the same way (bricks are made which are equal to the half and fourth part) of an *adhyardha* brick."[81]. In other words, just as the *ardhya* and *padya* bricks referred to are made by diagonally dissecting the *pancami* bricks, so also by diagonally dissecting an *adhyardha* brick, we get new types of triangular bricks. But the *adhyardha* being oblong in shape, the similar process of diagonally dissecting it yields altogether three—instead of two—triangular shaped bricks which, in the Sulva terminology, are *ardha-adhyardha, dirgha-adhyardha* and *sula-padya.* Thus :

6. *Ardha-adhyardha.* The *adhyardha* brick being oblong having its two sides as 24 and 36 *angula*-s, the simple division of it into two by drawing a diagonal gives us two triangular bricks, the sides of each of which are 24, 36, 43.266 *angula*-s. Its area = $\frac{1}{2}$ of 864 = 432 sq. *angula*-s.

However, dividing the same oblong brick into four by drawing two diagonals on it, we get two types of equal traingular bricks. As Thibaut explains, "As the *adhyardha*-brick is not a square, but an oblong, it is divided by its two diagonals into four triangles of which two and two are equal to each other. The technical names of these two different kinds of triangular bricks are *dirghapadya* and *sulapadya*".[82] Incidentally these two names do not occur in the text itself and are found in the commentaries. However, for our purposes, the point to be noted is that by cross-division of the oblong of 36×24 *angula*-s by two diagonals, we get two pairs of bricks —each in the pair having the same measure. These are :

7. Dirgha-padya = the sides of which are 36, 21.63, 21.63 *angula*-s.

8. Sula-padya = the sides of which are 24, 21.63, 21.63 *angula*-s.

Area of each Dirgha-padya brick = 216 sq. *angula*-s
Area of each Sula-padya brick = 216 sq. *angula*-s.

9. *Ubhayi*, literally made of parts of both. The text describes it as follows : "The eighth part of a *pancami* and of

81. *Ibid* iii. 121.

82. Thibaut on *Baudh. Sul. Su.* iii. 121.

an *adhyardha* have to be combined in that manner that three corners (i.e. a triangle) are the result"[83]. Thibaut explains : "For this purpose the *padya* of a *pancami* is divided by a line drawn from the apex to the middle of the base ; the result is a rectangular triangle the sides of which are 12 *angula*-s, 12 *angula*-s and 16 *angula*-s 33 *tila*-s. This is the eighth part of the *pancami*. In the same way, the *dirghapadya* of an *adhyardha*-brick is divided, the result is the nominally eighth part of the *adhyardha*—a rectangular triangle the sides of which are 12 *angula*-s, 18 *angula*-s and 21 *angula*-s 21 *tila*-s. By putting together the equal sides of these two triangles, we get a scalene triangle the sides of which=16 *angula*-s 33 *tila*-s, 21 *angula*-s 21 *tila*-s and 30 *angula*-s. The name of a brick of this kind is *ubhayi*". (Its area = 180 sq. *angula*-s).[84]

10. *Astami*, literally one-eighth ; "And bricks which are equal to the eighth part of a *pancami*".[85] Following the general principle of dividing by drawing diagonals, we get four triangular parts of a *pancami* brick and then by drawing a line from the vertex to the middle of the base of each of these four triangles, we get altogether eight triangular bricks, each of which will be one-eighth of a *pancami* brick in areas, The three sides of each of such eight bricks will be 12 *angula*-s, 12 *angula*-s and 16.97 *angula*-s. Its area $= {}^1/_8 \times 576 = 72$ sq. *angula*-s. Such then, are the ten varieties of bricks required for making the structure called *vakrapaksa syenaciti* of the second type. As the *Baudhayana Sulva-sutra* sums up : *iti dasa*—"These are altogether ten different classes of bricks".[86]

Before passing on to survey the technique of laying such varieties of bricks discussed by our text for the construction of the structure covering the total area of 7½ square *purusa*-s resembling a falcon with stretched and bent wings and made of five layers each with 200 bricks in such a manner that the joint of two bricks in one layer does not overlap the joining of bricks in the layer immediately above it, we may as well note another point. The ten varieties of bricks just enumerated are required only for the construction of one type

83. *Baudh. Sul. Su.* iii. 122.
84. Thibaut on *Baudh. Sul. Su.* iii. 122.
85. *Baudha. Sul. Su.* iii. 123.
86. *Ibid.* iii. 124.

of structure. To the technologists and mathematicians of the Sulva texts, however, came down the problem of constructing not only this type of structure. They were required to construct various types of structures besides. Understandably the need was felt also for other types of bricks for the construction of structures of other shapes. In the Sulva texts, accordingly, we come across specifications of bricks of various other shapes and sizes, all bearing technological-mathematical names. We may have some idea of this from the following list, though for the sake of simplicity we give here only the shapes and sizes of some of these bricks, without their technical names :

1. Square bricks, the side of which is one-fourth of a *purusa i.e.* 30×30 *angula*-s.
2. Square bricks the side of which is one-sixth of a *purusa,* i.e. 20×20 *angula*-s.
3. Square bricks the side of which is one-tenth of a *purusa,* i.e. 12×12 *angula*-s.
4. Oblong bricks of 24×12 *angula*-s in size.
5. Oblong bricks of 20×10 *angula*-s in size.
6. Triangular bricks sides of which are 30, 30, 42.42 *angula*-s.
7. Triangular bricks obtained by dividing a 30×30 *angula*-s square brick into four parts by drawing two diagonals on it ; i.e. each of the four such triangular bricks has for its three sides : 30 *angula*-s, 21.21 *angula*-s and 21.21 *angula*-s.
8. Bricks of the shape of a trapezium obtained by adding an oblong of the size 15×7½ *angula*-s to a triangle on it having the three sides as 15 *angula*-s, 15 *angula*-s and 21.21 *angula*-s.
9. Bricks of the shape of an irregular pentagon—the sides of which are 30 *angula*-s, 7.5 *angula*-s 21.21 *angula*-s, 21.21 *angula*-s and 7.5 *angula*-s—with the total area of 450 square *angula*-s (i.e. half of the area of 30×30 *angula*-s). The name given to the type of brick is "swan-beaked" or Hamsamukhi.
10. Trapezium-shaped bricks with sides measuring 6, 12, 18 and 16.97 *angula*-s.
11. Trapezium-shaped bricks with sides measuring 24, 24, 48, 33.94 *angula*-s.

12. Rectangular bricks measuring 38 *angula*-s 25 *tila*-s by 19 *angula*-s 12½ *tila*-s.
13. Triangular bricks the measurement of which is obtained by diagonally bisecting the above rectangular bricks.
14. Triangular bricks the measurement of which is obtained by quartering the above (i.e. No. 12) rectangular bricks with two diagonals and having for its base the longer side.
15. Triangular bricks the measurement of which is obtained by quartering the above (No. 12) rectangular bricks with two diagonals and having for its base the shorter side.
16. Square bricks with an area of one-thirtieth of a square *purusa,* i.e. square bricks with an area of 480 square *angula*-s, or 21.9×21.9 *angula*-s.

The list may indeed be made much longer. But that is not necessary for our purpose. Readers interested in a more exhaustive list of the brick types mentioned in the Sulva texts, may profitably look up *The Sulbasutras* by Sen and Bag. For the purpose of our own discussion, two points are specially to be noted in this connection.

First, the impression of brick technology we have from the Sulva texts is, to say the least, extremely sophisticated. Evidently enough, it presupposes a class of highly skilled craftsmen whom it may not be an error to consider as whole-time specialists or at least those that had inherited the tradition of generations of specialists. Judging from the minute details of ritual prescriptions recorded in the *Yajurveda* and *Brahmana*-s, the Vedic priest-craft, too, cannot but presuppose similar whole-time specialists, though devoted to matters other than technology in its real sense. This suggests that the class of brick-technologists and Vedic priests could hardly belong to the same class of people. If, in spite of this, it is argued that a section of Vedic priests chose to specialise in brick-technology, it would be necessary to admit that they did it not in the strict capacity of Vedic priests but somehow deviating from their priest-craft.

Secondly, the brick-types mentioned in the Sulva texts are, from the very nomenclature used for these, unmistakable pointers to technological-mathematical temper caring above all for the accuracy of measurement, which, in its turn, evidently presupposes precision of observation. Such a theoretical temper is totally wanting in—and, as a matter of fact, even contrary to—not merely in the general trend of thought we find in the *Yajurveda* and the *Brahmana* texts but also in the very nomenclature of the brick-types in the priestly literature in its strict sense. Thus the mere fact that the *Sulva-sutra*-s come down to us as appended to the priestly literature cannot be viewed as a decisive evidence for the Vedic priests themselves having been the makers of mathematics as codified in the Sulva texts.

16. MASONRY AND ARCHITECTURE

Equipped with such a wide variety of brick-types the masons and architects in the *Sulva-sutra*-s proceed to the physical construction of the brick-structures, showing again their basic interest in mathematical questions combined with superb craftsmanship—or, perhaps as better expressed, in the mathematical questions arising out of the requirements of their craftsmanship. Thanks to the pioneering work of Thibaut and others, it is not necessary for us to try to re-explore the details of this aspect of the Sulva texts. Depending on the critical evaluation of the major commentaries on the texts, they have analysed the cryptic aphorisms and reconstructed from these with excellent diagrams the technique followed for the physical construction of the different brick-structures with different names. For our own purpose of meeting the masons and architects in the *Sulva-sutra*-s, it may be enough to follow Thibaut, though readers interested in more details may profitably consult also the recent work by Sen and Bag.

Our present purpose being to have some idea of how the Sulva mathematics is inextricably related to the work of the masons and architects, we are not required to discuss here the construction of all the brick-structures. It may be adequate, instead, to follow Thibaut's review of the technique followed for the construction of the comparatively complicated brick-structure, the *Vakrapaksa-syena-citi*—a brick-structure having the shape of a falcon with outstretched and bent wings. The Sulva texts discuss the construction of different forms of such a brick-structure, of which we shall quote here what is said about the two types.

We have seen the diagrams of the two different layers of the brick-arrangements of the first kind of Vakrapaksa-syena brick-structure prepared by Thibaut on the basis of the *Apastamba Sulva-sutra.* Explaining these Thibaut[87] observes :

The following extract contains Apastamba's rules for the first kind of the *Vakrapaksa-syena* :

'He who wishes for heaven, may construct the altar shaped like a falcon ; this is the tradition.

'His wings are bent and his tail spread out.

'On the west side the wings are to be drawn towards the east, on the east side towards the west.

'For such is the curvature of the wings in the middle of the birds, says the tradition.

'Of the whole area covered by the seven fold *agni* with *aratni* and *pradesa,* take the *pradesa,* the fourth part of the *atman* (body without head, wings, and tail) and 8 quarter bricks ; of these latter, 6 form the head of the falcon ; the remainder is to be divided between the two wings'.

The *sutra* determines what portions of the legitimate area of the *agni* have to be allotted to the different parts of the falcon construction. The whole area of the *saptavidha agni* is seven *purusa*-s with the addition of the two *aratni*-s on the wings and the *pradesa* of the tail, altogether $7\frac{1}{2}$ *purusa*-s. Now the fourth part of the atman (of the primitive *syena-citi*) one *purusa* and the *pradesa,* i.e., an oblong of 120 *anguli*-s by 12 *anguli*-s$=1/_{10}$ square *purusa* and eight quarter bricks, (i.e., square bricks the side of which is equal to the fourth part of a *purusa*=30 angulis, so that they cover together an area of $1/_2$ square *purusa*) are given to the wings in addition to the area which they cover in the primitive *agni,* only they have to cede in their turn three of the eight quarter bricks, which are employed for the formation of the head. The original area of both wings together being $2^2/_5$ *purusa*-s, their increased area amounts to $2^2/_5+1^3/_5-3/_{16}=3^{13}/_{16}$ square *purusa*-s, for one wing to $^{129}/_{32}$ square *purusa*-s.

"Nine and half *aratni*-s (=238 *anguli*-s) and three quarters of an *anguli* are the length of the wing.'

The breadth of the wing is the same as in the primitive *syena,* i.e., =1 *purusa*=120 *anguli*-s. Dividing the area of the wing mentioned above by the breadth we get the length. Up to this, the wing has the shape of a regular oblong ; the following rules show how to produce the curvature :

'Make ties at both ends of a cord of two *purusa*-s length and a mark in its middle.

'Having fastened the two ends of the cord at the two western corners of the oblong forming the wing, take it by the mark and stretch

87. Thibaut in SHSI II. 449-53.

it towards the east ; the same is to be done on the eastern side (i.e. the cord is fastened at the two east corners and stretched towards the east). This is the curvature of the wings.'

By stretching the cord, fastened at the west corners, a triangle is formed by the west side of the oblong and the two halves of the cord, and this triangle has to be taken away from the area of the wing. In its stead the triangle formed, when the cord is stretched from the eastern corners, is added to the wing.

'Thereby the northern wing is explained'.

The curvature is brought about in the same way :

'The *atman* is two *purusa*-s long, one and a half *purusa*-s broad'. This is not the final area of the *atman,* as we shall see further on ; but an oblong of the stated dimensions has to be constructed and by cutting pieces from it we get the area we want.

'At the place of the tail, stretch a *purusa* towards the west with the breadth of half a *purusa*'.

That means : construct an oblong, measuring one *purusa* from the east to the west, half a *purusa* from the north to the south.

'To the south and to the north of this oblong, construct two other oblongs like it, and dividing them by their diagonals remove their halves so that half a *purusa* remains as breadth at the jointure of *atman* and tail.

The result is the form of the tail which we see in the diagram.

'At the place of the head a square is to be made with half a *purusa,* and from the middle of its east side cords are to be stretched to the middle of the northern and the southern side.

The triangles cut off by these cords are to be taken away from the area of the head.

'Then the four corners of the *atman* are cut off in the direction towards the joining lines. This finishes the measurement of the *syena.* Its four corners are cut off by four cords connecting the ends of the lines in which the *atman* and the wings touch each other with the ends of the lines in which head and tail are joined to the *atman*'.

17. AN EXAMPLE OF MATHEMATICAL EXCELLENCE

We propose to add only one point to the above.

It is concerning the mathematical excellence implicit in what is just said. This is evident from one special brick-type mentioned for the construction of the structure. As described by the *Apastamba Sulva-sutra,* "One class of bricks has the length of the fifth of a *purusa,* the breadth of a sixth bent in such a way as to fit (the place in which they are to be employed)".[88] Thus, in other words, these bricks have the size of 24 *angula*-s by 20 *angula*-s. But these are not rectangular bricks, as indicated by the word "bent". What, then, is meant by "bent" here? Thibaut explains: "By *nata*, 'bent', the Sutrakara means to indicate that the sides of the brick do not form right angles. The shape of the brick is of a parallelogram the angles which the sides form with each other, are the same which the wings of the *syena* (falcon) form with the body".[89]

Let us now note the special mathematical interest of the use of this type of bricks.

In the first layer of the brick-structure under consideration, as can be seen from Thibaut's diagram, this type of bricks is used only for the construction of the wings—60 such bricks in each wing, i.e. 120 such bricks in all. The total area to be covered by each wing is 27,450 square *angula*-s, i.e. 54,900 sq. *angula*-s for the two wings together.

Thus, in other words, the total area to be covered by 120 bricks each of which is 24 *angula*-s in length and 20 *angula*-s in breadth is to be 54,900 sq. *angula*-s. Had these bricks been rectangular, the area of each would have been 24×20=480 square *angula*-s, i.e. the total area covered by 120 rectangular bricks would have been 57,600 sq. *angula*-s, instead of the prescribed total area for the two wings being 54,900 square *angula*-s. Hence is the crucial importance of the expression of *nata* or 'bent' in connection with such bricks. It means that the area covered by a parallelogram-shaped brick 24 *angula*-s long and 20 *angula*-s broad is not 480 square *angula*-s but evidently 457.5 sq. *angula*-s. One hundred and twenty such parallelogram-shaped bricks thus cover the total area of 54,900 sq. *angula*-s, which answers to the area prescribed for the two wings together. This, in short, means that the technician-mathematicians of our text *knew the way of calculating the area of a parallelogram the length of the two sides of which are given (in this particular case the actue angle between the two sides of the parallelogram being 72.3875 degrees) and that this area is different from that of a rectangle with the same length and breadth.* For a

88. Quoted by Thibaut, *Ibid*. 453.
89. Thibaut *Ibid*.

mathematician today, the calculation implied in this difference may be somewhat familiar. But to meet in the ancient Sulva texts mathematicians having precise knowledge of this difference—or, more simply, the technique of precise calculation of the area of a parallelogram with given sides as distinguished from a rectangle of the same given sides—is, to say the least, highly remarkable from the historical viewpoint. The step taken by the Sulva mathematicians must have been a prodigious one. From the statement that 120 parallelogram-shaped bricks—the two sides of which are 24 and 20 *angula*-s —should cover the total area of 54,900 square *angula*-s, it can be reasonably thought that they were also aware of the necessary angle in which such bricks were to be "bent", i.e. the fact that the area of a parallelogram depends on the exact angle between two of its adjacent sides or on its altitude. In the present case had the acute angle of the parallelogram been greater or lesser than 72.3875 degrees, the total area to be covered by 120 such bricks too, would have been greater or lesser, or the total number of bricks required would have been different.

18. TERMINOLOGICAL PRECISION

The making of an exact science presupposes the creation of a whole host of terminologies, each to be used in a precise sense. In the names of the brick-types improvised in the Sulva texts we already have some idea of how this was done by the makers of mathematics in ancient India. But terminological precision is not confined in the Sulva texts only to such brick-names. The need was also felt to use precise terminologies specifically in matters mathematical. Old words or words already in circulation were sometimes retained no doubt. However, connotation of these are firmly fixed, generally in the interest of mathematical precision. Only a few examples of this may be enough for the purpose of appreciating the general theoretical climate of the *Sulva-sutra*-s. Here is a list of some such terminologies from the *Baudhayana Sulva-sutra* :

Amsa : Literally, shoulder. Technically, two eastern corners of a square brick-structure I.34.

Aksnya : A technical word introduced in Sulba-sutra in the specific sense of 'diagonally'. Thus, *aksnya-rajju* means a diagonal line, a cord stretched across a square or oblong. I.46.

Prauga : Literally, forepart of the shafts of a chariot. But technically in the specific sense of isosceles triangle. Thus, *ubhayatah-prauga* (lit. having an isosceles triangle on both sides) means a rhombus. I.57.

Rju-lekha : Straight line. II.32

Karani : Literally, an instrument. But technically in the specific sense of a side (of a square). I.55.

Caturasra : Ordinarily, four-cornered. But, technically, a square (= *sama-caturasra*) or rectangle (= dirgha-caturasra). I.22.

Tiryak : Ordinarily, lying crosswise, obliquely, transversely. But, technically, breadth. I.46. Thus, *tiryan-mani* means breadth of shorter side of an oblong. I.38, I.48.

Nyanchana : A mark determined by the following procedure : Make a mark of the western half of the cord less the fourth part (of the half). The name of this mark is *nyanchana* I.32-33.

Tri-karani and Tritiya-karani (I.47) : Thibaut explains : "It seems that, in order to find the *tritiya-karani*, we are directed to find at first the *trikarani ;* the third part of the *tri-karani* is the *tritiya-karani* ; for it is the side of a square the area of which is one-ninth of the square of the *tri-karani* and consequently, one-third of the given square".

Pramana : Measure. I.3.

Parikarsana : Literally, dragging about. But, technically in the sense of 'circle'. III.210.

Parsvamani : Horizontal side, i.e. longer side of an oblong. I.48.

Viskambha : Literally, a prop, support. But, technically, in the sense of 'diameter'. I.25 & 26.

Savisesa : a ratio : diagonal of a square divided by the side of the same square= $\sqrt{2}$=1.414256. I.62.

Vrdhra
(Vrddha) : A technical term, meaning, cut-off piece. I.50.

19. POLITICAL PHILOSOPHY OF THE SATAPATHA BRAHMANA

From what is so far discussed about the Sulva texts—specially the one associated with the name of Baudhayana—one point is on the surface. It is the importance sought to be attached to accuracy—accuracy of the table of linear measures, in the enunciation of the geometrical propositions required although the technological problems to be confronted for construction of brick-structures of various shapes, in the determination of the shapes and sizes of the different brick-types, in improvising a whole host of technical terms without which exact science is inconceivable. Judged specially in the ancient context, this zeal for accuracy in the Sulva texts cannot but appear to us to be remarkable, even where the texts fail to have exact solutions of some of the problems these are bold enough to confront, e.g., in the case of squaring a circle or circling a square.

What, then, is the basis of this emphasis on accuracy? The quest for theoretical consistency—and sometimes perhaps also deductions though implicit—need not be overlooked. These seem to be specially conspicuous in the case of the geometrical propositions, with the enunciation of which our text opens. Nevertheless, even in these cases we meet the manual workers with their strings and poles, drawing in their own ways the basic geometrical figures and discussing their properties with very strong empirical bias. This is indicative of the actual source of the demand for accuracy of the Sulva texts. What we are trying to emphasise, in other words, is that the zeal for accuracy has its real moorings in actual observation, connected with manual operation. It is this that forms the very basis of the mathematical science in the *Sulva-sutra*-s.

This point is of material significance for what we are trying to argue, because it is an index to the general theoretical temper of the makers of mathematics in ancient India. Before hasitly concluding that they were the Vedic priests, we cannot evade one question: How far this theoretical temper can at all be imputed to these priests? How far, in other words, is it legitimate to assume that the Vedic priests were actually keen on the accuracy of observation with a distinct mooring in manual work? In order not to be arbitrary, we have to seek the answer to the question from the priestly literature in its strict sense, specially the *Brahmana*-s which are overtly interested in ideological questions. In the special context of our discussion, it may not be a methodological error to concentrate mainly to the internal evidences of the *Satapatha Brahmana*, which, among the *Brahmana*-literature, discusses the question of the fire-altars most elaborately.

What, then, is the theoretical temper—or, more broadly, the characteristic ideology—of the Vedic priests of which we have indications in the *Satapatha Brahmana?* And, how far is it compatible with the zeal for direct observation having a distinct mooring in manual operation?

Before we proceed to seek an answer to these questions, it is necessary to note one point. Though essentially a ritual text, the *Satapatha Brahmana* is very definitely committed to a political position—to the validation of a social norm—and the general theoretical temper of the Vedic priests is inextricably connected with it. In other words, the characteristic ideology of these priests and their theoretical temper are in need of being discussed in the general background of this political commitment.

What, then, is the nature of this political commitment? Our text does not mince words about it. Its essence is to establish the political power and privileges of mainly the Ksatriyas—"the nobility" or "the lordly power"—who, in those days, could be the *yajamana*-s or rich patrons financing the sacrifice, though, along with it, the superiority of the Brahmin priests,—"holy power"—subsisting almost exclusively on the *daksina* or fees for performing the sacrifice.

Let us begin with an account of a controversy about altar-construction we read in the *Satapatha Brahmana.*

While the Sulva texts specify the shape and size of every brick-type and generally give the brick-names in terms of these, the *Satapatha Brahmana,* being a ritual text after all, is naturally interested primarily in their magical efficacy in terms of which the brick-types are frequently named. One such name is *lokam-prna,* usually translated as the "space-filling" ones. It seems that from the view-point of one theological trend, which is associated with the name Tandya, the *lokam-prna* is a "technical term for those bricks which have no special prayers belonging to them, but are piled up with a common formula beginning with *lokam-prna chidram-*

prna, 'fill the space, fill the gap' ".[90] These are supposed to be contrasted with "the *Yajusmati* (prayerful) bricks which bear special names, and have special formulas attached to them".[91] Apparently sharing this view, the theologian Tandya declared :[92]

Surely the bricks possessed of prayers are the nobility, and the space-fillers are the peasants, and the noble is the feeder and the peasantry the food ; and where there is abundant food for the feeder, that realm is indeed prosperous and thrives : let him therefore pile up abundant space-fillers.

But the text refuses to accept it as "the standard practice". Hence it elsewhere declares.[93]

And, again, as to why he lays down a *lokam-prna*—the *lokam-prna,* doubtless is the nobility (or chieftaincy), and these other bricks are the peasants (or clansmen) : he thus places the nobility (or chieftain), as the eater, among the peasantry. He lays down in all the layers : he thus places the nobility, as the eater, among the whole peasantry (or in every clan).

In his notes Eggeling[94] explains the controversy :

At VI. 1. 2.25, Tandya was made to maintain that the Yajusmatis, or bricks laid down with special formula, were the nobility, and that the lokamprnas, laid down with one and the same formula, were the peasants, and as the noble (or chieftain) required a numerous clan for his subsistence, there should be fewer of the former kind of bricks, than the established practice was. This view was however rejected by the author of the *Brahmana,* and here, in opposition to that view, the *lokamprna* is identified with the nobility, and the Yajushmatis with the clan.

To the modern readers the controversy over the question whether the *lokam-prna* (space-filling) and *Yajusmati* (prayerful) bricks are to be identified with the peasantry or the nobility may appear to be totally arbitrary. But not the political view on which the representatives of both the theories fully agree. According to it the "noble" is to be made supremely powerful and the peasantry completely subservient, by the magic power of the ritual. This political norm repea-

90. Eggeling, SBE XLI. 153 n.
91. *Ibid.*
92. *Sat. Br.* vi. i. 2.25 Tr. Eggeling.
93. *Ibid.* viii. 7.2.2 Tr. Eggeling.
94. Eggeling SBE. XLIII, 132 n.

tedly occurs in the *Satapatha Brhmana* even in connection with the theme of altar-construction. Thus :[95]

And inasmuch as, in going from here, the horse goes first, therefore the Ksatriya, going first, is followed by the three other castes ; and inasmuch as, in returning from there, the he-goat goes first, therefore, the Brahmin, going first, is followed by the three other castes. And inasmuch as the ass does not go first, either in going from here, or in coming back from there, therefore the Brahmin and Ksatriya never go behind the Vaisya and Sudra : hence they walk thus in order to avoid a confusion between good and bad. And, moreover, he thus encloses those two castes (the Vaisya and Sudra) on both sides by the priesthood and the nobility, and makes them submissive.

Another ritual brick-name in the text is *rtavya* usually rendered as "seasonal bricks". And the text declares :[96]

And, again, as to why he lays down seasonal (bricks), the seasonal (ones) are the nobility and these other bricks are the peasantry ; he thus places the nobility as the eater among the peasantry. He lays down (some of) them in all the layers : he thus places the nobility as the eater among the whole people.

Again :[97]

And the seasonal (bricks), indeed, are also the nobility ; by the (different) layers he thus builds up the nobility above (the peasantry)...Let him not thereafter place over them any other brick with a sacrificial formula, lest he should place the peasantry above the nobility.

And so on. Many more passagees like these can be quoted from the *Satapatha Brahmana.* But that is not necessary. What is necessary is to note that the ritual placing of the bricks in the text is designed also to serve a frankly political purpose. For the brick technology itself the priests could have depended on the Harappan survivors, because we have so far no knowledge of any other tradition of highly sophisticated brick technology ante-dating or contemporaneous with the *Satapatha Brahmana.* What seems to be distinctive of the text, however, is the addition to it, and that includes this political philosophy, because the text—like the literature of the Vedic priests in general—belongs to the period of the rising

95. *Sat. Br.* vi 4.4.13 Tr. Eggeling.
96. *Ibid* viii. 7.1.2. Tr. Eggeling.
97. *Ibid* viii. 7.1.12. Tr. Eggeling.

political power of the nobility on the ruins of the ancient tribal organizations of the early Vedic peoples. In such a period, the strengthening of the political power of the nobility —though by mgical rituals—was only likely to be a paying proposition for the priests.

Eggeling in his own way notes this political motivation in what he calls the "sacrificial metaphysics" of the *Satapatha Brahmana*, or, more specifically, in those portions of the text discussing the question of the altar-construction. We may profitably quote him at some length :[98]

> The theologians of the Brahmanas go, however, an important step further by identifying the performer, or patron, of the sacrifice—the Sacrificer—with Prajapati ; and it is this identification which may perhaps furnish us with a clue to the reason why the authors of the Brahmanas came to fix upon 'Prajapati' as the name of the supreme spirit. The name 'Lord of Creatures' is, no doubt, in itself a perfectly appropriate one for the author of all creation and generation ; but seeing that the peculiar doctrine of the Purusa-sukta imparted such a decisive direction to subsequent dogmatic speculation, it might seem rather strange that the name there chosen to designate the supreme being should have been discarded, only to be employed occassionally, and then mostly with a somewhat different application. On the other hand, the term 'Prajapati' was manifestly a singularly convenient one for the identification of the Sacrificer with the supreme 'Lord of Creatures' ; for, doubtless, men who could afford to have great and costly sacrifices, such as those of the Srauta ceremonial, performed for them—if they were not themselves Brahmins, in which case the term might not be inappropriate either—would almost invariably be 'Lords of Creatures', i.e. rulers of men and possessors of cattle, whether they were mightly kings, or petty rulers or landed proprietors, or chiefs of clans. It may be remarked, in this respect, that there is in the language of the *Brahmanas* a constant play on the word 'praja' (progenies), which in one place means 'creature' in general, whilst in another it has the sense of 'people, subjects,' and in yet another the even more restricted one of 'offspring or family'.

Such, then, is the political philosophy of the *Satapatha Brahmana*. It is the political philosophy of a split society in which the powers and privileges belong to the "nobles" or "kings", though secondarily also to their ideological apologists, the priests, who take upon themselves the task of

98. Eggeling in SBE Vol. XLIII, intro. xv-xvi.

stabilizing with their magical rituals the power of the nobility, or, as Renou observes, "Vedism was in charge of a priestly elite who served a military aristocracy".[99] The details of such rituals often appear to us to be rather trivial, often making the texts almost unreadable even in standard translation. But that is not the impression the priests intend to give to their patrons. For the purpose of impressing the nobility of the efficacy of these trivialities, the essential points of this political philosophy is sometimes projected back on the ancient Vedic mythology which, as we shall later see, usually gives us the impression of a different political temper altogether—the political temper characteristic of the tribal collectives. For the priests of the *Brahmana*-literature, however, that does not matter. They freely tamper with what they themselves claim to inherit. Thus they project back their political views on the ancient Vedic pantheon itself and want their patrons to see in it the fulfilment of their desired reality. Class difference among the mortals is introduced among the ancient deities. The group of Vedic gods called the Maruts, for example, are now made to stand for the common people, while the despotic power of the nobility is represented by gods like Indra and Varuna. Here are typical passages illustrating this :

Varuna, doubtless, is the nobility, and the Maruts are the people. He (the priest) thus makes the nobility superior to the people. And hence people here serve the Ksatriya, placed above them.[100]

He muttered that verse addressed to Indra and referring to the Maruts. Indra indeed is the nobility, and the Maruts are the people... 'They shall be controlled', he thought, and therefore that verse is addressed to Indra.[101]

Now some, on noticing any straw or piece of wood among the Soma-plants, throw it away. But let him not do this; for the Soma being the nobility and the other plants the common people, and the people being the nobleman's food—it would be just as if one were to take hold of and pull out some food he has put in his mouth, and throw it away.[102]

99. L. Renou, RAI 60.
100. *Sat. Br.* ii. 5.2.6.
101. *Ibid* ii. 5.2.27
102. *Ibid* iii. 3.2.8.

Some ritual details are sought to yield the symbolic interpretation of what "make the Ksatra superior to the people. Hence the people here serve, from a lower position, the Ksatriya seated above them."[103] Similarly, other ritual details are interpreted to show how "the Ksatriya, whenever he likes, says, 'Hallo Vaisya, just bring to me what thou hast stored away.' Thus he both subdues him and obtains possession of anything he wishes by dint of this very energy."[104]

20. GENERAL THEORETICAL TEMPER

With this political philosophy in mind—which, incidentally is characteristic of the *Brahmana*-literature in general—let us turn to the question of the general theoretical temper of the priests. For our present purpose, it is not necessary to attempt any exhaustive survey of the question. Our purpose here is only to ascertain how far is it credible to accept the Vedic priests as the makers of mathematics embodied in the Sulva texts. For this purpose it is necessary to take note of their theoretical temper, because the most outstanding characteristic of the Sulva texts is the insistance on accuracy of observation and this as usually having moorings on manual operation. If we ignore or overlook this characteristic, little or nothing of genuine mathematical interest is left for us in the *Sulva-sutra*-s. Hence it is impossible to evade the question : How much of of the zeal for accuracy of observation can be imputed to the Vedic priests, so that they can be credited with the making of this mathematics ?

From what is already noted about the political philosophy sought to be validated in the *Satapatha Brahmana,* it should be obvious that direct observation—and specially the accuracy thereof—cannot be a strong point of Vedic priests, notwithstanding their insistance on the meticulous observance of ritual details. In fact, all this ritual details have one purpose and that is the creation of awe and wonder, which, as Eggeling rightly observes, had always been the technique for promoting hierarchical aspirations. As he comments, "even practical Romee" could not do without it : "The Roman statesmen sub-

103. *Ibid* i. 3.4.15.
104. *Ibid* i. 3.2.15.

mitted to these transparent tricks rather from considerations of political expediency than from religious scruples; and the Greek Polybius might well say that 'the strange and ponderous ceremonial of Roman religion was invented solely on account of the multitude which, as reason had no power over it, required to be ruled by signs and wonders."[105] The same is perhaps all the more true of the Vedic priests who seem to have the additional anxiety of impressing the rich patrons financing for expensive sacrifices that these assured for them all sorts of material prosperity, besides absolute political power of keeping the direct producers—primarily the peasants—complete control.

The key, therefore, to the general theoretical temper of the Vedic priests is the technique of creating awe and wonder, of marvel and mystery, supported at best by weierd analogy and palpably absurd symbolism. What goes completely against all this is the description of things as they are, which, in its turn, evidently enough, results from direct observation. From the priestly point of view, therefore, direct observation specially with an interest in precision is on the whole undesirable. What is desirable instead is a mystical veil on things, under the cover of which they can best operate as effective ideologues.

Interestingly enough, the priests whom we meet in the *Brahmana*-literature are aware of this and they repeatedly express it, though in their own way. One of their favourite formulas is: "the gods love the mystic"—*paroksa-priyah iva hi devah* or *paroksa-kamah iva hi devah.* It is true that in the *Brahmana*-literature this dictum occurs mainly in justification of all sorts of fanciful etymologies concocted by the priestly corporations. But we shall see that in the *Brhadaranyaka Upanisad* —which, according to the Vedic tradion, is directly appended to the *Satapatha Brahmana*—Yajnavalkya himself wants to he very clear and categorical about its generalised philosophical implications.

But let us first have some idea of how—and how frequently —this dictum proclaiming the divine love or divine sanction for purposive mystification occurs in the *Brahmana*-s. We quote here some of the passages from the *Satapatha Brahmana* and this in the course of the discussion of the altar-construction :

105. Eggeling SBE XII. Intro p. x.

Now the embryo which was inside was created as the foremost (*agri*) : inasmuch as it was created foremost (*agram*) of this All, therefore (it is called) Agri : Agri, indeed, is he whom they mystically call Agni ; for the gods love the mystic. And the tear (*asru*, n.) which had formed itself become the *'asru'* (m.) : *'asru'* indeed is what they mystically call *'asva'* (horse), for the gods love the mystic.[106]

This same vital air in the midst doubtless is Indra. He, by his power (*indriya*), kindled those (other) vital airs from the midst ; and inasmuch as he kindled (*indh*), he is the kindler (*indha*) : the kindler indeed—him they call 'Indra' mystically (esoterically), for the gods love the mystic.[107]

And as to why it is called 'Ukha' ;—by means of this sacred performance and this process the gods at that time dug out these worlds ; and inasmuch as they so dug out (*ut-khan*), it (the pan representing the worlds) is called *utkha,'—'utkha* being what they mysteriously (esoterically) call *'ukha',* for the gods love the mysterious.[108]

And, again, why he puts it on the lotus-leaf. When Indra had smitten Vrtra, he, thinking that he had not laid him low, entered the waters. He said to them, 'I am afraid : make ye a stronghold for me' ! Now essence of the waters there was that they gathered upwards (on the surface), and made it a stronghold for him ; and because they made (*kar*) a stronghold (*puh*) for him, therefore it is *'puskara' ; 'puskara'* being what is mystically called *'puskara'* (lotus-leaf), for the gods love the mystic.[109]

He said, Verily this one has lifted me from out of all evil ; and because he said 'he has lifted me ont (*udabharshit*)', hence (the name) *'udumbhara' ;—'udumbhara'* doubtless being what is mystically called Udumbara, for the gods love the mystic. 'Wide space (*uru*) shall it make (*karat*) for me'! he said, hence *'urukara'; 'urukara'* doubtless being what is mystically called *'ulukhala'* (the mortar) ; for the gods love the mystic.[110]

Like practically all arguments in the *Brahmana*-literature the etymologies suggested are weird and absurd and, in any case, the text itself does not want us to take these seriously from the

106. *Sat. Br.* vi. i.i. 11.
107. *Ibid* vi. i.i. 2.
108. *Ibid* vi. 7.1.23.
109. *Ibid* vii. 4.1.13.
110. *Ibid* vii. 5.1.22.

philological standpoint, because all these are intended only to cater to the divine taste for deliberate mystification after all. At the same time, it will be an error for the historian of science today to ignore or overlook the possible consequences of the dictum for the development of science and scientific temper among the priestly corporations. Among the Vedic scholars there is a tendency no doubt to read in the dictum some spiritual significance eluding a superficial understanding of it. But the great idealist philosopher of the *Brhadaranyaka Upanisad,* Yajnavalkya, wants in his own way to put a stop to such speculations. Evidently he feels that the dictum is in need of some explication. While, therefore, delivering a grand metaphysical discourse to Janaka, the king of Videha, he comes out with it and declares :

paroksa-priyah iva hi devah, pratyaksa-dvisah[111]—"the gods are fond of the mystic (or mysterious or obscure) ; *they detest direct knowledge.*"

The brief expression *pratyaksa-dvisah*—conveying the divine distaste for direct perception—can have no purpose here but to bring out the full philosophical implication of the brief priestly dictum. And this seems to speak volumes. If the gods themselves detest direct knowledge or perceptual knowledge, the mortals can develop any craze for meticulous observation only by flouting the divine.

It may be objected that even in the *Brhadaranyaka Upanisad,* the philosopher Yaynavalkya wants to retain the implication of the priestly dictum restricted to the etymological questions.[112]

But are we really justified in taking Yajnavalkya's reiteration of the priestly dictum with the explicatory addition *pratyaksa-dvisah* as intended to be restricted to mystical etymology, i.e. of rejecting the possibility of it being a pointer to the general theoretical climate of the priestly corporations and their philosopher-successors ? The answer seems to be clearly in the negative. The entire structure of the grand world-denying metaphysics propounded by Yajnavalkya is, in an important sense,

111. *Br. Up.* iv. 2.2.
112. Samkara's commentary on the passage may give us such an impression.

based on the dictum of detesting direct perception, which, if allowed any validity, runs the risk of taking the material things of the world on their face-value or as real materials things. Here is only one example of how his philosophy depends on the censorship of direct knowledge. In the *Brhadaranyaka Upanisad* Yajnavalkya declares :[113]

> Verily, where there seems to be another, there the one might see the other ; the one might smell the other ; the one might taste the other ; the one might speak to the other ; the one might hear the other ; the one might think of the other ; the one might touch the other ; the one might know the other. An ocean, a seer alone without duality becomes he whose world is Brahman... This is a man's highest path. This is his highest achievement. This is his highest world. This is his highest bliss. On a part of just this bliss other creatures have their living.

We shall later have the occasion to discuss this philosophy in some detail. But we have already seen that *acarya* Prafulla Chandra Ray as a working scientist could see that the power and popularity of this philosophy had been an important factor accounting for the decline of the scientific spirit in India. For the present, let us return to the point we have been trying to drive at.

We have now some idea of the general theoretical temper of the Vedic priests not only from the political philosophy which they want mystically to validate but also from their oft-repeated dictum of the divine fondness for the mystic or the deliberately obscure and, therefore, by implication their distaste for direct observation. Before hastily assuming that these same priests were the actual makers of mathematics embodied in the Sulva texts, we cannot thus evade the obvious question : How far such a general theoretical temper can at all be conducive to the making of this mathematics ?

There seems to be only one answer to this question. And that is in the negative. All that was really life-giving to the Sulva mathematics must have been accuracy of observation and this as frankly connected with manual operation. Therefore, notwithstanding the fact that the *Sulva-sutra*-s come down to us as appended to the priestly manuals, there are considerations that make us very strongly hesitate to follow the gene-

113. *Br. Up.,* iv. 3.31-32.

rally accepted view that the Vedic priests were themselves the makers of this mathematics. The most overriding consideration for this hesitation is of course the fact that this mathematics presupposes a highly sophisticated brick technology, which, archaeologically speaking, is just inconceivable among the Vedic priests. To this is to be added the evidence of the general theoretical temper of these priests, which is incompatible with that which imparts real significance to this mathematics.

21. UNSOLVED PROBLEMS AND POINTERS TO FURTHER RESEARCH

Yet the mathematics in the Sulva texts is a fact, and, as a fact requires some explanation from the historian of science. We have failed, because of reasons explained, to subscribe to the facile view that it was created by the Vedic priests to meet the requirements of their sacrificial ritual. Hence we are required to look for some alternative explanation for its origin. At the present stage of research, however, such an attempt leads us at best to the realm of conjectures. This, in other words, means that the problem is in need of further investigation and research, and the conjecture we shall venture would be ventured only with the hope of drawing the attention of the more competent historians of science to its general plausibility, At the same time, we shall be obliged to mention the number of yet unsolved problems that such a conjecture ushers in.

The conjecture we are drawn to by the data available so far is as follows. Though codified at a much later period, the mathematics in the Sulva texts came down from a very ancient period and one is tempted to presume that it could be the period of the First Urbanization. The strongest point in favour of such a presumption is the imposing brick technology of the Harappan Culture, i.e. what still survives of it in spite of the well-known accounts of brick-robbery in a big scale. The mathematics in the Sulva texts, as we have repeatedly asserted, is inconceivable without the tradition of highly sophisticated brick technology which, from the archaeologists' viewpoint, is absent throughout the period of the *Yajurveda, Satapatha Brahmana* and the *Kalpa-sutra-s*. Even admitting a very late date of the codification of the *Sulva-sutra*-s (which are appended to the *Kalpa-sutra*-s), it is hard to believe that this could happen much later than *c.* 300 B.C., i.e. what the archaeologists call the second phase of the NBP period when brick technology is reintroduced in some scale in Indian history. But the sophistication of brick-technology, the development of geametrical propositions wanted to meet its theoretical needs, the evolution of a whole host of terminologies to meet the requirements of this mathematics—all these and many other characteristics of the Sulva mathematics which still amaze us—evidently presuppose a very long time of development. That could perhaps be a number of centuries, though how many centuries remains anybody's guess. But it seems impossible to bring down the date of the Sulva texts many centuries later than the 300 B.C. Therefore, the only tradition of sophisticated brick-technology ante-dating the Sulva texts, as far as our knowledge goes today, is that of the Harappan culture. Since the Sulva mathematics is inconceivable without the tradition of a sophisticated brick-technology, it is difficult to reject outright the presumption that it developed during the period of First Urbanization though, as *somehow* transmitted to a much later period, it was codified in the form of the *Sulva-sutra*-s. We have emphasised the word *somehow,* because at the present stage of our research we have no knowledge of *how* exactly this could be transmitted. From the standpoint of our presumption, this therefore is a matter of further research and investigation.

In this connection, two points may not be irrelevant. We have in the *Satapatha Brahmana* and also elsewhere in the Vedic literature a list of the succession of teachers (*vamsa*-s). Max Muller[114] and other specialists in Vedic literature are, generally speaking, not inclined to look at the list as purely fictitious. The list is a long one and it wants us to take back to a very ancient period or a remote antiquity. Modern Vedic scholarship is yet to be clear about the exact implication of the list of succession of teachers. Here, therefore, is a point that may be further investigated from the viewpoint of our presumption.

Secondly, there seems to be a growing interest among our archaeologists to explore and excavate the sites of post-Harappan culture and they are giving us more information about

114. Max Muller HSL. 438 *Seq. cf.* also Eggeling SBE, XII, intro. pp. xxxiii f.

what is called the overlap of late Harappan Culture with cultures succeeding it. What is so far achieved by our archaeologists in this direction may not be very imposing. But the direction itself appears to be promising and it discourages the tendency to think that round about 1750 B.C. the Harappan Culture came to a complete extinction, leaving later Indian culture to inherit nothing from it.

It seems unfortunate that A. Ghosee, one of our ablest archaeologists, in the last chapter of his otherwise brilliant book *The City in Early Historial India,* appears to give us such an impression. Criticising specially the view expressed by S. R. Rao in *Excavation at Rangpur and other explorations in Gujarat* (*Ancient India,* 18 & 19, 1963 pp. 4-207), Ghosh[115] observes :

> It is now certain that there was a wide gap of seven to nine hundred years between the disappearance of these chalcolithic cultures and the emergence of the historical period in the regions of their occurrence, so that the chances of the former having anything to do with the letter are extremely remote. And even more remote, actually non-existent, is the likelihood of any Harappan urban tradition filtering through them into northern India, where the historical cities sprang up not earlier than 600 B.C.

Again[116] :

> Archaeologists and anthropologists must recognize the phenomenon of the loss of ancient cultures. For example, what does survive of the Sumerian civilization in present-day Iraq, or of the dynastic civilization of the Nile in modern Egypt, or of the Sabaean-Himyaritic civilization in the contemporary life of south-Arabian peninsula ? In every civilization there is a point from which cultural traits can be traced onwards down to contemporary times, and so far as the Indian civilization is concerned, the Indus civilization is not that point.

Hence is his conclusion : "In building their cities in the early historical period the people were writing on a clean slate, with no Harappan nor any other mark on it."[117]

All this if intended to mean that the Harappan Culture came to an absolute dead end—or to something like an absolute zero

115. A. Ghosh, CEHI 77.
116. *Ibid* 85.
117. *Ibid* 89.

—with the final decline or destruction of its imposing cities, would hardly be accepted by most of the serious archaeologists. On the contrary, beginning from the days of R. P. Chanda and John Marshall, we have very significant light being thrown on later Indian religion by the material relics unearthed at the Indus sites, and, as for ancient Egypt and Mesopotamia we have before us the writings of Gordon Childe showing how the scientific achievements of these two other primary centres of "the urban revolution" are inherited and absorbed by later European culture. There seems to be no *prima facie* ground to argue that the third primary centre of "the urban revolution" was bound to be an exception to this. What is true, of course, about the observations of Ghosh just quoted is that there is a gap *in our knowledge* about the content and mode of survial of the Harappan achievements in later Indian culture; but this present *ignorance* is only *our ignorance* and to draw any positive conclusion from it about what objectively happened in Indian history would be no more logical than to argue that our failure so far to decipher the Indus script indicates that the Indus peoples themselves were pre-literate. Incidentally, it must be mentioned here that A. Ghosh himself, in some of his other writings—specially in his editorials in *Ancient India*, Nos. 10 and 11—does not give the impression of totally negating the Harappan "survivals" in later Indian culture.

We have argued all these in order to emphasise only one point. In default of any better hypothesis of the making of mathematics in ancient India, the presumption based on certain circumstantial evidences that it might have taken place in the ancient Harappan period cannot be summarily rejected, though how it came down to the later Vedic priests and could be adopted by them to add a new awe-inspiring dimension to their sacrificial ritual remains a matter of further research.

Such a presumption evidently takes it for granted that mathematics—and, for that matter, quite a sophisticated form of it—did develop in the Harappan culture. This, we are going to argue in the next chapter, is not an undue assumption. We propose to wind up the present discussion only with one point.

22. FUTURE OF SULVA MATHEMATICS

The mathematics of the Sulva texts, it is only reasonable to think, could not have come into being abruptly. The presumption, on the contrary, is that it must have taken a long time—how long we do not know—to reach the stage of development it did. This, in other words, means that we have to infer some process of growth or development in the making of mathemetics in ancient India—a growth in which the theoreticians had largely to draw on the active experience of the technicians—specially of the brick-makers, brick-layers, architects and engineers.

What is apparently strange, however, is that historically speaking this mathematics came to an abrupt end. The mathematics of the *Sulva-sutra*-s—particularly the geometrical aspect of it—had really no future. Instead of developing, it simply withered away, leaving practically no legacy for later mathematical acivities in India, except perhaps, as Thibaut shows,[118] in the survival of some stray Sulva terminology—like *varga* and *karani*—though with new arithmetical-algebrical significance infused into them and this by pushing out the original geometrical interest of the Sulva texts. In any case, the positively geometrical interest of the Sulva mathematics lacks continuity of further development. As Thibaut puts it, "Clumsy and ungainly as these old *sutra*-s undoubtedly are, they have at least the advantage of dealing with geometrical operations in really geometrical terms, and are in this point superior to the treatment of geometrical questions which we find in the *Lilavati* and similar works."[119]

It is generally assumed that in the history of mathematical activity in India, after the Sulva texts the next mathematical work found so far is the *Bakshali Manuscript*—so called because it was discovered in a village called Bakshali (near Peshwar). Hoernle, to whose pain-staking work for years we owe a systematic restoration of the text, wants to put it in the third or fourth century A.D.—a date on the whole endorsed by serious schools like Buhler, Datta and others. Admitting this we are to place it in the advanced stage of Second Urbanization, with a gap of many centuries from the Sulva texts. In any case there is absolutely nothing to show any vestige of Sulva mathematics in this work. Found near the crossing of many important trade-routes of the time, the contents of this

118. Thibaut in SHSI II. 476 ff.
119. *Ibid* 472.

work take interest in mathematics mainly in the sense in which the commercial communities care for their calculations. But more of this work later.

The real landmark of mathematical work in Indian history after this is the *Aryabhatiya*—the work of Aryabhata I who was born in A.D. 476. Though really a work on astronomy, it takes keen and skilled interest in mathematics, specially in its second part called *Ganitapada*. The reason for its interest in mathematics is quite obvious. Precise astronomical calculations are impossible without precise mathematical knowledge. In Aryabhata's work, we indeed come across a renewed interest in geometry. But this is geometry not in the sense in which we have it in the Sulva texts, for the simple reason that in the *Aryabhatiya* geometry moves away from the earth to the sky, i.e. is concerned with the positions, movements, etc. of the heavenly bodies. In short, Aryabhata shows hardly any legacy of the Sulva geometry presumably because by his time it had already become a dead curiosity of the past.

There were, of course, a considerable number of commentators on the Sulva texts. However, strangely enough, as Thibaut shows, they often misunderstand or misinterpret the mathematics of the Sulva texts. "Trustworthy guides as they are in the greater number of cases, their tendency of sacrificing geometrical construction to numerical calculations, their excessive fondness as it might be styled of doing sums renders them sometimes entirely misleading."[120] The most important reason for this, as Thibaut wants us to understand it, is that "they represent the later development of Indian mathematics."[121]

This last point is very clearly illustrateed by B. B. Datta. Rama, a commentator on *Katyayana Sulva-sutra,* quotes from the work of the later mathematician Sridhara (c. A.D. 750);[122] Sivadasa, a commentator on *Manava Sulva-sutra* quoted the second Bhaskara (*c.* A.D. 1150) ;[123] Karavindasvami, a commentator on *Apastamba Sulva-sutra,* quoted certain passages

120. *Ibid* 473-74.
121. *Ibid* 473.
122. B. B. Datta, SS, 11.
123. *Ibid* 12.

from *Aryabhatiya* ;[124] so did Dvarikanatha in his commentary on the *Baudhayana Sulva-sutra.*[125]

What else could the commentators do than go in for later predominantly algebric methods of demonstration when the genuinely geometrical promise of the Sulva texts came to some kind of dead end or, at any rate, failed to develop furher ?

Here, then, we have another problem which the historian of science in India cannot afford to ignore or overlook. Why was it that the Sulva geometry failed to have any further development ?

From the viewpoint of our own presumption the answer to this is not so difficult. From the circumstannce of this mathematics being inextricably related to brick technology—and failing to find any tradition of sophisticated brick technology antedating the Sulva texts except in Harappan Culure—we have presumed that this mathematics could have developed in Harappan Culture, notwithstanding there being various unsolved problems at the present stage of our research specially about its transmission to the later period. We have further presumed that in this later period, the Vedic priests wanted somehow to utilize this brick-technology perhaps from the stragglers of Harappan technicians, employing them for the purpose of building awe-inspiring sacrifical altars ; along with this brick-technology the mathematics—which was the outcome of the theoretical requirement of the technology—could have come down to the Vedic priests which they codified in the *Sulva-sutra*-s.

One advantage of this presumption, however, seems to be that it may throw some light on the apparent peculiarity of this mathematics having once a period of growth or creative development and yet having practically no future in the subsequent history of mathematical activities in India. This follows from the socio-economic function of the brick technology in the Harappan period and this as contrasted with the later Vedic period.

In the ancient Harappan period, the technology of making and using burnt bricks did have a very positive socio-economic function. Apart from being essential for defence against floods

124. *Ibid* 16.
125. *Ibid* 18.

—to which, as we have already seen the Allchins have very legitimately drawn our attention—practically everything that appears to be so spectacular about the Harappan Culture largely depended on brick technology : the houses, the citadels, the granaries, the drain-system, the dock-yards are just a few examples that should substantiate our point. The socio-economic function of brick technology in Harappan Culture is in fact so obvious that it requires not much elaboration. When a technology serves such a positive socio-economic function the theoretical requirements of it—in the present case, the mathematical knowledge required by it—acquires nourishment for creative development.

But what is the function of the same technology—or its survivals—as shifted to the later Vedic period ? Its socio-economic function is not only just nil but definitely negative—because it is now harnessed to total economic waste—the building of some complicated structures for assuring the patrons financing the sacrifice that these would magically ensure for them the fulfilment of their desires like gaining cattle and food, annihilation of foes and quick transit to heaven. Beyond creating such apparently absurd and false assurances for the *yajamana*-s —and, incidentally, validating in the course of the performance of the rituals the despotic power of the "nobles" of the age —brick structures of the Vedic priests had no function at all, or the only other function that these had was ensuring for the parasitical priest class their livelihood in the from of *daksina*-s or sacrificial fees which, according to the Dharmasastra norm, was about the only legitimate source of income for them. Harnessed to such total economic drainage, neither the brick technology nor the theoretical outcome of it—namely geometry—could have any future. And the fact is that these had none in the Vedic tradition itself.

C. G. Kashikar argues, of course, that it is wrong to think that bricks were used during the comparatively later Vedic period exclusively for ritual purposes. "It cannot be said," he observes, "that the bricks which then presumably formed an essential material for house-building, were employed only for the ritual purpose."[126] But what is the basis of this as-

126. C. G. Kashikar, in *ABORI* 1981, 126

sumption that bricks were then used for house-building? Surely not any archaeological evidence, because the archaeologists have so far not been able to find any relic of brick-built house in the Vedic settlements proper usually assumed to be the PGW sites. In default of archaeological evidence, Kasikar has recourse only to literary evidence. But as we have already seen, the only literary evidence he manages to cite is as follows: "Bricks as house-building material are mentioned in Kesava's *Paddhati*."[127] Though claiming to belong to the Vedic tradition—for it is a commentary on a *Srauta-sutra*—the date of the *Paddhati*, according to Kashikar himself, is 13th century A.D. Quoting the evidence of such a late text in substantiation of what is alleged to be true in a period anterior to the fifth or sixth century B.C. seems hardly to be a way of writing history, specially when the thesis built on it flouts an overwhelming mass of anchaeological data.

127. *Ibid* 126 n.

CHAPTER 7

SCIENCE IN FIRST URBANIZATION

1. PRELIMINARY REMARKS: MATHEMATICS

The difficulties in discussing the mathematical achievements during the period of Frst Urbanization, though well-known, may be briefly recapitulated. We have no literary evidence testifying to this—nothing comparable to what is available in the other two primary centres of "the urban revolution", like the clay tablets of Mesopotamina or the papyrii of Egypt.

It is difficult, of course, to reject outright Childe's conjecture of possible mathematical documents of the Harappan Culture being lost with the unknown perishable materials on which these could have been recorded, specially because of his two arguments. First, the Harappan script, which survives for us mainly on the "seals", could not possibly have been invented just for the purpose of inscribing on these; there are therefore grounds to presume that the script was invented to meet some other requirements of the urban centres. Secondly, on the analogy of Mesopotamia and Crete, it is not *prima facie* absurd to think that the main impetus leading to the invention of the script in the urban centres of the Harappan Culture could be accounts keeping. This must have been required not only for the city administrators, say in charge of the huge granaries from which were catered the basic needs of the whole-time urban specialists and others who were themselves not direct producers, but also for the traders of merchants of the Harappan cities, about whose existence there can be no doubt. At the same time, there is a limit beyond which a hypothesis can be built up merely on the evidence of the unknown. In any case, even admitting the possibility of mathematical documents of Harappan Culture perishing with the unknown materials on which these might have been documented, it is impossible for us today to form any definite and systematic idea of the nature of the mathematical activities in the period of First Urbanization.

But all this does not at all mean that we are totally debarred from forming any idea whatsoever about the possible mathematical activities in the Harappan Culture. Though without any direct literary evidence, there are a number of indisputable archaeological data which want us very strongly to *presume* that there had been considerable mathematical activity in the Harappan Culture ; we have even very reasonable grounds to infer the nature of the mathematical activities, though at the present stage of research, only in scrappy outlines and perhaps somewhat tentatively. Fortunately, some highly competent scholars have already applied their minds to the subject. They include V. B. Mainkar, M. N. Deshpande

2. THE ARCHAEOLOGICAL DATA

and others, on whom we shall freely draw.

What, then, are these archaeological data ?

To begin with, a number of actual mathematical instruments are found among the ruins of the Harappan sites. Deshpande[1] gives the following list of these :

1. A neatly finished piece of shell from Mohenjo-daro showing in its extant form nine divisions ; it is unanimously accepted by the archaeologists as the remain of a measuring scale.
2. A bronze rod from Harappa, also viewed as the remain of another measuring scale.
3. A small measuring rod of ivory from Lothal.
4. 'A peculiar object also from Lothal was probably used as a compus for measuring angles.'
5. "Terracotta plumb-bobs of different sizes from Lothal with or without vertical rods are also reported from Mohenjo-daro and Harappa."
6. "A graduated scale intended perhaps for measuring has also been reported from Kalibangan."

Before passing on to discuss these in some detail, it is necessary to add that these are not the only archaeological data for inferring mathematical activities in the Indus Valley Civilization. To the above list are to be added the following :—

1. Units of weights uniformly in circulation throughout the vast area covered by the Harappan Culture. We are indebted to Mainkar for a brilliant systematisation of these along with the mathematical deductions from their evidences.
2. Standardization of the brick-measures and the geometrical principles followed in their making, to which may be added

1. M. N. Deshpande, IJHS, VI, 1. 10-11.

the geometrical principles usually followed by the "seal"-cutters.

3. The monumental remains of fortifications, granaries, public-baths, roads, house-blocks excavated at Harappa, Mohenjo-daro and recently also at Kalibangan and the remains of a dockyard at Lothal. Geometrical knowledge apart, we are obliged to deduce from these the knowledge of mathematics in other forms, like calculations of the number of workers, engineers, architects required for their construction and hence the amount of food etc. necessary for their subsistence.

4. Certain designs on Harappan potteries, like "squares inscribed in compus-drawn intersecting circles" appear to be pointers to the theory and practice of geometry.

Thus, notwithstanding the fact that no written document testifying to the mathematical activities of the Harappans is available for us, we have many evidences that remain inexplicable without the admission of mathematical activities on their part. It remains for us to see if any light is possibly thrown by what we propose to call "the method of retrospective probing", *i.e.* deductions from the mathematical documents of later times which are evidently in need of the assumption of antecedent historical development.

3. THE SCALES OF LENGTH MEASURE

We begin with a brief account of the scales or instruments for measuring length, fragmentary remains of which are so far recovered from the ruined sites of the Harappan Culture.

These, as we have already said, are three. Of these, one was first reported by Mackay in 1938 in his *Further Excavations at Mohenjo-daro*[2], the second first reported by Vats in 1953 in his *Excavations at Harappa*[3] and the third first reported in 1973 by Rao.[4]

2. Mackay FEM, Pl. cvi. 30 ; also pp. 404-5.
3. Vats EH Pl. CXXV, 39a and p. 365.
4. S. R. Rao, in IAR, 59-60, pl. xiii. b., also p. 17. cf. also S. R. Rao, LIC 105 : the one "found at Lothal is said to be more accurate, as the divisions marked on it (Fig. 28, pl xxxii A) are smaller than those marked on the scale from Mohenjo-daro. A graduated scale is reported from Kalibangan too, but details are not known".

According to the sites in which these are found, therefore, these three scales may be referred to as the Mohenjo-daro-scale, the Harappa-scale and the Lothal-scale. The physical descriptions of these scales, though oft-repeated in recent archaeological literature, may briefly be recapitulated.

The Mohenjo-daro-scale is made of shell, the Harappa-scale of bronze and the Lothal one of ivory. The selection of the materials has its own interest, not merely because of their durability but also of their comparative resistance to easy contraction or expansion due to temperature variation. Much more important for our own discussion, however, is the mathematical aspects of the scales—the unis of linear measures suggested by these.

The Mohenjo-daro scale first[5] :

> It has nine graduations and there is a hollow circle on one graduation, while on the fifth graduation therefrom, there is a large circular dot. This sequence would indicate that a hollow circle would occur five graduations after the dot and so on. The graduation lines are made with a fine instrument and are uniformly thick and properly graded in length. The distance between two adjacent lines is, on an average, 6.7056 mm. The accuracy of the graduations is very high, as the mean error of graduation is 0.075 mm. Thus the length contained between five graduations is 33.528 mm and the length between one hollow circle and the next would be 67.056 mm. This length of 67.056 mm apparently constitutes the major graduation of the scale. Since the sub-division of the major graduation of the scale is decimal, we may reasonably assume that the original scale had ten major graduations on it. This assumption leads us to the conclusion that the length of the total scale could be 670.56 mm. This is a very interesting result, as the total length of the scale is practically equal to two-thirds of a metre. It is also noteworthey that the length-scale appears to be decimally divided like the metre.

We shall presently see that the evidence of the series of weights also found in the Harappan Culture corroborates the decimal system having been current there. But let us first note the system of length measures as evidenced by the scales. Mackay, to whom we owe the first report on the Mohenjo-daro scale, also drew our attention to the decimal system on which it was fashioned, and observed :[6]

5. Mainkar in FIC 146.
6. Mackay FEM, 404-5.

The decimal system of liner-measure is known in Egypt as early as the Fourth Dynasty, and a decimal division of the cubit in the Twelfth Dynasty has been noted by Kahun. Both the decimal and the sexagesimal systems were in use in early Sumer, though it is not yet known which came first. According to Langdon, both systems were in use in Jemet Nasr; and on the Fara tablets, also, which must be dated to the Early Dynastic period, the two systems were used. We are told, however, that a purely decimal system is found on the Proto-Elamite tablets; and it may be that it was from Elam that the system was introduced into N.W. India, though on the basis that every man has ten fingers it seems to me that the decimal system should be more primitive than the sexagesimal, and that it may have had independent origins.

In view of the trade relations between the Harappans and Mesopotamians, diffusion of scientific ideas need not be necessarily ruled out, though in the absence of any positive evidence in favour of it, it seems safer not to indulge in conjectures. In any case, the systematic and uniform use of the decimal system in the Harappan Culture—about which we shall presently see more—is a prominent feature of the period of First Urbanization and it could have interesting consequences in the comparatively later period of Indian culture. For the present, however, let us pass on to the other scales found among the Indus ruins.

S. R. Rao, as we have already noted, has reported a truncated piece of ivory scale at Lothal:[7]

The Lothal scale is 15 mm broad, 6 mm thick and has a length of 128 mm. Only 27 graduation lines can be easily seen, the length over which these lines are spread being 46 mm. The average distance between the graduation lines is, therefore, 1.704 mm. The sixth and the twentyfirst graduation lines are longer than the rest. The length between these two graduation lines is, by calculation, 25.56 mm.

Is it, then, indicative of some different unit of length measure than was current in Mohenjo-daro? Mainkar[8] has answered this question and we may quote him at some length:

The Mohenjo-daro scale and the Lothal scale are apparently different but on analysis they prove to be practically equal. It will be noticed that 20 divisions of Lothal scale are equal to 34 mm, which is almost equal to the distance between the hollow circle and

7. Mainkar in FIC 146.
8. *Ibid.*

the circular dot on the Mohenjo-daro scale, namely 33.53 mm. This fact establishes that the two scales are related, but their division into smaller graduations are different. The smaller graduations of the Mohenjo-daro scale are at every 6.706 mm, while those of the Lothal scale are at 1.7 mm, the ratio being 4 Lothal graduations are equal to one graduation of the Mohenjo-daro scale. The smallness of the Lothal scale-graduation indicates that it was used for finer measurements.

One example of the need for such finer measurement is suggested by S. R. Rao: "The Indus seals can be measured more accurately in terms of the smaller divisions of the Lothal scale than in terms of the divisions of the Mohenjo-daro scale."[9] "This", observes Mainkar, "is a very plausible and satisfying explanation."[10]

Much more important for our immediate discussion are two other points sought to be established by Mainkar. These are: (1) His view of the possible use of the Indus Valley scales of linear measurement in brick-making and architectural techniques in the period of First Urbanization, and (2) His attempt to correlate the units of length measures with the same in later Indian history.

Before passing on to these—particularly the first of the two points—we may as well note the very legitimate caution with which he begins this discussion:[11]

> While comparing the theoretical and actual length of linear measures, it should be borne in mind that, even under modern conditions, it requires very costly machinery to draw lines to a thickness of 0.050 mm. Further, it is difficult to read, with the naked eye, graduation lines separated by a distance of less than half-a-milimetre. It is, therefore, necessary to understand that when the calculated length of say, 68 mm, is equated with an actual distance of, say 67 mm, we may not be introducing a substantial error in our argument. Such equations have always to be treated as guidelines rather than strict mathematical formulations.

With this caution in mind, let us see how Mainkar wants to show that "the three scales discovered in three widely separated centres of the Indus Civilization are interrelated, and can

9. S. R. Rao, in LIC 107.
10. Mainkar in FIC 146.
11. *Ibid.*

explain statisfactorily the dimensions of the bricks, the baths, the dock and the like."[12]

The bricks first. Mainkar considers the following five "nominal sizes" of the bricks found in the different centres of the Indus Civilization :

1. Mohenjo-daro : 225×115×57 mm.
2. Lothal and Mohenjo-daro : 250×125×60 mm.
3. Lothal : 280×140×65 mm.
4. Kalibangan and Mohenjo-daro : 300×150×75 mm.
5. Kalibangan : 400×200×100 mm.[13]

We have already noted that the standardization of the ratio of the three dimensions of the bricks—the length, breadth and thickness—as 4 : 2 : 1 is extremely significant from the view-point of efficient bonding, and it reminds us of the ratio of the sides of the standard bricks being used in our times. For the present let us note Mainkar's analysis of how the determination of the brick-sizes mentioned above appear to be indicative of the use of the Indus Valley scales. As he observes :[14]

It is obvious that the longest side of the Mohenjo-daro brick, namely 225 mm. is nearly 9×25.6 mm. i.e. 9 times the major graduation of the Lothal scale. The longest side of the Lothal brick, namely 250 mm, is equal, within limits of error to 150 small graduations of the Lothal scale (1.7 mm. of each) or 10 major graduations of 25.56 mm. each, while the longest side of the larger brick, namely 280 mm is made by the addition of one major graduation or 15 small graduations on the Lothal scale, i.e. 11 large graduations (25.56×11 mm).

The brick with 300 mm side is a further extension of the above principle and comes to 180 Lothal small graduations or 12 large Lothal graduations (25.56×12 mm). In like manner the 400 mm brick would appear to be equal to 16 larger divisions of the Lothal scale (25.56×16 mm).

This analysis shows that perhaps, bricks were made in dimensions which were integral multiples of large graduations of the Lothal scale, namely 25.56 mm. The other dimensions of the bricks, being in the ratio of $1 : {}^1/_2 : {}^1/_4$ also fall into a rational number of Lothal graduations.

12. *Ibid* 147.
13. *Ibid.*
14. *Ibid.*

Admitting this, we have to accept that brick-making according to the application of some definite scale which we come across practically throughout the *Sulva-sutra*-s is indicative of a very ancient tradition, inasmuch as this tradition goes back to the period of the First Urbanization. This, in other words, means that the relation of mathematical calculation with brick-technology has a hoary past.

In view of the large number of brick-types mentioned in the Sulva texts each with very specific measurement in terms of the units of linear measures accepted by the texts, it would be a laborious process to try to assess the measurements of the Sulva bricks in terms of the scales of the Harappan Culture. Besides, that is not necessary for our main argument, namely that it is not *prima facie* impossible to try to trace the tradition of the application of mathematical calculation to brick technology to the ancient Harappan Culture. This tradition, once accepted, may explain the meticulous care taken by the brick-makers of our Sulva texts to be specific or accurate about the measures of the brick-types, a large number of which they had to improvise in order to meet the requirements of the peculiar structures they were asked to execute. Incidentally, this technique of improvising new and newer brick-types, too, could have its roots in the Harappan culture, where, apart from the standardised bricks, we also meet with various other brick-types, like the T-shaped one assumed as needed for covering the drains and the wedge-shaped bricks used for the construction of wells, drains or the grinding floor of the granaries.

But there is another point of considerable interest which may as well be noted in this connection. In spite of various conjectures, the fact remains that we have no definite knowledge of the language of the peoples in the Indus Valley Civilization. It is, therefore, futile to speculate on the possible terminologies used by the Harappan peoples for the units of length measures. In the history of Indian culture, the earliest evidences for such terminologies are to be found in the *Sulva-sutra*-s and *Arthasastra*. In both, the basic unit for length measure is called an *angula*, literally 'the finger'. For the sake of precision, however, the *Arthasastra* defines it as "the maximum width of the middle (part) of the middle finger of a middling man."[15] Whether the unit *angula* of the *Arthasastra* is exactly the same as understood in the *Sulva-sutra*-s may be open to some discussion, for the *Arthasastra*[16] proposes to measure it in terms of eight *yavamadhya*-s (the width of the middle of eight *yava*-s) whereas the *Baudhayana Sulva-sutra*[17] conceives it in terms of fourteen grains of the *anu* plant (understood by Thibaut as *panicum miliaceum*). But we may note here one point of some interest. According to both the texts[18] the longer unit called *aratni* (losely translated as 'cubit' by Kangle) is conceived in terms of 24 *angula*-s and it is also the same according to Yallaya's explanation of Aryabhata[19] though the latter uses the word *hasta* instead of *aratni* (literally the length from the elbow to the tip of the little finger). In any case, the fact is that the term *angula* stands for the basic unit of length measure in later literature, inclusive of the Sulva texts and there are at least some hints suggesting correlation between the *angula* of the *Baudhayana Sulva-sutra* and of the *Arthasastra* as well as of much later astronomical works.

Earlier writers like J. F. Fleet[20] were satisfied by roughly equating the *angula* to 3|4th of an inch, which makes it 19.499 mm. On the basis of a more meticulous calculation, however, Mainkar equates the length of the *Arthasastra angula* to 17.78 mm. This gives a very interesting clue to correlate the basic unit of later linear measure, viz. the *angula,* to the length measure of the Lothal scale. As Mainkar[21] puts it :

> The author has shown, (in his articles) tracing the development of length and area-measures in India, that the *angula* which is the basic unit of length measures, mentioned in the *Arthasastra,* is 17.78 mm. This value is so nearly equal to the value of ten small graduations of the Lothal scale (1.703×10 mm), that they may be considered as being practically equal. If this is accepted, and Rao agrees with it, the entire series of length-measures specified in the *Arthasastra* falls in a pattern with the Indus scales. The author has shown in his articles mentioned above, that the length-measures used in India throughout later periods were related in some manner or other, with the length-measures specified in the *Arthasastra.* It is,

15. *Arthasastra,* ii. 20.7.
16. *Ibid* ii. 20.5.
17. *Baudh Sul Su.* i. 4.
18. *Arthasastra,* ii. 20. 12 ; and *Baudh. Sul Su.* i. 16.
19. Shukla and Sarma, *Aryabhatiya,* intro xliii.
20. J. F. Fleet in JRAS, 1912. 233.

therefore, possible to assert that the Indus length-measures had a very profound influence on the length-measures used in India up to a few years back.

S. R. Rao wants to go a step further :[22]

> It appears that both 'foot' and 'cubit' were treated as units for linear measures. The 'foot' is said to be of 13.2 ins. (33.5 cms.) and the 'cubit' varying between 20.3 and 20.8 ins. (51.5 and 52.8 cms). The houses in Lothal can be measured in terms of complete units of 'foot', e.g. House No. 159 (phase IV A) measures 40×20 units, and warehouse 117×123 units, the unit in each case being 13.2 ins.

But before passing on to see more of the application of mathematics to the brick-structures of the Harappan Culture, we may ask ourselves a simple question. Could it be that the correlation of the *angula* of the later texts inclusive of the Sulva-s with the linear measure of the Indus scales be itself an indication that wants us to seek the roots of the Sulva mathematics in the mathematical activities in the First Urbanization?

4. BRICK-TECHNOLOGY AND MATHEMATICS IN FIRST URBANIZATION

While analysing the *Sulva-sutra*-s we were led to the view that the mathematics codified in these texts is inconceivable without the tradition of highly sophisticated brick-technology. The texts give us the impression that this mathematical knowledge was above all the outcome to meet the theoretical requirements of the brick-makers, brick-layers, architects and other technicians, who were required to execute the construction of certain specified forms of brick-structures. At the same time, we were confronted with an apparently anomalous situation. The texts cannot but be placed in a period which, archaeologically speaking, was unaware of any sophisticated brick-technology. Hence we were led to raise the question concerning the possible roots of this mathematics in the mathematical activities of the First Urbanization, one of the most conspicuous features of which had been highly sophisticated brick-technology. But the first point that requires to be established before answering the

21. Mainkar in FIC 147-48.
22. S. R. Rao, LIC 107.

question is that we have definite evidences indicating mathematical activities in Harappan Culture. We have just seen that the linear measures of the broken scales found among the ruins of the First Urbanization appear to foreshadow the basic unit of the linear measure assumed by the Sulva texts. We now pass on to discuss how far we are obliged to presume the application of mathematical knowledge from the remains of the brick-structures of the First Urbanization.

In default of anything directly documenting mathematical activities during the period of the First Urbanization, it is impossible of course for us to hope to have any systematic knowledge of Indus mathematics.

Nevertheless, it is important to note that serious archaeologists and other scholars have felt obliged to argue that the brick-structures of the First Urbanization remain unexplained without the assumption of the application of a good deal of mathematical knowledge.

Here is how M. N. Deshpande[23] puts the general argument :

> The monumental remains of fortifications, granaries, public-baths, roads and house-blocks excavated at Harappa, Mohenjo-Daro and recently at Kalibangan and remains of dockyard at Lothal imply a good deal of arithmetic and knowledge of geometry. Keeping of accounts for the construction of public buildings such as of labour and material would entail complicated calculations. Unfortunately, direct evidence of such accounting is not available. As regards the knowledge of geometry besides the few measuring rods and other instruments which have come to light... we have largely to depend for such deductions on the data supplied by the buildings themselves. It is obvious from the meticulous care the Harappans took in planning the city with well laid-out streets that they knew fundamentals of surveying. This would include knowledge of levelling as without detailed measurements it would not have been possible to plan the sewage system. The use of standardized bricks having plain rectangular faces, parallel sides, sharp, straight, right-angled edges including wedge-shaped bricks in the construction of circular wells so as to produce the inner and outer circumference would presuppose knowledge of geometry of parallels and circles.

Mainkar with his illuminating analysis of the correlation of the standard brick-sizes of the Indus ruins with linear mea-

23. M. N. Deshpande, in IJHS VI. No. 1. 9.

sures of the Indus scales, gives us some specific examples of the mathematical calculations implied by a number of the brick-structures of the First Urbanization. We quote him at some length :[24]

Turning to other sources of measurement the 'Great Bath' at Mohenjo-daro had average measurements of 11.89 m. × 7.01 m. with a depth of 2.44 m. Adjoining the 'Great Bath' were smaller rooms of 2.9 m. × 1.8 m. each. It is apparent that each of the smaller rooms had approximately ¼ of the average measurements of the Great Bath' (2.9×4=11.6 and 1.8×4=7.2). In their turn, the measurements of the smaller bathrooms are related to the bricks having 300 mm as the longest dimension. Thus, the longer side of the bathrooms is 10 times and the smaller side 6 times the longest side of the 300 mm brick. The longer side of the 'Great Bath' is equal to a length of 40 bricks and the smaller side to a length of 24 bricks the depth being equal to the length of 8 bricks.

Rao has recorded that the average measurements of the dock at Lothal are 214 m × 36 m, the foundations being 1.78 m wide and the walls above the ground 1.04 m. These measurements are interesting because 1.04 m is equal to 40 large graduations of Lothal scale (25.56 mm), while 1.78 m is equal to 1000 times the small graduations of (1.7 mm) of the Lothal scale or $66^2/_3$ large graduations of 25.56 mm. The major dimensions of the dock are in the proportion of 6 : 1. The dimension of 36 metres is 20 times the width of the foundations, namely 1.78 m. The latter value is in turn related to the small graduation of the Lothal scale by 1,000 times.

The width of the doors in the houses at Mahenjo-daro was about 1.02 m.; this is 40 times 25.56 mm. the large division of the Lothal scale or 15 times the large division of Mohenjo-daro scale, namely 67.06 mm.

There are numerous other dimensions and measurements which could be analysed. But, it is considered that the examination carried out so far is adequate for indicating that the three scales discovered in three widely separated centres of the Indus Civilization are inter-related, and can explain satisfactorily the dimensions of the bricks, the baths, the dock and the like.

All this may be taken as pointers to a very interesting line of research requiring the co-operation of archaeologists, architects and mathematicians for determining the nature of mathematical knowledge necessarily presupposed by the structural remains as well as by the decorative designs of the Indus relics. R. P. Kulkarni of the Maharashtra Enginnering Research

24. Mainkar in FIC, 147.

Institute, seems to be taking some step in this direction in his article *Geometry as known to the People of Indus Civilization.*[25] From the designs and figures on some "seals" corroborated by certain structural remains at Mohenjo-daro and Harappa, he argues that "the people of Indus Civilization might be knowing the following postulates and geometrical constructions" regarding "properties of rectilinear figures".[26]

"Postulates : (a) The diagonals of a square (or rectangle) divide it into equal areas. (b) The diagonals of a square (or rectangle) are of equal length. (c) The bisectors of the sides of the square cross each other at right angles.

"Constructions : (a) To divide a line of a given length into two equal parts. (b) To divide a line of a given length into any number of parts of equal length. (c) To divide a line of a given length into different lengths as required. (d) To construct a square|a rectangle of sides of given lengths and with accurate right angles. (e) Construction of dividing a given quare|rectangle into small squares|rectangles of equal areas. (f) To divide a rhombus (and, therefore, may be a triangle) into rhombuses (triangles) of equal area. (g) To draw two lines parallel to each other." Admitting all this, we are obviously reminded of some of the basic propositions discussed in the Sulva geometry.

Similarly, Kulkarni proposes to read the knowledge or the potentials thereof of "properties of a circle" and the "knowledge of area and volume of rectilinear figures" as forming part of the geometry as known to the people of the Indus civilization. Much of what he claims may appear to us as he himself admits, to be more or less conjectural[27] and hence may fail to create conviction of the mathematicians today. Besides, for the archaeologists, too, his article is likely to appear as but based on rather inadequate data, inasmuch as he depends for these mainly—even exclusively—on Marshall's *Mohenjo-daro and the Indus Civilization* and Mackay's *Early Indus Civilization*, showing no awareness at all of the spectacular results of the later archaeological works done at sites like Lothal,

25. R. P. Kulkarni IJHS, 1978, Vol. XIII No. 2. 117-124.
26. *Ibid* 119 ff.
27. *Ibid* 122.

Kalibangan etc. Such limitations of his article are, however, not to be misunderstood or misinterpreted as limitations of the basic direction of approach. On the contrary, it may be taken as a pointer to very fruitful research though requiring the collective efforts of working engineers, architects, mathematicians and of course the archaeologists. In any case, the evidence of the *Sulva-sutra*-s of the formation of geometry to meet theoretical requirements of the brick-technology cannot but invite us to explore the possibility of studying more intensively the geometrical knowledge of the Indus period, one of the most prominent features of which was very imposing brick-technology, though in default of any literary evidences it may be imposible for us to know if the mathematical activities during the period of First Urbanization was ever systematically codified as was done in the Sulva texts.

Incidentally the codification of the *Sulva* texts in their present form cannot but raise a number of highly intriguing questions. Even if we are allowed to postulate the codification of geometrical knowledge in the Harappan period on some unknown perishable materials, the fact remains that we do not yet know with certainty the language of the Harappans. In his recent work on *A Comprehensive Sulvasutra Word Index,* Axel Michaels gives us the impression that a considerable number of the technical words in these are apparently non-Sanskrit. Admitting this, we have no ground to argue that these non-Sanskrit words have roots in the Harappan language. Besides, a very large number of *Sulva* words are positively Sanskrit, as positive is the fact that the language and style of the composition of these works are Sanskritic. Since it will evidently be too much to assume that the *Sulva* texts were later adaptations by the Vedic priests of some unknown—or purely hypothetical—Harappan texts, the problem of the codification of the Sulva texts in their present form remains an open question. Perhaps only this much may be asserted that since the basic geometrical knowledge in these texts is inconceivable without the tradition of very sophisticated brick technology (which was there in the Harappan culture but totally absent in the period usually assigned to the *Sulva* texts), their mathematical core presumably came down from the Harappan past. Towards the beginning of the period of Second Urbanization some people belonging to the circle of the priestly elites somehow took an absorbing interest in this ancient mathematics and codified it in Sanskrit, though freely drawing on "loan words" specially in matters concerning technological and terminological. However, there are likely to be many objections intertwined with unanswered questions even against such a hypothesis. Hence it seems reasonable to leave the question of the codification of the *Sulva* texts as an open one, requiring a good deal of further investigation.

5. SYSTEM OF WEIGHTS

At the present stage of research, whatever we may reasonably presume about the geometrical knowledge and its systematisation in the Indus Civilization, there are also other grounds to infer mathematical minds working during the period of First Urbanization. We shall mention here one of these, namely the system of weights.

We are indebted to A. S. Hemmy, who, in his chapter on *System of Weights at Mohenjo-daro* in Marshall's *Mohenjo-daro and the Indus Civilization*, initiated the analysis of the subject. Later archaeological work specially at Lothal has revealed more data related to the subject, necessitating a revised understanding of the system of weights in the Indus Valley Civilization. We are indebted specially to Mainkar who, in his paper on *Metrology in the Indus Civilization*, has given us a systematic account of the system of weights in Harappan Culture based on up-to-date evidences. It may be enough for our own purpose to follow the discussions of Hemmy and Mainkar to have some idea of the mathematics underlying the system of weights in the period of First Urbanization.

Before proceeding to this mathematics, it may be useful to note two points. First, what cannot but amaze the archaeologists today is the uniformity of the system of weights that prevailed unchanged for over five hundred years throughout the vast area covered by the Harappan Culture—an area which, as we have already said, is roughly estimated at 500,000 square miles. "Such uniformity of weights over such vast spans of distance and time is a unique feature in the history of metrology."[28] Secondly, the objects used for weighing pur-

28. Mainkar in FIC 141.

poses are "rectangular blocks, mostly of a tawny or light grey banded chert, but also of other hard rocks such as gneiss. In one on two cases their form is cylindrical, but for the most part it is cubical. The blocks are well finished and polished, and are generally in a good state of preservation... The results obtained by weighing these blocks show conclusively that they are weights belonging to a definite system."[29] This, again, appears to be something remarkable, particularly when compared to the weights in the other two primary centres of the Urban Revolution. As Mainkar[30] observes :

> The available information (Berriman 1953) shows that there were many systems of weights and measures prevalent in Egypt, and there was no uniformity in them even remotely comparable with that noted in the Indus valley. The Sumerians also could not claim comparable uniformity in the field of weights and measures. Even the shapes of the weights used in the Egyptian and Sumerian civilizations were entirely different. Some Sumerian weights were made in the shape of animals. The shapes of the Indus weights were highly practical.

Analysing the weights found in Mohenjo-daro and Harappa, Hemmy prepared a Table of the Denomination of Weights, which suggests that the system followed was partly binary and partly decimal. In 1973, Rao discovered at Lothal another system of weights which when tabulated, appears to be binarily arranged. The classification of Indus weights into partly binary and partly decimal does not appear to be logical to Mainkar[31] who, therefore, argues as follows :

> The practical approach would appear to be to take all these weights together and divide them afresh into two or more logical groups. It is obvious, to any observer that the decimal system is primarily used in the progression of the Indus weights. The principle used for the progression of the denominations of the metric weights is to multiply and divide the decimal unit by two. For example, take the unit of 100 g. By multiplying it by two, we get 200 g. dividing it by two gives 50 g. If the same operation is repeated with other decimal units the series of denominations obtained is 1, 2, 5, 10, 20, 50, 100, 200, 500 and so on. Thus the weights form a series in the ratio of 1 : 2 : 5 and decimal multiples of each of the numbers in this

29. Hemmy in Marshall's MIC 589. Mackay FEM 601 ff. who adds the analysis of 220 weights found by subsequent excavations.
30. Mainkar in FIC 141.

ratio. Applying this principle to the totality of the weights recorded by Hemmy and discovered by Rao, two distinct and well-knit series of denominations are obtained, as shown in (the following) Tables.

First Series of Weights

Mean Weight (Approx.)	Ratio (Approx.)
1.2184	0.05
2.285	0.1
5.172	0.2
13.792	0.5
27.584	1
55.168	2
137.90	5
271.33	10
546.70	20
1417.5	50
2701.4	100
5556.0	200
10865.0	500(?)

Second Series of Weights

Mean Weight	Ratio (Approx.)
0.871	0.05
1.770	0.1
3.434	0.2
8.5753	0.5
18.1650	1
33.3052	2
174.50	10
6903	500(?)

These two tables leave out only two weights of 6.829 grammes (or 6.896 grammes of Rao) and 4.3370 grammes, which happen to be approximately a quarter of 27.405 grammes and 18.165 grammes respectively and may have been used for special purposes. These two tables are formulated by the application of the same logical principle as has been used 4000 years later in the progression of the denominations of weights of the metric system.

It may also be noted that the weights of ratio unity in the two tables are rationally related to each other. The unit weight of 27.584 grammes of the first series, is 50 per cent higher than the unit weight of 18.1650 grammes of the second series.

Thus the units of weights so far discovered in the Indus Valley Civilization, as systematically arranged and analysed, cannot but give us the impression of trained mathematical mind working behind their improvisation. Even in default of any document directly testifying to the mathematical activities in the period of First Urbanization, therefore, we are in possession of many circumstantial evidences indisputably indica-

31. *Ibid* 142-43.

ting the development of mathematical knowledge of the Harappans.

6. MATHEMATICAL INSTRUMENTS?

It will be relevant to mention here the discovery of an interesting article among the Indus ruins which many archaeologists are inclined to view as a mathematical instrument—probably a compass used for surveying purposes. We are indebted to S. R. Rao not only for its discovery at Lothal but also for first drawing our attention to its possible use or function. Here is how he puts all this :[32]

> Town-planning presupposes a careful survey of the land and measuring of angles for determining the alignments of streets, drains and houses. For this purpose an instrument made of shell was used at Lothal. It is a hollow cylinder with four slits on each of the two edges. When placed on a horizontal board it can be used almost as a compass in plane table survey for fixing the position of a distant object by viewing it through the slits in the margins. The lines so produced pass through the central point. If the opposite slits on one side are joined by cords they cut at right angles, and if all the cords passing through the slits on both the margins are drawn on the came plane as shown in Plate XXXII. B they intersect one another at the centre and the angles so formed by the eight lines measure exactly 45° each. Obviously this instrument must have been used in land-survey and for fixing alignment of streets and houses. Marshall considered similar objects found at Mohenjo-daro as personal ornaments, but they are too thick to serve as finger-rings and have too many slits to serve as pendants.

There may be scope for further discussion about the possible use of this apparently peculiar object. In the meanwhile it may be noted that serious archaeologists and scientists are inclined to view it as possibly connected with the mathematical activities in the Indus Valley Civilization. Deshpande, for example, observes that it "was probably used as a compass for measuring angles."[33] D. P. Agrawal endorses Rao's view of it having been a surveying instrument and adds, "Recent explorations have produced this instrument from Pabhumath and other sites of Saurashtra.[34]

32. S. R. Rao, in LIC 107.
33. M. N. Deshpande, in IJHS VI, 1. p. 11.
34. D. P. Agrawal, in Possehl's IA—NP 90.

7. ASTRONOMY IN FIRST URBANIZATION

To begin with, let us recapitulate here one point already mentioned. We began with Childe's understanding of "the Urban Revolution," who notes that one of the achievements of this revolution was the creation of exact and predictive sciences. In this connection, he mentions two sciences, namely mathematics and astronomy which were brought into being by the inherent requirements of the revolution. However, of the three primary centres of the urban revolution, only two—namely Egypt and Mesopotamia— have furnished the archaeologists with direct documentary evidences of the making of mathematics and astronomy. The third primary centre of the urban revolution, namely the Indus Valley Civilization, has not provided the archaeologists with similar documents. Nevertheless, as we have tried to argue, as far as mathematics is concerned, there are many circumstantial evidences which cannot be explained without the admission of the making of mathematics in this third primary centre of the urban revolution.

First, certain archaeological relics—ranging from the broken scales, the units of weights etc. to the magnificent brick-structures like the Mohenjo-daro bath, the granaries, the Lothal "dock-yard", residential quarters and the superb town-planning remain totally unexplained without the presumption of the development of mathematical knowledge. To these may also be added certain motives of decorative art specially on the potteries which suggest geometrical knowledge. We have already given some specimens of this.

Circumstantial evidences in the second form wanting us to trace the roots of classical Indian mathematics in the achievements of Harappan culture—on which we have tried to put special emphasis in our discussion—is bound to prove highly controversial. We have described it as based on the 'method of retrospective probing.' Broadly speaking, the method is as follows. We come across certain mathematical works as belonging to a period the technological and other peculiarities essentially presupposed by which do not at all agree with archaeologically ascertained facts. Nor does the theoretical temper forming the background of this mathematics agree with the essential ideas and attitudes of those to whom this mathematics is customarily attributed. In such circumstances—and in view of the palpable absurdity of the assumption that the mathematics was codified *before* its creation—we felt obliged to move backwards in Indian history and see if there was any period which fully answered to the technological and other requirements presupposed by the making of this mathematics, and we came across such a period in the mature Harappan Culture. Hence we wanted to presume that this mathematics was created in that period though at the present stage of our research we are unaware of how its tradition was transmitted to the much later period and the conditions of its codification in the form in which it reaches us. This, in short, is the essence of what we have proposed to call the methodology of retrospective probing and it remains for us to see its relevance for the inference of scientific activities in other forms in the period of First Urbanization.

With this recapitulation of what we have argued so far, let us pass on to the question of the other exact and predictive science, namely astronomy, which Gordon Childe argues developed to meet the requirements of the Urban Revolution. And our question is : Is there any sufficiently sound ground to presume the making of astronomy in the third primary centre of the urban revolution ? Here, again, we are confronted with formidable problems. Notwithstanding the recent tendency of some scholars to read astronomical knowledge in the Indus "seals"—a tendency to which we shall presently return—the fact remains that we have no direct documentary data to prove it, not to speak of delineating the nature of this astronomical knowledge even admitting its making in the Mature Harappan culture.

In such circumstances, we are left to raise only two questions. First, are there real circumstantial evidences wanting—or even compelling—us to presume that astronomical knowledge did develop in the Indus Valley Civilization ? Secondly, how far does our method of retrospective probing help us to infer—or at least to guess—the development of astronomical knowledge during the period of First Urbanization ?

The first of the two questions is answered in very broad terms by one of our senior archaeologists, M. N. Deshpande, who starts from the obvious fact that "the entire civilization flowered forth as a result of surplus agriculture economy."[35] With this point in mind, let us quote his observation[36] on the possible development of astronomical knowledge in Harappan Culture :

As agriculture must have largely depended on the rains, the knowledge of seasons and a calendar cannot be ruled out. The regularity of the movement of the heavenly bodies could not have escaped the attention of the Harappans. The priest dominated political hierarchy postulated in respect of this civilization may have strengthened its position on this specialized knowledge. The priests may have kept accurate astronomical records and forecast celestial events and advised the farmers on the time of sowing, harvesting and perhaps the flooding of the rivers so as to forewarn them of the possible catastrophe, which in the end is supposed to have contributed to the destruction of Harappan cities. It is, however, not possible to say with certainty whether they followed the solar or the lunar calendar, though the meagre evidence at our disposal would suggest the adoption of the solar calendar, for the representation of the moon is conspicuous by its absence while a solar symbol is present in the form of painted motif on pottery and in the form of urus-like animal which according to Mackay may have been a solar deity with the head of the beast taking the place of the sixth ray. In the absence of depiction of the moon and stars Mackay had suggested that inhabitants of Indus valley—an agricultural people—did not pay that attention to stars which, according to Smith, is a distinguishing feature of the agrarian population. Such an inference would not hold ground if we consider the following facts : The Harappans must have possessed the knowledge of stars for, without it, it is difficult to account for the rigid north-south and east-west orientation of the streets and lanes. In the burials the bodies of the dead were oriented : head pointing to north and feet towards south. The identification of the pole-star for this purpose has to be credited to the Harappans. Besides, the depiction of a boat on one of the seals and a potsherd and the recent evidence of a dockyard at Lothal would definitely point that the Harappans were navigators and ships laden with goods plied along the western coast as far as Bahrein Islands and perhaps beyond. In the circumstances we may have to conclude that the Harappans possessed sufficient knowledge of astronomy involving angular measurements of heavenly bodies.

I am aware of the fact that many writers on the India Valley Civilization would resent more or less strongly to some assum-

35. M. N. Deshpande, in IJHS Vol. VI No. 1. 6.
36. *Ibid* 8-9.

ptions on which Deshpande's observations are based. There is, for example, literally a wilderness of conjectures today concerning the socio-political organisation of the Harappans ; so his subscription to the postulate of the "priest dominated political hierarchy in respect of this civilization" is not likely to be smoothly accepted. At the present stage of our discussion, however, let us not digress into this controversy, though we shall be compelled to return later to it and shall see why many prominent features of the Indus Civilization remain unexplained without it and also the flimsiness of the other postulates proposed as alternatives to it. Nevertheless, it is not easy to question the main points of Deshpande. A civilization thriving on vast argicultural surplus cannot do without a developed calendrical system. And the development of a calendrical system is not conceivable without astronomical knowledge. This is part of common knowledge, which E. N. Fallaize sums up as follows. Astronomy is not "called for by the practical requirements of a population which lives chiefly by hunting. For an agricultural people, however, foreknowledge of the recurring seasons becomes essential, and it is necessary that some means should be found to mark the proper seasons for performing the operations to ensure the food supply."[37] If such be the essential need for a calendrical system—and therefore also of astronomy—even at the early stage of the development of agricultural technique, it would be too bold to imagine that the vast area under Harappan culture thriving basically on agricultural surplus could do without it, particularly when we remember that the Indus Valley Civilization was further confronted with the problem of flood and inundation of the rivers and therefore also the agricultural operations of the region required the know-how of predicting or anticipating these.

The other circumstantial evidences mentioned by Deshpande wanting us to presume astronomical knowledge in Harappan culture appear also to have substantial weight. Thus, for example, the accurate or near-accurate knowledge of north-south and east-west directions followed by the city-planners of the Indus Valley Civilization and also the burial custom of the Harappans implying the knowledge of cardinal directions could be possible on the basis of the identification of some "fixed

37. E. N. Fallaize, in ERE XII. 62.

star", just as in default of any archaeological evidence recovered so far suggesting the mariner's compass or its prototype, the navigational activities of Harappan traders following sea routes seem to suggest dependence on the konwledge in some form of the heavenly bodies.

All these are presumptions no doubt—presumptions that lead us at best to some sort of a vague assumption of astronomical knowledge in some form during the period of our First Urbanization. But are we supposed to leave the matter like that? Or is there any other way of moving towards a more substantial knowledge of the development of astronomy in the Indus Valley Civilization?

8. SPECIMEN OF WILD CONJECTURE: "GREAT BATH" AN ASTRONOMICAL OBSERVATORY?

In the absense of definite document, the archaeologists are often left to conjecture no doubt. At the same time, there should be some difference between a reasonable conjecture and a just wild one, with not even the semblance of real data supporting it. It is our misfortune that the extensive literature on the Indus Civilization is not without such wild conjectures. We shall mention here one example of this, because it professes to throw very important light on astronomy of the period under discussion. Nor can we just afford to ignore it, because the fact is that somehow it has found place in an otherwise very prestigious publication.

I have before me a very imposing volume with the title *Frontiers of Indus Civilization* edited by B. B. Lal and S. P. Gupta—both respected names among Indian archaeologists today. It contains contributions from eminent archaeologists of India and abroad and some of these contributions are of outstanding importance. But it also contains an article by Brij Bhusan Vij on "Linear Standard in the Indus Civilization". The most startling conclusion the author wants to arrive at in this is that the Great Bath of Mohenjo-daro was an astronomical observatory. As he puts it, "The 'great bath' of Mohenjo-daro was scientifically constructed to calculated dimensions, indicating the keen sense of mathematics and astronomy. This was therefore an important observatory, apart from being a 'community bathing|swimming pool.'"[38]

38. Brij Bhusan Vij in FIC 156.

The very idea of an important observatory being also a swimming pool may appear to be extraordinary, and it seems that the author himself is not serious about it. In the same article he observes, "The 'bath' can be interpreted to have been specifically constructed for use as a 'possible observatory' rather than a mere 'community swimming or bathing pool'."[39] He repeats, "The Great Bath at Mohenjo-daro was constructed to meet specific intention|requirement, possibly to make an astronomical observatory, and that it was not just a 'community swimming pool'".[40] So the emphasis is obviously on its having been an astronomical observatory.

Such a view is startling, of course, and we are not aware of any other serious archaeologist suggesting such a possibility, notwithstanding the fact that the architectural skill and mathematical calculations evidently required for its construction cannot but be considered as an archaeological marvel. But had the Great Bath been an astronomical obseervatory—or, had our author been able even to adduce any really plausible consideration in favour of the hypothesis—the problem of the making of astronomy in the Indus Civilization would not have remained at a rather groping stage, as it is today. So the crucial question is : How far has our author been able to substantiate the possibility ?

Leaving aside some modern scientific jargons with which he begins the discourse, like the energy levels of the krypton .86 atoms and hyperfine frequency of Hydrogen etc.—the bearing of which on the Great Bath having been an astronomical observatory or not is obviously irrelevant—let us note what he has to say about the Bath itself. Referring to his own diagram, he observes :[41]

ABCD represent the base plan of the 'great-bath' with its broader side BC—23 feet or 7.01 metre ; and the diagonal AC—46 feet or 14.02 metre. If the standard diamensions of the bricks used during that age is accepted as 300 mm × 150 mm × 75 mm, it will be seen that the length, breadth and depth needed about 40, 24 and 8 'brick-lengths' respectively. A total of 61,444 such standard bricks would be needed to fill the volume of the 'great-bath'. This suggests that

39. *Ibid* 154.
40. *Ibid* 155.
41. *Ibid.*

the idea of a right-angled triangle, i.e. the Pythagorus (*sic !*) theorem, and hence *pi*/2 goes back to antiquity ; at least to the Indus Valley Civilization.

To this he adds[42] :

The longer side, AB, thus worked to the measured dimension of 39.78 feet or 12.124 metres (i.e. 3 × 7.01 metre). Not only this, the depth of the 'great-bath' was very carefully chosen to represent an angle *pi*/18 between the base diagonal and the 'skew diagonal', i.e. the angle CAE in the figure. The measured height (from base upward) represents the linear distance 8 feet or 2.44 metre, which make Sine CAE – Sine *pi*/18 = 2.44/14.02 = 0.174 037 conforming the statement that Sine CAE = Sine 10° = 0.173 6 (from common log tables). Hence, the idea of angles and their application was definitely known in the antiquity of past India". [From these data] the author therefore, concludes : The Great-Bath at Mohenjo-daro was constructed to meet specific intention/requirement, possibly to make an astronomical observatory ; and that it was not just a 'community swimming pool.'.

I have myself failed to see in all this anything substantial from the astronomical viewpoint specially as indicative of the Great Bath having been an astronomical observatory. However, not being an astronomer myself, I had the feeling that I might have overlooked or failed to understand something of genuine astronomical significance without which it is not easily conceivable how such an article could find place for itself in the imposing and prestigious publication. With appropriate humility, therefore, I passed on the article to two of my friends who are themselves astronomers and have very kindly agreed to advise me on astronomical matters in preparing the present study. They are Ramatosh Sarkar and Apurva Kumar Chakravarty. Both of them scrutinised the article and told me that from the astronomical point of view at any rate they could read nothing really substantial in the article itself. Whether the Harappans developed any astronomical science at all is a different question. As far as this is sought to be substantiated by the article under consideration and specially the startling claim that the Great Bath at Mohenjo-daro was actually an astronomical observatory, R. Sarkar fully agreed with A. K. Chakravarti and advised me to ignore this article altogether as sub-controversial.[43]

42. *Ibid.*
43. See Appendix.

I have mentioned all this only to emphasise one point. We are not obliged to impute importance to everything written about the Harappan Culture howsoever imposing—and even otherwise important—a volume may be in which it finds place, as is the *Frontiers of Indus Civilization.* Even completely beseless conjectures may creep into such volumes, somehow eluding the critical judgement of its responsible editors.

9. ASTRONOMY AND THE INDUS "SEALS"

Are we, then, left with nothing of the nature of real archaeological data indicating the making of astronomy in the Indus Civilization ? A number of serious scholars both in India and abroad want us to answer the question in the negative. There is a growing tendency among them to see in the Indus "seals" indications of a calendrical system and other astronomical data.

We begin with an observation of Asko Parpola who has given us an excellent summary of this. Discarding the tendency of some scholars who "have proposed various Sanskrit solutions to the problem of the Indus script" on certain grounds,[44] Parpola has assumed that the language of the Indus people was presumably of Dravidian affinity—a presumption in favour of which Iravatham Mahadevan in his *Study of the Indus Script Through Bi-lingual Parallels*[45] has very strongly argued. Arguing on this presumption Parpola is inclined to agree with the Soviet scholars working on the Indus script who claim to read in the sign 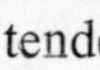 "as *min* 'fish'=*min* 'star' and therefore the sign  as six stars, which is the "Old Tamil name of Pleiades, aru-min, 'six star' ".[46] Assuming this to be some sort of evidence for his view, Parpola proceeds to argue :[47]

44. Asko Parpola in EIP 167.
45. I. Mahadevan in Possehl's ACI 261ff.
46. Asko Parpola in EIP 178. 1. Mahadevan TIS 717 : the fish sign alone occurs 381 times (p. 718) ; the sign of six lines occurs 38 times and the compound sign of fish with six lines occurs 16 times (p. 732). The inscription numbered 2128, e.g., contains this compound sign. (p. 61).
47. *Ibid.* 179.

The above mentioned interpretations of the fish signs agree with the factual information from later times. The names of planets and constellations read in the Indus texts are actually attested in historical documents and are of genuine Dravidian etymology. The fish signs which we read as planetary names make the most important group of those signs that we concluded from the contextual evidence to be probably god's name. The worship of such astral deities seems to have survived in the *navagraha* worship of later Hinduism ; in the Harappan religion the five planets proper seem to have represented gods who in later times emerged as leading Hindu deities : Brahma, Rudra/Skanda, Kala (Yama), Krsna and Balarama.

Like Father Heras (1953) and our Soviet colleagues (*Proto-Indica:* 1968, *Proto-Indica* : 1972) in their own ways, I have also followed up clues which the fish signs of the Indus script give by suggesting that ancient Indian astronomy and star lore may be largely of Harappan origin. Since the present book is intended for students of Indian archaeology, I want to conclude by referring to some important results, which in addition to giving further confirmation to the above hinted interpretations are of archaeological interest. While absent in the Avesta and the older books of the Rgveda (the latter composed on or soon after the arrival of the Rgvedic tribes to India), the names of the lunar asterisms (Sanskrit *naksatra*) appear in a complete list in *Atharvaveda.* It seems quite certain that the Aryans learnt the naksatras in India. While such luni-solar calendars as that connected with the naksatras are not created by primitive or nomadic peoples (Nilsson, 1920), they form an essential element in all early urban civilizations (Steward, 1955 p. 194 ff). The date when the *naksatra* calendar was complied can be determined by means of astronomical evidence inherent in it : it was in all likelihood around the 24th century B.C. (Needham, 1959 p. 246 ff). which coincides with the peak of the Harappan urbanization, adding thus one more means for defining its chronology. The orientation of the Harappan cities according to the cardinal points, which could be accomplished by astronomical means only, is tangible evidence for Harappan practice of astronomy. The Dravidian origin of the *naksatras,* on the other hand, seems proved by the long unexplained word *b(h)ekura* or *ori* used as their appellation in the *Brahmana* texts, which appears to be of Dravidian etymology (cf. Tamil *vaikuru-mtn* and Burrow & Emeneau, 1961, nos. 4570, 608).

Two points in this observation are in need of clarification. First, about the team of Soviet scholars working on the Indus script, with which Parpola's own team is in substantial agreement. Secondly, about Parpola's reference to Needham's authority which is really based on many complex considerations, i.e. is not as simple as Parpola's own reference is likely to give us an impression.

As for the Soviet scholars, Parpola[48] himself explains :

Only after the publication of our first preliminary report in 1969 we came to know the methods and results of a team of Soviet scholars who happened to start working on the Indus script simultaneously with us in 1964 (*Proto-Indica* : 1968). Headed by Yurij V. Knorozov, well known for his earlier computer work on the Mayan script, they, too have made use of the computer. Even otherwise our independent researches are in concord in several important respects. Soviet scholars have come to the same conclusion about the type of the Indus script, including the use of the rebus principle, attempted a division of the inscriptions into words, tried to identify grammatical elements, and to interpret individual signs from their contexts. Moreover, we agree also in considering the Indus language to be of Dravidian affinity.

Without trying to go into the methodology of the Soviet scholars, the full report of which is yet to reach us, we may quote here from a recent Soviet publication the general conclusions they are driving at. In *The Image of India : The Study of Ancient Indian Civilization in the USSR* by G. Bongard-Levin and A. Vigasin we read :[49]

Research into proto-Indian writing enabled Soviet scholars to reveal certain features of the religious-mythological concepts of the inhabitants of Harappan settlements, first and foremost various kinds of cult objects (anthropomorphic, zoomorphic, sacred trees, "deified vessels"), but the most important was to establish the general character of proto-Indian cosmogonic concepts and the calendar system.

Soviet scholars came to the conclusion that the inhabitants of Harappan settlements divided the year into three large and six small seasons. The symbols of the small seasons were representations of animals, the aurochs (the unicorn ; it was also the symbol of the year), goat, tiger, shorthorned aurochs and the bull. The zoomorphic symbols also, apparently, denoted the large seasons : the season of overflowing of rivers was "transmitted" by a representation of the crocodile ; the year began with the season of the aurochs ; the zebu and the scorpion symbolished the vernal and autumnal equinoxes. The sixty-year cycle (the cycle of Jupiter), which was followed in ancient India also in a later period, obviously arose in the Harappan era. The "Proto-Indians" divided this cycle into five twelve-year periods.

It is obviously premature for us to comment on this. But it may be incidentally mentioned here that already there is a

48. *Ib.* 178.
49. G. Bongard-Levin and A. Vigasin 194.

tendency among some scholars to trace the source or origin of the alphabet to the names of the heavenly bodies. As J. Needham sums it up, "Moran (H. A. Moran, *The Alphabet and the Ancient Calendar Signs*, Calif. 1953) has sought to show that the letters of the earliest alphabets were derived from the symbols of the 28 lunar mansions, and thus to explain the fact that most sets of alphabetic letters have from 25 to 30 components (D. Diringer, *The Alphabet : a Key to the History of Mankind*, N. Y. 1948). A full phonetic alphabet needs 46. The thesis is original, the hypothesis even seductive, but the presentation marred by too much special pleading".[50] There is thus some *prima facie* support for those scholars in India and abroad to read in the Indus "seals" some calendrical|astronomical sings, though it is needless to add that what is inferred from the Indus signs will be convincing to the extent to which it coheres with the other evidences, inclusive of archaeological and—if possible—literary ones.

Secondly, as we have already said, there is another point requiring clarification about the general observation of Parpola we have quoted. It is concerning his mention of Needham's authority about the dating of the *naksatra* system, which, as is well-known, has a very important place in Indian astronomy. To repeat what Parpola says : "The date when the *naksatra* calendar was complied can be determined by means of the astronomical evidence inherent in it ; it was in all likelihood the 24th Century B.C. (Needham, 1959, pp. 246 ff)." Lest it gives the wrong impression that Needham himself arrives at this date, it may be useful for our purpose to be clearer about this point.

Not that Needham does not discuss the *naksatra* system, usually rendered as the system of "lunar mansions" or "lunar zodiac". But the real reason for this is his primary interest in the history of astronomy in China, where we come across a strikingly similar system, which is called that of *hsiu*. We also come across an Arabic parallel of it, called the system of *al-manazil*. Hence, there is a good deal of controversy among the scholars about the exact origin of these strikingly similar systems. As Needham[51] sums up :

50. Needham SCC III. 239 note-g.
51. *Ibid*. III. 252-53.

And this is where we come to the problem which has caused so much controversy, namely, the relation of the Indian *naksatra* and Arabic *al-manazil*, 'moon-stations' to the Chinese *hsiu*. First brought to the attention of Western scholars by Colebrooke in 1807, many lists of the *naksatra* are available, while Biot analysed the data on them which al-Biruni had collected in the + 11th century. Lists of the Arabic *al-manazil* will be found in Higgins and others. They were certainly pre-Koranic, and the Hebrews knew them as the *mazzaloth*, while they even got into Coptic (Chatley). Iranian versions of the moon-stations are also known. The earliest reference outside Asia is in a Greek papyrus of the + 4th century analysed by Weinstock.

The common origin of the three chief systems (Chinese, Indian and Arab) can hardly, be doubted, but the problem of which was the oldest remains. That of the *manazil* is not a competitor, but the other two have elicited from time to time remarkable displays of vicarious chauvinism on the part of indianists and sinologists.

Needham himself seems to attach much importance to the view of Biot, though also differing with him rather strongly on an important point. By analysing the Chinese calendrical system of the "lunar mansions", Biot tried to establish two main points. First, from the internal evidence of this calendrical system, it can be inferred that it originated in the 24th century B.C.—or, to be more precise in 2357 B.C.[52] Secondly, the actual place where this system originally developed was China. Needham seems to be in favour of the first point and scrap the second. Referring to the paper "On the origin of the Twentyeight Mansions in Astronomy" (*Popular Astronomy*, 1947, pp. 62 ff.) by Chu Kho-Chen, who wants to support Biot's second point and summarises most of the evidences currently put forward for the view that already in the third millennium B.C. the *hsiu* system came into being in China, Needham comments rather sharply, "The great difficulty about this is that all the archaeological and literary evidence is against so early a date."[53] Discarding the theory of the Chinese origin of this system, Needham himself is inclined to think "that it may be possible to find a common origin in Babylonian astronomy for all 'moon-station' systems."[54] Such

52. *Ibid*. III. 177, note-c.
53. *Ibid*. III. 249.
54. *Ibid*. III. 254.

a view was first suggested by A. Weber in 1852;[55] it was as strongly supported in 1909 by H. Oldenberg[56] though vigorously contested by many scholars.[57]

The origin of the *naksatra* system being still highly controversial, we have found it preferable to discuss this separately in the Appendices. Biot's view, viz. it dates back roughly to the 24th century B.C., is reasserted by my advisers on astronomy from evidence of *Brahmana* texts. For the present we may mention only some other points.

Admitting Biot's dating of the formation of the original nucleus of the system and encouraged by the recent tendency of reading some of the *naksatra* names in the Indus "seals", it may not be *prima facie* impossible to think that the progress of our knowledge of the Indus civilization may in the future reach a stage showing the prevalence of the *naksatra* system (or something closely resembling it) already, during the period of our First Urbanization. Admitting this,—and we shall presently see some more substantial evidences supporting its possibility—we have to accept Parpola's claim that the Vedic peoples—originally unaware of any astronomical knowledge worth its name—eventually acquired (and used scraps of it) from the tradition coming down from the Harappan period.

10. METHOD OF RETROSPECTIVE PROBING : CHRONOLOGICAL POINTER

While discussing the mathematical knowledge contained in the *Sulva-sutra*-s, we have already mentioned and depended upon what we have called the method of retrospective probing. The assumption on which it is based may be briefly reiterated. When we come across some materials codified in the literature of a period which, on various considerations, has got to be considered as later than the data itself, there remains the possibility of moving background to some period in which such

55. A. Weber's *History of Indian Literature* (first German Edition, 1852, p. 21), and the first of his *Essays on the Nakshatras*, 1860, passim.
56. H. Oldenberg, in JRAS 1909, 1095 ff.
57. See G. Thibaut, in JASB 1894, 144-163; W.D. Whitney, in JAOS 1864, 1-94.

data is logically conceivable. Hence the presumption is that though the data come down to us as forming part of the literature of a later period their origin is to be traced to the earlier period from which they are somehow inherited.

Though without formulating this method and certainly not accepting its real implication, H. Jacobi[58] and B.G. Tilak[59] independently of each other followed it unconsciously, as it were, and opened for us the possibility of determining the antiquity of Indian astronomy. What they did was to recover from the Vedic literature certain references to basically astronomical observations. These observations, analysed according to the modern methods of astronomical calculations, want us to go back to hoary antiquity—to the third or fourth millennium B.C., though, according to Tilak, the period could be even earlier. From this, they argued straightway that the date of the Vedic literature itself is to be pushed to the same antiquity, without hesitating at all to *assume that the actual date of the observation must be the same as that of the literature in which it is referred to or mentioned in some form.* But such an assumption is not really obligatory. On the contrary the possibility remains that certain traditions actually come down from a hoary antiquity and accepted as authentic without verification by later writers. In the case of the Vedic literature—and more particularly in the early *Samhita*-s and the comparatively later *Brahmana*-s—such a possibility becomes quite strong sepcially because of a number of reasons.

First, the astronomical data found in the *Samhita*-s and *Brahmana*-s are so desultory—and what is wrose, so deeply embedded in discussions concerning ritual trivialities and theological disputations—that it requires much more than average proficiency in Vedic literature to locate these and disentangle the astronomical data from the framework of highly quaint logic seeking support from these for their rituals or

58. H. Jacobi, in *Festgruss on Rudolf von Roth,* Stuttgart, 1893, 68-73 ; in NGWG, 1894, 105-116 ; OC 1894, I. 103-108 cf. also H. Jacobi in JRAS 1909, 721-726.

59. B.G. Tilak *The Orion,* Bombay, 1893 ; for a brief review of the controversies over the views of Jacobi and Tilak, see Winternitz I. 293 ff., and for the different views expressed on the point see specially note 1, Winternitz, I. 295.

somehow connecting their own rituals with these data. It is indeed difficult to imagine people with any genuine interest in astronomy talking of astronomy in such a casual manner and using astronomical knowledge for this kind of mystery-mongering. On the contrary, it is quite conceivable that the priests whom we meet in the *Yajurveda* and the *Brahmana*-s—interested as they were above all in their *daksina* or sacrificial fee—could be trying to use every scrap of astronomical data coming down from a hoary past to add to their sacrificial rituals an awe-inspiring appearance *without bothering to verify these* astronomical data by the direct observations. We have added the last clause, because as we shall presently see the astronomical references in the *Brahmana*-s etc. do not really fit in with spatial or temporal contexts of these texts, though these can be taken as pointers to regions other than those of the Vedic settlements proper and to a period much anterior to that of the *Brahmana*-s.

Secondly,—and this is really a much more serious reason why the conclusions of Tilak and Jacobi could not and did not have much impact on Vedic scholarship,—there are other and sounder ways of dating the Vedic literature and these prevent us from pushing back the date of the literature to the hoary antiquity, as Jacobi and Tilak proposed. It is evidently outside the scope of our present discussion to try to enter into the technicalities of the question of determining the date of the Vedic literature. What is possible for us here is to quote some authoritative opinions expressed on the question. R.S. Sharma observes. "The French scholar Louis Renou, a life-long student of the Vedic texts accepted the view of Max Muller that the Aryans appeared in India around the fifteenth and sixteenth centuries B.C. and placed the hymns of the *Rg Veda* around this date."[60] The *Yajurveda* texts are much later. As Sharma continues, "On the present showing, the use of iron in the Indo-Gangetic divide and the Upper Gangetic basin, in which the *Yajus* texts and the *Brahmana*-s and *Upanisad*-s were compiled, cannot be taken back earlier than 1000 B.C., for this metal is known to several texts. Renou thinks that the *Brahmana*-s should be placed between the tenth and seventh centuries B.C."[61]

60. R.S. Sharma, MCSFAI 168.
61. *Ibid.* 169.

Accepting these chronological as well as geographical views, we may first briefly discuss some of the astronomical data to be found in the *Brahamana*-s and in texts usually accepted as belonging to this tradition. The main purpose of our discussion being only to show the possibility of the making of astronomy during the period of our First Urbanization, we propose to select here from the Vedic literature primarily two data, which, though without agreeing with the time-space contexts of the Vedic literature, can reasonably be accepted as pointers to the Indus Valley Civilization. After discussing these, we shall return to the views of Jacobi and Tilak, mainly to show the somewhat necessary limitation of their time which appears to have been at least one reason that obliged them to push back the dating of the Vedic literature to an unacceptable antiquity.

One of these astronomical data is concerning a certain asterism or *naksatra* called Krttikas, with which incidentally all the lists of the 27 *naksatra*-s mentioneed in Book XIX of the *Atharvaveda* as well as the different recensions of the *Yajurveda* like the *Taittiriya-Samhita*, *Kathaka-Samhita* and *Maitrayani-Samhita* begin, though of course the last mentioned text mentions 28 *naksatra*-s by way of adding a certain Abhijit to it. All the important Vedic data about the *naksatra*-s as well as the views of the modern scholars on these, are to be found in the *Vedic Index* by Macdonell and Keith though strangely enough the French scientist Biot—a predecessor and in some sense preceptor of Louis Pasteur—is mentioned here as a Chinese scholar ![62]

Jacobi argued that the Krttikas were counted the first *naksatra* because it then coincided with the vernal equinox ; from this he wanted to conclude that the Vedic culture was already in existence by 3000-2000 B.C. when this astronomical phenomenon did occur. How far this argument still holds good is a separate question. But certain other things said about the Krttikas have obvious chronological implications.

Such a statement is to be found in *Satapatha Brahmana* ii.1.2.1-5.[63] As is only to be expected in the *Brahmana* literature, the statement forms part of a theological controversy. The controversy is concerning the setting up of the two fires—

62. Macdonall and Keith VI. I. 409-430.
63. Tr. Eggeling SBE XII. pp. 282-3.

called Garhapatya and Ahavaniya—under the asterism considered most desirable for the purpose. Two views are mentioned in this connection : 1) that the fires are to be set up under the Krttikas and 2) these are not to be set up under the Krttikas. The ground adduced for the second view is that originally the Krttikas were the wives of the Seven Rsis (Saptarsi = Ursa Major or the Wain) and since the latter rise in the north and the Krttikas in the east they were precluded from having intercourse with their husbands, and the same will result from the act of setting up of the two fires under the Krttikas. The authority in our *Satapatha Brahmana*, however, rejects this view and offers a series of arguments in favour of the recommendation of setting up the two fires under the Krttikas. These arguments are :

(a) The Krttikas are Agni's asterisms, so that if he sets up his fires under Agni's asterisms, he will bring about a correspondence between the fires and Agni's asterisms.

(b) While the other lunar asterisms consist of one, two, three or four stars, the Krttikas are the most numerous of asterisms (consisting of seven, or according to others, six stars) ; hence he, by setting up the fires under the Krttikas, obtains an abundance.

(c) These (i.e. the Krttikas) do not move away from the eastern quarter whilst the other asterisms do move from the eastern quarter ; thus his two fires are established in the eastern quarter.

The general framework of the theological disputation concerning the desirability or otherwise of placing the sacrificial fires under the Krttikas can obviously have no interest for the historians of science. Nevertheless, one point mentioned in this connection cannot be ignored or overlooked. It is the statement that the Krttikas do not move away from the eastern quarter, while the other asterisms do move from the eastern quarter. Our text is quite firm on this point : *eta ha vai pracyai diso na cyavante|sarvani ha va anyani naksatrani pracyai disas cyavante.*[64] Commenting on the words *pracyai na cyavante* (does not swerve from the east) Sayana obser-

64. *Sat. Br.* ii. 1.2.3.
65. P. C. Sengupta, in IHQ 1934, 536.

ves *niyamena suddha-pracyam eva udyanti* (invariably rises in the exact east). Following this P. C. Sengupta seems to be justified in saying, "This means that the Krttikas rose exactly at the east"[65] and this is viewed as a distinctive peculiarity of the asterism, i.e. as contrasted with the others. Eggeling also in his translation of the *Satapatha Brahmana* accepts such an understanding and writes, "The Seven Rsis rise in the north and they (the Krttikas) in the east."[66]

Accepting this meaning of the statement we have a very interesting astronomical information here. It is the observation of the fact of the Krttikas rising exactly at the east. Depending on astronomical works like the (modern) *Surya-siddhanta,* eminent Indologists like Burgess, Whitney and others identify the Krttikas with the asterism Pleiades, with *Eta* Tauri as its determinative star. If, therefore, we take this as an actual piece of observation (somehow) codified in the *Satapatha Brahmana,* it is possible for the astronomer-mathematicians to calculate and determine the date when this observation actually occurred. So I passed on this data to the astronomer-mathematicians advising me on the present work, namely, Apurva Kumar Chakravarty and Ramatosh Sarkar, and both of them arrived at the view that the astronomical observation under consideration wants to take us back to the middle of the third millennium B.C.—a result agreeing with the calculations of P. C. Sengupta.

To avoid digression into the technicalities of the calculations involved, I have preferred to use their discussions of the questions in the form of separate Appendices. For the purpose of our present discussion we shall mention here only a few points based on these.

The *Eta* Tauri can rise exactly in the east only when its declination is nil, or, as Sarkar shows, when its celestial longitude is zero and celestial latitude is negligible, i.e. roughly coinciding with the vernal equinox.

But the celestial longitude of the star in 1985 is 59° 47′ 24′.

Calculating on the basis of the precessional rate of 1° for every 72 years on an average, we have to go back to 2334 B.C. to reach the period when the celestial longitude of *Eta* Tauri would have been thus nil.

66. Eggeling in SBE XII. 283.

Calculating on the same basis, however, by 1000 B.C.—i.e. roughly the time beyond which the *Satapatha Brahmana* (with all its inner complexities) cannot be pushed back by responsible Vedic scholars—the celestial longtitude of *Eta* Tauri would have been 18° 28′ 12″.

All this means that the statement that the Krttikas rise exactly in the east could by no means be based on an actual observation of the period of the *Satapatha Brahmana* though the date of the event amazingly coincides with the peak period of the Harappan culture when, therefore, it could form part of real astronomical observations.

Could it, then, be that the statement under consideration, viz. that the Krttikas rise exactly in the east, actually formed part of the astronomical knowledge which really developed in the Harappan Culture and it "somehow" came down to the authors|compilers of the *Satapatha Brahmana* who accepted and codified it without bothering to verify it by actual observation just to add some weight—or, perhaps better, some mysterious awe—to their ritual prescriptions? Astronomical contents developing in a very ancient period receiving condification centuries later is not *prima facie* impossible. Here is an example we quote from Needham: "Assyriologists have long been familiar with a number of cuneiform tablets which were preserved in the library of King Assurbanipal (Ashur-bani-apli, 668 B.C. to 626 B.C.) at Nineveh, but which date as to contents from the late 2nd millennium B.C."[67] These tablets could, of course, belong to an earlier date. But the point is that the preservation of these in the royal library of much later date presumably meant that importance was still being attached to these in the comparatively later period when their astronomical contents must have substantially changed to tally with actual observations of the time of their careful preservation. Thus, acceptance of ancient data without fresh verification is not *prima facie* impossible. In any case, the *Satapatha Brahmana* shows that what the priests say about the Krttikas has not even the semblance of interest in actual observation of the phenomenon referred to: it is just mentioned as one of the many grounds favouring the view that the sacrificial fires were to be placed in the east, because the Krttikas—un-

67. Needham SCC, III 254-56.

like the other asterisms—rise in the east, without in the least bothering to observe that in their time the Krttikas did not *actually* rise exactly in the east. To the astronomers of the Harappan Culture, it was presumably different. During their time the Krttikas did rise exactly in the east and in all likelihood they actually observed it.

The admission of such a possibility, seems to indicate some presumptions of far reaching consequences. Evidently enough, persons interested in astronomical observations could not be satisfied in noting a single phenomenon in its isolation like the rising of the Krttikas, specially when it is added that this is a distinctive peculiarity of the Krttikas as contrasted with the other asterisms. On the contrary, the presumption is that the observation under consideration formed part of a system of astronomy, only scraps of which reached the Vedic priests who wanted to use these in connection with their sacrificial cults. One is thus tempted to raise the question : Could it be that the entire *naksatra* system—which we find referred to in Book XIX of the *Atharvaveda* and the different recensions of the *Yajurveda* without ever giving us the impression of any systematic interest in astronomical observations and always arbitrarily connected with nothing more than theological trivialities—had its real roots in the astronomy of the Harappan Culture ?

The question is, of course, highly complex and too many views are already expressed on the astronomical data in the Vedic literature to allow any discussion of the subject without entering into a good deal of polemics. At the present stage of our discussion we may as well avoid it and to have separate appendices for the more technical details. Instead of that we shall note here only a few points of obvious interest.

The refences to the *nakasatra*-s apart, we come across in the Vedic literature certain statements with obvious astronomical interest. Among the modern scholars, Tilak and Jacobi are the more eminent ones who first put much emphasis on these. Rejecting the tendency of brushing these off as sheer priestly nonsense, they took these as genuine pieces of observation. Applying the modern methods of calculating the possible dates when such astronomical data could form part of actual observation, they felt compelled to go back to an early antiquity —far beyond the time usually admitted by the Vedic scholars as that of the composition of Vedic literature. From this they argued that the dating of this literature was in need of serious revision, or that the actual date of the making of this literature was to be pushed back to a great antiquity. Though some of the prominent scholars like Valle Poussin, Barth and Winternitz felt that it will be wrong to ignore "Jacobi's great chronological argument"[68] the majority of Vedic scholars wanted to reject this as just absurd. Not that the astronomical data —and these as pointers to much antiquity—were necessarily questioned. What was questioned was the argument that since such data are found recorded in the Vedic literature, these are to be taken as indicative of the date of the literature itself. What is ignored is the possibility that the date of observation of certain phenomenon is not necessarily the same as that of their codification, that the tradition of certain facts of observation may come down from a hoary antiquity though only to be used in codified form at a much later period by people with scanty enthusiasm for direct observation and therefore without the need felt for verifying the data with direct observation of their own. This was perhaps not so in the case of the priests during the First Urbanization, when as "organisers of production" they had to take actual astronomical observations with much seriousness. Cut off from this social function, the Vedic priests became on the whole social parasites subsisting only on the *daksina*-s or sacrifical fees received from the rich patrons financing the sacrifices. Accordingly, it was only natural for them to develop a different theoretical temper, which, as we have already noted, was basically opposed to the spirit of direct observation, inasmuch as this went against the tendency of deliberate mystification : *paroksa-priyah iva hi devah, pratyaksa-dvisah.* At the same time, certain scraps of knowledge about the celestial phenomena forming part of ancient astronomical lore could somehow come down to them and could be effectively used by them to add grandeur to their sacrificial cults, which, therefore, they could as well graft in their own literature. Thus in short, there are grounds to think that the actual date of observation of the astronomical phenomena mentioned in the Vedic literature may be much more ancient than their mention or codification in the Vedic literature. From

68. See Jacobi in JRAS 1909 721.

this point of view, the dating of the Vedic literature from the astronomical data contained in these may be more or less fallacious.

At the same time, we have to note a somewhat obligatory limitation of the conditions under which Tilak and Jacobi had to work. During their time nothing was known about Indian Culture pre-dating the Vedas ; for them the Vedas were necessarily the starting point of understanding Indian culture. In short there was nothing for them on the basis of which to think that certain astronomical data coming down from the pre-Vedic period could somehow find place in the Veda.

It is interesting to note that Jacobi understood this limitation, though in his own way. In 1909, he said that if "we were quite sure that Vedic culture was not older than 1200 or 1500 B.C." we would be obliged to seek other explanations of the astronomical data contained in these. However, referring to the usually accepted date of the Vedas he added :[69]

> As long as this fact remains in suspense, either my arguments or these three subversive interpretations given to them by my opponents will appear plausible in accordance with the estimated age which critics assign to Vedic culture. When the new theory on the antiquity of the Veda was first discussed, I made this same statement to Mr. Tilak, who wished to enter upon a campaign against all opponents. I told him that the discussion would have no definite result *unless excavations in ancient sites in India should bring forth unmistakable evidence of the enormous antiquity of Indian civilization.*

Dramatically enough, hardly within two decades when Jacobi wrote this, the archaeologists' spade did prove the enormous antiquity of Indian culture. But it also relieved the scholars from the earlier but somewhat necessary limitation of equating the antiquity of Indian culture with that of the Vedic culture. Thus scraps of astronomical data that found place in the Vedic literature remained. So also the possible date of their actual observation. However, it became a chronological pointer to the Indus Civilization and not to Vedic Civilization.

We have noted here only one example of this. The Krttikas rise exactly in the east, asserts the *Satapatha Brahmana*. Though today they do not rise exactly in the east—and though even during the period of the *Satapatha Brahmana* they did not

69. *Ibid* 722, emphasis added.

rise exactly in the east—they did so during the peak period of the Harappan Culture, which moreover, based as it was on vast agricultural surplus, did need an astronomical-calendrical system, and therefore, also required the observation of the heavenly bodies as accurately as was possible in those days.

11. ARBITRARINESS OF THE INTEREST IN ASTRONOMY OF VEDIC PRIESTS

We have thus seen that there are some astronomical data recorded in the comparatively later Vedic literature that cannot be based on actual observation during the time of the composition or compilation of these texts. On the contrary, depending on mathematical-astronomical calculations we are led to accept the view that the actual observation on which these are based chronologically agrees with that of the Harappan period. We are inclined to accept these as chronological pointers to the making of astronomy during the period of our First Urbanization. We are inclined to presume further that though in the comparatively later Vedic literature, such propositions with astronomical interest are extremely desultory—mentioned here and there in the context of sacrifical rituals without showing any intrinsic connection of these with the sacrifices—there is no reason that the astronomical propositions recorded in these texts were originally only a bundle of stray observations. On the contrary the presumption is that these originally formed part of a coherent astronomical system, only fragments of which came down to the Vedic priests or were sought to be used by them for adding some kind of aura to their sacrificial craft. In substantiation of this argument we shall mention here only one example of how casual had been the astronomical interest of the Vedic priests.

In the *Satapatha Brahmana,* just after mentioning certain consideration for the desirability of setting up the sacrificial fires under the Krttikas—one of these being that the Krttikas rise in the east—the text passes on to mention various other *naksatra*-s or asterisms under which the same fires could be placed, offering arguments in favour of each, none with any astronomical interest. Thus :[70]

70. *Sat. Br.* ii. 1.2.6-13.

He may also set up his fires under (the asterism of) Rohini. For under Rohini it was that Prajajati when desirous of progeny, set up his fires...

He may also set up his fires under (the asterism of) Mrgasirsa. For Mrgasirsa, indeed is the head of Prajapati; and the head (*siras*) means excellence (*sri*); for the head does indeed mean excellence; hence they say of him who is the most excellent (*srestha*) of a community, that he is the head of a community....

He may also set up his fires under the Phalgunis. They, the Phalgunis, are, Indra's asterism, and even correspond to him in name; for indeed Indra is also called Arjuna this being his mystic name, and they (the Phalgunis) are also called Arjunis....

Let him set up his fires under the asterism Hasta whoever should wish that (presents) should be offered to him: then indeed (that will take place) forthwith; for whatever is offered with the hand (*hasta*), that indeed is given to him.

He may also set up his fires under Citra....

So from the viewpoint of the Vedic priests, the recommendation of setting up the fires under the Krttikas is not to be taken with a great deal of seriousness and so also the astronomical consideration mentioned in favour of it. There are various other alternatives to it based on alternative considerations, which, from the priestly viewpoint at any rate, are equally sound. What is indisputable about this discourse is that certain names of the asterisms came down to the priests of the *Satapatha Brahmana* and they somehow tried to connect these with their sacrificial cult. Indeed the discourse read as a whole—specially its concluding portion—seems to give us the impression that, though aware of these *naksatra* names, they were little concerned with understanding what these actually were. They suggested for the word the absurd origin from the sense of "powerlessness" and even went to the extent of imagining that the sun was the best or most desirable or most powerful of the *naksatra*-s! Before imputing any genuine astronomical interest to the Vedic priests we have therefore to take note of the concluding portion of the discourse which reads as follows:[71]

Originally these (*naksatras*) were so many different powers (*ksatra*), just as that sun yonder. But as soon as he rose, he took from them (*a-da*) their energy, their power; therefore he (the sun) is called Aditya, because he took from them their energy, their power.

71. *Ibid* ii. 1.2. 18-19.

The gods then said, 'They who have been powers, shall no longer (*na*) be powers (*ksatra*).' Hence the powerlessness (*na-ksatratvam*) of the *naksatras*. For this reason also one need only take the sun for one's *naksatra* (star), since he took away from them their enregy, their power. But if he (the sacrificer) should nevertheless be desirous of having a *naksatra* (under which to set up his fires), then assuredly that sun is a faultless *naksatra* for him; and through that auspicious day (marked by the rising and setting of the sun) he should endeavour to obtain the benefits of whichever of those asterisms he might desire. Let him therefore take the sun alone for his *naksatra*.

Thus whatever expectation in the astronomical interest of the Vedic priests might have been aroused in us by the statement that the Krttikas rise in the east is washed away by this almost total fantasy about the *naksatra*-s with which the discourse concludes. Apparently, the statement about the Krttikas that we read in the *Satapatha Brahmana* came down to the priests from some other source.

That the astronomical scraps in the priestly literature were borrowed from other peoples whom the Vedic priests considered as aliens and opponents seems to be faintly suggested by the *Taittiriya Brahmana*[72] which refers to a class of *asura*-s named *kalakajna*-s. Since *kalakajna* literally means 'the knowers of time', it may not be totally impermissible to read in it the suggestion of conversance with calendrical science. If so, the mention of them as *asura*-s is significant, for in the Vedic literature it is one of the typical words referring to the rivals and aliens—the most outstanding opponents of the gods. It will no doubt be a mark of impermissible haste to jump at the conclusion only from this evidence that we have here a reference to the Harappans with their calendrical system, for there is a good deal of controversy among the modern scholars about the actual people referred to in the Vedic literature as the *asura*-s. At the same time it may be another error to ignore completely the suggestion of the *Taittiriya Brahmana* that in ancient times there were *kalakajna*-s—presumably astronomers or experts in the calendrical calculations and interestingly enough, they belonged outside the circle of the Vedic peoples themselves—in fact considered as the aliens and

72. *Tait. Br.* i. 1.2. 4-6.

opponents of the Aryans so-called. From the scraps of astronomical data recorded in the Vedic literature which, chronologically speaking, are pointers to the period of the Harappan Culture, it may be a mark of hasty historiography to dismiss outright the possibility of these *kalakajna*-s having been the astronomers of the Harappan period.

12. METHOD OF RETROSPECTIVE PROBING : GEOGRAPHICAL POINTER

If the later Vedic literature contains astronomical data which, when analysed, appear to be chronological pointers to the Harappan period, there are also data in the same literature which appear to be geographical pointers to the Harappan region as the place of astronomy in its making. We shall concentrate here mainly on one of these. It is to be found in *Vedanga Jyotisa*, literally 'astronomy as a limb of the Vedas or Vedic studies'.

Vedanga Jyotisa is a brief text that have come down to us in two versions—one claiming to belong to the tradition of the *Rgveda* and the other to that of the *Yajurveda*. The former is called *Arca-jyotisam* and the latter *Yajusa-jyotisam*. The text is very brief : in the printed edition we propose to follow here,[73] the first contains thirtysix verses and the second fortyfive. In both versions, the language used is classical Sanskrit, which is now current and hence from this viewpoint the work cannot be very ancient. But that is no index to the astronomical contents of the work, because the author|authors of both its versions claim that they simply present the views of a certain ancient authority called Lagadha[74]—a circumstance which accounts for the basic similarity in the astronomical content of the two versions. We have really nothing that can be considered as historical knowledge about this Lagadha, beyond perhaps the bare fact that the astronomical views he stood for is traceable to considerable antiquity. There is a tendency among the modern scholars to determine the date of

73. *Vedanga Jyotisam* ed. S. C. Bhattacharyya, Calcutta, 1974.

74. RV-J. 2 ; VJ-Y 44. Interestingly the name appears to be quite peculiar and it seems that it is not Sanskritic at all. We are aware of no scholar throwing light on this peculiar name.

Lagadha from certain astrononiical data recorded in the text. As T. S. Kuppanna Sastry has recently claimed :[75]

> Verses 6, 7 and 8 of the Yajur Vedanga Jyotisa (Y-VJ) show that at the time of Lagadha the winter solstice was at the beginning of the asterism Sravistha (Delphini) segment and that the summer solstice was at the mid-point of the Aslesa segment. It can be seen that this is the same as was alluded to by Varahamihira in his *Pancasiddhantika* and *Brahatsamhita*. Since VM (Varahamihira) has stated that in his own time the summer solstice was at Punarvasu $\frac{3}{4}$, and the winter solstice at Uttarasadha $\frac{1}{4}$, there had been a precession of $1\frac{3}{4}$ stellar segments, i.e. 23° 20″. From this we can compute that Lagadha's time was $72 \times 23^{1}/_{3} = 1680$ years earlier than VM's time (*c*. A.D. 530), i.e. *c*. 1150 B.C. If, instead of the segment, the group itself is meant, which is about 3° within it, Lagadha's time would be c. 1370 B.C.

Admitting the accuracy of the observation of the phenomenon under discussion and admitting further that the actual observation is to be attributed to Lagadha himself, we have to admit that he lived roughly in the 14th century B.C.—i.e. in a period shortly following the end of the Mature Harappan Culture when, in the post-Harappan sites excavated recently by the archaeologists much of the traits of the Harappan Culture was still alive, and in any case, a period ante-dating by several centuries the making of the priestly literature of the Vedic people—the *Yajurveda* and the *Brahmanas*. This by itself casts doubt on the usual assumption that the *Vedanga Jyotisa* embodies an astronomical system evolved by the Vedic priests to meet the requirements of their priest-craft. Whitney strongly argues that this Jyotisa has no "relation to the Vedic ceremonial".[76] Besides, the possibility of the *Vedanga Jyotisa* incorporating into its astronomical system ideas coming down from a remote past cannot be totally ruled out, specially in view of the modern scholars failing to read in it a clearly consistent astronomical system : much of this brief text remains obscure for us and, as Thibaut has very convincingly shown, these remained obscure even to the most famous commentator

75. T. S. K. Sastry, in IJHS Vol. 19, No. 3, Suppl. 1984, p. 13 ; For discussion of the date of the observation under consideration, see also Pillai, *Indian Ephemeris* Vol. I, pt. i. 444-45.
76. Whitney OLS, Second Series. 384.

of it, namely Somakara, who, therefore, felt obliged to suggest fanciful meaning to many of its verses.[77]

As for the antiquity of the astronomical contents of the *Vedanga Jyotisa,* it may not be irrelevant here to mention another point. Varahamihira (*c.* A.D. 530) in his work 'the five astronomical systems' or the *Pancasiddhantika,* begins with an account of what he calls *Pitamaha Siddhanta* or the astronomical system of the grandfathers—evidently a figurative way of indicating its hoary antiquity. Compared to the other systems of astronomy discussed by him, this is considered by Varahamihira as crude and undeveloped, and hence he gives very meagre information about it.[78] But the interesting point is that Thibaut, comparing these information with some of those of the *Vedanga Jyotisa* feels "that the astronomical book quoted by Varahamihira as *Pitamaha Siddhanta* must have been either the Jyotisa itself or a work very much like it."[79] We may perhaps have in this a clue to the apparent obscurity—often amounting to almost illegibility—of the extant versions of the *Vedanga Jyotisa,* because, as we have already seen, the classical Sanskrit used in the texts indicates that these must have been written or codified in a period much later than that of the formation of the astronomical core of the *Vedanga Jyotisa,* and, as such, much of it could not be properly understood—or was even misunderstood—by the authors of the *Jyotisa* texts.

With this point about the antiquity of its astronomical content, we now pass on to consider an information contained in the *Vedanga Jyotisa* which, though astronomical in nature, helps us to locate the region in which this astronomy presumably developed. We are going to see that this again seems to be a pointer to the frontier of the Indus civilization.

Let us first try to be clear about the data itself.

According to *Vedanga Jyotisa*, one solar year consists of 366 civil days (a civil day meaning the time from one sunrise to the next, i.e. a day-sum-night or nycthemeron, consisting

77. G. Thibaut, in SHSI II, 480 ff.
78. *Panca Siddhantika* Verse-4. See also S. B. Dikshit, BJS pt. II, p. 3.
79. G. Thibaut, in SHSI II, 488.
80. VJ—J, verse 28.

of our 24 hours).[80] One solar year consists of two equal *ayana*-s,[81] i.e. in each *ayana* there are 183 nycthemeron. The two *ayana*-s are : 1) *uttarayana,* i.e. the period beginning with the winter solstice and ending on the day just preceding the summer solstice, and 2) *daksinayana,* i.e. the period beginning with the day of the summer solstice and ending with the day just preceding the winter solstice. The text further asserts that during the *uttarayana* each day (from sunrise to sunset) increases by one *prastha* (the meaning of which we shall presently see) and the night decreases by one *prastha* ; while during the *daksinayana* the opposite takes place.[82] Further the maximum time difference between a day (i.e. from sunrise to sunset) and night (i.e. from sunset to sunrise) is 6 *muhurta*-s.

What then, are meant by *prastha*-s and *muhurta*-s ?

The contraption presupposed for time-measurement was called *jalayantra*[83] or a vessel with a hole for discharging water from it, and the water discharged from it was measured by units called *prastha.* Accordingly, the word *prastha* was also used as a unit for time-measurement, i.e. the time taken for the discharge of one *prastha* of water from the vessel. According to *Vedanga Jyotisa,* 15¼ (15.25) *prastha* of water is discharged during the time period called one *danda.* Hence was the practice of referring to one *danda* as equal to 15¼ or 15.25 *prastha*-s. Thus, one *prastha* approximately=0.06557 (or $^4/_{61}$) *danda*-s. Hence, in one *ayana,* the total increase of a day or a night=183 *prastha*-s=183×$^4/_{61}$ *danda*-s = 183× 0.06557 *danda*-s, i.e. 12 *danda*-s. Now, 2 *danda*-s (also called 2 *nadika*-s=1 *muhurta*),[84] i.e. 12 *danda*-s=6 *muhuta*-s. The *Vedanga Jyotisa* calculates one *savana* day (i.e. one sunrise to the next or our 24 hours) as consisting of 30 *muhurta*-s (i.e. 1 *muhurta*=our 48 minutes).

Now, as already stated, for *Vedanga Jyotisa,* the maximum time difference between one day (sunrise to sunset) and one night (sunset to sunrise) is 6 *muhurta*-s. On the starting day

80. VJ-Y, verse 28.
81. *Ibid.*
82. VJ-Y, verse 8.
83. Fleet in JRAS. 1915. 213-230. S. B. Dikshit BJS. 224ff.; S. R. Das in IHQ. 1928. 256-69.
84. VJ-Y, verse 32.

of the *daksinayana* (i.e. summer solstice), the day is longest and the night shortest. Hence, on the summer solstice, the day is 18 *muhurta*-s (=our 864 minutes or 14 hr.s 24 min.s) and the night 12 *muhurta*-s (=our 576 minutes or 9 hours 36 minutes). Similarly, on the winter solstice, the day=12 *muhurta*-s and the night 18 *muhurta*-s.

Such then, is the data we have in the *Vedanga Jyotisa*. And it is exceedingly important for our discussion. According to the astronomical knowledge we now possess, such relation between the longest and shortest day of the year cannot and does not hold good all over the globe. On the contrary, it holds good only for some specified latitude. The point is already noted by earlier scholars. Siteshcandra Bhattacaryya, for example, on whose edition of the *Jyotisa* we have depended here, observes that this phenomenon holds good only for latitude 34°45′ north, though unfortunately he does not give us the calculation on which he depends.[85] Besides, the calculation has to be based on the exact meaning assumed by the text of the concept of sun-rise and sun-set and also certain other points. So I have passed on the data to Ramatosh Sarkar and A. K. Chakrabarty for more precise calculation of the possible latitude from which the observation under consideration could be actual. Taking into consideration various alternative possibilities, Sarkar arrives at the conclusion that the most plausible alternatives are: (1) latitude 34.3° (34°18′) North, or (2) 34.5° (34°30′) North. His calculations and observations are to be found in Appendices. Remarkably enough, these come rather close to the view expressed by Siteshcandra Bhattacaryya and Sarkar explains in the Appendices why his calculations differ marginally from that of Chakrabarty because of some differences in their presuppositions.

With these points in mind, we may return to our main discussion. The astronomical data concerning the longest and shortest day is a pointer to some geographical location. Significantly enough, this geographical location falls within the cultural frontier of the Indus civilization, the latitude indicated being somewhat more to the north of Harappa whose latitude is 30°38′ North, though much nearer to Taxila.

85. S. C. Bhattacharya, *VJ* 18.

It is also significant to note from the point of view of our discussion that the place of observation of the longest and shortest day of the *Vedanga Jyotisa* cannot be true of the regions where the Vedic peoples eventually settled and produced their ritual literature—the *Yajurveda* and the *Brahmanas*. The latitude of the Vedic settlements could not be more than 28° N.[86] Therefore, we are inclined to conclude that in the *Vedanga Jyotisa* we have an important clue which, geographically speaking, indicates that its astronomical contents were presumably based on the observations of the Harappans, though it came down to the much later period and to a different region altogether, where the Vedic priests wanted somehow to connect it with their sacrificial ritual, branding it as a "limb of the Vedas" (*Vedanga*), perhaps without understanding and certainly not verifying the astronomical contents that came down to them. This, for us may be a clue also to the unintelligibility of the text in its codified form.

The observation of the longest and shortest day under discussion gives us only an idea of the latitude and, as such, it holds good of the entire belt of the globe of this latitude, inclusive of Mesopotamia, the southern capital of which, namely Babylonia, has the latitude 32°33′ N. This might have been one of the reasons why it was suggested that the astronomical calculation under consideration developed in Babylonia and the Indians borrowed it from the Mesopotamians. However, apart from the possibility of independent and parallel developments, we have to consider here another point. The *Vedanga Jyotisa* view, as we have seen, is based on certain detailed calculation and also on the use of a certain apparatus for measuring the time-unit. So long, therefore, we do not come across the relevant calculations and the relevant instrument for measuring time-unit in Mesopotamia, the mere fact that the latitude roughly agrees with that of Mesopotamia—combined of course with the fact that the ancient Mesopotamians also developed their system of astronomy—cannot prove that the Indians borrowed the astronomy of *Vedanga Jyotisa* from

86. The Aryavarta of the Dharmasastras is the middle Gangetic zone, extending originally from Kuruksetra to Allahabad.

them. This is a point already argued by Whitney and Thibaut. As the latter put[87] :

> Regarding the disputed point whether the rule fixing the length of the shortest and longest days of the year has been borrowed by the Indians from some foreign source, for instance from Babylon, or sprung up independently on Indian soil, I am entirely of the opinion of Prof. Whitney who sees no sufficient reason for supposing the rule to be an imported one. It is true that the rule agrees with the facts only for the extreme north-west corner of India ; but it is approximately true for a much greater part of India, and that an ancient rule—which the rule in question doubtless is—agrees best with the actual circumstances existing in the North West of India is after all just what we should expect.

The last point of Thibaut is in need of some comment. What did he mean by saying that the observation "agrees best with the actual circumstances existing in North West of India" and it is "after all just what we should expect" ? Why in particular, it is just what we should expect ? The answer appears to us to be as follows. Thibaut's article on the *Vedanga Jyotisa* appeared in 1877, when the only valid starting point for understanding the cultural history of India was the *Rgveda* and when it was practically unanimously assumed that the Vedic people entered India from the North West. Thus, it seems, that Thibaut tacitly assumed that while entering India from the North West the Vedic people actually observed that the longest day consisted of 18 *muhurta*-s and the shortest of 12 *muhurta*-s, which therefore formed part of their assumptions. However, there are many considerations against such a possibility. If the Vedic people actually observed such a phenomenon while they were enterning India from the North West, it is only reasonable for us to expect some reference to it in the earliest strata of the *Rgveda*. But the fact is that the entire *Rgveda* is totally unaware of any observation even remotely suggesting this. Secondly, there is little scope to doubt that when the Vedic people entered India, they were on the whole nomadic pastoral people, for whose economic life astronomical knowledge could not be a necessity, though when they settled down many centuries later in the "Aryavarta" (or "Madhyadesa"), astronomical knowledge could be—and perhaps did—form part of the

87. G. Thibaut, in SHSI, II. 488.

need of their economic life, though it seems that they depended for this purpose more on surviving hearsay than on actual observations. Thirdly, from what we actually read about the technological development of the Vedic people in the *Rgveda* itself, it requires a great deal of Aryan chauvinism to imagine that during this period of the oral composition of this vast literature they could by any chance improvise the time-measuring instrument—something quite sophisticated as judged in the ancient context.

But what could Thibaut do in 1877 when absolutely nothing was known about the Harappan civilization in North West India with its imposing technological development and producing the vast agricultural surplus on which the civilization thrived? The strong presumption is that their socio-economic life did require an astronomical system, though at the present stage of research, in default of any direct documentary evidence of this astronomical system, we are somewhat obliged to follow the method of retrospective probing to understand it. Fortunately, the method seems to yield some positive result, for in the literature of the Vedic peoples—which is much later than the Indus civilization—we have both chronological and geographical pointers to some system of astronomy already developed in the Indus civilization.

CHAPTER 8

POTTERY, TRANSPORT, TEXTILE AND OTHER TECHNOLOGIES

Santanu Maity & Sujata Maity

1. PRELIMINARY REMARKS

We have so far argued that notwithstanding the serious difficulties caused by the lack of any direct document testifying to mathematics and astronomy in the ancient Indus Civilization, there are circumstantial evidences wanting us to presume the making of these as "conscious" sciences during the period of our First Urbanization. It remains for us to discuss the question of the possibility of the emergence of conscious science in other forms—specially what we call nature sciences these days like chemistry, biology, physics, etc.—during the same period.

Our starting point, however, is formed of the archaeological data concerning the main technologies of the Harappan culture, because we began with the proposition that conscious science—whatever form it may eventually acquire—is necessarily rooted in technique. Without an overview of the more prominent technologies of the Harappan period, therefore, we cannot proceed to the question of the possibility of other conscious sciences in the Harappan civilization. The present chapter as well as the one following it are intended to give such an overview. In the first of these two chapters, a sketch of some prominent technologies as pottery, transport, textile, etc. are given, while that of metal technology in the second. The reason for having a separate and somewhat detailed discussion on metal technology is that among the various techniques it has the greatest potentials of moving towards what is sometimes described as the approximation to international science.

I am indebted to my young colleagues Santanu Maity and his wife Sujata Maity—both of whom are working archaeologists—for sketching out the present chapter. For the next one I am most grateful to D.P. Agrawal—one of our leading archaeological scientists—for the permission of using his article published sometimes back in the *Indian Journal of the History of Science* : D.C.

2. POTTERY AND CERAMIC

In archaeology pottery is regarded as the index to culture. To an archaeologist pottery assumes the importance of the alphabet. Though profusely found in archaeological excavations and explorations in the cultural frontiers of ancient India most of these are not tested in the laboratory. Generally archaeologists are inclined to the typological study of the sherds and their probable correlation with others. In most cases technological analysis is not properly done. But the technological study in details—such as the composition of clay, tempering agents, nature and components of the slips or glaze, nature of firing, porosity test and different chemical analyses—may throw light on the technological advancement of a culture. Not only that, the scientific study of pottery "throws light on almost every aspect of the past—social, cultural, religious, economic, political and what not."[1] Though the importance of pottery for archaeological reconstruction of past cultures is immense, yet "Archaeologists, in India no less than other parts of the world, have often been imprecise in the terms they employ in describing pottery, and no single system of description has so far been accepted."[2]

Each and every object, found either in excavations or in explorations has its own history of technology. But their finders are not always in a position to determine their technological "make up", either due to ignorance or sloth. But so far as the understanding of the Harappan Ceramic technology is concerned, we are indebted mainly to E.J.H. Mackay, who has discussed it elaborately, both in Marshall's book and his own.[3] Not only pottery ; Mackay has also discussed the techniques of manufacturing other objects, which are on the whole accepted by the scholars of the succeeding period. I have before me a book with the title : *A study of Harappan Pottery* by O. Manchanda.[4] In this book what is discussed about 'General Characteristics' of Harappan pottery—seems to be an echo of Mackay. Even what Sankalia discussed in his *Some Aspects of Prehistoric Technology in India* is no exception to this. Mackay is there-

1. B. B. Lal in PAI 3.
2. B & R Allchin BIC 287.
3. J. Marshall MIC Vol. I ; Mackay FEM.
4. O. Manchanda SHP.

fore the pioneer in scientific approach to the subject. In our own discussion we have generally tried to follow him.

Since the twenties of our century innumerable Harappan sites have come to light in a widely distributed area, thanks to the field-work of the archaeologists of many countries. Recently the homogeneity of the Harappan culture in various aspects has been challenged by some scholars on the basis of fresh archaeological data.[5] In fact, regional diversity of any culture is an accepted phenomenon in the field of archaeology. The Harappan culture is no exception to this. But the pottery technique of the Early, Mature and Late Harappan phases was basically the same. It is from this technique and its products that the dynamism of a culture either in its mature or decadent forms can be easily ascertained though along with its regional diversities. While discussing pottery of the Harappan period, the Allchins in their earlier work observed :

> The extra-ordinary self assurance of the pottery of the Harappan period calls for comparison with the crafts and arts of the Roman empire or of Victorian Britain. There is extreme standardization, and technical excellence goes hand in hand.... Even after allowing for the increasing 'Indian-ness' of the craft, its roots in the west, in Baluchistan and Iran, are undeniable.[6]

But in their subsequent work, they seem to put greater emphasis on the more "Indianness" of the technique : "Mature Harappan pottery represents a blend of the ceramic traditions of Baluchistan on the one hand, and those of India east of the Indus (as exemplified in Fabric A at Kalibangan) on the other.. it developed its own somewhat stolid character...competent and self-assured."[7]

. It reminds us of Marshall who said[8] :

> The very multiplicity and variety of its shapes—most of them peculiar to the Indus Valley and quite distinct from those of Persia and Mesopotamia—are evidence enough that the craft of the potter had been practised there from time immemorial.

In the same volume E.J.H. Mackay[9] observed :

5. B. K. Thapar in B. B. Lal (ed.) FIC 14-15.
6. B & R Allchin BIC 288.
7. B & R Allchin RCIP 197.
8. J. Marshall MIC I, I. 28
9. Mackay in Marshall's MIC, I, 287.

Little of the pottery from Mohenjodaro resembles in shape that hitherto found in Mosopotamia and Elam. The wares even of adjacent countries are seldom much alike in form; each race evolves such shapes as are suitable and uses them with but slight modifications over long periods. It will be seen that most of the pottery forms are far removed from being primitive, showing that the potter's craft was well advanced.

On the basis of the above observations, we are inclined to suggest that for the history of origin and technique of Harappan pottery, we need not look abroad. It had its own technological development in Indian subcontinent, neither in Egypt and Iran, nor for that matter in the "fertile crescent".

The Harappan pottery is generally called as the Black on Red Ware, because this constitutes the majority of the entire assemblage. But excavations and explorations at major Harppan sites have yielded many other varieties, such as Red Ware, Buff Ware, Gray Ware and Black and Red Ware etc. Of course, such categorization is based on broad divisions and is not strictly applicable. These potteries are either plain or of the painted varieties; except at Kalibangan nowhere the percentage of painted and unpainted varieties has been absolutely detected. The potteries are of two fabrics—fine and coarse—and the section varies from thin to thick according to their shapes. It has already been mentioned that the shapes of the Harappan pottery are quite different from those of Iran, Mesopotamia and Elam. Nigam[10] likes to classify the shapes into the following groups:

(A) The shapes which are met only in the Harappan civilization. These include goblets, beakers, perforated jars, certain types of jars, and vases etc. (B) In this group can be placed such shapes which occur in the pre-Harappan cultures and continue in the Post-Harappan period in association with other cultures. Here are such common shapes viz., dish-on-stand, cup-on-stand, bowls, dishes, basins, caskets, vases, jars, lids and ring-stands etc. (C) Zoomorphic containers, and (D) The Black and Red Ware.

Such considerable varieties of shape would not have been possible if the potter's craft was less advanced. Technological inheritance for generations was the prime force behind this perfection.

10. J. S. Nigam, in EIP 136.

The quality of pottery depends to a great extent on the selection and preparation of clay. The textural integrity and evenness of the surface of the pottery is obtained by means of selecting fine clay devoid of rough and coarser elements. The potters of the Harappan period had the opportunity of utilising and employing local alluvial clay for pot making. In fact, this was easily available from the bed of the rivers and was not much laborious for levigation. Well levigated clay was essential for pottery. On the basis of the practical study of the nature of the clay, the Harappan potters would have selected the natural alluvial clay. Not only that ; they might have their efficient process of levigation of clay for the production of the pottery in a mass scale, though the archaeological evidences for this are still lacking. Evidences for the preparation of clay, one of the most vital aspect of potting technique, are also still lacking. Yet most probably in ancient Sind the process was the same as it is today. The potters were content merely to beat the clay dry, and to strain it through the canesieve. But another easy method of collecting dry clay, picking up and throwing out the rough and coarser agents and thoroughly kneading it with the required water quantity might have also been followed by them.

On analysis it is found that Harappan Wares contain sands and lime, either singly or both ; these were mixed with the clay of both painted and unpainted pottery. Sand is regarded by the potters as the 'bone' of pottery. Mica has also been traced as a tempering agent or degraissant along with sand or lime, and so far mica is concerned the mixing was probably not deliberate, because mica is naturally found in the sand on the banks of the Indus. Both sand and mica are essential for the smooth handling of clay on the wheel and they also save the pottery from cracking while being dried in the sun. Sometimes smaller jars were made of a special paste, devoid of sand and lime. However, when burnt the clay generally turned pink or light red and this was due to the presence of iron compounds in the clay. For Gray Wares, according to Mackay, a paste of slate colour—in some cases approaching black—was employed. In his view :[11]

11. Mackay FEM 174-75.

Clay of this kind apparently required no tempering, and only in exceptional cases was any other substance mixed with it for this purpose. Whether the colour of the pottery made from it was natural or was darkened by the admixture of some carbonaceous material with the clay has yet to be determined. I am inclined to think that as this pottery varies considerably in tint something was added in varying proportions with the purpose of darkening it.

Hamid suggests that this was due to the presence of ferrous iron in the clay.[12]

Potteries are shaped either by hand or on wheel or by moulding. A few handmade sherds are found in the entire assemblage of Harappan pottery so far and the hand-modelling was also rather crude, which could be the work of children. No moulded variety has yet been reported by the excavators. In fact, almost all the pottery assemblage of Mohenjo-daro, Harappa and other sites are wheelmade; but scholars are not unanimous about the wheel on which these were shaped, because the material representation of an actual potter's wheel has not yet been found. But both Mackay and the Allchins, studying the evenness and striation marks on the pottery have suggested the use of footwheel, which is still found at Beluchinstan and Sind. Mackay likes to give credit to the Harappan (and not to the Aryans) for introducing footwheel in the region. The Allchins[13] observe:

The foot-wheel is still in use in Sind and parts of the Punjab, Saurashtra, and the North-west Frontier province, in contrast to the Indian spun-wheel found east of the Indus, and may be taken with some degree of certanity to be a legacy of this period. The modern foot-wheel closely resembles those which are found right across Iran and into Mesopotamia, both today and probably in ancient times.

Though the use of the foot-wheel has been suggested by Mackay and the Allchins, the possibility of using hand-wheels for potting cannot be altogether ruled out, because hand-wheels have also produced wares not inferior to the Harappan pottery. Without the actual evidence of the wheel used in that period, we may suggest the simultaneous use of both types of wheel. The advantages of a particular type over the other cannot always completely oust the less advantageous one; it is rather likely to be dominant.

12. Marshall, MIC, I, 311 note-2.
13. B & R Allchin RCIP 199.

The majority of the vessels of Harappan period have a flat base, yet round-bottomed or pointed based potteries are not altogether rare. The flat base of a pot was more convenient for flat brick-paved floors. These flat bases have groovings which indicate that the pots were removed from the wheel by a string or grass in slow motion. The string was held either between two hands of the potter or one end of the string was tied to the little finger of the potter and the other stuck again on the base of the pot which was to be removed from the wheel, so that while the wheel was in motion the jar was automatically cut off. The bases of Harappan potteries are not well finished in comparison to the upper portion. Trimming on the lower portion is a common feature of the Harappan pottery. It was done in vertical direction by a knife or some similar instrument.

Sometimes the jars having angular shoulders were at first made in two pieces which were fitted together when wet and again they were put on the wheel for the final touch. But these fitting points are not easily detectable due to their obliteration by trimmings both inside and outside. Offering stands other than the jars were also made in two pieces—the first part is that of base and stem and the second part is the pan. It is likely that after the modelling of the pots they were dried in a steady temperature, and to reduce porosity burnishing was done by mechanical friction on the surface in a 'leather-hard' stage.

The natural surface of the pottery was hidden, because slip was used on its surface before firing, which added to the pot a polish, lustre and bright hue. Not only that. It was also used as a preventive against the porosity of the pottery with the aid of a mop. Sometimes slips were applied to the pot more than once according to necessity. The slips were of various colours, viz., cream, white, black, chocolate, purple, pink etc. Slips of red oxide for better class pottery were used. Black slip was probably obtained by using lamp black or powdered charcoal added to the levigated clay. The juice of Tuthi (*abutilon Indicum*) was probably applied to give a black gloss on the surface.

For decorative devices some other treatments were done before firing: cording, incising, scoring, perforating or by employing impressions on the pottery. For cording, a cord was wound around a vessel when it was on the wheel, either in slow motion or in a stationery state. Incising was generally confined to the bottom of the pans, and it was probably done by a pointed tool or metal comb. Scoring was executed generally on the upper part of the pottery with a very sharp tool, most probably a metal comb. Many perforated cylindrical vessels and potsherds are found at several Harappan sites, the purpose of which is still undetermined. Mackay observes[14]:

The pattern was scratched on the unbaked clay of this perforated ware to guide the work of the cutter; some of these marks remain, though they should all have been removed in the process of cutting the incisions.

Impressions were executed probably by a wooden stamp before the baking of the pot. But another type of incising needs to be mentioned. These are the graffiti marks on the pottery. This was generally done with the aid of a pointed tool after the baking. One such sherd bears the representation of a boat. Such graffiti marks have also been found on the gray sherds.

Paintings were drawn on the potteries before or after baking. Paintings which were executed on the pottery before baking were composed of inorganic substances and due to chemical changes in firing a considerable colour range was achieved. It is observed in the *Early Indus Civilisation* that "The slip used for a great part of the painted ware is red ochre,... After a jar had been shaped on the wheel, the ochre was painted thickly over it, ... The designs....were painted on this red surface with a brush before firing, the material used being a thick, black, or purplish-black, paint made from manganiferous haematite".[15]

Baking of pottery is one of the most vital stages of that craft—the stage in which the Indus potters showed their excellent skill and capacity. Mackay[16] observes :

The pottery of Mohenjo-daro was well baked. The resulting fabric was hard enough to stand a considerable amount of knocking about, and its uniformity of colour shows that the potter had considerable

14. Mackay FEM 181.
15. Mackay EIC 109.
16. Mackay FEM 176.

control over the heat of his furnace........ We have found comparatively few over-baked vessels that had been put into use; nor do I remember a jar or potsherd that could be said to be underfired, though some mistakes must have been inevitable.

Due to firing—either in kilns or in open fire—some chemical changes occurred in the clay and these changes depended mostly on the composition of the clay, the temperature of firing etc. As Mackay[17] says:

> The firing of a jar appreciably altered its colour, changing a light grey clay to a pinkish-red shade owing to the checimal action of the small amout of iron in it. Those very few specimens which are definitely over-fired have a distinctly greenish hue, which is due to the presence of a complex ferrous compound in whose formation lime plays an important part.

In Marshall's *Mohenjo-daro and Indus Civilization* vol. 1., Mackay, due to the lack of evidence reported that probably the potteries were burnt in open fires. But in his own later work he has changed his opinion considerably due to the findings of the potter's kilns in subsequent exacavations. Thus, he observes[18]:

> In the more recent excavations, however, we have found quite a number of kilns, which judging from the masses of potsherds around them were exclusively used for the baking of pottery. The remains of two such kilns were unearthed.

Following Mackay, Hegde has discussed the kilns in his article *Ancient Indian Pottery-Kilns*. The description of the kilns, in Mackay's[19] words, is:

> The main features.....are a pit for the wood or reed fuel and a domed compartment above to hold the vessels to be baked, communication between the two being effected by round holes in the floor of the upper chamber.

Certainly kiln baking had some advances over the open firing method, yet Hegde is probably right when he says[20],

> It is doubtful if much of the Harappan pottery was baked in kilns.... The rarity of ancient Indian pottery-kilns in the archaeo-

17. *Ibid.*
18. *Ibid* 177.
19. *Ibid.*
20. K. T. M. Hegde "Ancient Indian Pottery-Kilns" in *Puratattva* No. 9, 1977-78, 110.

logical record suggests that more than 99 percent of the ancient Indian pottery was baked in open fires. But unfortunately for the archaeologists, this baking device does not leave behind tangible evidence.

Actual observations have shown that potteries produced by open-fire is not inferior to the kiln-burnt ones. The large scale demand of the Harappan population inclusive of trade certainly required a mass production of pottery which could not be fired only in the constructed furnaces or kilns. It is probable that some special class of pottery for which special care in baking was needed was burnt in the kilns, while the others were prepared by open fire method.

After baking, a pot is ready to serve its purpose. But sometimes paintings are executed on it for decorative purposes. Harappan potteries are inclusive of painted varieties. These are either monochrome or polychrome. It has already been suggested that the clay and other components of plain and painted wares were almost the same. Sometimes a raised effect on the surface was offered by means of applying paints on the already polished slip by hair brushes of varying fineness or by a reed pen. For polychrome wares red ochre and a green pigment "terre verte" were used; but in polychrome wares painting was generally executed after baking. Innumerable designs were painted: intersecting circles, triangles, squares, lozenges, floral and animal designs etc. It is most probable that some of the designs not only required knowledge of geometry but also some instruments such as the compass and scale. Not only that; some anatomical data appear to be embodied in depicting animals and human motifs.

Lastly, we may have some words on the so-called glazed wares or Reserved Slip Wares and Black and Red Wares. Practically speaking, there is little difference between glazed and Reserved Slip Wares. Their process of manufacturing is almost identical. F.R. Allchin has expressed his doubt whether these were truly glazed or had only a surface gloss due to polish.[21] Mackay himself was doubtful also in describing

21. Personal discussion of F.R. Allchin with D.K. Chakrabarti, which is mentioned by Chakrabarti in this article 'Reserved Ship Ware in the Harappan Context", in *Puratattva,* No. 8, 1975-76, 159:

these sherds as 'truly glazed'.[22] Chakrabarti thinks that for these as 'reserved slip wares' *no extra-Indian origin can be traced*. Whatever may be the case, according to Sankalia five stages were required to prepare this type of pottery. These are :[23]

1. Coating of the pot, after modelling with a coloured clay.
2. Burnishing of the coated surface with a blunt instrument.
3. Application of a slip and drying of the pot in the sun.
4. Removal of a part of the slip by a comb-like tool, thus creating a pattern or design.
5. Firing of the pot at a high temperature.

These wares are not frequently found in the Harappan period. The so-called glazed wares were "made of a light gray clay of medium thickness, well-baked and fairly tough. After the application of a purplish black slip probably composed of manganese, a thorough burnishing was done on the

"F.R. Allchin points out that it is not clear from the published description whether the pottery was truly 'glazed' or as seems probable, only with a surface gloss resulting from the fusion of particles of the dressing. There should be a glass surface in a truly glazed pottery and this surface can be achieved through a number of techniques. Accidental fusion of particles of the dressing need not be called a true glaze."

22. In Marshall MIC II, 577-78. Mackay under the heading 'Glazed pottery' has actually described two glossy terracotta (pottery) beads and sherds which look very like copies of Mosaic glass along with the result of chemical analysis of the beads by Hamid. More interesting is the foot-note no. 2 on page 578 of the same volume. "The term 'glaze', here as well as in some other places, is used rather loosely for want of better name for this substance, which though glassy in appearance is not a true glass—ancient glasses being essentially soda-lime silicates."

In FEM 188, Mackay observes : "This glazed ware is so like mosaic glass that one feels that some craftsman experimenting in this direction perhaps produced a vessel composed entirely of ropy glaze which had fused throughout and could, therefore, not be truly termed as glass."

Thus it is clear that Mackay was a bit hesitant to accept these sherds as glazed wares. However, if we accept these as glazed wares, the Harappans may be credited as the first users of these sherds in the world.

23. H.D. Sankalia SAPTI 12.

surface. After that a glaze was applied to the surfaces but certainly before firing a portion of both glaze and slip were removed by a comb-like instrument in such a way so as to leave a straight or wavy decorative mark on them.

Black and Red wares are found in the Harappan context mainly at the Saurashtra area. The pottery is black in the interior and red on the surface. H.N. Singh has discussed this ware elaborately in his book *History and Archaeology of Black and Red Ware.*[24] His work is indeed an impressive effort and deserves special mention. It is interesting to note that scholars are not unanimous regarding the technological make-up of the Black and Red Ware. Scholars in India and abroad have advocated divergent theories, which more or less concentrate on the technique of firing. More scientific study and research is likely to throw light on other aspects of this subject, such as the preparation of clay and application of slip etc. However, we may at the present state of research accept Singh's[25] comment :

> The technological personality of Black and Red Ware is poignantly diverse for divergent views have been expounded in respect of the technological make up, for example, inverted, single and/or double firing technique of manufacture. There may be still more complex methods of its manufacture which, however, are not firmly tested. This shows a changing technological personality of the Black and Red Ware.

3. TERRACOTTA

Generally baked clay objects are regarded as terracotta and so it is reasonable to connect these with potteries. Terracottas have been found in many perhistoric sites throughout the world and Indian sub-continent is no exception to this. In many prehistoric sites older than those of Harappan we have come across varieties of terracotta in considerable number. The Harappan sites have also yielded the same in great number. The occurrence of a large number of terracotta specimens in Harappan sites is due to the easy availability of alluvial clay —the core material of terracotta—and its plasticity helped a good deal in modelling. A careful study may reveal a techno-

24. H.N. Singh HABRW 1982.
25. *Ibid* 65.

typological continuation of this craft, somehow in a reformed way till now.

The Harappan terracottas include many varieties, such as human and animal figurines, gamesmen and toys and many other models.

The paste of clay used for the modelling of terracotta was virtually the same which was used for making the potteries. This clay burned to a light red colour or dark pink, according to the degree of heat. Probably the source of clay for the potters and terracotta-modellers was the same and it is likely that potters also had some share in modelling other terracotta objects, along with the specialists of this craft. The same 'degraissants', viz., lime and mica have been found mixed with the clay, used for the modelling of other terracotta objects. All the specimens are modelled generally by hand and only a few of them have been found as shaped in mould, viz., the masks and the fine bull. The human figurines, most of which are broken, were made solid. Their physical description is unnecessary for our discussion. These were very simply modelled by hand and some 'applique' ornamentation was executed on them. Male figures are rare and female figurines mostly of "Mother Goddesses" are conspicuous in their typical representation. The eyes were represented by pellets of clay, usually oval in shape. The mouth was portrayed by elongated pellets with horizontal incised line to demarcate roughly the lips. The nose was made prominent by simply pinching-up of a portion of the clay.

In modelling of the animals, however, a different method was followed. Mackay's observation on these may usefully be quoted here :[26]

> Unlike the human figures, many of the larger-made animals are hollow inside. Some of them must have been made on a core, but of what material the core was, it is as yet impossible to say, for the inner surfaces of the broken figures are uniformly smooth, though uneven. The core was clearly combustible, since it leaves no trace behind. There are always vent-holes in the unbroken figures, evidently made to permit of the escape of the gases formed in burning the materials of the core. Other figurines were made in a mould. It is easier to press a thin sheet than a thick mass of clay into the crevices of the mould. With the exception of the mask-like faces and the

26. Mackay in Marshall MIC I. 349.

fine bull......which were certainly all made in a mould, the pottery figures of both humans and animals were entirely modelled by hand.

Sometimes the eyes of the animals were represented by a deep incision in the clay and small round pellets were inserted into it to represent the pupil. Sometimes only the pellets were inserted without any incised border line. It is interesting to note that wrinkles in the skin are portrayed by means of incised lines, and heavy folds by the addition of strips of clay to add a 'realistic' touch. In some case the animals stand on a flat pottery base. The animals, represented as models, are dove, peacock, squirrel, mongoose, hare, elephant, monkeys, dogs, pigs, ram, bull, rhinoceros etc.

Sometimes red painting was applied on the models, though the majority of them are devoid of any paint. Better specimens were treated with a cream coloured slip or were washed over with a dark red paint.

Mention should also be made of some terracotta toys like rattles, figures with movable heads, etc. The rattles were roundish in shape with small terracotta pellets inside. These were all made of light red ware. Probably some combustible elements with some already baked pellets were used as the core which was covered with clay and the modelling of the rattle was efficiently done by hand. After firing, the combustible elements were turned into gases leaving the baked pellets inside. We have not found any vent holes for the passing of the gases after the combustion. Probably the gases would pass through the porous fabric of the pottery and to maintain the porosity no slip was probably applied on the rattles. Whatever might have been the case, all the terracotta models were generally well-fired.

The toy animals or human figures with movable heads or hands were separately made and then attached to each other by fastening a string or stiff hair in between.

Like the pottery the terracotta models of the Harappan period betray a sophisticated knowhow. In fact, technical process of the terracotta modelling was not very much different from that of pottery. The basic difference was that the potteries were generally made on wheel while the terracotta models were shaped by hands and sometimes by moulds.

4. TEXTILE

Textile specimen of the Harappan period is extremely scanty due to their perishability. A few specimens which are available were due to their preservation by being impregnated with metallic salts—either of silver or of copper. Instead of conjecturing on what may possibly be found by future excavations we have at present to depend only on what is already found, though this gives us only a very inadequate idea of the textile technology of the Harappan period.

It is almost certain that spinning and weaving were known to the Harappans, as is attested by the discovery of actual woven fabrics, spindle whorls, bobbins, etc.

For spinning of threads, spindle whorls are necessary and considerable numbers of these spindle whorls—chiefly of terracotta and also of expansive faience and shell—are found from almost every Harappan site. Marshall is of the opinion that spinning was a household craft practised both by the well-to-do and the poor alike.

Fragments of cotton textile, dyed red with madder have been found in Mohenjo-daro. Marshall observes :[27]

> The cotton resembles the coarser varieties of present-day Indian cottons, and was produced from a plant closely related to *Gossypium arboreum* or one of its varieties.

This type of cotton was not obtained from any wild species but rather from cultivated plants. However, Mackay has cited the results of the examination of the specimen by James Turner, Director of the Technological Research Laboratory, Bombay. Turner[28] in his preliminary report remarks :

> The fibre was exceedingly tender and broke under very small stresses. However, some preparations were obtained revealing the convoluted structure characteristic of cotton. All the fibres examined were completely penetrated by *fungal hyphae*.... As this examination has been confined to a fragment measuring 0.1 inch in one direction by 0.3 inch in the other direction, these results can only be regarded as tentative.
>
> (1) Fibre : cotton.
> (2) Weight of fabric : 2 Oz. per square yard.
> (3) Counts of Warp 34's. Counts of Weft : 34's.

27. Marshall MIC I, 33.
28. *Ibid*, II. 585.

(4) Ends (Warp threads) : 20 per inch Pock (Weft threads) : 60 per inch.

The results of the examination of other specimens have also been incorporated in Mackay's *Further Excavation at Mohenjo-daro.*[29] The examination of one such specimen on copper razor by A.N. Gulati by means of visual inspection, examination of the fabric with magnifier and a microliner counter, etc. has yielded the following results.[30]

> The weave of this fabric was plain, i.e. with an almost equal number of picks and ends per unit length.... The mean of these observations gives 44 picks and 43 ends per inch. The diameter of the yarn varies between 1/50th and 1/80th of an inch...... That the counts of the yarns used in the manufacture of the fabric under examination lie between 13's and 17's. Taking 15's as the average value, the weight of the fabric would be 4 ozs. per square yard. It is therefore to be concluded that the material used in the manufacture of this fabric was cotton.

The use of bast which is highly fragile for weaving has not yet been clearly found from the examined specimes but this was used as string or cord on some metal objects. These were not of convoluted structure and are not properly identified. These were either of jute or flax or of straw or of some wild grass.

Some scholars have suggested that wool was probably used for clothing ; but there is no evidence for this so far unearthed.

Probably the Harappans were the first in the world to utilise cotton for manufacturing their clothes and garments.

From no Harappan sites is so far discovered any trace of a loom by which weaving was practised. Probably these were made of perishable materials. Whether these were simple horizontal ground loom or the vertical two-beamed loom or the warp-weighted loom is not known. But textile technology was advanced enough to meet the growing need of consumers. The recovered specimens of textile from Mohenjo-daro are enough to give a fair idea of the development of textile technology in the Harappan period. As the Allchins[31] comment :

> That woven textiles were already common in the Indus civilization, and that the craft for which India has remained famous was already

29. Mackay FEM 591-594.
30. *Ibid* 592-93.
31. Allchins BIC 293.

in a mature stage of development, must be inferred from this single find, and from occasional impressions of textiles upon earthen ware, pottery and faience from the Harappan sites.

Therefore, the history of textile technology in India may be traced back to a period approximately 4500 years back. This is one example of how the Harappans surpassed the Sumerians or Egyptians in many respects of technology.

5. TRANSPORT

The invention of the wheel is well-known for its revolutionary consequences. We shall discuss here mainly how it facilitated transport in the Harappan culture.

Growing archaeological evidences, though indirect in nature, throw ample light on the transport technology of the Harappans which they effectively devised to suit both land and water.

We may first discuss the land transport. For inland transport bullock-carts were used—a considerable number of model representations of which in terracotta have been found in many Harappan sites. From these models along with their present equivalents still functioning in the Sind region one can easily reconstruct the actual specimen which was most frequently used. As Piggott observes[32] :

> For slower and heavier transport, the ox-cart was extensively used in the Harappa culture. Models of carts in clay are among the commonest antiquities on the prehistoric sites of the Punjab and Sind, and the type represented is exactly that which today cracks and groans with its ungreased, nearly solid wheels in the villages round Mohenjodaro, ... The type, we now know, is unchanged even in its wheel-base, for the recent Harappa excavations have revealed cart-ruts belonging to an early phase of the city's occupations having a width of some 3 feet 6 inches, which is that of the modern carts in Sind. Such things have a long survial-value—the standard gauge of modern English carts was already fixed by the third century B.C., in these islands.

S.R. Rao[33] attempts to reconstruct the models of the carts in more detail :

> Three main types from Lothal are reconstructed with the help of the toy wheels and cart frames found in excavation. The first type has a solid chassis, which is concave or flat. The second and third

32. Piggott 176.
33. S.R. Rao LIC 123.

types have a perforated chassis, but the latter has, in addition, a detachable cross-bar. On such a chassis wooden posts were fixed to form a boxlike frame..... The wheels of Lothal carts were attached to the free projecting ends of the axle which itself was secured with leather straps to the main frame. Lynch-pins seem to have held the wheels in position. The carts with a detachable cross-bar and those with a chassis made up of two curved bars were confined to Saurashtra. Perhaps the latter were meant for carrying light loads, while the other two types carried heavy loads.

In this context Gordon Childe's[34] reconstruction of Harappan cart may also be mentioned :

In the Indus Valley, numerous models illustrate the under-carriages of the carts of the Harappa civilization, which were constructed in just the same way as the contemporary village carts of Sindh, whose axle turns with the wheels. The frame consisted of two curved beams set parallel, and joined by two to six cross-bars. The pole might run under the cross-bars or be mortised into the foremost. Two or three corresponding holes in each of the side-beams must have held upright poles to contain the box of the vehicle, presumably of wickerwork. A pair of holes at the centre of each side-beam would have held pegs projecting downwards to fit on either side of the axle. Thanks to this arrangement, it is easy to dismantle the vehicle by simply lifting the frame off the axles.

In fact, Childe's reconstruction appears to be more scientific than that of others.

Mention may also be made of another model resembling modern Ekka, the miniature bronze and copper models of which have been found from Harappa, Chanhu-daro etc. This model looks like a little covered trap with a canopy and curtains of fabrics set up on four poles above the cart-frame. In Harappan model is a seated driver, well forward between the shafts and observed as holding the reins with one hand. Gordon Childe also has accepted these as passenger carts. As Childe describes[35] :

A chassis of the usual type supports a light body covered with a gabled roof. The passengers sit back to back, and in one case the driver sits on the front cross-bar of the frame. Similar vehicles can still be seen in India.

34. Childe in HT Vol. I 717-18.
35. *Ibid* 719.
36. S.R. Rao LIC 123.

Rao[36] conjectures that these were the same as the horse-drawn chariots, described in the *Rgveda,* for which direct evidence at our disposal is at present indeed meagre. Whether spoked wheels were in use or not is a matter of controversy and we have to wait for more evidence.

The long distance journeys through wooded or undulating tracts must have been conducted by caravans of pack-oxen.

For long distance trade and commerce, land transport only was not sufficient and to supplement the land transport an elaborate water-transport network was definitely devised by the Harappans. It was equally or more important than land-transport. In this case also our evidences are not direct, though it is generally admitted that boat and ships were built in this period. Their shapes and technological setup can be reconstructed mainly on the basis of their representations either on seals or on potsherds or even from the terracotta models. S.R. Rao claims that Lothal was a boat building centre too.[37]

From the roughly sketched boat on a potsherd from Mohenjo-daro the following technical characteristics may be discerned[38] :

> It has a sharply upturned prow and stern and is apparently controlled by a single oar. The mast may possibly be a tripod, and one or two yards are shown, or, conceivably, one only, the second line representing a furled sail.

This type was probably river-boats, from which goods were loaded on a shelving bank. On a reused seal the representation of another boat has been found, which is 'mastless with a cabin and a man at oar'. From this representation it is presumed that the real one was probably lashed together at both bow and stern—perhaps an indication of their making of reeds, like primitive Egyptian boats. In the centre there is a hut-like representation, the prototype of a modern cabin, which was also probably made either of wood or of reed. A steersman could sit at a rudder or steering oar on a raised platform over the bow. This was also used probably for riverine traffic. From the terracotta models of boat from Lothal three major types of it are distinguishable ; of these, two types had sails and the other was without sails. In a complete terracotta model of

37. *Ibid* 124.
38. Mackay FEM 183.

boat, according to Rao,[39] a sharp keel, pointed prow and a high flat stern are observed along with three blind holes, one near the stern the other near the prow and the rest near the edge of the boat for fixing the mast, for fastening the ropes of the sails and for supporting the oar respectively. In other models pointed keels, raised sterns and prows are noticeable while in the third type flat base and pointed prow are prominent with the absence of provision for mast.

Rao is of opinion that "the first two types of sharp-keeled boats with provision for the mast must have sailed on high seas, whereas the third type resembling canoes was used in the estuary only."[40] On two potsherds found at Lothal painted depiction of multioared boats have been found, which were most probably used in ancient Sind.

More direct evidences for the anchorage of vessels have also come from Lothal. These are actual stone anchors, circular or triangular in plan with rectangular section and one perforated right across for passing the ropes. "Blind-holes" are also bored on them for anchor-fixing. The identification of these circular and triangular perforated stones as anchors, as is proposed by Rao,[41] is not free from doubt. More scientific and serious investigation may determine their true nature.

The transport technology of the Harappan period, in all probability, has somehow been transmitted to the succeeding period without little change, the traces of which have already been cited by eminent scholars.[42] In comparison to present technology the transport technology of the Harappan period may well be regarded as pre-historic, but in that period it was no less advanced than that of the contemporary Egyptian or Mesopotamian civilization.

39. S.R. Rao LIC 124.
40. *Ibid*.
41. *Ibid* 125.
42. Piggot, (PI. 176) ; Allchins (BIC 272-73) ; Childe (HT, Vol. I, 717-19) ; and Rao (LIC, 123-25), etc. are inclined to draw the analogy of the Harappan bullock carts with the carts in Sind of the 20th century. They also like to see similarity of modern 'ekka' with that of the Harappan models. Rao and Allchins are of the opinion that the present-day country boats and the Arab Dhows have probably some similarities with the Harappan boats.

6. TECHNOLOGY OF SOME STONE OBJECTS

Besides the use of metals like copper and bronze, the Harappans did not give up stone to meet their daily necessities ; rather they had exploited it to the fullest for various purposes. Varieties of objects made of either hard or soft stone were used by them.

Of these we shall discuss mainly the technology of stone blades, boxes, vases, statues and some other minor objects. The technology of stone beads would be discussed separately. Like their ancestors the Harappans were accustomed to the making of stone implements and articles in such a way as to transmit the technology to the Late Harappan Period or even to other chalcolithic peoples after dispersal from their original centres.

The Allchins observe, 'In spite of the commonness of metals, stone was not abandoned, and chert blades were prepared from cores which in turn had probably been exported from such great factories as that at Sukkur (Sakhar)'.[43] Not only this ; the stone blades were imported 'to form a uniform item of equipment at Harappa, Mohenjo-daro, Lothal and Rangpur, Kot-Diji and Kalibangan.'[44]

These blade-flakes of flint|chert were long and parallel-sided, and the technology followed to prepare them was but an inheritance from the preceding periods. Sankalia[45] is of the opinion that this technique was discovered in the upper palaeolithic period. The blade-flakes were long and narrow with more or less parallel edges. These blades were made from a carefully prepared core either by "pressure flaking" technique or by "crested ridge" technique—probably the latter.[46]

In this technique, all the irregularities or whatever could be easily removed from a chalcedony nodule were first removed by a round stone hammer. Secondly, a ridge was prepared along the length of the prepared core by alternate flaking. This ridge is supposed either to guide the regular removal of parallel-edged flakes or it creates a line of weakness which makes it easy for the removal of the first series of flakes.

Such crested ridge flakes and cores still having a ridge on them are found in the Harappan and later chalcolithic cultures.

43. B. & R. Allchin BIC 134.
44. *Ibid,* 268-269.
45. H. D. Sankalia SAPTI 3.
46. *Ibid* 5.

At Mohenjo-daro, flint blades have been found from almost every house. From the discovered specimens the following characteristics have been deduced by the excavators : (i) The flakes were not short ; (ii) The form of flint at Mohenjo-daro was a 'brownish-grey chert' which was often mottled ; (iii) No trace of patina is observed either on core or on flake.[47] But the observation of Mackay in the second phase of excavation work at Mohenjo-daro is worth adding to this. He says, 'We have in recent years unearthed comparatively few flint flakes or the cores from which they were struck. Those that have been found, whether in the early or later levels, are simple, long, thin, rectangular specimens, trapezoidal in section, with untouched edges and only occasionally with the bulb or percussion at the end. In a great number of cases the bulb seems to have been deliberately removed by snapping it off, with the idea of making the blade as uniform as possible throughout its length... doubtless these ribbon flakes were mostly used as knives.'[48] Mackay is also of the opinion that the flakes were prepared from the 'imported core' and 'no great degree of skill' was required for flaking from the core ; only a little practice was enough for the purpose. He also indicates that the possible sources of flint was 'probably the lime stone outcrops at Sukkur' which is supported by the recent observations of the Allchins :[49]

> At Sukkur blades of the type found in Mohenjo-daro and other settlements were also made in large quantities at extensive working areas using the same sources of material as Palaeolithic craftsmen had done earlier. The Harappan craftsmen cleared the earlier materials and flint nodules that had suffered long exposure to the elements, and piled them up in long ridges, or threw them down the sides of convenient hills. They then used the freshly exposed nodules of flint for blade production. The craftsmen appear to have sat cross-legged at chosen spots in the cleared area, and the actual places where they sat could still be recognized when we visited the site in 1976—some four thousand years later ! Small kidney-shaped areas were completely cleared of flints down to the residual soil, and concentrations of waste materials, cores and blades could be seen beside them. Studies of the stone industries from settlements of the Mature Indus Period show a distinct and probably rapid development and

47. Mackay in Marshall MIC II. 485.
48. Mackay FEM 395.
49. B. & R. Allchin RCIP 196-97.

convergence from the more varied, individual assemblages recorded at Early Indus and all earlier settlements. The range of artefacts made from blades (in which the local and regional distinctions were seen), drops away almost entirely. What is left is an industry consisting of long, regular blades made from carefully prepared cores of very high quality material obtained from outstandingly good sources of supply. The blade industry of this period is highly professional, and this craft shows an effortless competence, without apparently any desire to produce novel or special results. This is a clear example of the kind of craft specialization that took place at the beginning of the Mature Indus urban period.

A more recent study on the functional aspect of the chipped stone tools has been made by J. M. Kenoyer on the basis of the fresh recovery of the implements from the excavation at Mohenjo-daro by Dales in 1964-65. Jonathan M. Kenoyer has made a careful scientific study of the techno-functional aspect of the tools and his study reveals that stone tools were indispensable, more economical and in some cases more efficient than metal tools.[50]

Bridget Allchin in her 'Substitute Stones' has also thrown some light on the functional aspects along with the technological make-up of some stone objects of Harrapan period. In her concluding paragraph she observes: "the Harappans, far from being a non-violent people, or being without any system of military organization other than that needed to defend their major cities, as has sometimes been suggested, must have had an effective army."[15] Such a conjecture seems to require further confirmation by evidences—direct or circumstantial.

The discovery of a considerable number of saddle querns and muller stones show that argiculturally the Harappans were solvent. These querns with a convex base were crudely made of hard gritty igeneous rock such as basalt and granite and also of sand stone. Three types of querns have been unearthed. On one type a small stone was pushed or rolled and in the other a stone was used as pounder. Eventually there would be a cavity on the second type due to overuse. A rotary quern has also been found from Lothal which is described by Rao :[52]

50. J. M. Kenoyer, "Chipped stone tools from Mohenjo-Daro" in FIC. 117-129.
51. Bridget Allchin, "Substitute Stones" in Possehl HC 237.
52. S. R. Rao LIC 150.

A unique contribution made by the Lothal craftsmen is the invention of a rotary mill in stone which required less manual labour for grinding corn in large quantities than was the case when a flat pedestalled quern was used. The mill consists of two circular pieces. The upper one has a circular feeding mouth and a square opening below the neck to fix a horizontal stick.

The other querns were probably made by cracking apart a small quarzite boulder and then rubbing together the two new surfaces till they became flat. For stone-rubber a hard white crystalline stone was preferred. The base of the rubber was flat and the upper surface rounded. It was used for preparing the skin. Hammer stones were made by trimming a natural nodule with a pointed tool.[53]

Some palettes of good workmanship are recovered from Harappan sites, which were probably used for rubbing haematite and other colours for cosmetics or for varieties of pigments. These were made from a very dark coloured slate or from dark grey semihard stone. Some stone-built burnishers have also been unearthed from the Harappan sites. These were generally made of chert with a high polish, sometimes bearing oval or triangular section ; these were probably used by the metal workers and potters. The drills, built of hard greyish black stone resembling a fine grained chert were also used by the Harappans for fashioning the interior of the stone bodies. The upper side of the drill was roughly concave and the bottom was convex and the two ends were levelled. It was used by fixing a upright forked stick. Varieties of mace heads—of alabaster, sandstone, cherty limestone etc.—were used, probably as weapons. These were either pear shaped or lentoid and even oval with a conical perforation for fixing a stick. The mace heads were lashed all over probably with raw hides to be firmly secured to a wooden stick.

Harappan sites have also yielded a large number of weights, both large and small with varieties of shapes, which are already discussed by us.

The Harappan stone vessels were generally made of soft stone like alabaster which could only contain dry or oily substances. The technique was not an advanced one—all the products look very rough and clumsy. Perhaps with the help of a tubular drill

53. Mackay FEM 393.

the core was bored. In some cases—specially in the faience vessels—it is found that a specimen was made in three or more separate pieces and then they were cemented together to allow larger cavity in the body ; at the same time narrow neck was provided to prevent the drying of the contents of the jar. Mackay says :[54]

> Most of these tall stone-jars were made on a lathe, though owing to their corroded surfaces it is difficult to prove it ; in shape they are so remarkably regular that it seems the most likely mode of manufacture.....
>
> The interior of each dish was probably hollowed out by means of a specially shaped borer....The exterior was then probably first roughly shaped and afterwards turned on a lathe, or the dish was laid upside down on a revolving horizontal wheel for the final trimming.

No complex technical method was applied to carve the statuettes, mostly made of soft stones like alabaster, limestone etc. 'Formally and conceptually too, they do not reveal any great imagination or technical skill.'[55] It is also observed by Mackay that 'Other features indicate the very primitive nature of the statuary of Mohenjo-daro, but, none the less, the art of sculpture had so far advanced as to separate some of the limbs from the body'.[56] The heads were crudely carved with the hairs portrayed neatly coiled up in a knot, sometimes hanging in a long plaited knot or even sometimes tied in a fillet. The ridge of the nose was portrayed in a line with the forehead ; the eyes were sometimes inlaid in a refined way. With a thick, short sturdy neck the statues indeed betray no technological mark. However, two statues from Harappa show better technical skill. The Allchins observe : 'no comparable sculptures are known from North India of the early Historic period, and secondly both have drilled sockets to take dowell pins to attach head or limbs, a technique not found in later sculptures.'[57] It is not yet certain whether painting was thoroughly executed on the statue or not. On the trephilo-design of a *shawl* (?) on the famous statue of Mohenjo-daro some

54. *Ibid* 317.
55. Amita Roy, "Harappan Art and Life : Sketch of a Social Analysis" in HS (ed. D. Chattopadhyaya) 177 ff.
56. Mackay in Marshall's MIC I. 362.
57. Allchins RCIP 205.

sort of red pigment is noticed, which suggests the possibility of applying pigment on the statues. The animals were however, carved from a block of stone of which a solid part remained as base.

Thus we see that technology of stone objects, in spite of the profuse use of metals, was not obliterated. In fact the Harappans could not discard centuries of experience due to habit on the one hand and economy and efficiency on the other. When they technically became more habituated to the use of metal they only reduced the stone artefacts but did not abandon it altogether.

7. SEAL CUTTING AND ENGRAVING

Perhaps the most interesting of all objects that have been unearthed at Harappa, Mohenjo-daro, Lothal, Kalibangan, Chanhu-daro, Rupar, and many other sites, are the seals, impressed with some kind of undeciphered pictographic writing, theriomorphic, anthropomorphic and therio-anthropomorphic representations. "The number so far discovered in excavation must be around 2000."[58] These types of seals were first noticed by Sir A. Cunningham as is attested by his report on Harappa in the *Archaeological Survey of India : Report for the years* 1872-73 (*pp.* 105-108). Yet even now the Harappan seals are still somewhat of an enigma ; in spite of computer-study recently undertaken for understanding the legends on these we remain far from being exact about what was written on them.

Regarding the functional aspects of these seals, scholars are yet to be unanimous. The theories they propound are but conjectures or a synthesis of various opinions. Roy has aptly remarked : "One cannot be too sure about the social function or purpose of these seals. But it is not unlikely that they served as magical amulets or as sealings of particular religious sects or cults or of economic groups. A deciphering of the writings alone may perhaps provide an answer."[59]

The list of the types of the seals, proposed by Mackay is as follows : (1) cylinder seals ; (2) square seals with perforated boss on reverse ; (3) square seals with no boss and frequently inscribed on both sides, (4) rectangular seals with-

58. *Ibid* 209. But see I. Mahadevan TIC for more exact number.
59. Amita Roy, in HS (ed. D. Chattopadhyaya) 121.

out boss ; (5) button-seals with linear designs ; (6) rectangular seals with perforated convex back ; (7) cube seals, (8) round seals with perforated boss ; (9) rectangular seals with perforated boss ; (10) round seals with no boss and inscribed on both sides etc.[60]

The seals made for impressing on a soft material like clay and bitumen : These were mostly made of steatite and also of agate, chert, copper, faience and even terracotta. Some of these have triangular or pentagonal or plano-convex section. Mackay discusses in detail the method of manufacturing the Harappan seals, which seems to form the basis of the later discussions of it by Rao and the Allchins.[61]

The seals were first cut into shape and size by means of a saw, which is evident from the saw-marks on the unfinished seals. When the seal was cut into the required size and shape, in most cases a boss was added to it. This boss was at first made roughly by horizontal and vertical cutting on the back. It was then rounded off by a knife and finished with an abrasive. Then a hole was bored through the boss either from the opposite side or horizontally through the bores to take a cord. Generally one third of the area on the back of the seal was covered by a boss. It is likely that before the finishing of the boss the engraving was done ; because with a raised projection on the back (i.e. the boss), the seal-cutter would find it a bit difficult to engrave on the opposite surface. According to Mackay '..the device and characters on the seal were cut either before the preliminary shaping of the boss or before it was rounded off and finished'.[62]

The designs were nicely carved by a sharp-pointed burin or by a small chisel. Pointed drills were not used to outline, the figures first ; rather it was used only when more details to the carved figures were added. It is interesting to note that the Harappan seal cutters were so skilful that before or without outlining a figure they could start engraving any part of the body with perfection.

Inscription to the seal was added after the carving of the animals or other motifs. According to Mackay, '...the croo-

60. Mackay in Marshall's MIC, II, 370-71.
61. S. R. Rao LIC 95 ; B. & R. Rllchin BIC 294 and RCIP 201.
62. Mackay in Marshall MIC II, 378.

kedness of some of the characters in the inscriptions suggests that they were added later. . . . The seal-cutters probably kept a stock of seals by them and added the inscriptions as required.'[63] Could it be that the inscriptions were outlined later by the scribes for the actual seal-cutters to execute on the seals? The inscriptions were generally arranged in a single row at the top of the seal. But sometimes a second row was added and in that case the letters were smaller due to the lack of space.

Next was the task of applying a coating on the seal, which produced a white hue and gloss on the surface. But certainly no colour was applied. Mackay was at first of the opinion that the coating and baking of the seal was done before the engraving of the motifs and inscription. As he put it: 'Possibly, before a seal was considered ready for engraving, it was coated in order to conceal blemishes and then was baked in a kiln. . . . . Besides whitening the surface of a seal, the baking would materially assist in hardening it owing to the loss of water that would result. The process appears to have been carried out before the seals were engraved.'[64] But the fact seems to be that the seals were probaly engraved before the coating and firing. The reasons are not far to seek. If a soft stone like steatite was baked it would have been too hard to be easily and smoothly worked out by the seal-cutters. Secondly, if the motifs and inscriptions were engraved first and then coated and fired, the edges of the engraved devices would have been more hard and compact than those which were cut after baking the seal. In his later work, Mackay seems to revise his earlier view and observes: 'There is no doubt that after being cut and engraved these seals were treated in some way to produce the white coat that covers them.'[65]

The chemical analysis of the coating on seals was first done by Sanaullah who identified the substance as steatite, saying, 'This surface substance is steatite or talc that has been deprived of the greater parts of its water, which is only possible by ignition.'[66]

63. *Ibid*.
64. *Ibid* 379.
65. Mackay FEM 346.
66. Report of Sanaullah in MIC II. 379.

The chemical analysis is as follows :[67]

	Percent
Silica	61.2
Oxides of aluminium and iron	2.4
Lime	Nil
Magnesia	34.6
Water (by difference)	1.8
Total=	100.00

Mackay adds : 'Mr. Horace Beck after microscopically examining this white coating has come to the conclusion that it is not a slip, but was made by painting the surface with an alkali and then subjecting it to heat. Sometimes this coat shows a certain amount of lamination caused either by the alkali being applied in more than one coat or by the overlapping of brush marks.'[68]

Both Rao and the Allchins agree to the view that the coatings were later applied on the seals. It is, however, unfortunate that chemical analysis of the coating on seals of other Harappan sites has not been incorporated in the excavation reports of the sites.

The Allchins have briefly discussed the method as follows : 'The seals were intaglios, made of steatite, first cut to shape with a saw ; the boss was then shaped with a knife and bored from either end. The carving of the animal motif was done with a burin, probably of copper, and at some stage, generally before carving, the seal was baked to whiten and harden its surface. An alkali was probably applied to the surface before firing to assist in the whitening and to glaze it'.[69] In their later work they have the same thing to say, though with the omission of three words 'The carving of the animal motif was done with a burin, probably of copper, and at some stage the seal was baked to whiten and harden its surface.'[70] The omission of some words—'generally before carving' after

67. *Ibid.*
68. Mackay FEM 346.
69. B. & R. Allchin BIC 294.
70. B. & R. Allchin RCIP 201.

'at some stage' is very interesting. It seems that the Allchins are hesitant to state that the seals were coated and baked before engraving which they proposed in their previous work.

Rao, however, has laid special stress on the Lothal seals. According to him :[71]

> The various stages of producing steatite seals can be followed by a careful study of the seals available from Lothal. In the first instance lumps of steatite were sawn off into tablets with the help of a wire-saw and subsequently chiselled with a stone or bronze chisel. The blocks so produced were reduced to the required size and shape by rubbing on a stone. After cutting the boss on the back, the animal motif was engraved on the face of the seal with the help of a beaked engraver of shell or chert. The pictographic script was subsequently engraved. The perforation of the boss was done by means of two flanged drills of copper driven one each from both sides of the boss. Finally the seal, if made of soapstone, was dipped in an alkali and heated to produce a shining creamy surface and to render the substance harder.

Thus we see that Rao and the Allchins have basically followed Macky in the technological interpretaion of Harappan seals.

8. BEADS

The technological history of sophisticated Indian jewellery and ornaments can be traced back to the Harappan period. The women of the Harappan period were fond of wearing varieties of ornaments, which not only show the technical skill and excellence of the craftsmen but deliberate sophistication of the crafts also. Varieties of innumerable beads, the chief repertoire of the Harappan ornaments were their valued possessions which certainly earned for them not only name and fame but also wealth.

Bead-making is an ancient craft in India. Mackay says: 'There is every probability that in India bead-making was one of the most ancient arts, since in most places the requisite materials were ready to hand.'[72] The Allchins also agree to this.[73]

71. S. R. Rao LIC 95.
72. Mackay in Marshall's MIC II. 510.
73. Allchins BIC 295.

The technique of making beads of semi-precious stones must long have remained a special feature of Indian craftsmanship. Archaeology is able to supply a mass of evidence from sites from the earliest neolithic and chalcolithic settlements in many regions to testify to its longevity. Another, most comprehensive, body of evidence comes from a bead-worker's shop at Ujjain, actually dating from about 200 B.C., which was excavated by N. R. Banerjee. Rao also thinks in the same way when he says : 'It is clear that bead-making has been an important industry during the last four thousand years or more successively at Lothal, Nagra and Cambay—all situated at the head of the Gulf of Cambay. The present day bead-makers follow the same process which the Harappan lapideries had evolved.[74]

On the basis of the above observations we may assert that the Harappans—with some inheritance of their immediate ancestors, whoever they might have been—developed a magnificent technological infrastructure of unprecedented advancement ; some traits of this, though suffering stagnation and conservatism for centuries, had somehow been transmitted to the succeeding cutters of Indian subcontinent, often modified in diffeent set-ups.

However, bead-making was a profitable industry for the Harappans. Rao has aptly remarked that 'the jewellers and bead-makers vied with one another in adorning the person of Harappan women with the choicest product of their trade.'[75] Varieties of beads such as a cylindrical beads, beads of cog-wheel type, fluted tapered beads, long barrel-cylinder beads, barrel shaped beads, long barrel beads, short barrel beads, disc shaped beads, globular beads, segmented beads, pottery beads, etched carnelian beads, composite beads, immitation etched carnelian beads, wafer-beads, tubular denticular beads etc. are found from Harappan sites.

The materials used for making beads also varied. These were steatite, faience, a hard viterious paste (allied to faience), pottery, shell, agate, carnelian, jade, quartz, limestone, jasper, lapis lazuli, green felspar or amazonite, onyx, gold, silver,

74. S. R. Rao LIC 103.
75. *Ibid*, 102.

copper or bronze etc. This shows that the Harappans were very keen in selecting the materials for making beads. At the same time they must have been quite familiar with the fibres of the stones which were not always locally available and hence imported from the far off places. The Harappans imported the raw materials and exported the finished goods and thus technology went hand in hand with production, consumption and export.

Mackay has done a comprehensive study on the process of bead making in ancient Sindh on the basis of his discovery at Chanhu-daro. Rao has also done the same on the basis of his discovery at Lothal in Gujrat. Wheeler, the Allchins etc. have basically followed the line of Mackay. According to Wheeler:[76]

> The Harappan beads are abundant, varied in form and material, and important historically.....The processes of sawing, flaking, grinding and boring the stone beads are well illustrated at Chanhu-daro, where a bead-maker's shop was found. The technique was a laborious and skilful one. The stone (agate or carnelian) was first sawn into an oblong bar, then flaked into a cylinder and polished, and finally bored either with chert drills or with bronze tubular drills. Alternatively, almost incredibly minute beads of steatite paste seem to have been formed by pressing the paste through fine gauge bronze tubes. The stone drills were very carefully made with tiny cupped points to hold the abrasive and water that gave the drill the necessary bite.... No site has produced so many of them as Chanhu-daro, and the possibility of an export-trade in beads from the Indus is worthy of consideration.

We shall discuss here the observations of Mackay. According to him the nodules of stones were at first split along the longer axis to produce some roughly made square or rectangular slips or rods but the easier method was to cut the rods or slips longitudinally from the core nodules by a saw together with fine abrasive. This saw was probably a toothless metal one. Rao[77] is of opinion that before the preparation of the rods or slips the stone was cooked in small earthen bowls. Then the coarse stone was split by giving heavy blow and it was thus reduced to the required size. However, later on, by careful minute flaking uneven angular surfaces were removed to make the slips or rods somewhat round. Then the round

76. Wheeler IC 98.
77. S. R. Rao LIC 103.

shaped rods or slips were ground and rubbed on a sand stone. For the barrel shaped beads one half of the bead was first rubbed and then the other half. The ends were also ground to flat. One of the most crucial parts of bead making was the boring of holes. Each end of the bead was roughened so that the drill could not slip at the time of boring. This boring of the holes was done by stone drills. Mackay says :[78]

> That stone and not metal drills were used in boring the hard stone beads of Chanhu-daro is now proved by a large number of stone drills being found there....The business end of these stone drills is rounded, with a slight depression in the centre. A micro-photograph of the end of one (greatest diameter 0.12 in.) clearly shows the concentric markings formed by an abrasive in its rotation against a hard substance, or, alternatively, the rotation of a hard stone against it. All these drills, whether black or dark brown in color, were made by roughly flaking the stone into a rod-like shape, and then grinding them in much the same way as the beads. They are never of the same diameter right through, but narrow slightly just above the working end and then thicken towards the butt, which more often than not has slightly faceted sides, doubtless to prevent its turning in the handle or chuck in which it was fixed. It might not at first appear possible to use so brittle a drill with the hardness of 7 against a material of similar hardness, such as agate or carnelian. By itself the drill would have made little or no impression on these stones, but the use of a fine abrasive with it, such as emery or crushed quartz, would entirely alter matters. C. H. Desch, Director of the National Physical Laboratory,.....to whom I submitted some of these drills for experiment writes as follows : 'I think that the depression at the end of the drills is intended to hold the abrasive under the drill and prevent it from escaping. I mounted one of the stone drills in a small Archimedean brace, which I held vertically. The action would be just the same as that of a bow drill. Using 120-mesh emery and water, I found that it took about 20 minutes to drill a depth of a millimetre in one of the rough pieces of carnelian. A small depression must have been made to locate the drill, as on a flat surface the drill wanders around before getting to work. The drilling is certainly quite practicable. Failing emery, possibly even sand may have been used. The wear on the drill is very slight.

But Rao comments : 'Drilling was done by the Harappans by means of two flanged drills of bronze placed pointed towards the centre at either end of the bead. Chert drills do

78. Mackay in JAOS, 1937, 6-7.

not appear to have been used for this purpose as suggested by Mackay.'[79]

Accepting the views of both Rao and Mackay, it may be noted that the Harappans used both stone and metal drills at two bead making centres viz., Chanhu-daro and Lothal. However, a bow was probably used to hold the stone drill at the time of working. Pump-drill might also have been used. Every hard stone bead was bored from both flat ends and thus the drill holes met approximately in the middle. Due to the use of fine abrasive the bore holes look very polished. In the Cambay region boring was done at last, whereas in the Sindh region boring was done before the final finishing. Then a fine polish was executed on the beads but the means are not yet clearly known. Rao says: 'Finally, the bead was heated once again to get the necessary glow...'[80] But Mackay does not mention it.

This, in brief, is the general process of making beads in the Harappan period. But mention should also be made of some special types of beads for which technical skill, patience and painstaking labour were certainly needed. One of these types was microsteatite beads. These beads were so minute that their boring was a real problem on the part of the bead-makers. Let us quote Mackay on this point:[81]

> Mr. Horace Beck, to whom I have submitted samples, writes as follows: 'The smallest bead which I have measured from Chanhu-daro is 0.0268 in. in diameter, which means 37.3 to an inch. It is stated that some specimens from the Harappa site are as small as 0.021 in., but the smallest that I have measured was 0.025 in. The perforation of the Chanhu-daro beads is approximately 0.01 in. An average of the perforations of the six beads measured was 0.0098, the maximum being 0.0107, the minimum 0.0083. After carefully examining a modern watch maker's drill which is capable of drilling holes 0.01 in. (0.25 mm.) in diameter, it is difficult to believe that either the drill or the bead could have been held by the hand, so I suggest that some form of lathe or jig must have been used'....
>
> Every one of this type of bead was bored, and, as the illustration shows, they had once been strung together. How such minute beads could be bored puzzles Mr. Beck and myself.

79. S. R. Rao LIC 103.
80. *Ibid.*
81. Mackay in JAOS 1937. 10-11.

To this Mackay adds :[82]

The beads were in fact made as long tubes, and the spiral furrowing...suggests that these tubes were shaped by squeezing a composition through an aperture; though how tubes rather than solid rods were produced it is difficult to see. Where two beads were found almost united and with their markings continuous, it seems that they must have passed after one another through the same apparatus.... At any rate these beads show craftsmanship of the highest order combined with extraordinarily good eyesight.

Mackay's observation appears to be convincing when he says that these were at first produced by extrusion process as a uniformed tube from an aperture. But about the type of aperture from which the tube was extruded, Mackay is somewhat silent. However, a great deal of information has been supplied to us by K. T. M. Hegde and others. They have put forward a convincing hypothesis of the material and technology of Harappan micro-beads after the recovery of such beads at Zekda. According to them these beads were not made from pure talc (4 Mg O5 SiO_2 H_2O) which is a hydrated silicate of magnesium and the presence of alumina indicates the presence of Kaolinite (Al_2 $O_3$2 SiO_2 2H_2O) in the composition of the original material and the Harappans deliberately mixed talc and Kaolinite in the ratio of five to one. But the natural formation of this material is also found as talcose steatite which may also have been used by the bead-makers of Harappan period.[83]

On the technology of these beads Hegde and others conclude :[84]

We therefore suggest, keeping in view the known Harappan technological infrastructure, the following simple process as a plausible method for the manufacture of these micro-beads. All that is needed is a circular copper or bronze disc with a few one millimeter diameter perforations near the centre. Each perforation must have a copper or bronze wire of 0.5 millimeter diameter with one end soldered or riveted near the perforation and the other end bent and positioned to be at the center of the perforation... The disc has fine holes at its periphery which allow it to be stitched all around to a well-knit piece of cloth. If into this device a paste of finely ground talcose

82. *Ibid* 12.
83. K. T. M. Hegde *et al* in Possehl HC 241.
84. *Ibid* 242-43.

steatite is put and the cloth gathered together and squeezed by hand, talcose steatite paste emerges through the perforations as tubes. These soft tubes can be cut as they emerge, at quick intervals, to convert them into microbeads.

Three skillful persons are necesary to do the job : one for squeezing the paste, the second for cutting the soft tubes and the third for collecting the soft microbeads, over a layer of fine ash on a dish, to avoid damaging them. The beads can then be baked at 900 to 1000° C in a kiln to harden them. We have adequate evidence to believe that Harappans had the necessary infrastructure to attempt such a process.

They had copper and bronze discs. There is even evidence of perforated examples. They knew both soldering and rivetting. They sometimes soldered the rivet by pouring molten copper or bronze around its base for additional firmness (Marshall 1931 : 489). They had cloth (Marshall 1931 : 585-86). The soft tubes appear to have been cut by a thin, sharp device to convert them into microbeads. A simple device to serve this purpose is a horse hair. Harappans had horses (Marshall 1931 : 653-54 ; Joshi 1972 : 135). The fact that they could melt copper at 1080°C, clearly shows that they had suitable kilns or open fires with a forced draught to raise the temperature of the fire to 1200°C so as to bake the beads at 1000°C.

It appears probably therefore, that the Harappans microbead makers produced these beautiful, tiny beads by means of simple process. But it was, nevertheless, a laborious, painstaking, unhurried job.

Larger steatite wafer beads were[85] at first shaped as a long rod and subsequently cut up into segments with the aid of a saw. The perforation of these beads were done by a fine metal point. But we have come across small wafer beads also, Probably these were made by means of cutting up steatite into thin plates and were rubbed down to required thinness. Then these plates were cut into blanks approximately to the size required for the bead. Then these were rounded off, perforated and heated at 900°C for hardening.

Another type of beads are the carnelian beads which were in all probability exported by the Harappans.[86] Whether they were the first discoverers of these beads is yet a question not fully answered.

85. Mackay in JAOS Vol. 57, 1937. 12.
86. *Ibid* 14.

On the basis of the detailed review of Dikshit, Sankalia has described three techniques of manufacturing etched carnelian beads :[87]

Technically, there are three types of etched beads.

Type I : White patterns on red background.

Type II : Black patterns on whitened surface of the stone.

Type III : Black patterns etched directly on the stone.

Beads of type I are most common ; of type II rare, and that of type III almost negligible, still all of them go back to the Indus or the Harappan Period, and thus the techniques should have been in existence by at least 2300 B.C.

Technique :

1. The white patterns on the red surface were prepared by making a thick liquid of potash, white lead, and the juice of *Kiral bush* (*Copparis aphylla*). This was then applied directly with a pen on the carnelian. Heated on a charcoal fire the design became permanent. Microscopic analysis of these beads showed that the etching produced a number of minute spots, under different coefficients of expansion. The white layers do not affect the extreme surface of the stone.

2. In type II beads, the white surface is first prepared with alkali. On these are drawn lines in black, prepared from metals like copper and manganese. The effect is sometimes purplish.

3. In type III, a pattern in black is etched directly on the original surface of the bead.

It is also further noted that the three techniques do not always appear exclusively. Combinations of Types I & II (Variety A) and of types I and III (Variety B) are at times noticed.

From the above one is tempted to think that the Harappans had some understanding of chemistry, though it remains for us to see whether chemistry as a "conscious" natural science could emerge among them.

87. H. D. Sankalia SAPTI 40-41.

TABLE I

(Tables of Chemical Analyses)

Chemical analyses of Ceramic materials found at Mohenjo-daro

(After Marshall, 1931, Vol. II. p. 689)

Specimen	Silica	Alumina	Ferric Oxide	Ferrous Oxide	Manganese Oxide	Lime	Mangnesia	Alkalies	Copper Oxide	Water	Analyst
1. Black bangle	54.28	19.63	—	8.70	0.13	9.63	4.39	3.43	—	—	Mohd. Sana Ullah
2. Greenish pottery	52.39	17.03	5.30	2.29	—	15.78	4.45	1.71	—	1.05	do
3. Faience Vase (bluish-green)	89.76	3.86	0.93	—	—	0.88	tr.	4.07	0.50	—	do
4. Faience bangle (bluish-green)	88.12	3.02	1.82	—	—	1.26	—	—	0.46	—	do
5. Faience tubuler bead (chocolate)	91.07	2.44	1.15	—	tr.	1.28	tr.	2.08	Cu 20 1.98	—	do
6. Steatite disc.	57.99	4.85	—	—	—	4.31	27.20	3.54	1.09	2.01	M A. Hamid
7. Faience Statuatte	57.23	3.69	—	—	—	6.39	28.99	1.88	0.46	1.36	do
8. Slip from steatite Seal	61.2	2.4	—	—	—	—	34.6	—	—	1.8	Sana Ullah
9. Steatite flat beads	63 65		—	—	—	—	83.80	—	—	1 09	do

TABLE—2

(*Analyses of the glazed Ware*)

(After Marshall, 1931, Vol. II, p. 578)

Analyst. Dr. M. A. Hamid

Pottery base of bead.	Per cent.	The White glaze.*	Per cent.
Silica	71.12	Silica	86.28
Alumina	9.27	Alumina and Iron oxide	7.78
Iron oxide	10.91	Lime	2.35
Lime	1.77	Magnesia	0.61
Magnesia	1.55	Soda	1.21
Soda and Potash	3.77	Potash	Nil
Loss on ignition	1.69	Loss of ignition	2.05
Total	100.08	Total	100.28

*The term "glaze," here as well as in some other places, is used rather loosely for want of a better name for this substance, which though glassy in apperance is not a true glass—ancient glasses being essentially soda-lime silicates.

CORRECTIONS TO TABLES I & III

Table I

Copper Oxide in Column 5 : Cu_2O, 1.98
Alumina in Col. 6 : 4
Ferric Oxide in Col. 6 : 0.85
Alumina in Col. 7 : 3
Ferric Oxide in Col. 7 : 0.69
Alumina in Col. 8 : 2
Ferric Oxide in Col. 8 : 0.4
Alkalies in Col. 4 : Na_2O, 4.50 ; K_2O, 0.65
Magnesia in Col. 9 : 33.80

Table III

Magnesium Carbonate to be transferred under the heading "Carbonate of Lime"
SK Site to be read as DK Site

TABLE III

Mortars (After Marshall, Vol. II / p. 689)

Locality	Gypsum	Carbonate of lime	Sand	Alkaline Salt	Moisture	Analyst
Wall (HR Site)	74.12	2.50	20.41	1.18	1.79	Mohd. Sana Ullah
do	63.25	0.66	31.61	3.47	1.01	do
Tank (SD Site)	43.75	13.78	38.04	2.47	1.96	do
Drain (DM Site)	56.73	24.87	16.64	—	1.76	do
Vat (HR Site)	Nil	69.58	21.71	5.44	3.27	M. A. Hamid
Drain and Cesspit (SK Site)	Nil	39.96	46.74	0.74	3.74	Mohd. Sana Ullah
Magnesium Carbonate	Nil	8.82	Nil	Nil	Nil	Nil

2. Nodules of gypsum mortar analysed by H. P. Plenderleith (after Mackay, FEM, p. 598)

Moisture	3.60%
Water (Combined)	0.01%
Silica	23.05%
Iron oxide and alumina	3.43%
Calcium sulphate	55.67%
Carbonates	Nil
Magnesia and alkalies	not estimated :—

"Lime mortar (Mackay, FEM, p. 598).

Calciumcarbonate	39.96%
Magnesium carbonate	8.82%
Gypsum	traces.
Water	3.58%
Sand, clay, etc	47.64%

TABLE—4

(After Marshall Vol. II pp. 689-690)

A black coal-like substance found at Mohenjo-daro has been identified by the writer as *Silajit* or *Shilajatu,* an ancient Indian Medicine. It occurs as an exudation on rocks in the Himalayas, and is popular with the physicians following the old school. . . .The composition of Mohenjo-daro specimen (M) is shown against those of four specimens analysed by Hooper (JASB.72 (1903), 98-103) in the table below :

Analysis of Silajit

	M	I	II	III	IV
Water	15.99	9.85	15.90	11.15	10.99
Organic Matter	55.24	55.20	49.86	51.55	56.86
Ash	28.77	34.95	34.24	37.30	32.15
	100.00	100.00	100.00	100.00	100.00
Ash :—					
Silica	8.23	1.35	1.62	18.10	10.15
Alumina	2.43	2.24	1.08	6.00	4.64
Ferric Oxide	1.44				
Lime	7.31	4.36	3.96	3.86	3.88
Magnesia	0.32	1.50	0.52	0.15	1.34
Alkalies	9.04	13.18	14.32	4.78	6.91
Carbonic acid, etc.	Not determined	11.51	12.13	3.69	4.83

(Analysis of M. by Dr. Hamid)

TABLE—5

Lollingite

(After Marshall, Vol. II, p. 690)

"Specimens of lollingite or leucopyrite also deserve special mention as they bear evidence of having undergone ignition. Under the action of strong heat, these minerals give off arsenic ; or its white oxide, when roasted in air. It is, therefore, highly probable that these minerals were employed for making arsenious preparations, either for medicinal purposes or for destroying life...The composition of a natural lollingite specimen (B) is given in the table below along with those from Indus Valley sites."

Analysis of Lollingite

	B.	I.	II.	III.	IV.	V
Iron	27.14	54.55	49.3	45.63	51.7	48.1
Arsenic	72.17	34.02	43.6	47.12	43.9	48.6
Copper	—	0.92	0.7	—	—	—
Sulpher	0.37	1.38	0.16	—	—	—
Water	—	7.68	4.7	—	—	—
Insol	—	1.45	0.8	—	—	1.9
Sp. gr.	—	4.0	5.6	—	—	—

(Analysis of the natural specimen (B) is by Brevik and those of the Indus specimens by the writer).

CHAPTER 9

METAL TECHNOLOGY OF THE HARAPPA CULTURE

D. P. Agrawal

[*Reprinted from IJHS v, No. 2, 1970, pp. 238-252*]

1. INTRODUCTION

For many millennia man struggled with stones to make his tools. The Stone Ages were long periods of stagnation and imperceptibly slow growth. Stone as a material for tools had its obvious limitations, so also wood and bone. The advent of the metal was a real breakthrough in man's technological progress ; it was not only tougher and susceptible of a finer edge and more durable but also fusible, malleable and ductile. Worn out tools could be recast and moulds could be improved to obtain any desired shape.

The mastery of the metal was a big boost to the social development. It improved the productivity of the society, led to the development of specialized crafts and provided knowledge of chemical and physical laws. Given the right type of ecology and social set-up, metal technology accelerated the pace of urbanization and led to a host of technical innovations.

Metallurgy had its various stages. In the beginning native copper was used only as 'stone' and was chipped to shape tools.[1] Later, it was hammered and cut to shape. But real metallurgy started with the 'Ore-stage' when man learnt to extract metal from its ores by smelting. This involved complicated processes. All complex processes had their origins at only a few centres which provided the optimum conditions for such discoveries and innovations. Such techniques could not originate independently everywhere, but were diffused from a few primary centres.

The mountain region extending from Anatolia, Armenia to Afghanistan is rich in metal minerals, especially the eastern flank. This zone is the home of wild pistachio and *Haloxylon amodendron*—as shown by recent pollen studies—which pro-

1. Forbes, R. T., *Studies in Ancient Technology*, 9 ; Leiden, E. J. Brill, 1964.

vide superb charcoal for metallurgical purposes.[2] Recently the importance of Kerman range in south-east Iran has been emphasized in the development of early metallurgy. Discovery of smelting equipment from Tal-i-iblis, in Mashiz Valley, datable to *circa* 4000 B.C., does indicate the probability of its being one of the earliest smelting centres.[3] It is quite likely that for the eastward diffusion of metal technology this site had a crucial role to play. In fact, evidence of contact between Rana Ghundai II and Daruyi near Tal-i-Iblis is reported.[4]

Nearer home, Deh Murasi[5] and Mundigak[6] provide evidence of metallurgy. The volute headed pin, bent blades of lance-head, shaft-holed axe may provide links between Mohenjo-Daro and Mundigak.

Pre-Harappan cultures it appears were not producing their own metal, as it is extremely rare, e.g. only a bangle at Kot Diji I, no metal at Siah Damb and Anjira and only an indeterminate object in Kalibangan I. It may, however, be noted that from Nal, knives, axes, saws, etc., quite a few objects from Mehi and knife blades from Damb Sadaat II and III are reported. In contrast, there is a sudden efflorescence of metal in the Harappa culture. For example, Mackay[7] reported from the further excavations at the DK-mound of Mohenjo-Daro alone, 14 spearheads, 64 knives, 23 axes, 2 swords, 53 chisels, 11 fish-hooks, 2 saws, 18 razors, 17 arrowheads and many other artifacts ! This richness in metal has a far-reaching socio-economic importance.

In trying to seek the reasons for the metal prosperity of the Indus people we will be unravelling several other important as-

2. Wertime, T. A., *Science,* 146, No. 3469, p. 1257, 1964.
3. Caldwell, J. R., and Shahmirzadi, S. M., *Tal-i-Iblis : The Kerman range and the beginning of smelting* ; Springfield, Illinois State Museum, 1966.
4. Lamberg Karlovsky, G. C., *American Anthropologist,* 69, No. 2, p. 145, 1967.
5. Dupree, L., Deh Morasi Ghundai : A Chalcolithic Site in South-Central Afghanistan ; *Archaeological Papers of the Museum of Natural History,* 50, pt. 2, New York, 1963.
6. Casal, J. M., Fouilles de Mundigak : *Memoires de la Delegation Archeologique Francaise en Afghanistan,* 17, Paris, 1961.
7. Mackay, E. J. H., *Further excavations at Mohenjodaro* ; New Delhi, Government of India Press. 1937-38.

pects of the Harappa culture too. Because metallurgy provides the technological base—the means of production—of a given society, the change in the means of production does determine to a considerable extent the superstructure. In the case of the Harappan urbanization also, metallurgy has played a crucial role.

2. THE PROBLEMS

To define the technological status of the Harappa culture in the fields of metallurgy and metal forging as also to understand its socio-economic implications, we will tackle the following problems :

(i) Did the Harappans use smelted copper ?
(ii) What types of mineral ores were used ?
(iii) Can we locate the sources of these ores ?
(iv) Did they practise deliberate alloying ?
(v) What metal-forging techniques they used ?
(vi) Does typological analyses show affinity of the Harappans with the Copper Hoards or with the other Chalcolithic cultures ?
(vii) What is the role of metal technology and the ecology in the urbanization of the Harappa culture ?

These are big problems and let it be admitted that there are no easy answers. But the chemical and metallographic analyses of our samples, as also previous work of other workers, do throw important light on the state of metallurgy in the Indus. In the light of chemical, metallographic, typological, as also ecological data we can now make an attempt to seek answers* for the problems posed above.

3. CHEMICAL ANALYSES

Before we spell out our approach it will be relevant to review the previous work. Desch started the pioneering work in this field under the aegis of the Sumerian Committee. He was in a way obsessed with nickel and went to the extent of seeking Transvaal copper ores (which contain nickel) as the

* The conclusions drawn here can only be tentative in the present stage of knowledge. But more analyses and data are being processed which would lead to firmer conclusions.

sources for Mesopotamian copper! Sanahullah[8] sought to distinguish the Harappan artifacts from the Mesopotamian ones on the basis of absence of arsenic in the latter. But Table I will show that arsenic and nickel both were present in west Asian artifacts and therefore arsenic is in no way distinctive of the Indus.

TABLE I

Sites	Total number of tools	With Ni	With As
Egypt ..	30	8	22
Khafaje ..	16	14	13
Ur ..	16	15	14
Kish ..	10	10	1

(Based on the data given by Burton Brown[9])

Problems of ore-correlations are quite controversial and complicated. In the West, Coghlan, Oldberg, Pittioni, Junghans and Sangmeister have done important work in this line, though Thomson[10] has challenged the very fundamentals of their work. However, Coghlan[11] has defended Pittioni and others' approach of comparing the impurity patterns of the ores and the artifacts for ore-correlations. Even Thomson[12] admits that the use of pyritic ores can be discerned from the increase of impurities (Table II) like tin, arsenic, iron and nickel. We observe an increase in these impurities in Mohenjo-Daro samples too (Tables III and IV), indicating the use of sulphide ores. Presence of sulphur too was frequently detected.

8. Sanahullah, *In* Vats, M., *Excavations at Harappa,* New Delhi, Government of India, 1940.
9. Brown, Burton T., *Exavation at Azarbaijan*; London, Murray, 1951.
10. Thomson, F. C., *Man,* 58, p. 1, 1958.
11. Coghlan, H. H., *Viking Fund Publications in Anthropology,* 28, 1960.
12. Thomson, F. C., *op. cit.*

TABLE II

[The use of pyrite ores from the late third millennium B.C. With the use of sulphide ores the impurities record a sharp increase (after Thomson 1958)]. n.d. = not detected spectrographically. Figures for copper in brackets are by difference.

Percentage composition

	Copper	Tin	Iron	Nickel	Lead	Arsenic	Zinc	Bismuth	Antimony	Silver
Native copper	(99·96)	0·0005	0·0002	0·03	n.d.	n.d.	n.d.	n.d.	n.d.	0·0025
Early third millennium (malachite ?)	(99·43)	0·02	0·3	0·005	0·002	0·1	0·005	0·0005	0·03	0·1
Middle third millennium (malachite ?)	(99·7)	0·03	—	0·002	0·01	0·2	0·005	0·002	0·03	0·01
Later third millennium (weathered pyrite ore ?) ..	(98·98)	0·5	0·1	0·03	0·03	0·3	—	0·0006	0·05	0·01
Later third millennium (pyrite ore ?)	(97·41)	0·87	0·02	0·16	0·56	0·20	—	—	—	—

Site: Moherjo-Daro, from lower levels

Percentage composition

Sample details	Ag	Fe	As	Sb	Pb	Bi	Cu	Sn	Ni	S	Rel. prob. % I	II	III	Reference
Pl. CXXVI, 5 DK 7535	—	0·33	0·66	0·25	0·59	—	91·01	6·14	0·48	0·12	0	18	82	(Mackay 1937–38)
Pl. CXXVII, 2 DK 7854 Axe	—	0·50	—	0·43	0·95	—	90·98	7·66	0·20	0·07	0	40	60	,,
Pl. CXXVI, 2 DK 7856 Chisel	—	0·51	—	1·25	0·39	—	75·25	7·84	0·61	—	0	23	77	,,
Portion of Axe DK 7861	—	0·10	—	0·14	0·22	—	88·49	9·88	0·30	0·06	0	44	56	,,
Pl. CXXVIII, 1 DK 5486 Axe	—	0·34	2·10	Tr	0·20	—	80·56	1·76	0·58	—	1·0	54	45	,,
Pl. CXXXIII, 4 DK 6043 Chisel	—	0·02	1·58	0·54	Tr	—	86·92	8·56	0·68	0·07	5·0	15	80	,,
Pl. CXXVIII, 15 DK 5360 Copper frying-pan	—	0·33	0·80	0·18	0·05	—	81·94	0·37	0·21	0·14	0	38	62	,,
Pl. CXXVII, 14 DK 7343 Bolt	—	0·29	0·24	—	0·81	—	97·23	—	0·89	0·10	0	69	31	,,
Pl. CXXVI, 4 DK 7853 Axe	—	0·28	0·40	0·06	0·71	—	94·64	0·31	0·33	0·69	0	6	94	,,
Pl. CXXXI, 32 DK 7859 Ingot	—	0·56	0·24	—	0·82	—	95·23	—	0·41	0·48	1	83	16	,,

TABLE IV

Percentage composition

Site: Mohenjo-Daro

Sample details	Ag	Fe	As	Sb	Pb	Bi	Cu	Sn	Ni	S	O_2	Rel. prob. %			Reference
												I	II	III	
Copper lump	—	0·03	0·15	0·88	0·02	—	96·67	—	1·27	0·98	—	0	19	81	(Marshall 1931)
,, ,,	—	0·49	—	0·98	Tr	—	97·07	—	0·31	1·15	—	5	31	64	,,
,, ,,	—	—	—	—	0·09	—	96·42	—	0·35	0·36	2·78	1	98	1	,,
,, ,,	—	1·51	1·30	Tr	Tr	—	92·49	0·37	1·06	Tr	1·01	4	53	43	,,
Fragments of implements	—	0·12	0·74	0·72	1·58	—	95·80	—	0·25	0·61	0·18	0	6	94	,,
Celt	—	0·15	4·42	—	0·26	—	94·76	0·09	0·14	—	—	0	3	97	,,
Copper chisel	—	0·59	3·42	0·10	3·28	—	92·41	—	0·15	0·05	—	0	18	82	,,
Bronze rod	—	0·15	1·96	1·15	0·17	—	91·90	4·51	—	0·16	—	0	1	99	,,
Bronze buttons	—	0·29	Tr	2·60	—	—	88·05	8·22	Tr	0·84	—	5	11	84	,,
Bronze chisel	—	0·35	—	0·35	0·70	—	86·22	12·38	—	—	—	0	34	66	,,
Bronze slab	—	0·42	1·17	0·33	0·11	—	82·71	13·21	0·56	—	1·49	0	5	95	,,
Bronze chisel	—	0·18	0·07	Tr	Tr	—	85·37	11·09	0·16	0·11	3·02	6	61	33	,,
Bronze lump	—	—	—	Tr	0·17	—	83·92	12·13	0·17	—	3·61	5	84	11	,,

TABLE V

Percentage composition

Site : Rangpur

Sample details	Period	Ag	Fe	As	Sb	Pb	Bi	Cu	Sn	Ni	O_2	Rel. prob. %			Reference
												I	II	III	
1. No. 324 Celt	IIc	—	—	—	—	Tr	—	91·2	2·6	2·1	4·1	47	51	2	(Rao 1963)
2. No. 663 Celt	IIa	—	—	Tr	—	Tr	—	91·35	4·09	Tr	4·6	,,	,,	,,	,,
3. No. 437 Bangle	IIc	—	—	Tr	—	Tr	—	86·4	11·07	1·8	0·73	,,	,,	,,	,,
4. No. 417 Knife	IIc	—	Tr	—	—	—	—	94·8	0·7	0·4	4·1	,,	,,	,,	,,
5. No. 330 Pin	III	—	1·88	—	—	—	—	91·8	0·6	5·88	—	8	84	8	,,
6. No. 442 Pin	IIB	—	1·86	—	—	—	—	96·6	Tr	0·8	0·74	,,	,,	,,	,,
7. No. 260 Bead	IIa	—	1·4	—	—	—	—	96·66	Tr	0·38	1·56	,,	,,	,,	,,
8. No. 635 Ring	IIa	—	0·45	—	—	—	—	96·1	Tr	0·2	3·25	52	25	23	,,
9. No. 169 Bangle	?	—	Tr	—	—	—	—	57·7	6·94	Tr	35·46	47	51	2	,,
10. No. 170 Amulet	IIa	—	0·57	—	—	—	—	77·6	Tr	0·1	21·73	52	25	23	,,
11. No. 141 Pin	IIa	—	0·24	—	—	—	—	65·4	6·78	0·51	27·08	26	56	18	,,
12. No. 526 Knife	IIc	—	Tr	—	—	—	—	59·0	5·28	Tr	35·72	47	51	2	,,
13. No. 525 Knife	?	—	1·08	—	—	—	—	59·6	2·69	—	36·63	8	84	8	,,

TABLE VI

Comparison of the impurity patterns of the copper artifacts from Ahar and the Khetri copper ore

Spectroscopic analyses

Sample description	Ag	Fe	As	Sb	Pb	Bi	Cu	Sn	Ni	Zn	Mn	Co	Au	Al	Cr	Mo	Zr	W	Ti	Mg	V	Gd	P	Si	Reference
Ahar axe ..	nd	+	+	+	+	+	+	nd	+	+	+	+	nd	+	+	+	+	nd	+	+	nd	+	+	+	(Hegde *in press*)
Ahar metal sheet ..	nd	+	+	+	+	+	+	nd	+	+	+	+	nd	+	+	+	+	nd	+	+	nd	+	+	+	,,
Khetri ore ..	nd	+	+	+	+	+	+	nd	+	+	+	+	nd	+	+	+	+	+	+	+	+	+	+	+	,,

Key: + = present
nd = not detected

We have , however, used the statistical approach of Friedman *et al.*[13] to determine the types of ores used. They analysed a large number of ore and artifact samples and arrived at the relative probabilities of occurrence of impurities in the three types of ores : native, oxides and sulphides.

Subjecting the more complete of the chemical data available for Mohenjo-Daro artifacts to statistical calculations (as per Friedman *et al.*) we arrived at the relative probabilities for the type of ores used as given in Tables III and IV. In the tables, relative probability I is for the native type, II for the oxidized type and III for the reduced type of ores. It it clear from Tables III and IV that there is a very high probability that sulphide ores were used by the Harappans from the beginning. On the other hand, at Rangpur fresh mining areas may have been tapped, as it appears that they used either native or oxidized copper minerals (Table V).

But once we go into the ore-correlations to determine definite mining areas which the Harappans used, the problem becomes more complicated. To get a definite answer to these problems many more analyses are needed ; but let us see as to what inferences are possible on the basis of the available data.

Table VIII shows our spectroscopic analyses of the various Indian ores and two Harappan artifacts (TF-Cu-3 and -6). It is evident from the table that only the Khetri ore has close correspondence with the impurity patterns of the Harappan artifacts. Hegde[14] too, following Pittioni, has compared the impurity patterns of the Khetri ore and the Ahar artifacts, and there is a surprising correspondence between the two (Table VI). Besides the common presence of several impurities in the Khetri ores and the Ahar tools, the absence of silver, gold and tin even in trace amounts is remarkable.

13. Friedman, A. M., Conway, M., Kastner, M., Milsted, J., Metta, D., Fields, P. R., and Olsen, E., *Science,* 152, No. 3728, p. 1504, 1966.
14. Hegde, K. T. M. (*in press*).

TABLE VIII

*Spectroscopic analyses**

Indian ores and artifacts

Sample description	Ag	Fe	As	Sb	Pb	Bi	Cu	Sn	Ni	Zn	Mn	Co	Au	Al	Cr	Mo	Zr	W	Ti	Mg	V	Gd	P	Si
TF-Cu-14a Madras pyrrhotite	nd	+	nd	nd	nd	nd	+	nd	+	+	nd	+	ns	nd	nd	nd	nd	+	nd	nd	nd	ns	ns	nd
TF-Cu-14b Madras pyrrhotite	nd	+	nd	nd	nd	nd	+	nd	+	+	nd	+	ns	nd	nd	nd	nd	+	nd	nd	+	ns	ns	nd
TF-Cu-5 Mohenjo-Daro galena ore	nd	nd	nd	+	+	nd	nd	nd	nd	nd	nd	nd	ns	nd	nd	nd	nd	nd	nd	nd	nd	ns	ns	nd
TF-Cu-15 Singhbhum chalcopyrite	nd	+	nd	nd	nd	nd	+	nd	+	+	+	+	nd	+	nd	+	ns	ns	+	+	+	ns	ns	+
TF-Cu-24 Khetri chalcopyrite	+	+	+	+	+	+	+	+	+	+	+	+	nd	+	+	+	ns	ns	+	+	+	ns	ns	+
TF-Cu-3 Chanhudaro celt	+	+	+	+	+	+	+	+	+	+	+	+	nd	+	+	+	ns	ns	+	+	+	ns	ns	ns
TF-Cu-6 Mohenjo-Daro spearhead	+	+	+	+	+	+	+	+	+	+	+	+	nd	+	+	+	ns	ns	+	+	+	ns	ns	ns

* My samples

Key : + = present

TABLE VII

Range	Elements
More than 1%	Si, Fe, Cu
0·1–1·0%	Al, Mg and Ni
0·001–0·1%	Zn, Mn, Co, Mo, Ti and V
Not detected	Sn, Pb, As, Au, Ag, Bi, Cr, Gd and Sb

As compared to Khetri ore, we got the following spectroscopic analyses for a chalcopyrite sample from Singhbhum :

The absence of Pb, As, Bi, Cr, Gd in the Singhbhum ore distinguishes it from the Khetri ore. But most of these elements are present, quite often in significant amounts, in the Harappan copper (Tables III, IV and V). This evidence suggests that the use of Rajasthan ores by the Harappans is more probable than the Bihar ores.

4. ALLOYING

Brown[15] has adduced evidence to show that tin alloying was known in the third millennium B.C. Childe[16] held that in Mesopotamia lead alloying for better moulding of copper was learnt by late Uruk times, whereas tin alloying appears in Early Dynastic times at the latest.

But in the Indus there is evidence for tin alloying right from the early levels (Table III), though in point of time later than West Asia. Of course, small amounts of Sn, Sb, Pb, Ni and Fe came from the ores only, as impurities, and were not deliberate additions. Tin-bronzes, with 8-11% tin, are the best for an optimum combination of strength, elasticity, toughness and the ability to stand shocks. More than 11% tin in bronze makes it brittle, yet is good for imparting it a bright polish.

Below we give the percentage of tools, out of 175 published chemical analyses studied by us, in different concentration-ranges of tin :

15. Brown, Burton T., *op. cit.*
16. Childe, V. G., *New light on the most Ancient East,* New York, Grove Press Inc., 1957.

Sn%	..	<1%	<8%	8-12%	>12%
Tool%	..	70%	10%	14%	6%

From the above, two points emerge : (1) most of the tools (70%) are of pure copper and (2) out of the bronzes only 14% fall within the optimum tin concentration of 8-12%. This indicates both scarcity of tin as also the inability of the Harappans to control the optimum range either due to the lack of knowledge or difficulties of correct mixing.

Data on arsenic alloying are not clear, as arsenic was not always looked for in the older analyses. But about 20 artifacts (out of the 175 mentioned above) show 1-6% concencentration of arsenic which could only be due to deliberate addition. Up to 4% arsenic in copper forms a solid solution, but increases the strength of the alloy in the cast condition only marginally.[17] However, it acts as a deoxidizer and is useful in closed-casting.

On the other hand, up to 8% tin forms a solid solution with copper and in annealed condition up to 16%. A 10% tin bronze will have a Brinell hardness of 135, as against 87 of pure copper, while retaining its ductility. Cold work on pure copper too can harden it comparatively, but will make it very brittle. Work-hardened bronze can achieve considerable hardness.

METAL FORGING TECHNIQUES

Most of the Harappan tools are of a simple type ; the sophistication of West Asia is absent here. Even such a utile device as a shaft-hole too was not adopted. All the Harappan axes are of a flat type only. Even the blades of knives and spearheads (with the exception of four mid-ribbed swords from Mohenjo-Daro)[18] are of a thin flat type. Amongst the distinctive Harappan types we may include the razors (especially the double-edged one), the blades with curved ends, chisels with broad rectangular tangs and narrow blades, barbed fish hooks, arrowheads with backward projecting bars. Fish hooks of course are pieces of superb craftsmanship and are unparalleled in any other Chalcolithic culture in India,

17. Tylecote, R. F., *Metallurgy in Archaeology,* London, Edwin Arnold, 1962.
18. Mackay, E. J. H., *op. cit.* 1937-38.

even in Mesopotamia and Egypt.[19] The cute figurines of the nude dancing girls from Mohenjo-Daro[20, 21] and Lothal are masterpieces of the Harappan bronze sculpture.

The dancing figurines mentioned above are not only fine sculptures, but are also the evidence of *cire-perdue* casting techniques used. Closed casting was a known art to the Harappans. Addition of small amounts of lead to enhance the fusibility of copper and of arsenic as deoxidizer may have been deliberate attempts for efficient casting. However, several specimens of axes[22, 23] with puckered surface and large blow-holes do indicate difficulties of casting. The large number of flat celts and other flat type of implements indicate a probable predilection for the easy open mould casting. Metallographic analyses indicate that cooling of the cast metal was slow and controlled.

Pots and pans were made using 'sinking' and 'raising' techniques. In some cases one can observe even the stake marks on the inside of vessels.[24] No wire drawing is attested, though wires were made by beating. Objects with thin sections, for example razors, arrowheads, even knife blades, were made by chiselling out from copper sheets.

Worthy of mention is the occurrence of a true saw from Mohenjo-Daro with unidirectional indentations ;[25] elsewhere it does not appear till the Roman times. The fine tubular drills[26] too are the earliest metal drills in the world.[27] They were probably made by beating over a mandrel. Mackay[28] suggested that they were perhaps used for making fine steatite beads.

19. Hora, S. L., *Ancient India,* 10 and 11, p. 152, 1954-55.
20. Marshall, J., *Mohenjodaro and the Indus Civilization,* Pl. XCIV, 6-8, Arthur Probsthain, 1931.
21. Mackay, E. J. H., *op. cit.,* 1937-38.
22. Mackay, *Chanhudaro excavations,* Pl. CXX, 27, New Haven, American Oriental Society, 1943.
23. Mackay, *op. cit.,* Pl. CXXXII, 36, 40, 1937-38.
24. Mackay, *op. cit.,* 1937-38.
25. Mackay, *op. cit.,* 1937-38.
26. Mackay, *op. cit.,* Pl. LXII, 1943.
27. Coghlan, H. H., *In* Singer, C., Holmyard, E. J., and Hall, A. R., *A History of Technology,* 1, Oxford, Clarendon Press, 1954.
28. Mackay, E. J. H., *op. cit.,* 1937-38.

Though soldering for gold was known to the Harappans, it appears that it was never used for copper. For metal joining the 'running on' and riveting techniques were used.[29] Lapping of tubular handles was also known.[30]

To determine if the Harappans knew the techniques of cold working and annealing, six samples were analysed from Chanhu-daro and Mohenjo-Daro. The metallographic analyses showed polygonal grains with twins indicating cold work and annealing. Lack of cracking and large pores in the case of a Mohenjo-Daro bowl (TF-Cu-7) may indicate slow cooling of the mould.

ECOLOGY AND TECHNOLOGY

We have tried to sketch above an outline of the Harappan metallurgical technology. A 2,000-year late date[31] for Harappan metallurgy than Iran and the evidence of a full-blown metallurgy from the start itself preclude any probability of independent origins in the Indus. Harappans were smelting quite pure copper from the sulphide ores and practised tin and arsenic alloying also. It is probable that they were using ores from Khetri area. They were relatively rich in copper and had learnt various techniques of metal forging. They knew 'sinking', 'raising', cold work, annealing, 'running on', riveting *cire-perdue* casting. They even had the earliest tubular metal drills and the true saw. We also noted the distinctive tool-types of the Harappa culture.

In comparison to the Harappans the other Chalcolithic cultures are poorer both in the quantity of the metal and the technology. The Copper Hoards are also rich in the metal, but they are comprised of unstratified collections from far-flung sites. Some of them may not belong to the same culture even. However, they have their own distinctive tool-types—the harpoon, the anthropomorph and the antennae sword—which are completely unrelated to the Harappans. The tool-repertoire

29. Mackay, *In* Marshall, J., *Mohenjodaro and the Indus Civilization*, London, Arthur Probsthain, 1931.
30. Mackay, *op. cit.*, Pl. CXXII.
31. Agrawal, D. P., *Science*, 143, No. 3609, p. 950, 1964.

of the Copper Hoards was specially adapted for a hunting-nomadic life in the thick primeval forests of the Doab.

The question arises, when the other protohistoric cultures too had the copper-metallurgy, why was it that only the Harappans ushered into urbanization. This brings in ecological factors.[32]

Ecology is the human habitat which is comprised of the animate and inanimate environment. Ecology provides the opportunities as also imposes limits on human societies. Let us try to reconstruct the ecology of the Harappan times and then study its influence on the society.

It was thought that the Indus area had greater rainfall in the past, had thick forests and swamps with fauna representative of hot and humid climate.[33] Fairservis[34] effectively countered these arguments and showed that the phenomenon of the Harappa Culture was more probable in a dry milieu. Since then Raikes[35, 36, 37] has proved that the climate was dry in that area, as it is today. Panjab and Sind fall in the rain contours of 10″ to 20″; in the lower Sind it is less than even 5″, as against 25″—40″ of annual rainfall in the Doab. The Indus has soft and pliable alluvial plains and a gallery forest. In contrast, the Doab was a thick monsoonal forest[38, 39]. The modern alluvial plains in this region are due to man-made deforestation.

The decline of Mohenjo-Daro has now been shown to have been due to the impounding of the Indus by tectonically caused mud-eruptions—at least four times during the life of the city.[40] But here we do not want to go into the causes

32. Agrawal, D. P., *Bull. Arch. Soc. India,* 1, p. 17, 1967-68.
33. Wheeler, R. E. M., *Early India and Pakistan,* London, Thames and Hudson, 1959.
34. Fairservis, W. A. (Jr.), *American Museum Novitates,* No. 2055, 1961.
35. Raikes, R. L., *American Anthropologist,* 66, p. 284, 1964.
36. Raikes, R. L., *Antiquity,* 39, p. 196, 1965.
37. Raikes, R. L., *Antiquity,* 41, No. 164, p. 309, 1967.
38. Calder, C. C., *In an Outline of Field Science in India,* edited by S. L. Hora, Calcutta, Indian Sci. Cong., 1937.
39. Stebbing, E. P., *The Forests of India,* Vol. 1, *London,* The Bodley Head, 1922.
40. Raikes, R. L., *op. cit.,* 1965.

and the controversies about the end of the Harappans. We are concerned here more about the origins of their urbanization.

One queer thing about the Harappans is that they impinge as a mature culture on the pre-Harappan cultures—the developmental stages are missing. The individuality of the Harappa culture goes against a foreign origin too. Was it, then, an 'explosive evolution'?

For the urbanization of a society the first requirement is the argicultural surplus. The Indus plains, with annually renewed fertile alluvium, could provide this surplus with little effort. The soil was so pliable that even wood and copper hoes were sufficient for cultivation. The surplus so produced brought the Indus society on the threshold of urbanization—because now the metal technicians could be provided for by the society. Childe said, 'The use of metal tools does not depend upon simply on techn:cal knowledge. A community can only use metal tools when it is producing an effective social surplus.' Metallurgy was a specialized craft and full-time specialists could be used only if a society had concentration and fluidity of the social surplus. Metallurgical know-how and a social-surplus could bring them to the threshold of urbanization. But why this metal efflorescence and full-blown urbanization appearing so suddenly?

It appears that the ecology (in the shape of vast alluvial pliable plains and a gallery forest) provided endless opportunities for increasing the production. What was needed was the tools for agriculture, transport and many other industries of an urbanized centre. Discovery of new mines (was it the Khetri belt?) may have led to sudden progress of the society. Technology, ecology and the abundance of the metal accelerated the pace of development. It may be suggested that the monopolists—discoverers of the mines—may have given the lead to the society and constituted the ruling elite of the Indus civilization. Sudden progress in technology, trade and agriculture could have led to the need of planned cities. If they are pre-planned cities, it is no use looking for their nebulous origins there!

In contrast, the Doab was a thick monsoonal forest. Its clearance needed the strength and the considerable abundance of iron. Limited clearance for agricultural patches was possible with copper. It was an ecology rich in the jungle game and river fish. The main tools of the Copper Hoards—the harpoon, the anthropomorph* and the antennae sword—are admirably suited for a hunting-nomadic life. So the ecology imposed severe limitations for any large-scale agriculture. There could be no great surplus and no urbanization till the advent of the Iron Age. The tools were probably made by the itinerant smiths who were at least economically released from the kinship bonds of their tribes.

Similarly, the ecology placed restrictions on the Chalcolithic cultures of Central India too. Agriculture on the sticky cotton soil without heavy iron ploughshares and coulters was not possible. With their meagre copper tools they could only cultivate the thin alluvial strips and thus could never produce sufficient agricultural surplus to come out of the village status.

Thus we see that a developed metallurgy, rich sources of copper, optimum ecological conditions and an efficient village society which could produce fluid social surplus—all contributed significantly to the first Harappan urbanization that India witnessed.

The facilities for the metallographic and spectrographic analyses were kindly made available to us at the Bhabha Atomic Research Centre, Bombay, by Shri Shivramkrishna and Shri D'Sylva to whom the author is thankful. For the encouragement and the guidance received, he is beholden to Prof. D. Lal.

* It was a definite missile and not a ritualistic human figure, as is shown by the sharp arms and the heavy blunted head.

Chapter 10

POSSIBILITY OF "CONSCIOUS" NATURE SCIENCE IN FIRST URBANIZATION

1. PRELIMINARY REMARKS

With this brief overview of the more significant technologies in the First Urbanization, we now pass on to the question of the possible emergence of the other nature sciences during the period. To begin with, however, we should like to draw the attention of the readers to one point in Gordon Childe's writings, whose lead we are trying basically to follow in our own understanding of the pre-historic period.

We have already seen that while discussing "exact and predictive sciences" ushered in by the Urban Revolution, he mentioned only arithmetic, geometry and astronomy, and not any other nature science, in spite of the spectacular progress in metallurgy and other techniques already on the eve of the full formation of the three primary centres of First Urbanization. Thus, we are inclined to understand that, in Childe's view at any rate, technology—though the most indispensable precondition for the making of science—does not by itself or smoothly pass on to "conscious" science. More factors are evidently needed for its making than are contained in the vast store of experience and knowledge acquired through mere technology. As he illustrates it with reference to metal technology[1] :

> The sciences applied in metallurgy are more abstruse than those employed in agriculture and even pot-making. The chemical change effected by smelting is much more unexpected than that which transforms clay into pottery. The conversion of crystalline or powdery green or blue ores into tough red copper is a veritable transubstaniation. The change from the solid to the liquid state and back again, controlled in casting, is hardly less startling. The actual manipulations themselves are more intricate and exacting even than those involved in pot-making, spinning, or boat-building...
>
> Metallurgical lore is the first approximation to international science. But it remains craft lore. All the practical science of the ancient smiths and miners was certainly embedded in an unpractical matrix of magic ritual. Assyrian texts, even in the First Millennium B.C., contain hints

1. Childe WHH 77-79.

of what such rituals may have involved—foetuses and virgins' blood. So do the remains of a bronze-workers' encampment in Heathery Burn Cave (Co. Durham) in England. Today barbarian smiths' operations are surrounded with a complex of magical precautions.

In the second place, the transmission of such lore by apprenticeship is largely imitative and therefore conservative...

Finally craft lore is liable to be secret. It is passed on from father to son or from master to apprentice. Craftsmen thus tend to form guilds or clans, which would guard jealously the mysteries of the craft.

2. ANCIENT TECHNIQUES AND MAGIC

Let us first try to be clear about one point. Early metallurgy, which of all the ancient techniques contained the greatest potentials for natural science, remained embedded in the matrix of magical beliefs and practices. So long as it was so, it remained more or less under the grip of superstitions, with its science-potentials crippled as jealously guarded secret craft-lores, disabling it to take the momentous step from pre-science to conscious science in the real sense. Why, then, was this matrix of magic?

It needs at once to be noted that the matrix of magic was not the characteristic of early metallurgy alone. It constituted, on the contrary, the limitation of ancient techniques in general and was necessarily so. Thanks to the lucid analysis of George Thomson, we can understand that its need varies inversely with the development of actual technique: the less developed is the actual technique the more is the need of magic. As he explains :[2]

It (magic) is an illusory technique complementary to the deficiencies of the real technique. Owing to the low level of production the human consciousness is as yet imperfectly aware of the objectivity of the external world, which accordingly it treats as though it were changeable at will, and so the preliminary rite is regarded as the cause of success in the real task ; but at the same time, as a guide to action, the ideology of magic embodies the valuable truth that the external world can in fact be changed by man's subjective attitude towards it. The huntsmen whose energies have been stimulated and organised by the mimetic dance are actually better huntsmen than they were before.

On the eve of the Urban Revolution, the metallurgists we are discussing were, of course, far more technologically ad-

2. G. Thomson SAGS I. 38-39.

vanced than the savage hunters. Hence is the normal expectation of them being free from the need of magic and moving towards an objective understanding of nature and natural processes, uninhibited by the limitations of magical fancies or superstitions. But such an emancipation from magic—and hence the move towards natural science—did not actually happen in history. There came into being other factors which, instead of emancipating people from superstitions, required its consolidation. Though apparently peculiar, such factors were included in the very preconditions for the Urban Revolution. This, as we shall see, is brilliantly described by Gordon Childe as "the dialectics of progress"—tremendous acceleration of man's move forward carrying also on its heels forces of retardation.

3. CHANNELISING SOCIAL SURPLUS

Let us try to be clear about this point. In view of the immense complexities in the metallurgical technique, it could not but be a full-time job. The smiths and miners, because of their specialisation, could not at the same time be the direct producers of their own means of subsistence—tilling the fields and minding cattle. They had, therefore, to live on the surplus produced by others, though obviously in exchange of their own products. It is presumed mainly on the ethnological evidences, that in the period shortly before the Urban Revolution, they lived as itinerant craftsmen ; as a single neolithic village could not, from its own surplus products, meet their requirements for the entire year, they had to move from village to village, exchanging their own products with the surplus food produced by each village and living on it for some period and then moved on to some other village with the same purpose.

An essential requirement of the Urban Revolution, however, was to have such full-time specialists settling permanently in the centres growing into cities, without which the requirements for the making of the cities could not be met. This means that the city-makers—or, more specifically, the organisers or rulers of the first centres developing into the cities—had to have at their disposal a somewhat vast amount of food-stuff for the annual maintenance of these specialist craftsmen. These governors, again, not being direct producers of food, had to depend on the channelisation of the surplus products of the peasants from near and perhaps also comparatively distant villages to the city centres, so that these could be stored in the city granaries for the full-time specialists. The improved tools of production on the eve of the Urban Revolution enabled the peasants to create the social surplus, just as the improved means of transport must have helped this process of channelisation to the city-centres. Still the problem remained of how to obtain or extract from the direct producers their surplus products, so that these could be funneled to fill the city granaries.

There were three conceivable alternative techniques that could make this possible: (1) direct plunder, (2) purchase and (3) persuasion by ideological devices. We are going to see why the third of these presumably best suited the city governors and that moreover we have the clue in this to the factor that must have acted as the most powerful agent for inhibiting the emergence of natural sciences from the metallurgical and other techniques that were developed enough to contain in pronounced form the potentials for the making of natural sciences: the ideological devices that could best suit the procurement of the social surplus from the peasants went against the basic requirement of the emergence of nature science from the technologies fully pregnant with its possibility.

Let us begin with the three possible alternatives for procurement.

The possibility of purchase of the surplus products of the direct producers by the city governors—perhaps inclusive of merchants or traders—is ruled out, because the proposition of purchase would presuppose accumulation of sufficient wealth (or in the ancient context, the accumulation of city-products with which to barter with the peasants), and this, in its turn, already employing full-time specialists by the merchants and traders to produce the typical urban products of commodities with which to buy or barter with the peasants. This, in other words, implies the accumulation of food stuff in the city granaries already accomplished.

We are thus left with two other alternatives, namely plunder and persuasion. Of these two, again, plunder is a cumbrous and complicated process: it presupposes the maintenance of a bureaucracy with at least a sizeably large armed force to coerce the peasants to part with their surplus products ; for their maintenance we have to presuppose again sufficient food stuff already accumulated in the city granaries. At any rate, compared to it, the method of persuasion is infinitely simpler and smoother. All that this method requires is the effective use of superstition. Thus, for example, once the peasants are made to believe that without offering a part of their products to the goddesses or gods—i.e. concretely to their earthly representatives, the priests or priestly corporations working for the god or god-king—they remain exposed to grave perils like draughts, plagues and disasters in many other forms, the peasants would willingly—and perhaps also eagerly—part with whatever they can afford from their own products. Besides, these priests or priestly corporations do not have to create such superstitions out of nothing : from the paleolithic age, people with extremely rudimentary understanding and control of nature were trying to supplement their real technique with the illusory technique, namely magic, from which the goddesses, gods and all sorts of supernatural agencies easily took shape. As George Thomson[3] very lucidly explains :

> The technique of magic is developed by the ruling class as a means of consolidating their privileges by investing them with supernatural sanctions. In this way the working class, being ignorant of the true causes of its subjection, is reconciled to its lot. This is the genesis of religion. Religion is an outgrowth of magic which emerges with the class struggle. It is an inverted image of social reality. Just as magic expresses primitive man's weakness in the face of nature, so religion expresses civilised man's weakness in the face of society.

This must have made the task of the governors of the early nucleus of cities all the simpler. They had only to systematise—to add awe and wonder—to a pre-existing system of beliefs and practices, and thus make the technique of persuading the direct producers to part with their surplus products to city rulers, from which to maintain the full-time specialists essential for the Urban Revolution.

But it will be wrong, of course, to take a merely negative view of the whole thing. The Urban Revolution was a momentous step forward in the history of mankind and the exploitation of the direct producers was a necessary precondition for

3. G. Thomson ER 9.

it, as was the use of superstition a precondition for this exploitation. We have, in other words, to understand the dialectics of progress which is summed up by Gordon Childe as follows :[4]

> Almost from the outset of his career, it would seem, man used his distinctively human faculties not only to make substantial tools for use upon the real world, but also to imagine supernatural forces that he could employ upon it. He was, that is, simultaneously trying to understand, and so utilize, natural processes and peopling the real world with imaginary beings, conceived in his own image, that he hoped to coerce or cajole. He was building up science and superstition side by side.
>
> The superstitions man devised and the fictitious entities he imagined were presumably necessary to make him feel at home in his environment and to make life bearable. Nevertheless the pursuit of the vain hopes and illusory short-cuts suggested by magic and religion repeatedly deterred man from the harder road to the control of Nature by understanding. Magic seemed easier than science, just as torture is less trouble than the collection of evidence.
>
> Magic and religion constituted the scaffolding needed to suport the rising structure of social organization and of science. Unhappily the scaffolding repeatedly cramped the execution of the design and impeded the progress of the permanent building. It even served to support a sham facade behind which the substantial structure was threatened with decay. The urban revolution, made possible by science, was exploited by superstition. The principal beneficiaries from the achievements of farmers and artizans were priests and kings. Magic rather than science was thereby enthroned and invested with the authority of temporal power.
>
> It is as futile to deplore the superstitions of the past as it is to complain of the unsightly scaffolding essential to the erection of a lovely building. It is childish to ask why man did not progress straight from the squalor of a 'pre-class' society to the glories of a classless paradise, nowhere fully realised as yet. Perhaps the conflicts and contradictions above revealed, themselves constitute the dialectics of progress.

4. NATURE SCIENCE AND URBAN REVOLUTION

To sum up the points we have discussed so far : The priests or the priestly corporations—which, at least in Egypt and Mesopotamia formed the nucleus of the ruling class even while acting on behalf of the god or god-king, and which, as we shall presently see, presumably also did in the Indus Civilization—played, historically a role infested with an inner contradiction.

4. Childe MMH 236.

On the one hand they acted as the *organisers of production.* On the other hand they had also to act as the *administrators of superstition.*

Before passing on to discuss how, in spite of this inner contradiction in their role, they could create—or at least encourage the creation of—arithmetic, geometry and astronomy as exact sciences, we shall try to understand why, because of this inner contradiction in their role—specially as administrators of superstition—they were also obliged to prevent the making of nature sciences based on reason and uninhibited experience, and this in spite of the development of the technological prerequisites for their making, as, for example, in the case of metallurgy.

While the Urban Revolution resulted from the series of spectacular developments in technology preceding it, and, as a matter of fact, it resulted in spectacular achievements in many forms, the practical needs of its new socio-economic structure also required vigorous consolidation of the magico-religious beliefs without which the basic need of the city life—the channelisation of the surplus products of the peasants to the city centres from which to maintain the whole-time specialists—is not easily conceivable. As a result, the original functional role of the system of magico-religious beliefs passed into its opposite. In the earlier stages of social development, these were illusory techniques supplementing real techniques : though without directly changing or controlling nature these could and actually did change the subjective attitude of the technicians themselves—infuse in them hope, courage and confidence—and thereby help them actually to better control nature, though indirectly.[5] In the hands of the city governors—the vigorous consolidation of the magico-religious beliefs, malignantly magnified, became the most effective instruments for controlling the people. In such circumstances, any attempt to understand nature and its laws depending on uninhibited reasoning and direct observation was sure to be frowned upon, if not actually prosecuted.

The question of mathematics and astronomy was, of course, different. The results of the former were too abstract to disturb the realm of the goddesses and gods and in any case it was necessary for the building of the gigantic ziggurats, temples and tombs of the god-kings, which, when built, could enhance

5. See G. Thomson SAGS I. 440.

the awe and grandeur of the priests and god-kings. Astronomy, besides having the obvious need for agricultural operations by way of preparing or regulating the calendar, had the added advantage of imputing supernatural knowledge and power to the priests and priestly corporations who incidentally, wanted to keep it a closely guarded secret : it enabled them to make the most wonderful predictions about the heavenly bodies as well as other predictable natural phenomena ; besides these heavenly bodies were viewed as veritable deities or at least as having some presiding deities, i.e. not as purely natural phenomena working according to just natural laws.

Thus the making of mathematics and astronomy in the primary centres of the Urban Revolution is not so difficult to understand. However what is also important to understand is the exclusion of the possibility of the making in these centres of other "conscious" sciences, aspiring to view nature as it is without any alien addition. This follows from the overriding need in these centres of another technique, which may be described as that of keeping the masses under control with fables and fears about the goddesses and gods—in short, the technique of using superstition. Nothing that disturbs this universe of superstitions could at all be tolerated. And hence there could be no scope in it for the making of nature science aspiring after an objective understanding of nature.

5. SUPERSTITION AND NATURE SCIENCE IN EGYPT : PLATO

The point we have been trying to make is not a new one. It was already noted by as sophisticated a philosopher as Plato many centuries back. We shall try to follow here George Thomson's[6] analysis of the point.

It needs first of all to be noted that besides being a very eminent philosopher, Plato was keenly interested also in politics. His earlier work the *Republic* already shows this and his latest work the *Laws* is frankly a work on politics . Politically speaking, his main problem was how to keep the masses —the slaves—under control. From this point of view, he was repelled—indeed horrified—by the earlier Ionian nature philosophers, who were trying to .understand nature in terms of

6. G. Thomson SAGS II (First Philosophers). Ch. XV.

nature itself ; from Plato's point of view, this had the most adverse effect for his political programme. Hence, in the *Laws*, he came out most sharply against the Ionian natural philosophers :[7]

—They say that earth, air, fire and water all exist by nature or chance, not by art, and that by means of these wholly inanimate substances there have come into being the secondary bodies—the earth, sun, moon and stars. Set in motion by their individual properties and mutual affinities, such as hot and cold, wet and dry, hard and soft, and all the other combinations formed by necessity from the chance admixture of opposites—in this way heaven has been created and everything that is in it, together with all the animals and plants, and the seasons too are of the same origin—not by means of mind or God or art but, as I said, by nature and chance. Art arose after these and out of them, mortal in origin, producing certain toys which do not really partake of truth but consist of related images, such as those produced by painting, music and the accompanying arts, while the arts which do have some serious purpose, co-operate actively with nature, such as medicine, agriculture and gymnastics ; and so does politics too to some extent, but it is mostly art ; and so with legislation—it is entirely art, not nature, and its assumptions are not true.

—How do you mean ?

—The Gods, my friend, according to these people have no existence in nature but only in art, being a product of laws, which differ from place to place according to the conventions of the lawgivers ; and natural goodness is different from what is good by law ; and there is no such thing as natural justice ; they are constantly discussing it and changing it ; and, since it is a matter of art and law and not of nature, whatever changes they make in it from time to time are valid for the moment. This is what our young people hear from professional poets and private persons, who assert that might is right ; and the result is, they fall into sin, believing that the gods are not what the law bids them imagine them to be, and into civil strife, being induced to live according to nature, that is, by exercising actual dominion over others instead of living in legal subjection to them.

—What a dreadful story, and what an outrage to the public and private morals of the young !

What, then, was to be done ? How to counteract this "outrage to the public and private morals of the young" with a natural view of nature ? For Plato himself there was frankly only one answer to it and that was to feed the young with lies

7. Quoted *Ibid* 323-24.

and falsehood, which, though deliberately fabricated, was, from the standpoint of Plato's politics, naturally also "beneficial". Already in the *Republic,* Plato recommended it without mincing words. Thus :[8]

—And even if this were not true, as our argument has proved it to be, could a legislator, who was any good at all and prepared to tell the young a beneficial falsehood, have invented a falsehood more likely to persuade them of their own free will to do always what was right ?

—The truth is a fine thing and lasting ; yet it is not easy to make people believe it.

—Well, was it hard to make people believe the myth of Kadmos, and hundreds of others equally incredible ?

—Which do you mean ?

—The sowing of the dragon's teeth and the appearance of the warriors. What an instructive example that is to the legislator of his power to win the hearts of the young ! It shows that all he needs to do is to find out what belief is most beneficial to the state and then use all the resources at his command to ensure that throughout their lives, in speech, story and song, the people all sing to the same tune."

Still the question remained about the feasibility of such a programme. Was it at all feasible to feed the people with lies and falsehoods, so that they were left with no enthusiasm at all for natural philosophy or the tendency to know nature as it is ? In the *Laws,* Plato argued that it was surely feasible, as was evidenced by the achievements of the ruling class of ancient Egypt. Thus :[9]

—What are the legal provisions for such matters in Egypt ?

—Most remarkable. They recognised long ago the principle we are discussing, that the young must be habituated to the use of beautiful designs and melodies. They have established their norms and displayed them in the temples, and no artist is permitted in any of the arts to make any innovation or introduce any new forms in place of the traditional ones. You will find that the works of art produced there to-day are made in the same style, neither better nor worse, as those which were made ten thousand years ago—without any exaggeration, ten thousand years ago.

—Very remarkable.

—Rather, I should say, extremely politic and statesmanlike. You will find weaknesses there too, but what I have said about music

8. Quoted *Ibid* 324.
9. Quoted *Ibid* 324-25.

is true and important, because it shows that it *is* possible for a legislator to establish melodies based on natural truth with full confidence in the result. True, it can only be done by a god or a divine being. The Egyptians say that the ancient chants which they have preserved for so long were composed for them by Isis. Hence, I say, if only the right melodies can be discovered, there is no difficulty in establishing them by law, because the craving after novelty is not strong enough to corrupt the officially consecrated music. At any rate, it has not been corrupted in Egypt.

So that is what Plato thought. In the capacity of a politician interested in the safety of the slave society he realised that it was dangerous to allow the natural view of nature to be encouraged, that superstition—though philosophically and scientifically a *falsehood*—is *beneficial* for keeping the people under control and that the feasibility of using superstition as an instrument of policing the state was already proved in ancient Egypt.

Another senior contemporary of Plato, Isocrates, said practically the same thing about the political function of superstition in ancient Egypt. As Farrington observes : "A sophisticated Greek of the fourth century B.C. cast a glance at the official religion of Egypt and detected its social utility. The Egyptian lawgiver, he remarks, had established so many contemptible superstitions, first, 'because he thought it proper to accustom the masses to obeying any command that was given to them by their superiors', and second, 'because he judged that he could rely on those who displayed their piety to be equally law-abiding in every other particular'."[10]

In such an atmosphere, it needs hardly to be added, people with a naturalistic view of natural phenomena could hardly be encouraged or even tolerated. Hence, in ancient Egypt, apart from certain officially approved sciences like mathematics and astronomy, there was no scope for the development of any nature science, and this in spite of the spectacular developments in technology containing potentials for the emergence of nature science. "There is an official mythology, transmitted in priestly corporations and enshrined in elaborate ceremonial, telling how things came to be as they are. There are no individual thinkers offering a rational substitute for this doctrine over their

10. B. Farrington GS 34-35.

own names."[11] In such circumstances the productive techniques continued to be handed down as craft-lores in the form of precepts and examples, and this among the members of the depressed classes, from whose actual experience and direct intercourse with nature it was too derogatory for the learned elites to learn anything and thereby to move to the actual knowledge of nature and its laws.

6. POSSIBLE ROLE OF SUPERSTITION IN INDUS ADMINISTRATION

Because of the concentration of political power among the priests and in the priestly corporations, what is discussed about Egypt was also true of Mesopotamia. But what about the third primary centre of the Urban Revolution, namely the Indus valley?

By analogy, one is tempted to argue that since it was necessary for the other two primary centres of the Urban Revolution to use massive superstition for policing the state, the same must have been true also of the third primary centre. Besides, as we have already seen, the accumulation of the surplus products of the direct producers in the cities, seems to be best explained by the presumption of the use of superstition.

Nevertheless, the assumption of the vast Indus "Empire" being ruled by the priests or priestly corporations cannot be a smooth one and there is literally a storm of controversy over the question of the actual socio-political organization of Harappan culture. The main difficulty here, as about many other questions concerning the Harappan Culture, is the want of direct literary documents attesting to some view or other. It is generally admitted no doubt that there must have been some strong centralised power enforcing its authority over the Harappan "Empire", because, without assuming it there is hardly any explanation of the manifold uniformity observed throughout it. But we have no direct knowledge of the nature of this centralised authority.

Many relics of the Indus Civilization are generally viewed no doubt as indicative of religion and religious beliefs.

From the time of the publication of the first full report on

11. *Ibid* 34.

the excavations of Harappa and Mohenjo-daro to that of the recent excavations at Kalibangan and Lothal, the archaeologists have shown how a large number of the relics of this forgotten civilization cannot but be understood as pointers to the powerful religious beliefs prevalent in the period. We need not try to prepare here a list of such relics ; readers interested in these may look up the recent book by the Allchins,[12] where most of these are very ably summed up. This does not mean, of course, that we have now a coherent and comprehensive understanding of the nature of Harappan religion and it is no use speculating on when and how any full account of it would be reconstructed. Nor is the basic fact to be ignored that among the eminent archaeologists controversies are still going on about its general outlines. However, all these do not materially affect our main argument, which requires only to be admitted that the Indus relics are unmistakably indicative of some presumably strong and widespread religious beliefs. If so, we have also to admit the custodians of these, namely the priests or priestly corporations, and, on the analogy of ancient Egypt and Mesopotamia, we are naturally tempted to assume that these priests and priestly corporations could remain quite aloof from the actual administration of the vast Indus "Empire". On the contrary, it would be logical to think that whatever might have been the actual nature of the social structure of the Indus Civilization, these priests or priestly corporations were only likely to have a large share—if not the decisive one—in the administrative machinery in the Harappan administration. From the remains of the imposing houses of the merchants in the Harappan cities, it is sometimes conjectured that these merchants could have formed the actual ruling class of the "Empire". But the obvious difficulty about such a conjecture is that it can hardly explain the eventual internal decadence of the Harappan civilization, about which the archaeologists are agreed. The merchant class—sensing as they do greater and still greater profitability—are drawn to improving and augmenting the production-process and hence the government under their rule is only expected to prosper rather than become a prey to eventual degeneration and decay. By contrast, a government under the

12. B & R Allchin RCIP 213 ff.

priest class is only likely to be under the grip of strong conservatism and hence also exposed to the possibility of eventual decline and decay.

Incidentally, D. D. Kosambi has advanced a new suggestion indicative of the effective use of religious superstition as an effective instrument for Harappan administration. As he puts it :[13]

Finally, the tools of violence were curiously weak, though nothing is directly known of their social mechanism for wielding force, which we call the state. The weapons found in the Indus cities are flimsy, particularly the ribless leaf-blade copper spearheads which would have crumpled up at the first good thrust. There is nothing like a sword in the main Indus strata. Archers occur in the ideograms, arrowheads of stone and copper have been discovered. The bow would be a survival of the hunting age. Of course, iron was not known, so that a few weapons in the hands of a small minority might have sufficed ; but the contrast with the excellent, sturdy though archaic, tools proves that the use of weapons was not very important. Therefore, the state mechanism, whatever it was, must have had some powerful adjunct that reduced the need for violence to a minimum. The cities rested upon trade, not fighting ; but if the army or police were not very strong, what helped the trader maintain his unequal sharing of profit ?

The answer seems to lie in religion. Though there are no great statues of the gods, what has been called the 'citadel' mound undoubtedly corresponds to the temple-zikkurat structures in Mesopotamia. The Harappan site has been devastated by brick-robbing, while at Mohenjo-Daro, what must have been the ruins of a major building in the sacred enclosure are covered by a Kusana stupa. But the adjacent 'Great Bath' at Mohenjo-Daro (filled with water drawn laboriously by hand from a special adjacent well, beautifully constructed with bitumen waterproofing between brick layers, a drain for emptying, and surrounded on three sides by cells) must have been a ritual tank, because of the beautiful and well-used bathrooms in every private house which distinguish the city from anything in proto-history, or Mesopotamia, or Egypt. Even a bather from the citadel could easily have descended the steps in the well which led down to the river. I have explained this as the prototype of the sacred lotus pond (*puskara*) ; which survived in later times.

Thus one of the main points stressed by Kosambi in favour of his view of the comparatively greater need of religious superstitions for policing the Harappan "Empire" is the flimsiness of the offensive weapons specially as contrasted with the ex-

13. D. D. Kosambi ISIH 59-60.

cellent, sturdy though archaic tools, which proves that the weapons usually needed to keep the people under control were not very important. To this may be added another point.

The weapons unearthed at the Harappan sites are indeed flimsy, indicating the lesser use of direct violence and hence comparatively greater use of religious superstitions. What could have perhaps made the argument stronger is another consideration.

Already in the introduction to the first full-length report on *Mohenjodaro and the Indus Civilization,* Marshall observes: "Their weapons of war and of the chase are the bow and arrow, spear, axe, dagger and mace. The sword they have not yet evolved: *nor is there any evidence of defensive body armours.*"[14] We have added emphasis on the last point, because, thanks to Needham's brilliant analysis, this seems to have very decisive importance for the point Kosambi has argued in favour of the relatively greater use of religious superstitions for the Harappan rulers. The offensive weapons mentioned by Marshall,—spear, axe, dagger and mace—are useful in hand-to-hand fight, or in cases of immediate confrontation of armed force with the exploited peasants when necessary. Besides these were flimsy after all. But not so are the bow and arrow, which are very effective long distance missiles. Now, as Kosambi admits, the bow and arrow are inherited from the hunting stage, there is nothing to prevent the assumption that the peasants were as much equipped with these as were the armed forces of the ruling class. And it is here that the *defensive body armour* has supreme importance from the viewpoint of military technology in the distant past. So long, therefore, the defensive body armour is not developed and used, some method other than direct violence is needed to keep the masses under control.

In the Chinese context, as Needham shows, the lack or inadequate development of defensive body armours, was one main reason for the ruling class patronising the Confucian views, preaching to the people to remain submissive to the lords. As he puts it:[15]

What was the situation, then, in ancient China? There the crossbow—a most powerful weapon—was invented centuries before any-

14. Marshall MIC I. Preface p. iv. Emphasis added.
15. J. Needham GT 168-9. Emphasis added.

where else. We know that the men of the feudal levies in ancient China (by that I mean between 800 and 300 B.C.) were armed with powerful bows. But at the same time protective armour was very little developed. The archaeologist Laufer has written a fine monograph on Chinese armour. It arises very late, and in early times you only get protective clothing made of bamboo and wood. Moreover, there are in the *Tso Chuan* countless stories of feudal lords being killed by arrow shots. *If the mass of the people as a whole were in possession of a powerful offensive weapon, and the ruling class were not in possession of a superior defensive means, one can see that the balance of power in society was different from what it was in, e.g., the time of the early Roman Empire, where the disciplined legions were rather well armoured, with bronze and iron. A slave population was possible because it was not in possession of the arms and armour of the legionaries,* nor did it have access to powerful bows. The principal Roman weapons were always the spear and the short sword. We know what troubles the slaves could give on the few occasions in which they did gain access to substantial stores of weapons, as in the revolt of Spartacus. In China, it was a different story, because from an early date the people had crossbows and the lords had poor defensive armour. If that was the case, it seems that the people in China had to be persuaded, rather than cowed by force of arms, and hence the importance of the Confucians.

For, as Needham had already shown, "During what may be called the high feudal period in China, which runs roughly from the eighth century to the third century B.C. the feudal lords were assisted and counselled by a group of men who afterwards became the school of philosophers which we know as the Confucian School."[16]

With these points in mind, we may now return to the question of the Indus Civilization...

Not that we know of there having been any philosopher in the Harappan culture, not to speak of any school of philosophy even remotely resembling the Confucian one. Notwithstanding Marshall's expectation to the contrary,[17] it can perhaps be safely asserted that there was nothing like that, or at least, even if there was any philosopher in the Harappan culture, we shall never know anything about him. What we do know, however, is that the *Harappans did not develop any defensive body ar-*

16. *Ibid* 155-56.
17. Marshall MIC I. 78.

mour. Even imagining that they developed something like that made of flimsy materials, these could be no good against the long-distance missiles—the bow and arrows—possessed in common by the army and the masses. It follows, therefore, that for keeping the masses under control it was essential for the ruling class to have something more than their military technology. This additional something had to have the efficacy of persuading the masses to remain as pious law-abiding citizens. In other words, in default of effective defensive armour they had to depend on ideological devices—or, to put it more bluntly, on religious superstitions on a really massive scale. Many stray objects found in the Harappan regions give us some glimpse of a rather imposing religion, though our archaeologists are yet to reconstruct any agreed view of this religion. But religion was there and so also there must have been its main accessory, namely, superstitions.

Could it, then, be that this religious ideology was the main instrument in Harappan culture for the purpose of policing the state? This brings us back to Kosambi's hypothesis though from a different premise altogether. What Kosambi argued from the flimsiness of offensive weapons seems to be strengthened by the lack of defensive body armour.

7. CONSCIOUS NATURE SCIENCE IN HARAPPAN CULTURE?

With these clarifications, we may now return to the main point we wanted to discuss. On circumstantial evidences we have been led to presume the making of "conscious" sciences in two forms—namely mathematics and astronomy—during the period of our First Urbanization. But the First Urbanization also witnessed spectacular achievements in many forms, of which specially metallurgy did contain the potentials for the making of several natural sciences. Nevertheless the question remains, could "conscious" natural sciences develop from these technologies? In default of any documentary data, we are obliged to try to answer this question on the basis of circumstantial evidences only. Two of such evidences are crucial for our present purpose. First, metallurgy in the ancient world, with all the spectacular potentials for the emergence of natural sciences, remains embedded in an essentially ascientific matrix of magic and rituals. Secondly, the consolidation of such myths and rituals leads to the formation of religious superstitions, and religious superstition is an effective factor contributing to the Urban Revolution itself, inasmuch as the channelisation of social surplus is necessary for the support of full-time specialists in the growing cities and religious superstitions best suit the purpose of persuading the direct producers to part with their surplus products. We have also seen that religious superstition could very possibly be the most effective instrument for policing the vast Indus "Empire". Our question, therefore is: Under such circumstances, could conscious nature science possibly emerge from their potentials implicit in the spectacular technologies witnessed by the Harappan culture?

The answer to this is evidently in the negative. The basic precondition for the formation of "conscious" nature science is to understand nature and the natural changes as these actually or objectively are, i.e. without the veil of magical beliefs on these and, of course, without imagining any supernatural agencies—the goddesses and gods—controlling or interfering with these. In short, it is the emancipation from magic and also from the negative aspect of magic taking the shape of organised religion. However, such organised religion was in all presumption a very effective instrument for Harappan administration. This could not allow the spirit of understanding nature and natural phenomena on the basis of uninhibited observation and reasoning, i.e. the very precondition for the making of conscious science. We are thus inclined to conclude that "conscious" science only in two forms, viz. mathematics and astronomy, could develop in our First Urbanization, as in the two other primary centres of the Urban Revolution. But the Urban Revolution also required an intellectual climate that precluded the possibility of developing conscious natural science from the series of spectacular technological innovations. For making of "conscious" natural science in its true sense, Indian history had to await the advent of the Second Urbanization which was no longer under the grip of priestly administration. But more of this later.

CHAPTER 11

END OF THE FIRST URBANISATION

1. PRELIMINARY REMARKS

The problem of the decline and end of the Indus Civilization is evidently one with which primarily the archaeologists are concerned. Still the historian of science cannot evade it. The reasons for this are obvious. First, in order not to be a mere inventory of the prominent achievements of the scientists, the history of science in India has to be related to the mainstream of Indian history—has in fact to be viewed as an important dimension of Indian history—more or less neglected though it may be by the histories of science in India so far. Evidently, the end of the Indus Civilization is a phenomenon too important to be ignored about ancient Indian history. Secondly, even from the restricted standpoint of the history of science in India, the end of the Indus Civilization with its imposing technological achievements and the possible making of arithmetic, geometry and astronomy coming to a virtual extinction with it means some kind of abrupt break in the scientific tradition, and it is incumbent for the historian of science to seek an explanation of it, for he is concerned not merely with factors that helped but also those that inhibited or disturbed its growth. And the fact is that from the technological-scientific point of view at any rate, the end of the Indus Civilization meant also the beginning of a fallow period that extended over many centuries. It is because of this that the archaeologists often refer to it as the "Dark Period" or "Dark Age", though, as we are going to see, from some viewpoint, this period is not totally dark after all. In spite of the loss of the technological magnificence and also the loss of script, it is the period of the creation of a vast body of literature—orally transmitted during many centuries—and also that of the making—or at least of the beginning of certain formal disciplines like linguistics, metrics, etc., required for the purpose of the meticulous preservation of this literature and such disciplines are not totally denuded of scientific interest. Further, a section of leading archaeologists at any rate are inclined to connect the end of the Indus Civilization with those people that created this vast body of orally composed literature and hence also initiated the formal disciplines required for its preservation. But more of this later.

2. DECLINE OF THE INDUS CIVILIZATION

That after a glorious career of about five hundred years, the Indus Civilization came under the grip of decline and degeneration is archaeologically indisputable. Depending mainly on Mackay's work, Gordon Childe gives us some typical examples of this : 'The last reconstructions of Harappan cities exhibit every sign of decadence. Old bricks were re-used for building mean houses on the sites formerly occupied by the spacious mansons of the bourgeoisie. The civic authority could no longer enforce the building regulations so strictly observed in more prosperous days so that the dwellings encroached upon the streets'.[1] Discussing the fate of Mohenjo-daro city, Wheeler observes[2] :

> One thing at least is clear about the end of Mohenjo-Daro : the city was already slowly dying before its ultimate end. Houses, mounting gradually upon the ruins of their predecessors or on artificial platforms in the endeavour to out-top the floods, were increasingly shoddy in construction, increasingly carved up into warrens for a swarming lower-grade population. Flimsy partitions subdivided the courtyards of houses. To a height of 30 feet or more, the tall podium of the Great Granary on the western side of the citadal was engulfed by rising structure of poorer and poorer quality. Re-used brick-bats tended to replace new bricks. The city, to judge from excavated areas, was "becoming a slum".

Why, then, was this decadence and degeneration after a splendid career of above five hundred years or more ? This is a question about which there is a good deal of controversy among the contemporary archaeologists. Perhaps it is premature at the present stage of research to expect any clear and definite answer to the question. What is possible nevertheless is to mention here some of the prominent views advanced with comments on their comparative plausibility.

To begin with, let us note that what was perhaps true of the Mohenjo-daro city was not likely to be true of the entire

1. G. Childe NLMAE 187
2. M. Wheeler IC (1979) 127

vast area covered by the Harappan culture. Nevertheless, we cannot possibly ignore the fate of the Mohenjo-daro city itself. At least one of the factors that contributed to the decline of this city was repeated flooding of it which, as the Allchins observe, 'has been long known and cannot be entirely discounted as a cause of local destruction'.[3] It is true that the Harappans could and did rebuild or repair the city repeatedly after the devastating floods ; but that must have sapped much of their energy and vitality, and hence also caused deviation from their main preoccupations. It is also tempting to conjecture that this could have further considerably weakened—or at least adversely affected—the prestige of the city rulers if they were priests, priestly corporations or priest-kings, because one source from which they were likely to create a belief of supernormal power in the popular mind could have been their capacity for predicting the coming of the floods based on their astronomical knowledge and calendrical science. If in spite of this, they failed to predict the coming of the floods caused mainly by natural catastrophes beyond the depth of their own understanding—not to speak of preventing these with their allegedly supernatural power—the city dwellers were likely to have raised awkward questions about their authority, which, as we have just seen, must have been considerably weakened, as a result of which the magnificently planned city was gradually reduced to some kind of slum.

So, the ravages caused by floods were likely to have been one of the factors that caused degeneration to the Mohenjo-daro city itself. But some of the archaeologists appear to go a step further and want to view the floods—along with other climatic and tectonic change—basically responsible for the degeneration and destruction of the Harappan culture as such. It is not necessary for our present purpose to attempt a re-examination of their views, because one of our most competent archaeologist-scientists, D. P. Agrawal, has already done it. We sum him up at some length.[4]

In 1956, M. R. Sahni suggested the possibility of explaining the end of Mohenjo-daro by the 'impounding of the Indus.'

3. B. & R. Allchin RCIP 224.
4. D.P. Agarwal AI 188-9

Raikes and Dales wanted to revive the theory more forcefully. According to them 'a tectonically caused large mud-extrusion impounded the Indus causing colossal silting', which eventually engulfed Mohenjo-daro. Lambrick, however, points to the difficulty of accepting such a theory. First, such a barrier would result in the shedding off of the silt-load by the river considerably upstream and not on the inner side of the dam. Besides, a permeable barrier could hardly stand the great impact of water coming at the rate of 2,270,000 litres per second : in 1819 a mud-extrusion caused by earthquake was washed away in 1826 by the first flood coming down the Nara. But Raikes and Dales came out with a renewed defence of their flood-theory, arguing that part of the water-discharge could take the course of the Nara. Agrawal, however, raises a number of basic questions against the view of Raikes and Dales : 'If there was such a vast lake around, how could any crops grow and therefore provide any surplus to sustain a city ? . . . . How could carts ply on a muddy road, if they were living in a quagmire ? How could the drains function at all ? . . . . Will not such extreme conditions kill the people *en masse* before any mud-lake engulfed them ?' And so on. Not that Agrawal denies that 'the Indus did affect the fortunes of the Harappans in more than one way. Frequent floods were sapping their energy. Their raised platforms and massive bunds indicate the severity of the problem. The Indus has been raising its flood level continuously necessitating safeguard measures, ... causing salinity increase and rendering vast tract useless for agriculture'. On the other hand, evidences from Rajasthan indicate fluctuations of wet and dry periods. The possibility of climatic changes cannot be fully ruled out. 'It is necessary to discuss in greater detail the recently discovered evidence of drastic changes in the palaeo-channel configurations, especially in Rajasthan. They did affect the Harappans as also the subsequent cultures in the area. Ghosh discovered a large number of chalcolithic settlements on the Ghaggar and Chautang in Rajasthan, now dried up. This raised questions concerning climatic changes in the area, as a result of which these rivers dried up. 'Our recent work, however, has emphasised the significance of the tectonically-caused environmental changes in north and west Rajasthan . . . . Dikshit has chronicled the various hypotheses in an historical sequence and several others have discussed the evidence of archaeological remains on these dried up beds.' On the evidences like these (to which are added a few more), Agrawal concludes: 'The Satluj once flowed into the Ghaggar following a path east of Roper, Sirhind, Patiala and Shatrana. The Ghaggar was a mighty river in the past and had on an average 8 km. wide bed. The Satluj, before assuming its present course, braided into a multitude of channels....As the enechlon faults controlled the river course of the Ghaggar, it was prone to drastic changes due to even minor tectonic movements. Both our Landsat imagery derived palaeo-channels and the field data given by Singh support the flowing of Satluj into the Ghaggar in the past. Some major easterly (from the Ghaggar) river of the past were changing its courses more frequently. We have been able to trace three such courses... Whereas these changes in the courses of various palaeo-channels are fairly clear, and also find support from the field data, the terminal course of the Ghaggar is far from clear....But when the Ghaggar was a perennial river, it is possible that it could have met the Nara and flowed directly into the Rann of Kutch, without meeting the Indus. The palaeo-channels beyond Marot do indicate such a possibility....As human settlements were thriving on their banks when they were alive, it would be easy to date these palaeo-channels. Archaeological explorations have thrown welcome light on these problems...The Ghaggar was alive during the pre-Harappan and Harappan times. The Painted Grey Ware (PGW) (*c.* 800-400 B.C.) mounds are located in the river-bed itself, probably indicating a much reduced discharge of the Ghaggar. The Chautang has a large number of Late Harappan sites....Considerable geomorphological and archaeological field work is required to document the vagrancy of these rivers. It has to be supported by chemical|minerological analyses of the core profiles from these river-beds to detect the signatures of the different relict rivers. Yet it is obvious that in north and west Rajasthan tectonically-changed palaeo-channel configurations were a major factor which affected the human settlements, perhaps from the pre-Harappan times onwards. Major diversions cut off the vital tributaries and growing desiccation, on the other hand, dried up the once mighty Saraswati and Drishadvati rivers. Around Lothal, the aerial survey carried out by us has indicated an annular pattern of drainage which points to tectonic disturbances. Either a tectonic uplift or an eustatic fall in sea-level was probably responsible for cutting off of the Lothal dockyard from the water channels and eventually from the access to the sea. Sea-level changes also need to be considered... The coastal-port sites of Sutkagendor, Lothal, etc. are far inland now but may have been connected with the sea in their heyday and perhaps fell out of use due to fall in sea-level. How actually did the sea-level changes affect the Harappan culture and its prosperity will require a more exhaustive study.'

Such, then, is an admirable summary of our present knowledge of the changes in physical geography affecting the fate of the Indus Civilization. At the same time, it needs to be noted that the decline and fall of the Indus Civilization must have been a complex phenomenon and to seek its full explanation only in these—or for that matter, exclusively in one type of causal factors—remains exposed to the fallacy of over-simplification. As Agrawal himself adds, 'It may, however, be emphasised that not a single cause but several contributed towards the decline and disappearance of the mighty Harappan civilization'.[5] This is a point on which a significant number of serious archaeologists concur. 'Just as the creation and maintenance of the system was the outcome of the successful combination of several factors', observe the Allchins, 'so too its breakdown could have been caused by the weakening of any one of these or the upsetting of their harmonious balance and interaction.'[6] Even Wheeler—about one of whose points there is a great deal of furore among our historians—observes :[7]

Let it be said at once that the factors instrumental in the dissolution of historic civilizations have never been of an uncomplicated kind. It can scarcely be supposed therefore that prehistoric or historic civilizations have endured simple destinies ; in other words, here too no single explanation can convincingly claim total truth. Over-ambitious wars, barbarian invasions, dynastic or capitalistic intrigue, climate, the malarial mosquito have been urged severally in one context or another

5. *Ibid.* 191
6. Allchins RCIP 191
7. Wheeler IC 126

as an over-all cause. Other theories have relied upon racial degeneration, variously defined or cautiously vague ; an enlargement, perhaps, of Samuel Butler's plaint that 'life is one long process of getting tired'. Recently, deep floods derived from violent geomorphological changes have been blamed for the end of the Indus civilization. In a particular context which has sometimes been amplified or decried without warrant, I once light-hearteedly blamed Indra and his invading Aryans for a concluding share in this phenomenon. The list need not be extended. It is safe to affirm that any one of these answers to the problem is far more likely than not to be fallacious in isolation. The fall, like the rise, of a civilization is a highly complex operation which can only be distorted and obscured by easy simplification.

Interestingly enough, among the causes of the decline and fall of the Indus "empire", Possehl mentions even the possibility of peasant revolt against the exploitors in the cities. As he puts it, 'Political conflict which could destroy the people producing the food and|or the productive potential of the land on which the non-agricultural urbanites ultimately depended could be another such explanation. Might we not even entertain such notions as peasant revolts against the ruling classes who may have been largely or even exclusively city based ?'[8]

Such then, are some of the conjectures concerning the decline and end of the Indus Civilization, though also with the caution sometimes expressed against putting an exclusive emphasis on any of these possible factors. The list of conjectures can indeed be enlarged adding to this such possibilities as the financial set-back resulting from the loss of foreign trade from Ur-III times when the 'Mesopotamians looked more towards Arabian and African markets than to Magan and Meluhha"[9]—Meluhha probably having been 'the generalized term for the Indus culture area'.[10] Another factor suggested is 'the "wearing out" of the land due to over-cultivation' though the Allchins observe that this 'seems unlikely as the population pressure can never have been very great, and in later times the land retained great fertility'.[11] At the same time, the Allchins

8. Possehl in ACI 188
9. S. Asthana in EIP 43
10. *Ibid.* 42
11. Allchins RCIP 224

observe, 'Another possibility which cannot be ignored is of epidemic diseases following in the wake of the floods, etc.'[12]

All these are conjectures, of course. Besides, it remains an open question whether these factors, taken severally or jointly, can explain only the internal decay of the civilization rather than its final destruction. Gordon Childe observes, 'This imposing civilization perished utterly as a result of internal decay accelerated by the shock of barbarian raids'.[13] He mentions the following archaeological evidences for the last point :[14]

Then the civilization was destroyed by barbarian invaders and the cities reoccupied by illiterate aliens. At Harappa these are represented only by extended and flexed burials in Cemetery H and the queer painted vases that accompany them. At Chanhu-daro and Jhukar, in Sindh, a distinct barbarian culture, the Jhukar culture replaced the Harappa civilization. Everywhere the literate tradition exemplified in the inscribed 'seal' was extinguished. But judging by the pottery from Cemetery H and Jhukar sites and by metalwork from the latter, some technical traditions were carried over. Presumably potters and smiths survived to work for new customers. Naturally, they produced quite novel objects to suit foreign tastes.

The pottery from Jhukar and Cemetery H is still wheel-made, painted and fired in the old techniques, but shaped quite differently and adorned with new designs. Jhukar smiths made shaft-hole axe-heads and probably axe-adzes and pins with swollen necks. Button or bead seals of stone, fayence or pottery replaced the rectangular glazed steatite 'seal' and where engraved with geometric designs, including the filled cross, or rarely with conventional beasts, instead of inscriptions and lifelike animals.

The button seal like the shaft-hole axe is plainly a north-western intruder in India. The closest parallels to the Jhukar seals, and an exact parallel to the axe-adze comes from Hissar III, in northern Iran. Putative intermediate links will be cited latter from Makran and Baluchistan. These agreements suggest that the barbarians who destroyed the Harappa civilization included at least invaders from north-western Iran. Wheeler has boldly suggested their identification with the Vedic Aryas. In any case, the rsis sang their Vedic hymns in a prehistoric night ; for the invasion completely broke the literary tradition, and there is no fixed point in Indian history till the reign of Darius.

12. *Ibid*. 225
13. Childe WHH 128
14. Childe NLMAE 187-88

3. ARYANS AND THE END OF THE INDUS CIVILIZATION

The view expressed by Wheeler and endorsed by Childe, namely that the decaying or already largely decayed Indus Civilization came to its final end by the invasion of the Indo-Aryan speaking peoples—or more simply by the Vedic peoples—has provoked a great deal of controversy and even much indignation among a section of our scholars.

Wheeler first expressed this view rather forcefully in 1947 in *Ancient India* No. 3, though in 1968 (reprinted in 1979), confronted with some strong criticisms and perhaps also taking note of other possible factors pointed to by other archaeologists as contributing to the final destruction of the Indus valley civilization, he gave the impression of withdrawing or somehow modifying his original emphasis and observed that he had once blamed Indra rather lightheartedly. The tone appears to be somewhat subdued and we shall try to see whether that is at all necessary. But let us begin by quoting his view as originally expressed. In 1947 he observed :[15]

> The Aryan invasion of the Land of the Seven Rivers, the Punjab and its environs, constantly assumes the form of an onslaught upon the walled cities of the aborigines. For these cities the term used in the *Rgveda* is *pur,* meaning a "rampart," "fort" or "stronghold". One is called "broad" (*prthvi*) and "wide" (*urvi*). Sometimes strongholds are referred to metaphorically as "of metal" (*ayasi*). "Autumnal" (*saradi*) forts are also named : "this may refer to the forts in that season being occupied against Aryan attacks or against inundations caused by overflowing rivers." Forts "with a hundred walls" (*satabhuji*) are mentioned. The citadel may be made of stone (*asmamayi*) : alternatively, the use of mud-bricks is perhaps alluded to by the epithet *ama* ("raw," "unbaked"). Indra, the Aryan war-god, is *puramdara,* "fort-destroyer." He shatters "ninety forts" for his Aryan protege, Divodasa. The same forts are doubtless referred to where in other hymns he demolishes variously ninety-nine and a hundred "ancient castles" of the aboriginal leader Sambara. In brief, he "rends forts as age consumes a garment."
>
> Where are—or were—these citadels ? It has in the past been supposed that they were mythical, or were "merely places of refuge against attack, ramparts of hardened earth with palisades and a ditch." The recent excavation of Harappa may be thought to have changed the picture. Here we have a highly evolved civilization of essentially

15. Wheeler reprinted in Possehl ACI 291

non-Aryan type, now known to have employed massive fortifications, and known also to have dominated the river-system of north-western India at a time not distant from the likely period of the earlier Aryan invasions of that region. What destroyed this firmly-settled civilization ? Climatic, economic, political deterioration may have weakened it, but its ultimate extinction is more likely to have been completed by deliberate and large-scale destruction. It may be no mere chance that at a late period of Mohenjo-daro men, women and children appear to have been massacred there (Mackay 1938b : 94f., 116ff., 172). On circumstatial evidence, Indra stands accused.

The evidence cited by Mackay of "men, women and children appear to have been massacred" at a later period of Mohenjo-daro is perhaps not so important for the central argument for viewing the Harappan cities being finally ransacked by the invading Aryans. As P. V. Kane[16] and others have shown, the presence of the few scattered skeletons at Mohenjo-daro can be explained by other hypotheses. Nevertheless the question of the Aryan invasion remains and it is as strongly supported by some archaeologists as bitterly contested by some others. We shall first mention some of the latter.

In 1964, G. F. Dales wrote on the "The Mythical Massacre at Mohenjo Daro."[17] Free use of innuendo and other rhetoric apart, the writer frequently refers to the authority of Marshall and Hargreaves, evidently ignoring the fact that some of their observations have become dated. However, from the archaeological viewpoint, his main or at least one of his main arguments is : '. . . what is the material evidence to substantiate the supposed invasion and massacre ? Where are the burnt fortresses, the arrowheads, weapons, pieces of armor, the smashed chariots and bodies of the invaders and defenders ? Despite the extensive excavations at the largest Harappan sites there is not a single bit of evidence that can be brought forth as unconditional proof of an armed conquest and destruction on the supposed scale of the Aryan invasion.' Depending mainly on his own theory of flood etc., Dales concludes : 'The enemy of the Harap-

16. P.V. Kane, *Presidential Address* : *Indian History Congress,* 16th Session, December 1953
17. G.F. Dales reprinted in Possehl ACI 293-296
18. *Ibid.* 294

pans was Nature. . . —Indra and the barbarian hordes are exonerated'.[19]

As if it is not enough to establish the innocence of Indra, K. M. Srivastava, depending mainly on the theories of natural calamities to explain the end of the Indus civilization, proclaims : 'Indra, therefore stands completely exonerated'.[20] However, he wants to go a step further and seeks clue to the genesis of Wheeler's theory of Aryan invasion. As he puts it :[21]

> In retrospect, when we look at Wheeler's career as an archaeologist in England and India and we see him as a Brigadier in the British army during World War II, we feel he could not interpret the dubious evidence of Mohenjo-daro and Harappa in any other manner. He started his real archaeological career from Maiden Castle excavations and soon emerged as an authority on Roman archaeology. His *Rome Beyond the Imperial Frontiers* clearly shows how deeply involved he was in the Roman art and architecture, Roman concepts of town-planning—citadels, lower towns, assembly-halls, etc. Thus when he was confronted with the twin-mound towns of the Harappans and their huge fortification walls and mud-brick platforms as well as the photographs of the so-called 'massacre' at Mohenjo-daro he was at once reminded of Roman history and archaeology. Marshall's (1931) and Vats' (1940) reports on Mohenjo-daro and Harappa, respectively, as well as Piggott's *Prehistoric India* (1950 : 244-248) provided him sufficient speculations on Aryan warfare and Indra's attack on Harappan towns. The common text-books on Indian history adorned his personal library. For the former army man and the Director General of Archaeology in India, as Wheeler was, the Aryan invasion of Indus towns (1961 : 249) was as simple as the Roman invasion of Britain and Turkey.

A Brigadier in the British army almost obsessed with the Roman model is naturally not expected to know much of the Vedas. Wherefrom, then, did he get the Vedic materials to substantiate his theory of the Aryan invasion? Srivastava has a simple answer to this : "In fact it was V. S. Agarawal who provided these references to Dr. Wheeler when the latter requested Professor Agarawal, then an officer in the Archaeological Survey of India, on tour at Harappa, although Dr. Wheeler never acknoledged it in any of his writings."[22]

19. *Ibid*. 296
20. K.M. Srivastava in FIC 441
21. *Ibid*. 442
22. *Ibid*.

All this is evidently imputing a certain dishonesty to Wheeler. Our point, however, is that if Wheeler is at all to be criticised for not acknowledging the real source of his information about the Vedic materials, we are to look elsewhere. The theory of the Aryan invasion of the Indus Valley Civilization—and this based upon a considerable amount of Vedic data (inclusive of those used by Wheeler)—was already advanced by R. P. Chanda in 1926 and 1929. Since these were published as *Memoirs of the Archaeological Survey of India* Nos. 31 and 41, Wheeler's want of acquaintance with these is not easily conceivable.

The fact that within a few years of the discovery of the Indus Valley Civilization, the sound Vedic scholarship of R. P. Chanda led him to the view that the ruin of the cities was finally due to the attack of the Vedic people under the leadership of their war-god Indra is, to say the least, a remarkable academic performance on his part. But most of the writers on Indian archaeology do not mention this and the hypothesis is generelly associated with the names of Wheeler and Piggott. Therefore we propose to reproduce here the writings of R. P. Chanda at considerable length.

4. R. P. CHANDA AND THE THEORY OF ARYAN INVASION

Here are some of the observations of R. P. Chanda published in 1926 in No. 31 of the *Memoirs of the Archaeological Survey of India* (pp. 1-5) :[23]

The archaeological discoveries at Harappa in the Punjab and at Mohen-jo-Daro in Sind have pushed back the monumental history of India from the third century B.C. to at least the beginning of the third millennium B.C. by one single stroke. A series of literary monuments, the Vedic *Samhitas*, the *Brahmanas*, and the *Sutras*, have long been known, the youngest in age among which is probably older than the third century B.C. But a wide divergence of opinion relating to the age of these works and particularly of the *Rgveda* among scholars renders their use as sources of history unsafe....

To facilitate the co-ordination of the data of Archaeology with literary evidences I propose to discuss in this paper some of the passages in the Vedic literature that throw light on the early history of the Indus valley.....

23. R.P. Chanda in MASI No. 31, 1-5

Many of the stanzas of the *Rgveda* contain references to Pura and Pur both of which terms mean *nagara*, 'town', in classical Sanskrit. In one stanza (7.15.4) an extensive (*satabhuji*) Pur made of copper or iron (*ayas*) is referred to. In another stanza (1.58.8) prayer is offered to Agni to protect the worshipper with Purs of *ayas*. In such passages *ayas* is evidently used in a metaphorical sense to denote strength. Susna, a demon, is said to have a moveable (*carisnva*) Pura (8.1.28). In the *Rgveda* Pura is much oftener connected with the enemies of the Aryas than with the Arya Rsis and warriors. Two of the famous Rigvedic kings, Divodasa, the chief of the Bharatas, and Purukutsa, the chief of the Purus, are found engaged in war with hostile owners of Puras. Divodasa was the son of Vadhryasva and grandfather of the more famous Sudas who defeated a confederacy of ten tribes including the Yadus, Turvasas and Purus on the western bank of the Parusni (Ravi). It is said (4.30.20) that Indra overthrew a hundred Puras made of stone (*asmanmayi*) for his worshipper Divodasa. The Puras that Indra overthrew for Divodasa evidently belonged to Sambara who is called a Dasa (non-Arya or demon) of the mountain (6.26.5). In one stanza (9.61.2), among the enemies of Divodasa are mentioned the Yadu (the Chief of the Yadus) and Turvasa (the chief of the Turvasas) with Sambara. The greatest feat that Indra performed on behalf of Purukutsa, the chief of the Purus, is thus described in a stanza (6.20.10), 'May we, O Indra, gain new (wealth) through your favour ; the Purus autumnal (*saradi*) Puras with thunder weapon, slew Dasas and gave autumnal (saradi) Puras with thunder weapon, slew Dasas and gave wealth to Purukutsa'. The epithet *saradi*, usually translated as 'autumnal,' is explained by Sayana in different ways. In his commentary on the above stanza he explains the term *saradi* as 'belonging to a demon named Sarat.' But in other places (1.131.4 etc.) he explains it as 'annual Puras of the enemies strengthened for a year with ramparts, ditches, etc.' The authors of the *Vedic Index* are of opinion that *saradi* or autumnal Puras 'may refer to the forts in that season being occupied against Arya attacks or against inundations caused by overflowing rivers.' The same exploit performed by Indra on behalf of the chief of the Purus is also referred to in certain other stanzas.....

The terms Pur and Pura mean *nagara*, 'city', 'town', and not fort. The Sanskrit equivalent of 'fort' is *durga* which also occurs in the *Rgveda* (5.34.7 ; 7.25.2). In one stanza (1.41.3) not noticed by the authors of the *Vedic Index* Durga and Pura occur side by side. Sayana here takes Pura as an epithet of Durga meaning 'neighbouring'. But if we can shake off our bias relating to the absence of towns in the Rigvedic period we can recognise in this stanza references to both fort and town. The recovery of the ruins of cities at Harappa and Mohen-jo-Daro leaves no room for doubt that the Rigvedic Aryas were familiar with towns and cities of aliens. It is futile to seek any more historical elements in the legends of Divodasa and Purukutsa than perhaps the names of these heroes. But if we eliminate the mythical and fanciful additions there is no reason to doubt the possibility of the nucleus. There existed and the folk memory remembered that there once existed Arya worshippers of Indra who waged wars against civilized aboriginal neighbours living in towns and fighting from within strong-holds. Who, then, were these enemies of the Aryas ? Do the hymns of the *Rgveda* give us any more information about them ?

It appears to me that the aboriginal towns-folk with whom the Aryas came into collision in the Indus Valley are called Panis in the hymns of all the books of the *Rgveda*. Yaska (*Nirukta* 6.27) in his comment on *Rgveda* 8.66.10 says, 'The Panis are merchants,' and in his comment on R.V. 10.108.1 (*Nirukta* 11.25) he calls the Panis demons. The distinction between the human and the superhuman Pani is also recognised by Sayana, the author of the commentary on the *Rgveda,* and the context justifies the distinction. The word Pani is evidently derived from *pana,* 'Price.' The human Panis of the *Rgveda* are wealthy merchants who do not offer sacrifice and do not give gifts to priests. In R.V. 1.124.10 the poet addressing Dawn says, 'Let the Panis who do not perform sacrifice and do not give gifts sleep unwakened (for ever)'. Another poet sings, 'Ye mighty ones (Asvins) what do you do there ; why do you stay there among people who are held in high esteem though not offering sacrifices ; ignore them, destroy the life of the Panis' (R.V. 1.83.3). A poet prays to Indra (1.33.3), 'Do not behave like Pani' (*ma Panihbhuh.*), which according to the scholiast means, 'Do not demand the price of kine.' Another poet, expecting a suitable reward for his offering of Soma drink, addresses the same deity as Pani (8.45.14). The Soma-drinker Indra does not like to make friends with the rich Pani who does not offer Soma sacrifice (4.28.7). A poet prays (3.58.2), 'Destroy in us the mentality of the Pani' (*jaretham asmat viPaneh manisam*). Sometime the Rsi (Poet) betrays a conciliatory mood. In one hymn (6.53) the god Pusan is repeatedly requested 'to soften the heart of the Pani' and make the Panis obedient. This hymn occurs in a book (6) of the *Rgveda* composed by Rsis of the family of Bharadvaja. In one hymn of this book (6.45.31-33) the poet, a Bharadvaja, praises Brbu, a Pani chief, for giving thousands and a thousand liberal gifts. Indian tradition long remembered this acceptance of gifts by Bharadvaja from the Pani Brbu as an exceptional case, an example of the special rule that a Brahmin who has fallen into distress may accept gifts from despicable men without being tainted by sin. We are told in the code of Manu (10.107), 'Bharadvaja, a performer of great austerities, accepted many cows from the carpenter Brbu, when he was starving together with his sons in a lonely forest.' (Buhler). Sayana in his commentary on R.V. 6.45.31 describes Brbu as the carpenter of the Panis.

It is evident from the hymns of the *Rgveda* that the Aryas were divided into two main classes, the priests and the warriors. Cattle breeding appears to be the main source of their livelihood, cows being the chief wealth. Agriculture was practised to a limited extent. A hymn (9.112) refers to the different professions followed and the crafts practised by the Aryas. Trade finds no place in the list. So the conclusion that the much maligned Panis were the representatives of an earlier commercial civilisation seems irresistible. Among the antiquities unearthed at Mohen-jo-Daro are coins with pictographic legends that indicate the very early development of commercial life in the Indus valley. The Panis probably represented this pre-historic civilisation of the Indus Valley in its last phase when it came into contact with the invading Arya civilisation. During the second millennium B.C. there occurred in the Indus Valley events analogous to those that occurred in the Aegean world at about the same time, that is to say, successive waves of invaders of Aryan speech poured from the north-west. These invaders who in the *Rgveda* call themselves Arya met in the southern part of the valley a civilised people who lived in cities and castles and mainly depended on commerce for their livelihood. The Arya conquerors who were inferior in material culture either destroyed the cities or allowed them to fall into ruin. Their great god Indra is called Puroha or Purandara, 'sacker of cities'. Like the pre-historic civilisation of the Aegean, the pre-historic civilisation of the Indus Valley also failed to survive the shock of the Aryan invasion.

5. ARCHAEOLOGY AN AID TO VEDIC STUDIES

Let us briefly put the point we have been trying to argue. The discovery of the Indus Valley Civilization, besides dramatically extending our knowledge of ancient Indian history, has also another dimension. It has become a tool for us for the interpretation of Velic literature—specially of the *Rgveda*. The Vedic scholars may profitably turn to the archaeological findings to see if these can provide us with any clue to certain otherwise unexplained—or at best fancifully explained—passages of the *Rgveda*. Thus, for example, the *Rgveda* speaks of a considerable number of cities in the Land of the Seven Seas and of the ransacking of these by the Aryans under the leadership of the war-god Indra. The whole thing cannot be brushed aside as a mere figment of imagination of the Vedic poets for the simple reason that those who have never seen any city cannot write about these : the Vedic peoples themselves could by no stretch of imagination be city-dwellers, it being overwhelmingly obvious from the internal evidences of the *Rgveda* that they were pastoral nomads after all. Therefore, before the discovery of Mohenjo-daro and Harappa—soon followed by the discovery of many other cities within the Harappan cultural zone—there could at best be some speculations about these *pura*-s or cities and of Indra's role as *purandara* or the sacker of cities in the *Rgveda*—speculations, some specimens of which are to be found in the *Vedic Index*. With the discovery of ruined cities in the Harappan cultural zone, Vedic scholars are relieved of the obligation of indulging in such speculations, notwithstanding the circumstance that many questions concerning the archaeological findings still remain controversial or perhaps not yet fully explained. Thus, for example, the real story of the scattered skeletons found in Mohenjo-daro may remain a problem for the archaeologists, so also that of the corpses of the Aryan soldiers and their arms still eluding the archaeologists' spade. We may hope that more digging and better inferences of the archaeologists would solve such problems. What in the meanwhile is gained by archaeology in the matter of throwing light on the *Rgveda* must not be ignored, for the Rgvedic references to the cities and of the ransacking of these seem to have no better explanation than is provided by the material remains in the Harappan cultural zone. To correlate archaeology with literary evidences is often considered to be a tricky and delicate matter. The importance of archaeology as tool for interpreting or understanding certain Vedic passages remains yet to be more adequately examined. Yet as a tool it sometimes proves to be of surprising importance. We shall mention here another example.

6. D. D. KOSAMBI AND THE VRTRA MYTH

Let us not forget, however, that the Vedic poets were poets after all. And as poets, they belonged to the ancient world. So they are not expected to understand and objectively describe the phenomena they record in their own way. We are thus bequeathed by them with the tricky task of disentagling reality from myth in their poetry. We shall mention here one example of how D. D. Kosambi attempts it.

Like Purandara or "ransacker of cities", Indra is often called in the *Rgveda* Vrtrahan or "slayer of Vrtra". For the Vedic poets, this was apparently one of his great achievements. As Macdonell observes, 'the chief and specific epithet of Indra is *Vrtrahan,* "Vrtra-slayer". It is applied about 70 times to him in the *Rgveda*'.[24] Vrtra is usually taken as the name of a dragon, destroying whom often with the aid of other gods, specially of the horde of semi-deties called the Maruts, was considered one of his major performances. But there are certain peculiarities about this performance which cannot be easily overlooked. Vrtra literally means the Obstructor, and is also described as *ahi*, literally 'serpent'. In other words, it is an obstructor, which also looked like a serpent. Many things are said by the Vedic poets about this serpent-like obstructor, among which some are extremely interesting. Vrtra, it is said, was complacent with the idea that its real vulnerable part was known to none ; however, along with the Maruts, Indra discovered its vulnerable part (iii.32.4 ; v. 32.5) and thus he struck and destroyed the 'obstructor' with such fierceness as to shake the heaven and world. And what was the result ? The whole area was flooded with water which was being obstructed by the serpent-looking obstructor. Here is how Macdonell gives some of the passages referring to the great exploits of Indra :[25]

> He smote Vrtra who encompassed the waters (vi. 20.2 ; etc.) or the dragon that lay around (*parisayanam*) the waters (iv. 19.2) ; he overcame the dragon lying on the waters (v. 30.6). He slew the dragon hidden in the waters and obstructing the waters and the sky (ii. 11.5); and smote Vrtra who enclosed the waters, like a tree with the bolt (ii. 14.2). Thus 'conquering the waters' (*apsujit*) is his exclusive attribute... For many dawns and autumns Indra has let loose the streams after slaying Vrtra (iv. 19.8)... He cleaves the mountain, making the streams flow or taking the cows (i. 57.6 ; x. 89.7), even with the sound of his bolt (vi. 27.1). When he laid open the great mountain, he let loose the torrents and slew the Danava, he set free the pent up springs, the udder of the mountain (v. 32.1-2). He slew the Danava, shattered the great mountain, broke open the well, set free the pent up waters (i. 57.6 ; v. 33.1). He releases the streams which are like imprisoned cows (i. 61.10), or which, like lowing cows, flow to the ocean (i. 32.2). He won the cows and Soma and made the seven rivers flow (i. 32.12 ; ii. 12.12). He releases the imprisoned waters (i. 57.6 ; i. 103.2), released the streams pent up by the dragon (ii. 11.2), dug out channels for the streams with his bolt (ii. 15.3), let the flood of waters flow in the sea (ii. 19.3), caused the waters pent up by Vrtra to flow (iii. 26.6 ; iv. 17.1).

And so on. The entire section of Macdonell dealing with Indra's exploit of destroying Vrtra—snake-looking obstructor—

24. Macdonell VM 60
25. *Ibid.* 59

needs to be read in full to see that according to the Vedic poets the main result of this was the release of pent-up water. Incidentally, Macdonell points to the use in the *Rgveda* of the word *arnas* or flood in this connection.[26]

Shorn of poetic imagination and the inevitable proclivity to myth-making in ancient poetry, what does all this really mean? D. D. Kosambi answers:[27]

Vedic Indra is described again and again as freeing the streams. This was taken as a nature-myth in the days of Max Muller, a poetic representation of the rain-god letting pent-up waters loose from imprisoning clouds. Recorded but ignored details of the feat make such an explanation quite impossible. Indra freed the rivers from the grip of a demon Vrtra. The word has been analysed by two most competent philologists [with full knowledge of Iranian (Aryan) as well as Sanskrit records] who did not trouble to theorise about the means of production. Their conclusion from purely philological considerations was that *vrtra* meant "obstacle", "barrage", or "bloquage", not a demon. The actual Rgvedic description independently bears this out in full. The demon lay like a dark snake across the slopes. The rivers were brought to a standstill (*tastabhanah*); when the "demon" was struck by Indra's shattering weapon (*vajara*), the ground buckled, the stones rolled away like chariot wheels, the pent-up waters flowed over the demon's recumbent body (cf. RV. 4.19.4-8; 2.15.3). This is a good description of dams (not embankments as Piggott would have it) being broken up, while such prehistoric dams, now called Gebr-band, are still to be found on many water-courses in the western parts of the region under consideration. The evidence for Indra's breaking up dams is not merely rationalization of the Vrtra myth. RV. 2.15.8: *rinag rodhamsi krtrimani*="he removed artificial barriers" makes this clear; *rodhas* means "dam" elsewhere in the RV, as in later Sanskrit. Indra is praised for restoring to its natural course the river Vibali, which had flooded land along its banks. That is, the pre-Aryan method of agriculture depended upon natural floods and flooding the lands on the banks of smaller rivers by means of seasonal (RV. 5.32.2) dams (without regular masonry), to obtain the fertilising deposit of silt to be stirred by the harrow. The Aryans shattered this dam system, thereby ruining the agriculture of the region and the possibility of continuing city life for long, or of maintaining the urban population. The fact of the ruin is undeniable; the causes have to be deduced from whatever data is available, which includes numerous heavy flood silt deposits that are visible in Mohenjo-daro excavations. The very floods

26. *Ibid.* 59
27. D.D. Kosambi ISIH 70-71

which endangered city and hamlet had made possible the argiculture which supported the inhabitants.

Kosambi's theory of the agricultural technique in the Harappan region depending on toothed harrow rather than plough has to be discarded, of course, specially after the excavation at Kalibangan revealed an actual ploughed field. However, at least among a section of archaeologists the "flood theory" is receiving increasing importance. If there is substance in this theory, it helps us to understand the Rgvedic passages related to Indra's exploit against Vrtra—the snake-like obstructor of water—much better than the earlier views wanting us to read this or that myth in these passages. Archaeology thus becomes a tool for the interpretation of Rgvedic passages which remain otherwise more or less mysterious for us.

7. SUMMING UP

One reason for the tenacious objection against the theory of the Aryan invasion as the final cause of the already decaying Indus civilization resulting from the cumulative effect of various possible causes seems to be frankly chauvinistic. Later Indian mythology wants to feed us with the belief that Indra was the king of the gods. Inputing to him such a negative activity as the destruction of the glorious civilization of ancient India can understandably hurt the feeling of some rooted primarily on religious convictions. An example of this seems to be the article on "The Myth of Aryan Invasion of Harappan Towns" by K. M. Srivastava already referred to. After attempting to refute elaborately the theory of Aryan invasion, he apparently feels the need of coming out sharply against Wheeler's statement that 'on circumstantial evidence, Indra stands condemned'. As we have already seen Srivastava passionately proclaims : Indra stands completely exonerated.

As against such passionate defence of the Vedic god, at least one point needs to be remembered. The *Rgveda* is a collection of songs and hymns of an ancient period after all, and hence it will be an anachronism to judge the standard of morality reflected in it by our contemporary standards. Sacking or destroying the Harappan cities may appear more or less deplorable if judged by the moral standards of our time. But it was not so to the Vedic poets, who saw in such actions the most magnificent feats of courage and strength, and, therefore, which in their standard must have been a highly laudable performance. That is why, they sang of the glory of Indra as *purandara* or destroyer of the cities. They would not have surely done it had they been under the influence of *our standard* of morality.

So the rather emotional statement completely exonerating Indra makes no real sense. To the ancient poets of the *Rgveda,* the personal bravery and strength of Indra was highly honourable, though these were often exhibited under the influence of alcohol or the intoxicating drink they called *soma.*

The point is that the standard of morality changes, not only from age to age but mainly because of the social conditions in which it is expressed. In the society in which the Vedic poets lived, certain acts considered most despicable by our standard of morality are actually praised by the poets in ways that appear to us to be most shocking. Here is an extreme example of this. In one hymn of the *Rgveda,* a certain poet praises Indra by way of addressing him as follows :

> Who has made thy mother a widow ? Who has sought to slay the sleeping and the waking ? What deity has been more gracious than thou, since thou hast slain the father having seized him by the foot ? (RV. iv. 18.12)

We have—from the standpoint of our own sense of morality —perhaps no adequate vocabulary to condemn such an act of abject patricide, particularly when it is associated with the story that Indra did it because his father had stolen some amount of the intoxicating stuff from Indra's stock (*Taittiriya Samhita* vi. 1.3.6), to which Sayana refers.

Can we, with our sense of morality, exonerate Indra from such an act ? We cannot. But such a question would have perhaps made no sense to the Vedic poet himself.

Wheeler's use of the words "Indra stands condemned" was perhaps intended to be rhetorical. But Srivastava's retort that "Indra stands completely exonerated" is just religious chauvinism.

CHAPTER 12

BETWEEN THE TWO URBANIZATIONS

1. PRELIMINARY REMARKS

In 1963, Leeonard Woolley describes the period intervening the two urbanizations in India as follows :[1]

> The Aryan conquerors were a simple, not to say a barbarous people, pastoral nomads for the most part, some perhaps petty farmers, who had not even a word for 'brick' in their vocabulary ; they could destroy, but they could not rebuild. Just as the Saxon pirates who settled in Britain shunned the Roman sites whose walls of massive stone seemed to their ignorance to be the work of devils, and when in due course began to build did so under the influence not of Rome but of the contemporary architecture of the Continent, so it was with the Aryans. Harappa and Mohenjo-daro were left to moulder into shapeless mounds. True, in the topmost levels are found shoddy huts built with re-used bricks above the ruins of the citadel, but the pots-herds prove that such were inhabited not by the invaders but by a remnant of the old servile population who now squatted on the sites where their masters had lived. For more than a millennium after their arrival the history of the Aryans is shrouded in utter darkness ; when at last, late in the fourth century BC., the veil lifts a little to show us northern India under the Mauryan Dynasty possessed of a great urban civilization, those cities have nothing in common with the old Indus capitals. The burnt-brick-and-bitumen construction which was the most striking feature of Mohenjo-daro architecture is a long-forgotten art ; in the earliest post-Harappa buildings of which we have any knowledge, those at Hastinapura, 80 miles northeast of Delhi, wood is the basic material, and the Mauryan palace at Patali-putra is also of timber. It would seem that when the Aryans began to build for themselves they made a fresh start ; the monuments which they set up were certainly not inspired by any memories of Mohenjo-daro. At Hastinapura and at Patalipura the remains are not sufficient to show the style of the buildings ; but the earliest buildings in stone are obviously based upon timber originals, and when we look at such a structure as the north gateway of the Great Stupa at Sanchi it is difficult to avoid the feeling that the inspiration comes from the wooden architecture of China.

We have quoted this because we feel that the statement "for more than a millennium after their arrival the history of the

1. L. Woolley in HMCSD VOL. I, Pt. ii, 458

Aryans is shrouded in utter darkness" seems to be in need of some modification, if not from the strictly archaeological viewpoint at least from the standpoint of the history of science and technology in India. It is true that the Vedic people—before being merged into Indian people among which the Aryan and non-Aryan elements are not very clearly distinguishable—left for us nothing spectacular from the strict point of view of material culture. But they created something most amazing in world history—a vast literature of over a thousand songs and hymns which is compiled as the *Rgveda-samhita.* There is no doubt that this has primary interest for the historian of literature. Nevertheless, as we are going to see, even as literature, this contains certain concepts with exceedingly interesting science-potential and even the mythological imagination embodied in it is not without interest specially when compared to the attitude to natural science that eventually developed among the ostensible followers of pure Aryan culture. Secondly, from the technique of retaining this vast literature in an almost incredibly meticulous form and also from the technique of understanding or interpreting its songs and hymns, there did develop certain formal sciences like those of metrics and linguistics, which, if totally ignored, cannot make the history of science in India sufficiently scientific. Considering these, we feel that the thousand years in Indian history after the end of the Harappan culture was not so much shrouded in total darkness as Woolley wants us to believe. Besides, the results of brisk work done by Indian archaeologists on the period intervening the two urbanizations appear to have no place in Woolley's observation, but these are not without interest whatsoever for understanding the achievements of the Vedic people. We begin with a brief account of this archaeological work.

2. RECENT ARCHAEOLOGY AND THE "DARK AGE"

"With the breakdown of the highly developed socio-economic system of the Indus civilization," observe the Allchins, "a major change took place, in that city life seems to have disappeared for several centuries before emerging afresh in the Ganges valley. At the same time a pattern of more or less uniform peasant agricultural settlements appears both inside the Indus region and beyond it. During the second millennium there is plentiful evidence of already developed regional cultures, frequently referred to as 'Chalcolithic' or 'Neolithic-Chalcolithic' in almost every part of the subcontinent".[2]

Any full review of all this falls outside the scope of our present study. What needs to be noted, however, is that the earlier idea of a total disappearance of the Harappan culture after the destruction or devastation of its major cities—particularly Mohenjo-daro and Harappa—is being increasingly modified by the more recent archaeological work. The need is thus felt to introduce new concepts as Late-Harappan and Post-Harappan culture, in which much of the Harappan traits are said to survive, though as co-mingled with locally developed new traits. Also are found certain sites described as those of overlapping and interlocking of Harappan or Late-Harappan culture with the newly developing local cultures.

The main indices to such sites are certain pottery types, the Black-and-Red Ware, Ochre Coloured Wares, Grey Ware and Painted Grey and Northern Black Polished Ware—the later phase of the last taking us to the fringes of the Second Urbanization. Of these we shall first have a few words on the Painted Grey Ware, because of some special interest attached to these.

3. PAINTED GREY WARE AND THE VEDIC PEOPLE

The sites associated with the Painted Grey Ware (PGW) have received much attention of our archaeologists, because, as A. Ghosh says, 'the geographical horizon of the Later Aryans is conterminous with that' and 'there is also a remarkable chronological proximity between the dates of the beginning of the Ware and the later Vedic age'.[3] Readers interested in the details concerning this, may go in for the monograph *The Painted Grey Ware : An Iron Age Culture of Northern India* by Vibha Tripathi (Delhi 1976). For our present purpose, however, it may be sufficient to quote some summary observations on the PGW sites by R. S. Sharma.

Here is Sharma's brief note on the geography of the PGW sites :[4]

2. B & R Allchin RCIP 229
3. A. Ghosh CEHI 6
4. R.S. Sharma MCSFAI 56

The later Vedic texts comprising the collections of the Yajus and Atharvan, the Brahmanas, and the Upanishads were composed in the land of the Kurus and Pancalas. This forms the major portion of western Uttar Pradesh, almost the whole of Haryana, and the neighbouring parts of the Panjab and Rajasthan. In geographical terms this area covers the Indo-Gangetic divide and the upper Gangetic plains. The divide includes the land between the Indus system and the Gangetic system, and covers a large portion of modern Panjab and Rajasthan and the whole of Haryana and the Delhi area. The Indo-Gangetic divide, if the northernmost portion of the Bari doab is included in it, is about 35,000 sq. miles.

At the same time, Sharma wants us to reemember that it would be wrong to connect the sites with the Vedic peoples alone :[5]

Painted Grey Ware sherds have been found in the same areas as are represented by the later Vedic texts. Although the PGW wares have been noticed in eastern UP and even in Bihar, their epicentre seems to be the upper Ganga and Sutlej basins. Nearly 700 PGW sites have been located in this region. They are in much larger numbers when compared with nearly 50 or so ochre coloured pottery sites, and there is no doubt that they indicate agrarian settlements on a large scale for the first time in this area. However there is nothing like an exclusive PGW culture because other wares such as black-and-red ware, black-slipped ware, red ware, and plain grey ware are also associated with them. Although very distinctive, the PGW sherds are not numerically predominant at any place. At Atranjikhera, where the PGW covers an area of about 650 sq.m., its incidence ranges between three and ten percent of the total pottery complex. Even where their number is fairly large, the PGW sherds may not exceed fifteen percent of the total pottery recovered from the PGW layers. Thus the PGW horizon represents a composite culture, just as the culture revealed by the later Vedic texts represents an amalgam of Sanskritic and non-Sanskritic, Aryan and non-Aryan elements.

Roughly speaking, the period covered by these sites is five hundred years—from 1000 B.C. to 500 B.C.

The PGW sites, generally speaking, are not indicative of advanced material culture except towards its end roughly in 500 B.C., which is also the beginning of the period of NBP Ware. Judged by the literary sources like the Upanisads and the Pali Tripitaka, we come across during this period the first foreshadowing of some characteristics of the urban or better proto-

5. *Ibid* 57

urban life, for the centres of these were still more of the nature of large villages than cities in the full sense.

But there is one point of immense technological significance about the PGW sites. Some of these are indicative of the earliest use of iron in northern India. How and wherefrom the earliest use of iron came into being is still a matter of much controversy. As Vibha Tripathi sums up some of the views :[6]

> Wheeler (1959) credited the Achaemenians with the introduction of iron in India in c. 600 B.C. Gordon (1958) thought that iron objects were not used in India prior to 500 B.C. M.N. Banerjee (1929) and Niyogi (1914), however, believed that iron was first brought into India by the Rigvedic Aryans. Forbes (1950) is also of the opinion that iron was known to the earliest Aryan settlers in India. Lallanji Gopal (1961), synthesizing archaeological and literary data, concluded that iron was introduced in the later Vedic period. He emphasized that the Rigvedic term *ayas* could not mean iron. Singh (1965) attached great importance to the pre-PGW level pieces found at Kausambi and on that basis dated iron to c. 1,000 B.C. But the late and degenerate nature of PGW at Kausambi (even the existence of a PGW deposit at the site has been questioned by Sinha) does not warrant the high antiquity claimed for iron at Kausambi. N.R. Banerjee (1964) believes that iron technology was introduced in India in c. 1000 B.C. by the Aryans. Tandon (1967-68) suggests that a chemical, technical and typological study of iron implements is required for understanding the multi-disciplinary problems posed by the Iron Age.
>
> These controversies do not allow any agreed dating and authorship of early iron. Meanwhile, further field work has produced evidence of a new kind.

Let us not, however, be involved here in such controversies. What interest us most in our study of the history of technology and science are the results reached so far by the field-work of our recent scholars. Here is how R. S. Sharma sums these up : 'In any case', he observes, 'the PGW phase marked by the use of iron in the Upper Gangetic plains and Indo-Gangetic divide can be reasonably dated to c. 1000-500 B.C.'[7] At the same time it needs to be noted that the use of iron in the earlier of these sites was extremely restricted. These 'mainly comprise spearheads, arrowheads, hooks, etc.'[8] This means that it would

6. Vibha Tripathi in RCIA 272
7. R.S. Sharma MCSFAI 58
8. *Ibid* 59

be fallacious to equate the PGW period with that of the introduction of iron technology in full sense. Apart from the fact that the limited use of iron artifacts are also found in certain other sites[9], one point needs to be specially noted: 'Broadly speaking iron technology made little impact on the economy of the PGW period. Mud or mud-brick houses with reed thatching, limited number of antiquities indicating restricted trade and commerce and absence of heavy argicultural implements during the PGW phase represent a typical picture of a village economy. 'The economy was simple. No doubt, in a slow-moving society the impact of iron was slow. The metal did not produce any spurt in the material prosperity of the society'.[10] To this may be added the further observations of R. S. Sharma :[11]

> Therefore till the sixth century B.C. northern India did not enter into a full-fledged iron age. Only in the second phase of iron associated with the NBP levels (500-200 B.C.) do we encounter more agricultural implements. This picture is consistent with the history of iron technology in Western Asia and elsewhere. In the first stage it was used for purposes of war, and in the second for handicrafts and agriculture. In the first phase in India, the use of iron could not be extended to production because of its paucity and primitive technology, but it may have helped the organizers of production in making their authority felt over the producers. However in this phase iron may have been used for clearance, for making wheels and the body of carts and chariots, and in the construction of houses because nails have been recovered from several PGW sites.

At the same time the iron-finds in the PGW sites associated with the Vedic peoples—in the kingdom of Kurus, of Pancalas, of the Matsyas, of the Madras,—and specially the fact that the 'largest deposit of iron weapons discovered so far belongs to Atranjikhera in the Pancala area'[12] are in no way to be overlooked. These evidences want us to think that the Vedic peoples knew iron, though without knowing more than its limited use for war purposes. One obvious reason for this must have been the relative scarcity of the metal: far-reaching conse-

9. Vibha Tripathi PGWIACNI 100ff.
10. *Ibid* 102. The quotation is from A. Ghosh CEHI, 10
11. R.S. Sharma MCSFAI 72
12. *Ibid*

quences of the use of iron technology were possible only when iron was found in abundance and hence became a "cheap" metal in the rich iron ore deposits in the eastern zone, which also became the main centres of second urbanization in Northern India. But the other important reason for the restricted use of iron by the early Vedic peoples before they got mixed up with the local peoples or "merged into Indian people" could have been their socio-economic organization : being nomadic pastoral people after all, they could hardly realise the importance of iron beyond the limited purpose of weaponary. However, before passing on to discuss the more spectacular results of the introduction of iron technology in an extensive scale in the comparatively later period, we may be permitted to digress a little and see if we can find anything deserving special notice from the viewpoint of the history of science as distinctive contributions of the early Vedic peoples.

4. THE RGVEDA-SAMHITA AND SCIENCE IN INDIA

If we have so far discussed mainly the destructive activity of the early Vedic people, we may as well note here some of their creative activities and see if there is something in these that interests us specially from the viewpoint of science in India.

By far the most outstanding creative activity of the early Vedic peoples is the *Rgveda-samhita* which comes down to us as a vast collection of 1028 songs and hymns in a total number of 10,552 verses or *rk*-s. Strangely, however, the makers of this vast literature were pre-literate people—or people without any knowledge of script—though the literature survive for us in its meticulous form by being transmitted through generations of sheer retentive memory ; it is thus called *sruti* or 'that which is heard'. Prompted apparently by chauvinistic enthusiasm, S. R. Rao has recently argued practically against the entire world of serious Vedic scholarship and observes : 'But the mere absence of written records during the Vedic period does not imply that writing was unknown to the Vedic Aryans. The Rigvedic Aryans might have used a perishable material for writing, as the Late Harappans did.'[13] No evidence of the Late

13. S.R. Rao LIC 160

Harappans using perishable materials for writing purposes is given by Rao. But the more important point is that the postulate of using perishable materials for writing is relevant only in case of the peoples who are otherwise definitely known to have the knowledge of the script. However, the ancient Vedic poets themselves would perhaps testify against such a possibility, inasmuch as they speak of "making songs by the mouth", or, in simpler language, what we call oral composition. Thus the poet Kanva, son of Ghora, addressed the Maruts :

'Make hymns by your mouth (*asye*) ; spread these like the cloud ; sing the *ukthya* (laudatory hymn) in Gayatri metre." (RV 1.38.14).

Incidentally, Sayana interprests *asye* as *svakiya-mukhe* or in one's own mouth, and thus leaves no scope for using any writing material—perishable or otherwise.

To this may be easily added certain other internal evidences of the Vedic tradition. In the *Nirukta,* Yaska uses the word *upadesena* (i. 20), which is understood straightway by Lakshman Sarup[14] as "oral tradition". Why does he do it ? The word *upadesa* is derived from the root *dis* meaning "to utter". *Upadesena* thus literally means "well uttered", easily brushing aside the possibility of recording in writing on any perishable material. Besides, the word usually rendered as "revelation" in the Vedic context is *amnaya*, as it is done for example in the *Vaisesika-sutra* (1.1.3). The word *amnaya,* derived as it is from the root *mna,* indicates oral communication : *mna kathane*. Here, again, the distinct implication is "to utter" or "to speak".

Thus Rao's idea of the Vedas having been once written on perishable material, goes against the Vedic tradition itself. But why does an archaeologist of his stature go in for such an extraordinary hypothesis ? Evidently, because he wants to revive and revitalise Ramachandran's thesis that the Vedic Aryans were the actual makers of the Indus civilization, one of the most formidable evidences against which is that these Aryans were pre-literate peoples. It may not, therefore, be irrelevant here to show how strong is the verdict of orthodox Vedic tradition itself against Ramachandan's view. One of the arguments of Ramachandran, as Rao says, is 'that the Indus seals depict

14. Lakshman Sarup NN Intro. 71

Vedic cult scenes'.[15] Ramachandran says, 'the Rgvedic idea that the bull does all the roaring to proclaim *dharma-vijaya* is caught up by the Mohenjo-daro seals and sealings representing the bull with its characteristic dewlap'.[16] As evidence of the bull "proclaiming *dharma-vijaya*" in the *Rgveda,* Ramachandran quotes the following from the *Rgveda* : "*tridha baddho vrsabho roraviti mahodevo martyanavivesa*", which Ramachandran translates : "The thrice-bent bull goes on roaring that the Great God (is no longer in some invisible heavens but) has actually completely entered the mortals."[17] The Rgvedic passage is all right (iv. 58.3), so also is the occurrence of the word *vrsabha* in it, as it also occurs in *Rgveda* iii. 55.17 ; vi. 73.1 ; vii. 101.1 etc. Only the "bull" is not there in these passages. Beginning from the *Mahabhasya* of Patanjali[18] *c.* 150 B.C. and *Nirukta* Yaska[19] *c.* 500 B.C. to Sayana[20] *c.* A.D. 1400, the word *vrsabha* in this context is derived from the root *vrs,* literally "to shower" and in the *Rgveda* it means "showering bounties."

So the bulls in the Harappan seals can be taken as evidences of the Aryan origin of the Harappan civilization only by flouting the authority of Patanjali, Yaska and Sayana who refuse to interpret the word *vrsabha* as the bull. Of course, any suggestion of *dharma-vijaya* is not there in the *Rgveda ;* it exists only in the imagination of Ramachandran and Rao. Besides as A. Ghosh points out that it is statistically impermissible to deduce the characteristic peculiarity of the Harappan civilization mainly from the evidence of the bulls in the Harappan seals. As he puts it, "The Unicorn which has no place in later Indian religious belief, heads the list [in the Harappan seals] with a frequency of 1,159 while the bull, in all its forms, is a poor second, with only 156."[21]

15. S.R. Rao LIC 137
16. T.N. Ramchandran *Presidential Address, Ancient India, Section* 1. quoted by Rao LIC 137
17. See S.R. Rao LIC 138
18. *Mahabhasya* Paspasa-ahnika
19. *Nirukta* Parisista xiii. 7
20. Sayana on iv.58.3 renders *vrsabhah* as *phalanam varsita,* i.e. showering results (bounties). Interpreting x.8.2 also he takes the word to mean *kamanam varsita i.e.* showering what is desired.
21. A. Ghosh in Possehl HC 321.

However the strongest evidence against the possible Aryan origin of the Harappan civilization is the socio-economic conditions of the early Aryans, who were nomadic pastoral peoples after all, still on the tribal stage of social development, and hence did not possess the tools or equipments to build up the highly developed urban civilization like the Harappan one.

Yet, as we are going to see, they had something significant to contribute to the history of science in India.

5. PREDOMINANTLY PASTORAL ECONOMY

The *Nighantu*—the earliest of the glossaries of Rgvedic words, which Winternitz is inclined to place sometime before 700 B.C.[22] mentions the word *go* or cow as one of the synonyms of *prthivi* or the earth (*Nigh*. i. 1). This justified Sayana to interpet the word *go*, as occurring in one of the comparatively later *rk*-s of the *Rgveda* (x. 31.10) as the earth. According to the *Nighantu*, again, the word *aditi* is the synonym of *go* as well as *prthivi* (ii. 11 and i. 1). Similarly, the word *ila*, besides being a synonym for *anna* or food, also stood for *go* and speech or *vac* (i. 1 ; i. 11 ; ii. 7 and ii. 11).

Such ambiguities tell their own story : the cow meant practically everything that the Vedic 'seers' or poets cared for—speech, food and in fact the whole world. The main theme of their hymns or songs was, of course, the praise of their gods. However, one of the seers went to the extent of singing the glory of the god because of being born of the cow (*gojata*) (vi. 50. 11). Even the *Pancavimsa Brahmana*—much later than the *Rgveda*—wanted to measure the distance between the heaven and earth by imagining a thousand cows placed on top of each other (xvi. 8.1 and xxi. 19). To all these R. S. Sharma adds : "Cattle were considered to be synomymous with wealth (*rayi*), and a wealthy person was called *gomat*. Terms for battle such as *gavisti, gosu, gavyat, gavyu* and *gavesana* were derived from cattle, which was the measure of distance (*gavyuti*) and also of wealth and wergeld. *Gopa* or *gopati* was the epithet given to the king. In the life of the family the importance of cow is indicated by the use of the term *duhitr,* one who milks, for daughter. Gods were divided into four categories, heavenly

22. M. Winternitz HIL I. 69 note

(*divya*), earthly (*parthiva*), cow-born (*gojata*) and watery (*apya*)."[23]

Further evidences are perhaps unnecessary. The poets or seers whose compositions come down to us as complied in the *Rgveda-samhita* were predominantly pastoral peoples. As for agricultural practices of the Rgvedic peoples R. S. Sharma gives us very interesting philological evidences. His starting point is the following argument :[24]

Terms for possessions found in the *Rg Veda* and having parallels in other Indo-European languages may throw light on the nature of property in the earliest Vedic society. Their number however is limited, and hence negative evidence becomes more important. It may be argued that terms for certain economic activities existed in pieces of Vedic literature which are not available, and that such terms should not be looked for in 'religious' literature. But the vocabulary of the *Rg Veda* is rich, and the main concern of the prayers is the material prosperity of the Aryans. Therefore the absence of words for some economic phenomena in the *Rg Veda* may be considered significant, especially when they are not found in other old allied languages. On the other hand we have also to take account of such terms as are found in Sanskrit and all other Indo-European languages and assign them to the earliest period of Vedic society, although all of them may not appear in the *Rg Veda*.

With this argument, he reviews the words related to agriculture and agricultural operations in the Vedic literature :[25]

It is significant that no common terms or cognates for cereals and cultivated plants are to be found in Indo-European languages, which indicates that cultivation made progress only when the Aryans settled in different countries. If we examine the various Indo-European terms for agriculture and vegetation listed in Chapter 8 of the *Dictionary* of Buck, it would appear that most Sanskrit terms for plough, furrow, cultivation, dig, spade, sickle, cereals, etc., do not have their cognates in Indo-European languages, although a few are found in Avestan. Although the term *ar* in the sense of plough is found in Indo-European languages, yet it has no linguistic parallel in Sanskrit. Linguistically this term cannot be reduced to *hala* (plough). All this would suggest that the Indo-Aryans took to agriculture in India, and to express this activity they adopted some local words.

23. R.S. Sharma MCSFAI 24
24. *Ibid* 22
25. *Ibid* 26

To these, it is tempting to add another point already mentioned by R. S. Sharma :[26]

> In ancient Indo-European languages there is no term corresponding to *krsti* in the sense of cultivator....Curiously enough Sayana understands *krstih* in the sense of *prajah* or people, though hostile. Hence the common notion that *krsti* may indicate cultivation in the Indo-European context has to be discarded. Similarly the use of the term *carsani* in the *Rg Veda* and the modern derivation *casa* (cultivator) in Bengali and *cas* (ploughing) in Maithili do not seem to indicate the original meaning of the Vedic term. It is suggested that the term is derived from *krs,* 'to plough' or 'to till'. But it is correctly thought that the term is derived from *car,* 'to move', and therefore the five *carsanayah* were five moving peoples who could be compared to the *pancajanah*.... The well-known term *hala* does not appear, but two other terms for plough, *langala* and *sira* occur in the earliest books ; the *varatra* or a leather strap of the plough is also mentioned. We hear of *phala* or ploughshare and furrows (*sita* and *sunu*) in Book IV, where a hymn is devoted to agricultural operation. However it has been argued by Hopkins that Book IV is the latest Family Book, and is as late as Book VIII.

6. RTA : THE PRIMORDIAL COMPLEX OF NATURAL LAW AND MORAL LAW

To sum up the discussion so far : The early Vedic peoples, specially judged by the older strata of the *Rgveda-somhita,* were predominantly pastoral tribes who eventually acquired some knowledge of agriculture perhaps from local peoples among whom they advanced and with whom they got increasingly mixed up, losing thereby their original racial identity. We are now going to see that for the key to their intellectual attitude it is desirable to begin with what is known in general about tribal peoples from unbiased ethnological studies. I have elsewhere[27] attempted to go into some detail of this question about the early Vedic people, highlighting the relics of the sense of tribal collectivity retained in the *Rgveda* : in inumerable passages of the *Rgveda* property is viewed as collectively owned by the kinsmen and shared out among them, significant decisions being taken by them in their tribal assemblies and individuality is recognised mainly in terms of special merit in war, crafts-

26. *Ibid* 25
27. D. Chattopadhyaya *Lokayata* Ch. 8

manship, healing technique, etc. These songs and hymns thus take us back to a general intellectual climate very different from that of ours.

In that intellectual climate we come across a concept with an exceedingly interesting science-potential. It seems to represent an instinctive awareness, as it were, of a primordial complex of natural order and social order—of natural law which is at the same time moral and social law—inviolable and obligatory even for the Vedic gods, who are at best sometimes conceived as its custodians but not its makers. Understandably, in the rather archaic songs of the *Rgveda* we do not expect the concept of the law of nature in a chiselled form. On the contrary, we come across it as largely mixed up with the mythological imagination of the Vedic poets. Nevertheless the possibility was not fulfilled. Rooted as it was in the instinctive morality of the collective life, its sense was lost to the later Vedic poets with their move forward to a social order based on class differentiation and early state power which, whatever might have been its exact nature, was dominated presumably by war-chiefs.

But let us first have some idea of the concept of *rta* in early Vedic consciousness.

The Angirasas whose spells and charms, along with those of the Atharvans, formed the original core of the *Atharvaveda,* were remembered also by the earliest stratum of the *Rgveda* as but hoary ancestors—'the fathers', 'our fathers', 'those ancient poets', etc. They were also said to have lived the ancient collective life and were hence free from jealousy in any form. All this, however, followed from their observance of the *rta* :

"Those ancient poets, the observers of the *rta*, were in joyful company of the gods ; these ancestors gained the secret lustre : with spells of truth they generated Usas. Being united with the common cattle, they became of one mind : they strive together, as it were, nor do they injure the rituals of the gods ; not injuring each other, they move with wealth." (RV vii. 76.4-5).

Such then was influence of *rta,* which, along with its derivatives and compounds, occur over 450 times in the *Rgveda.* It has naturally to be viewed as one of the key concepts of this vast collection of songs and hymns. It is also so archaic in content that it is not easy to suggest any fully satisfactory modern equivalent for it, and even the traditional commentator Sayana fails to explain it uniformly throughout the *Rgveda.*

Modern scholars have nevertheless attempted to translate it and they appear on the whole agreed to see in it some kind of a peculiar complex of natural as well as social law, or of the cosmic order which is at the same time the social order.

Winternitz used it in the sense of the "order of the universe."[28] Macdonell took it to mean the "physical and moral order",[29] though adding, "the notion of this general law, recognised under the name *rta* (properly the 'course' of things) we find in the *Rgveda* extended first to the fixed rules of the sacrifice (= rite), and then to those of morality (right)".[30] According to Keith, it was the term for the cosmic order as well as moral order : "The term cosmic order, *rta,* and its opposite *anrta,* express also moral order".[31]

As for the Vedic poets themselves, they wanted to raise this principle to the most exalted position. Thus :[32]

The dawns arise in the morning according to the *rta ;* the fathers have placed the sun in the heaven according to the *rta ;* the sun is the bright countenance of the *rta,* and the darkness of the eclipse is contrary to law, *vrata.* The year is the wheel of the *rta* with twelve spokes. The red raw milk, the product of the white uncooked cow, is the *rta* of the cow under the guidance of the *rta.* Agni, the fire, which hidden in the waters and the plants, is produced for man from out of the kindling sticks, becomes the shoot of the *rta,* born in the *rta.* The streams flow in obedience to the law of *rta.*

Radhakrishnan, apparently obsessed with the zeal to read his own idealistic outlook in the *Rgveda,* tries to give an idealistic twist to the concept of the *rta.* As he puts it,[33]

....*Rta* literally means 'the course of things'. It stands for law in general and the immanence of justice. This conception must have originally been suggested by the regularity of the movements of sun, moon and stars, the alternations of day and night, and of the seasons. *Rta* denotes the order of the world. Everything that is ordered in the universe has *rta* for its principles. It corresponds to the universals of Plato. The world of experience is a shadow or reflection of the *rta,* the permanent reality which remains unchanged

28. M. Winternitz HIL I. 154
29. Macdonell HSL 75
30. *Ibid* 67
31. A.B. Keith RPVU 248
32. *Ibid* 83
33. S. Radhakrishnan IP I. 78-9

in all the welter of mutation. The universal is prior to the particular, and so the Vedic seer thinks that *rta* exists before the manifestation of all phenomena. The shifting series of the world are the varying expressions of the constant *rta*, so *rta* is called the father of all.

The comparison of the Rgvedic *rta* to Platonic Ideas of which the worldly things are but shadows would have appeared frankly so incomprehensible to the Vedic poets themselves that it perhaps does not need any comment. Throughout the vast collection of primitive poetry, if we are at all justified in reading any central theme, it is the intense desire for this-worldly wealth, i.e. as far as the pastoral peoples are capable of thinking or imagining. On the contrary, the comparison of the Vedic *rta* is perhaps valid with certain other concepts of the ancient period. These are the concepts of *asa* in the Avesta and the *tao* of the Chinese Taoist. Keith observes :[34]

> In the physical world there rules a regular order, *rta*, which is observed repeatedly, and which is clearly an inheritance from the Indo-Iranian period, since the term *asa* (*urta*) is found in the Avesta, and has there the same triple sense as in Vedic India, the physical order of the universe, the due order of the sacrifice, and the moral law in the world.... The identity of the Vedic and Avestan expressions is proved beyond possibility of doubt by the expression 'spring of *rta*', which is verbally identical in *Avesta* and the *Rgveda*.

More interesting, perhaps, is the comparison of the Rgvedic *rta* to the *tao* of the Chinese Taoists, to which Cornford has drawn our attention.[35] But we shall return to *tao* and *rta* later.

Filliozat is inclined to see in the *rta* "the rudiment of the scientific notion" and observes :[36]

> The *rta* is the regular order in general, being the moral rectitude, the ritual exactitude and the true law of the universe ; in one word, it is the 'Norm'. Varuna is the Lord and the guardian of this 'Norm'. The concept of *rta* goes back to the Indo-Iranian period ; to the Vedic *rta* correspond exactly the Avestan *asa* and the Old-Persian *rta*. He has for basis the sentiment that all things are harmoniously regulated and are not left to chance in this universe.

34. A.B. Keith RPVU 83
35. F.M. Cornford FRP 172ff
36. J. Filliozat CDIM 91-92

The fixity of the stellar figures, the periodical return of the mobile stars to similar positions, the evident verification of all the principal and regular phenomena of nature have, from a very early date, imposed this notion. Among the peoples inclined to explain the organism by the cosmos and the cosmos by the organism, this notion must have quickly extended itself from the sky to the earth, from the world to the living being. Although victim of innumerable accidents, the existence of this last one could not be conceived as being outside the orbit of the great cosmic law, as the same body was a little cosmos. If this is not, properly speaking, a scientific concept of natural law, it is at least an idea quite near the same. The scope of this idea is absolutely general because there is nothing which is not governed by *rta,* including the moral behaviour of man. Sometimes the Indian and the Iranian minds are blamed for having thus mixed up the physical and the moral but it should also be recognised that this very mind has also raised above them a sufficiently large concept for covering both of them.

An interesting point—and, as we shall presently see, a point also of considerable interest—about the Rgvedic *rta* is its relation to the Vedic gods as conceived by the ancient poets. *Rta*—the primordial complex of natural law and social law—has *nothing to do with the will of God* or that of any Vedic deity : the deities are at best the custodians of the *rta,* but they have nothing to do with the making of it or bringing it into being by their will. Here are some of the passages from the *Rgveda* in rough English rendering illustrating the point :

'O Agni, your brilliance comes to us and you brought the cows of *rta* equally to us (*rtasya dhenah anayanta sa-srutah* : Sayana interpreted *sa-srutah* as *samanam gacchantyah*) (i. 141. 1).' 'O Mitra and Varuna, you bring *rta* for the *yajamanas* and let the *yajna* be bountiful (i. 151.3). 'As of old, O Indra, you remain the custodian of food and the custodian of the *rta* ; you help us in searching our cows and be friends with us (i. 132.3). 'O Mitra and Varuna, 0 Asuras—the possessors of *rta*—you proclaim loudly of *rta,* since you two are great experts of heaven ; do connect us with cow and water.' (i. 151.4) 'You two (Mitra and Varuna), possessors of *rta* are the foremost suppliers of cows in the *yajna*...'(i. 151.8). 'He who gives to the bright followers of the *rta,* and whom the Adityas increasing,—he, as the foremost, goes with wealth in a chariot to distribute wealth in the assemblies.' (ii. 27.12) 'The great Agni increased without any restraint in the expanse with water and food in the past ; it lay down in the source of *rta* (*rtasya yonau*), being of a charitable disposition and being friends with the water.' (iii. 1.11) 'Agni, the custodian of the *rta* and the possessor of the *rta,* is, like Bhaga, the leader of men.' (iii. 20.4) 'O Visvadevas, he who bears the *rta*—him, the seniormost, do you procure large number of cows.' (iii. 56.2). 'Usas, like the Dyavaprthivi (heaven-and-earth), the custodian of the *rta* and of wealth.' (iii. 61.6) 'O Dyavaprthivi, let your *rta* prevail that we may get wealth along with food. (iii. 54.3) 'As our fathers, (Angirasas), in the past, spread the *rta* so did they discover the tawny cows.' (iv. 2.16) 'From the ancient days, the *rta* has got many waters and much wisdom which discards the undersirables ; the verses of the *rta* opened the deaf ears of men.' (iv. 23.8) 'The holdings of the *rta* are fast, the manifold forms of the *rta* are delightful, the praisers desire abundant food of the *rta,* by virture of the *rta* cows are obtained and they (the cows) enter into the *rta.*' (iv. 23.9) 'Having pleased the *rta* the praisers gain strength and water ; the earth does yield the best cows only for the sake of the *rta* and it is vast and incommensurable because of the *rta*'. (iv. 23.10) 'In the ancient days, Usas-es were truths born of the *rta,* who gave wealth as soon as approached and, praising whom with *uktha* ready wealth was obtained.' (iv. 51.7). 'Usas-es, the deity who knew the abode of the *rta,* made the cows.' (iv. 51.8). 'Usas, the friend of the Asvins, was the mother of cows and the protector of the *rta.*' (iv. 52.2) 'Desirous of the *rta,* the ancients, in the past, praised you, O Agni, for their protection—you, Agni the Angiras, who are great delighter of the mortal, the giver of food (and the lord of the dwelling)'. (v. 8.1). 'O Mitra and Varuna, who, wishing for the *rta,* can get you? Protect us in the abode of the *rta ;* give animals and food to those who desire *yajna.*' (v. 41.1) 'You (Varuna and others) are the custodians of the *rta,* born of the *rta,* increasers of the *rta,* the terrible enemies of *anrta* (the opposite of *rta*) ; thus may we and other heroes remain with happiness and wealth in your abode' (vii. 66.13). And many more like these.

A superficial understanding of the exaggerated epithets used by the poets for their gods may mislead us into imagining that the gods were conceived as determining the course of nature. But the point is that the gods as gods were not doing it. On the contrary, they could do this only in the capacity of observers or guardians, or upholders, of the *rta.* The gods were even thought of as owing their very existence to the *rta,* for it was often said that they were born of the *rta* itself. All these are to be understood along with the fact that human beings, too, were similarly conceived by the Vedic poets ; the ancient Angirasas, as we have already seen, sat in the joyful company of the gods, and this, as observers of the *rta.*

Of all the Vedic gods, Mitra and Varuna—particularly the latter—were most often mentioned as the gods who determined the course of nature :[37]

> Varuna's power is so great that neither the birds as they fly nor the rivers as they flow, can reach the limit of his dominion, his might, and his wrath (i. 24. 6).... He embraces the all and the abodes of all beings (viii. 41.1 & 7)..... Varuna is omniscient. He knows the flight of birds in the sky, the path of ships in the ocean, the course of the far-travelling wind, and beholds all the secret things that have been or shall be done (i. 25. 7 ; 9 & 11) ; ....No creature can even wink without him (ii. 28. 6). The winkings of men's eyes are all numbered by Varuna, and whatever man does, thinks, or devises, Varuna knows (AV iv. 16. 2 ; 5).

All these are true. Still the question is : To what did Varuna owe all this stupendous power ? The Vedic poets left us with no uncertainty as to the answer. They were never tired of telling us that of all the Vedic gods Varuna (often, of course, along with Mitra) had the closest connection with the *rta.* Varuna, along with Mitra, was the guardian of the *rta—rtasya gopa*—and only in this capacity were they the rulers of the rivers and the bestowers of food and rain.[38] They were the revealers of the *rta* and the increasers (or upholders) of the *rta,* but all these, significantly enough, were accomplished by the aid of the *rta.*[39]

"O Mitra and Varuna, you touch the *rta* and increase the *rta* with the aid of the *rta* and spread yourselves for the purpose of increasing the *yajna.*"[40] Mitra and Varuna were seen

37. A.A. Macdonell VM 26
38. *RV* vii 64.2
39. *Ibid* i. 23.5.
40. *Ibid* i. 2. 8

in the place of the *rta* covered by the *rta*.[41] The epithet "observer of order" (*rtavan*), predominantly used of Agni, is also several times connected with Varuna and Mitra.[42]

Now if this was the nature of the relation of the *rta* with Mitra and Varuna, what reason have we to assume that the *rta* was dependent upon the gods, or that it was the expression of the will of the gods? There is none and even Keith had to admit this, though partially:[43]

> The streams go on their way according to the *rta* of Varuna, heaven and earth further the *rta* of Mitra, and the two gods appear as the lords of the *rta* the right. Yet on the other hand they are reduced to a lesser grade in that they appear also as the charioteers of the *rta*, the furtherers of the *rta*, the guardians of the *rta*, something which therefore exists apart from them.

The epithet for 'upholding the *rta* by the aid of the *rta*' though mostly applied to Mitra and Varuna, was also applied to the Adityas and to the gods in general.[44] The Adityas, as also Agni and Soma were looked upon as the guardians of the *rta*.[45] The gods were often described as being born of the *rta*: Soma, the king and god, was born of the *rta* and increased manyfold the *rta* with *rta* (*rtena yah rta-jatah vivavrdhe raja devah rtam brhat*).[46] No wonder, therefore, that Soma was identified with the *rta* itself.[47] Agni was born of the *rta* and hence was shining (for the poet).[48] This birth of Agni from the *rta* had interestingly also a reference to the hoary antiquity: a *rk* of the oldest stratum of the *Rgveda* referred to him as *rta-jatah purvih*.[49] Asvins, too, were born of the *rta*.[50] Being born of the *rta*, the *gana* (i.e. the tribal collective) of the Maruts was without blemish.[51] The gods themselves behaved according to the *rta*: Soma, shining with the *rta*, spea-

41. *Ibid* v. 62.1
42. Macdonell VM 26
43. A.B. Keith RPVU 84
44. Macdonell VM 26
45. *Ibid*
46. *RV* ix. 108.8
47. *Ibid* ix. 62.30
48. *Ibid* i. 36.19
49. *Ibid* iii. 20.2
50. *Ibid* iii. 58.8.
51. *Ibid*, v. 61.14

king the *rta,* was purified and flowed towards Indra.[52] The Visva-devas (all-gods) were upholding the *rta* (*rta-dhitayah*).[53] It was because of the *rta* that Agni obtained his immortality.[54] More examples are not necessary. Evidently the principle of the *ṛta* was to the Vedic poets much more fundamental than any hypothetical will of the Vedic gods.

Lest we are misled to view the *rta* as the will of the gods, it is worth remembering further that the *rta* was not connected with the gods alone. Even the cows, while lowing the *rta* obtained the technique of the *rta.*[55] Sarama, the dog, recovered the cows, with the aid of the *rta,*[56] just as the all-gods (Visva-devas) obtained the cows by increasing the *rta.*[57]

In the understanding of the Vedic poets, the *rta* is thus an independent principle operating—or desired to operate—by its own inner dynamism, and the Vedic gods are themselves supposed to observe it or act according to it, though in some poems the gods are also described as being the "custodians" of the *rta* or the "protector" of the *rta.* But nowhere in the *Rgveda* is there any suggestion of this primordial concept embodying the potentials of what the later scientists call laws of nature yielding to the will of God or of the gods.

The point is exceedingly interesting and to see its special interest we begin with two observations of Needham.[58]

> For without doubt one of the oldest notions of Western civilisation was that just as earthly imperial lawgivers enacted codes of positive law, to be obeyed by men, so also the celestial and supreme rational creator deity had laid down a series of laws which must be obeyed by minerals, crystals, plants, animals and the stars in their courses. Unfortunately, if one turns to the best books and monographs on the history of science, asking the simple question, when in European or Islamic history was the first use of the term 'laws of Nature' in the scientific sense, it is extremely hard to find an answer. By the +18th century it was of course current coin—

52. *Ibid* ix. 113.4.
53. *Ibid* v. 51.2.
54. *Ibid* i. 68.4.
55. *Ibid* x. 61.10.
56. *Ibid* v. 45.7.
57. *Ibid* iii. 56.2.
58. J. Needham. SCC II. 518.

most Europeans are acquainted with these Newtonian words of + 1796 :

> Praise the Lord, for he hath spoken,
> Worlds his mighty voice obeyed ;
> Laws, which never shall be broken,
> For their guidance he hath made.

Here evidently the conception of the laws of nature follows from the monotheistic conception, the roots of which in European thought are ultimately traceable to the Hebrew view. As Needham observes :[59]

> Far more certain as another contributory line of thought was that which emanated from (or was transmitted from the Babylonians by) the Hebrews. The idea of a body of laws laid down by a transcendent God and covering the actions both of man and the rest of Nature is frequently met with, as Singer and many others have pointed out. Indeed, the divine lawgiver was one of the most central themes of Israel. It would be difficult to overestimate the effect of these Hebrew ideas on all occidental thinking of the Christian era—'The Lord gave his decree to the sea, that the waters should not pass his commandment' (*Psalm* 104)—'He hath made them fast for ever and ever ; he hath given them a law which shall not be broken' (*Psalm* 148). Furthermore, the Jews developed a kind of natural law applying to all men as such, somewhat analogous to the *jus gentium* of Roman law, in the 'Seven Commandments for the Descendants of Noah' (*Isaacs,* I). This was liable to conflict with Talmudic law (*Teicher,* I).

How the conception of the laws of nature as understood in modern science got eventually disentangled from monotheistic theology is an exceedingly important question, and, as we shall see, it is relevant also for understanding the history of science in ancient India. Before passing on to all this, let us try to have a fuller idea of the conception of *rta* in the *Rgveda.*

The first point to be noted in this connection is that the conception of *rta*—though containing in embryo the idea of the laws of nature—could never have occurred to the Rgvedic poets as ordained by any omnipotent Divine Creator for the simple reason that the monotheistic theology was totally unknown to them. Whether, among some later Indian philosophers there was a move towards monotheism is, of course, besides the point, for we are discussing here the understanding

59. *Ibid* II. 536.

of the Rgvedic seers or poets and not the later philosophers ostensibly pledging allegiance to *Rgveda.* However, from the passages already quoted from the *Rgveda* it is obvious that specially in the verses in which the concept of the *rta* occurs there is not even any hint of monotheism and there is no hint whatsoever of this concept suggesting the primordial complex of natural order and social order having anything to do with the will of any omnipotent creator.

What, then, could possibly be the source of the concept of the *rta*? Since it represents a pre-monotheistic stage of thought, one way of seeking for its roots is to begin with the genesis of it as an aspect of the monotheistic conception itself. Following Henry Maine's *Ancient Laws*, Needham sums it up:[60]

> Scholars unversed in the history of jurisprudence turn naturally to the well-known book of Maine. He first explains that the earliest law was the case-law of unwritten custom in primitive societies. Their usages were not commands and there was little sanction save the moral disapproval of the society if they were transgressed, but gradually a body of judgements grew up after the differentiation of society into classes; the 'dooms' of Teutonic, or the themistes of Homeric, chieftains. With the growth of State power these judgements could more and more afford to overstep the bounds of the precepts which the society had formerly followed, and continued to follow, as being, for it, demonstrably based on universally acceptable ethical principles. And thus the will of the lawgiver could embody in codes of enacted statutes, not only laws which had as their basis the immemorial customs of the folk, but also laws which seemed good to him for the greater welfare of the State (or the greater power of the governing class) and which might have no basis in *mores* or ethics. This 'positive' law partook of the nature of the commands of an earthly ruler, obedience was an obligation, and precisely specified sanctions followed transgression.

To this may be added another point. Needham substantially agreeing with Edgar Zilsel in viewing the history of human law and the laws of nature[61] notes that among the ancient Greek thinkers the Stoics maintained that "Zeus (immanent in the world) was nothing else but *koinos nomos,* Universal Law"[62] and adds: "Zilsel, alert for concomitant social phe-

60. *Ibid* II. 519.
61. *Ibid* II. 533, note *a.*
62. *Ibid* II. 534.

nomena, notes that just as the original Babylonean conceptions of laws of Nature had arisen in a highly centralised oriental monarchy, so in the time of the Stoics, a period of rising monarchies, it would have been natural to view the universe as a great empire, ruled by a Divine Logos."[63]

With these points in mind we may return to our conception of the *rta* in the *Rgveda.* If the laws of nature, as originally understood, were viewed abroad as connected with the idea of the will of one Supreme God, i.e. as a corollary of monotheism, monotheism itself had its own history and was but an extension in the ideological sphere of the idea of a Supreme Ruler of the State, benevolent or otherwise. Such an idea of God could not and did not emerge among the Rgvedic poets, from whose consciousness was not washed out the memory of the tribal collectivity—the memory of property being collectively owned and shared out among the tribesmen (*bhaga* and *amsa*) and of society being ruled by the tribal assemblies (*sabha* and *samiti*).

What, then, could be the source of the conception of the *rta* ?

Thomson[64] has argued that "man's consciousness of the external world was determined from the outset, not by the relations between the individual and his natural environment, but by the relations which he had established with his fellows in the development of production... Only in this way is it possible to explain why the external world should appear so differently to peoples standing at different levels of culture.... Such developments only become intelligible when we understand that man's consciousness of the world around him is a *social* image, a product of society."

Proceeding on the basis of this we may, in accordance with our method, ask ourselves : what is known in general about the moral consciousness of the backward peoples surviving in the truly tribal societies, which, as tribal societies, must be basically similar to that of the early Vedic poets ?

The following passages are from Engels :

63. *Ibid.*
64. G. Thomson SAGS II. 45-46.

The grandeur and at the same time the limitation of the gentile order was that it found no place for rulers and ruled. In the realm of the internal, there was as yet no distinction between rights and duties; the question of whether participation in public affairs, blood revenge or atonement for injuries was a right or a duty never confronted the (Iroquois) Indian; it would have appeared as absurd to him as the question of whether eating, sleeping or hunting was a right or a duty.[65]

The tribe, the gens and their institutions were sacred and inviolable, a superior power, instituted by nature, to which the individual remained absolutely subject in feeling, thought and deed. Impressive as the people of this epoch may appear to us, they differ in no way one from another, they are still bound, as Marx says, to the umbilical cord of the primordial community. The power of these primordial communities had to be broken, and it was broken. But it was broken by influences which from the outset appear to us as a degradation, a fall from the *simple moral grandeur of the ancient gentile society*. The lowest interests—base greed, brutal sensuality, sordid avarice, selfish plunder of common possession—usher in the new civilized society, class society....[66]

The Vedic *rta*, in its aspect of the human relations, could have originally been what Engels called here the 'simple moral grandeur of the ancient gentile society': the laws regulating the relations of the members of the pre-class society were instinctively apprehended by them as sacred and inviolable, a superior power instituted by nature to which the individual remained absolutely subject in feeling, thought and deed. And if this was so, their consciousness of the external world, being a social image, could only be an extension or projection of the same. This could give the early Vedic poets their conception of the *rta*—an archaic complex of the physical and moral law, so sacred and inviolable that even the greatest gods were born of it.

That this was presumably so can perhaps be seen from the internal evidences of the Vedic literature itself. Most of the modern scholars discussing the world-outlook of the ancient Vedic poets have discussed the glory of the *rta* in the *Rgveda*. What needs to be added to it—and what is not generally discussed by the Vedic scholars—is the *sense of the loss of an-*

65. F. Engels OF 258.
66. *Ibid* 163; emphasis added.

cient rta in the Vedic literature itself, evidenty reflecting the later conditions.

To begin with, it needs to be remembered that the *Rgveda* as a vast collection of songs and hymns, was composed over many centuries. The internal chronology of the *Rgveda*, therefore, still continues to be a question of considerable controversy. What seems to be beyond controversy, however, is the assumption that throughout the vast period of the composition of the *Rgveda*, the Vedic society could not be a stationary one. The ancient collective life of the Vedic tribesmen—the relics and memory of which so abound in the *Rgveda*—could not and was not a permanent feature of the Vedic society. In short, the tribal collectivity had to disintegrate, mak'ng room for a class divided society. Here is how Engels describes this process of disintegration, depending mainly on the evidences about the Indo-European peoples available in his times :[67]

> The military commander of the people—*rex, basileus, thiudans*—became an indispensable and permanent official. The popular assembly was instituted wherever it did not yet exist. The military commander, the council and the popular assembly formed the organs of the military democracy into which gentile society had developed. A military democracy—because war and organisation for war were now regular functions of the life of the people. The wealth of their neighbours excited the greed of the peoples who began to regard the acquisition of wealth as one of the main purposes in life. They were barbarians : plunder appeared to them easier and even more honourable than productive work. War, once waged simply to avenge aggression or as a means of enlarging territory that had become inadequate, was now waged for the sake of plunder alone, and became a regular profession... Internal affairs underwent a similar change. The robber wars increased the power of the supreme military commander as well as of the sub-commanders. The customary election of successors from one family...was gradually transformed into hereditary succession, first tolerated, then claimed and finally usurped. The foundation of hereditary royalty and hereditary nobility was laid. In this manner the organs of the gentile constitution were gradually torn from their roots in the people, in gens, phratry and tribe and the whole gentile order was transformed into its opposite : from an organisation of tribes for the free administration of their own affairs it became an organisation for plundering and oppressing their neighbours ; and correspondingly, its organs were transformed from ins-

67. F. Engels OF 267-9.

truments of the will of the people into independent organs for ruling and oppressing their own people. This could not have happened had not the greed for wealth divided the members of the gentes into rich and poor; had not 'property differences in a gens changed the community of interest into antagonism between members of a gens' (Marx); and had not the growth of slavery already begun to brand working for a living as slavish and more ignominous than engaging in plunder.

What do we read in the *Rgveda* as the repercussion of this change? One aspect of it is already noted by Roth and Whitney:[68] the ancient glory of Varuna, the custodian of *rta par excellence*, was lost; he had to move to the background to make room for the supermacy of Indra, the war-god of the *Rgveda*. This is mythology no doubt. What specially interests us, nevertheless, is the sense of the loss of *rta* along with the degradation of Varuna. As Kutsa, whose songs we read in the admittedly latest portion of the *Rgveda* (namely the first *mandala*), cursed the new development with the sense of *rta* lost:

I ask thee, O *yajna*, the ancient one (*avama*: Sayana took this to refer to Agni, the first of the gods)! Let his (i.e. *yajna*'s) messenger speak with due consideration: Where is the *rta* of the past gone? Who is the new one (*nutana*) that holds it? Know this of me, O Heaven-and-Earth (i. 105.5).

[Sayana's commentary on the word *nutana* is worth mentioning: 'if there were such a (new) one, the present condition of mine would not have been; hence there is none such.' No less interesting is the challenging tone of the *rk*, a tone that runs through the other verses attributed to the same poet:]

All these gods, who are in the three spheres, where is the *rta* of yours gone? Where, again, the absence of the *rta*? Where, as of old, are the *yajna* (*ahuti*) of ours? Know this of me, O Heaven-and-Earth (i. 105.5).

Where, O gods, is the holding of the *rta*, where is the watchfulness of Varuna? Where, again, is the path of the great ways of Aryaman? And hence are we fallen in misery. Know this of me, O Heavan-and-Earth (i. 105.6).

We ask of Varuna, the knower of the path and the maker of food, —I utter this from my heart, let the *rta* be born anew (*navyah jayatam rtam*). Know this of me, O Heaven-and-Earth (i. 105.15).

Keith observed, "the idea of *rta* is one which, like the moral elevation of Varuna, has no future history in India."[69] Per-

69. A. B. Keith RPVU 35.

haps another way of putting it is that the ancient collective life with its "simple moral grandeur of the ancient gentile society" had no future in the history of India until recent times when the demand for it is reiterated though at an incomparably higher level.

It is no use speculating what would have happened in India had the Vedic pantheon of innumerable gods (and a few goddesses) developed into something like Christian monotheism? European parallels tempt us to think that the ancient concept of the *rta* would, in that event, have made room for something like the will of God imparting laws to nature. But the fact is that this did not take place. The sense of *rta* withered away from the Vedic consciousness, leaving the Upanisads only desultorily to mention the old word without its ancient grandeur. In the subsequent history of Indian thought, it was replaced in two ways. In the theologico-political circle, it made room for the law of *karma* and in the circle of scientist-materialists—often branded as abject heretics—it assumed the concept of *svabhava* which, though literally meaning "nature" perhaps carried the connotation of the "Laws of nature". But more of this later.

It is not the place for us to discuss how, in the history of European thought, the concept of the law of nature, originally understood as the will of God, gradually got disentangled from its theological association and assumed the secular form of natural science. Zilsel has already discussed it with whom Needham substantially agrees. Readers interested in the question would profitably go in for their writings. What concerns our immediate discussion is another point. Did the Vedic people contribute to the general fund of the scientific achievements of ancient India in any other important sense than the proto-scientific idea of natural law which they called the *rta* and which in the science circle anticipated the concept of *svabhava*?

7. WRONG WAY OF READING SCIENCE IN THE VEDAS

There are various attempts to read in the Vedas science or the potentials thereof. But there seems to be a wrong way as well as a right way of doing this.

68. See A. A. Macdonell VM 65.

The examples of the wrong way of reading science in the Vedas are indeed numerous. Such an attempt is perhaps also understood specially during the period of the struggle for our national independence. An unarmed people fighting colonial forces had somehow to boost up its morale by trying to overrate the achievements of our ancient *rsi*-s. The motivation of inflating the scientific achievements of the Vedic peoples, in other words, was to show that howsoever strong might have been the brute force of the colonisers, it must not be forgotten when their ancestors were at best but barbarians, our sages could intuitively attain knowledge in certain forms which took many many centuries for the Europeans to reach in the form of natural sciences.

With the achievement of our political independence, however, this way of boosting up the morale of the Indian people has lost its relevance. Since 1947, India has been trying to take steps to develop into a modern nation, and hence the need is felt for developing in India modern science and technology in their right sense. At the same time there remains the need of highlighting the importance of our national heritage—of reminding ourselves of the true glory of our past, so that the hangover of the old colonial mentality of at least a section of our people is fully overcome. But this must not be done in a wrong and perverted manner, one form of which is to claim that everything worthwhile in modern science was already achieved by the ancient sages of the Vedas.

Unfortunately such a tendency to overrate the scientific achievements of the Vedic people still persists in the country. We have already seen in our *Introduction* how a modern Indian scientist as eminent as M. N. Saha had to struggle against it. This, as he has argued, is a totally perverse way of glorifying our ancient culture, besides being a factor inhibiting the real development of modern science in India. In spite of all that Saha and the rightminded people like him did, the tendency continues in our country, which, to say the least, is unfortunate. It will perhaps require a separate book or booklet for a full-length survey of it as still persisting, though one expects that it should decline with the objective understanding of the Vedas.

In such circumstance, however, it proved specially annoying to see also in some of the "spiritual" leaders the zeal to overrate almost to a fantastic extent the science-contents of the Vedas. This has started creating considerable confusion among a section of our people without proper acquaintance either with Vedas or with modern science. A certain "great sage" or Maharshi known as Mahesh Yogi has started what he calls "Vedic University", with its main administrative centre somewhere in Switzerland. As advertised in a really big way in some leading Indian newspapers as well as scientific journals abroad, this university is "undertaking research study in the relationship of the *Rig Veda* to the latest discovery of modern physical and biological sciences." Recently, in 1985, the said "University" held "First European Conference on Vedic Science". I have before me copies of its proceedings. These open with the message of Mahesh Yogi the founder of the university 'via conference telephone from South American Continental Capital of the Age of Enlightenment, in Brasilia, Brazil' : "With the discovery of the unified field by modern science, we feel inspired to offer to the world the complete knowledge of the organizing power of nature. This is going to provide a new status to life on earth, giving rise to a unified field based perfect civilization."

The theory of Unified Field of modern science being outside the scope of my own specialisation, I passed on the Report of the Conference for its assessment to Partha Ghose, who has substantial reputation among our physicists. Following is the reply I received from him :

8th September, 1985

Dear Professor Chattopadhyay,

I was amazed and concerned to read Newsreport No. 1 of the Maharishi Vedic University which you gave me to read and comment on... The essence of the scientific method lies in the hard renunciation of the all too tempting desire to grasp the "ultimate truth" immediately and subjectively in a holistic flash of "revelation" or "enlightenment". In Einstein's words the road to the paradise of science is "not as comfortable and alluring as the road to the religious paradise" [A. Einstein in "Autobiographical Notes" in *Albert Einstein : Philosopher-scientist* ed. by P. A. Schilpp, Harper & Row], but it is more 'trustworthy'. There is no giving up of logic at any stage, however arduous the path might be, although logic itself keeps developing as new phenomena are discovered that do not fit the old concepts. In this way one gradually attains a surer, more and more unifying and deeper mental grasp of things. And what is more, such knowledge is communicable in mathematical language, publicly available and applicable.

The Vedas, on the other hand, are assumed by the Maharishi and his followers to contain a holistic, subjective knowledge of the absolute truth which is extra-logical. To talk of "Vedic science" and "holistic science" is therefore a contradiction in terms as far as the word goes.

The modern unified field theories of physics are very firmly rooted in the traditional scientific method, and in no way reflects the kind of "three in one" structure of the "Maharishi Technology of the Unified Field" (whatever that might be). From what I could glean there is some vague resemblance of the latter with the measurement problem in quantum mechanics of the late 1920's. But even there the subjective interpretation is by no means accepted universally. In fact, the majority of physicists hold to an objective interpretation, and surely nowhere is the subjective interpretation ever used in practice.

This kind of hasty identification, I find, usually stems from a lack of genuine interest and understanding of modern physics, and is not only pointless and superficial but misleading, particularly when it is made by a person like the Maharishi who is held in great esteem by a large number of devotees all over the world who are unable to evaluate the true worth of his statements and accept them uncritically.

The tendency to glorify the Vedas by trying to give them a garb of scientific respectability is pathetic. It only reveals a basic lack of respect for them and a failure to appreciate their true worth which, I think, lies in their exquisite poetry and occasional flashes of philosophical wisdom. Taking into account the age in which they were conceived and written, I find them extraordinarily rich. However, it is difficult to believe they contain the whole of science.

If they do and if the Maharshi claims to know it all, why is he making us poor mortals spend so much money and effort in rediscovering it ?...

I was amazed to read Dr. Geoffrey Clements' alleged statement that the "benefits of the Maharshi Technology of the Unified Field have already been validated by more than 300 scientific studies in all areas of physiology, psychology, sociology, and ecology, conducted at over 150 universities and research institutes on all continents." Anybody who makes such a sweeping statement without first facing professional scientific bodies and without giving references to authentic publications in respectable professional journals, only deserves to be ignored.

Two comments will be in order here. First, it was Einstein who first emphasised the need for a unified field theory. In this time he was thinking in terms of unifying electromagnetism and gravitation. He spent the last 30 years of his life trying to achieve this unification, but he failed; the problem remains unsolved even today. On the other hand, electromagnetism and the weak nuclear force (responsible for radio-activity) have been satisfactorily unified into a single electro-weak theory by Salam, Weinberg and Glashow (1967|68). Some interesting theories (called Grand Unified Theories or GUTS) of unifying this electro-weak force with the strong nuclear force have also been proposed, but there is as yet no conclusive experimental evidence in support of such a theory. Physicists have nevertheless gone ahead and proposed a unification of all the four fundamental interactions including gravity by using the idea of supersymmetry (the symmetry between fermions and bosons) and strings (in which the basic objects are string-like). But these are still in a very embryonic stage. To conceive of a grand idea of unification is one thing and a very creditable thing; to give it a concrete shape is quite another. If and when physicists finally succeed in constructing a satisfactory unification of all the fundamental forces, will it make sense to say that Einstein knew it all?

Secondly, these modern unified field theories only deal with the purely physical world: they have nothing whatsoever to say about physiology, psychology, sociology or ecology. Science is far, far away from anything resembling a unified theory of all its branches...

I hope the above comments will be of some use in clarifying the true position of modern physics.

With kind regards and best wishes,

Yours sincerely,
Partha Ghose

But let us leave the question of the so called "Vedic University" and turn to another recently published book which has created considerable sensation in recent time. It is called *Vedic Mathematics or 'Sixteen Simple Mathematical Formulae from the Vedas' (For one-line Answers to all Mathematical Problems)* by Jagatguru Swami Sri Bharati Krishna Tirtha Maharaja Shankaracharya of Govardhana Matha, Puri, originally published by the Benaras Hindu University in 1965, and, as reprinted in an inexpensive edition, is having very wide circulation these days.

The special difficulty created by the book is that its author is far from being an amateur as far as mathematics (specially arithmetic) is concerned. The book is full of highly clever algorithms, often showing amazingly simple ways of arithmetical calculations. This makes it impossible to brush off the intrinsic importance of book. What nevertheless is extremely deceptive about it is its title and major historical claim. The title wants us to believe that the algorithms are actually contained in the Vedas (or auxiliary Vedic works). To create such a make-belief, the author gives sixteen formulas in Sanskrit, on which all his calculations are said to be based, adding that all these are from the Vedas. But it has proved impossible to trace these formulas to the Vedas. In short, notwithstanding the mathematical excellence of the book, its title has no more worth than that of a fiction. It is thus another recent example for us of the wrong way of reading science in the Vedas.[70]

70. Reviewing the book in IJHS Vol. iii. No. 1, May 1968 (pages 59-60), A. K. Bag states : "The author admits that these sutra-s and corollaries have been derived from the *Atharvaveda,* till now no scholar has been able to trace this relationship." In the same journal (vol. xviii, No. 2, p. 223), R. C. Gupta observes : "from the language of the 16 *sutra*-s (formulae), it is clear that they are author's own composition in modern Sanskrit but employing the old *sutra*-style. Hence author's claim that they are 'contained in the *Parisista* (the Appendix-portion) of the *Atharvaveda,* (p. xv) can be justified only by regarding them, following a suggestion by V. S. Agrawala (see p. 6 of his *Foreword*), as a new Parisista added according to the tradition of formulating subsidiary apocryphal texts. It seems that the author attached the name of Vedas to his work in retaliation to one of his teacher's habit of refuting the opinion that even formula of modern mathematics are contained in the Vedas (see *Bulletin of the National Institute of Sciences of India,* No. 21, 1963, p. 253). In his *Preface* (pp. xiii-xxx), the author talks of 'The Astounding Wonders of Ancient Vedic Mathematics' and says that 'the Vedas should contain within themselves all the knowledge needed by mankind.' (p. xiii)". However, as an earlier reviewer (T. S. Kuppanna Sastri in VIJ vol. 4 pt. i, pp. 108-9) rightly observes : "One would expect from the title *Vedic Mathematics* that the contents are to be found in what are accepted as the Vedas, or in the *Parisista*-s belonging thereto intended for the elucidation of the Vedic rituals. But the work is not Vedic in this generally accepted sense, neither the so-called *sutra*-s nor the contents being found anywhere in the Vedic literature. Therefore, we can take it, with the General Editor (*Intro*. pp. 6-7), that the contents are not what are in the Vedas but what *should be* there on account of the Vedas being the source of knowledge, if only the author had not described the work, in his *Preface* (p. xiii), to be 'on the astounding wonders of *ancient Indian Vedic* mathematics', and inces-

8. RIGHT WAY OF READING SCIENCE IN THE VEDAS

We have mentioned examples of trying to over-rate science and science-potentials in the Veda not because it is typical of the modern scholars but because a considerable number of our scholars are not altogether free from this basic tendency.

As already observed, for the pioneers of Indian studies in modern times, the Vedas remained virtually the only starting point as sources for Indian history and hence they were under the peculiar obligation to seek the roots of all science and technology in the Vedas. With the discovery of the Indus Valley Civilization such an obligation has ceased to exist; yet the tendency to read the beginnings of technology and science in the Vedas continues at least among many scholars. There seems to be also a simple method to follow for the purpose. It is, in short, to consult the *Vedic Index of Names and Subjects* by Macdonell and Keith. Admitting the extraordinarily remarkable scholarship that has gone to its making, it may as well be remembered here that the work was first published in 1912, when research in what are called the loan words—particularly words of the local people entering into the Vedic vocabulary—was at best in its infancy. Secondly, the work is based on the survey of a vast literature beginning with the *Rgveda* and stretching upto the Upanisads—the latter foreshadowing the period of the second urbanisation, when the Vedic people lose much of their original identity, being largely mixed up with the local people and adopt much of their material culture. The older method is thus no longer tenable.

We have already had some objective idea of the material culture of the Rgvedic people, so a good deal of caution is necessary before reading in the *Rgveda* much of advanced technology and science. Unfortunately such caution is not always maintained and we frequently come across exaggerated acco-

santly throughout the work, tried to create the impression that what he gives are *from the Vedas themselves*. His effort at glorifying the Vedas and Hindu culture by these false claims will only create a revulsion of feeling when the truth is known. The pity is that what the author gives is *not even Hindu* classical mathematics."

unts of the technological and scientific achievements of the Rgvedic peoples.

What appears to be more unfortunate is that behind the anxiety to impute to the Vedic peoples undue achievements, what was really something of the nature of extra-ordinary importance about them from the viewpoint of the history of science is often obscured or ignored. The very composition of the *Rgveda*—embodying as it does 1,028 songs or hymns by pre-literate pastoral peoples—is itself a wonder of wonders. No less a wonder is the preservation of it by sheer retentive memory. The literary merit of this vast literature is discussed by others and falls outside the scope of our discussion. What is within the scope of our discussion—and without which no history of science in India can be adequate—is the technique developed for its preservation in memory, with meticulous care for rightly pronouncing each syllable in the vast literature, the understanding of the metre in which each hymn is composed, the grammatical syntax of their composition, their exact meaning, and so on, though not without internal controversies among later thinkers who got absorbed in such problems. Understandably, all this formed the starting point of a number of formal sciences, like phonetics, metrics, linguistics, etc.

Fortunately, I found my young colleague Navjyoti Singh of NISTADS, New Delhi, has for some years been working on this theme. What is all the more fortunate for me is that he has conceded to my request to contribute a chapter on this positive aspect of science and science-potentials in the Vedic tradition. With grateful thanks to him, I am using his contribution as the next chapter of the present study.

CHAPTER 13

LINGUISTICS AND ORAL TRADITION IN THE PERIOD BETWEEN THE DECLINE OF HARAPPAN CULTURE AND THE RISE OF MAGADHAN CULTURE

Navjyoti Singh

1.0 PRELIMINARIES

Study of history of science depends on what is thought as worthy of being called exact science. Obviously the understanding of what constitutes the subject-matter of exact science changes with history. In contemporary world, a rigid view of the boundary of natural and social sciences has prevented historians of science from giving attention to linguistic researches of the past.[1] Adequate grasp of the exact sciences in ancient India has suffered most because of this. This is so because of the unique and distinct position that the study of human language occupied in ancient India. Linguistics truly can be called a queen of sciences in ancient India. In this chapter we will give contextual outline of the development of linguistics and at the same time seek historical reasons for the same.

1.1 A HISTORICAL PARADOX

History of Indian subcontinent poses a peculiar problem in the period between the decline of Harappan Culture (*c.* 1500 B.C.) and sprouting of Magadhan Culture (*c.* 500 B.C.). This period witnessed general decline in material culture and technology but at the same time saw maturing of human expression in the composition of poetic hymns as well as in the formulation of cosmogonic ideas as is recorded in the Vedas.

1. An only comprehensive work on history of science in India, Bose et al (1971), does not even mention linguistics, except on pp. 25-26, where again importance of linguistics is brutally minimised by reducing the worth of its source materials to merely 'a veritable source of information of many scientific and technical subjects' (p. 26). And these subjects are everything other than linguistics.

The beginning of this period is marked by the decline of urban centres, degeneration of Harappan technology and artefacts and disappearence of Harappan script. Only towards the end of this period that the regeneration of urban culture begins and also Brahmi and Kharosthi scripts come into being.[2] In our view the most striking fact is that this period is characterised by the absence of writing and urbanity and at the same time much of the corpus of Vedic literature was composed and originally redacted in this period.[3] This is a historical paradox and calls for explanation.

Unlike Harappan script, which disappeared in this period, contemporaneous hieroglyphic script of Egypt (*c.* 2700 BC—) or cuniform of Sumer (*c.* 2500 BC—) or ideograms of China (*c.* 2500 BC—) got strengthened through this period. Strengthening and refinement of writing played an indispensable role in stabilisation, preservation and standardisation of language. Egyptian writings on papyrus (2700 BC—O AD) and Babylonian writings on clay tablets (2500 BC—O AD) stand in testimony. Writing in these cultures helped the fixation of language and thus long texts composed in the period could

2. Archaeological evidences point abundantly towards absence of writing on non-perishable material. Asokan inscriptions (3rd cent. B.C.) all around the country in Brahmi and Kharosthi scripts are the earliest evidence of revival of writing. Though few cryptic epigraphs in Brahmi have been dated earlier to Asokan inscriptions, even then they do not in any case go back to before 5th cent. B.C.
3. Literature which is composed in this period is vast and that too in diverse literary styles. The four Vedas (Rk, Yajus, Saman and Atharva) were composed in this period. *Rgveda* is the most archaic composition and came into being in the early part of the period. Much of Brahmana, Aranyaka and Srauta classes of literature were composed in the later part of the period. Some of Pratisakhya and Siksa classes of literature are also composed in this period. Much of the literature in the period adopts metric and poetic style but prose literature was also composed (*Yajurveda, Satapatha Brahmana* and *Nirukta* of Yaska). Even Sutra style literature (*Rkpratisakhya*) is born in the end of the period. Later Jaina and Buddhist literature also records non-Vedic aspects of oral compositions in the period.

be preserved. In fact, the possibility of composition of long texts itself is generally related to the invention of writing and its further refinement. Unwritten language is bound to fluctuate rapidly, thus making preservation of long composition virtually impossible. This absence of long texts from the past of most of the cultures in the world can be understood in the context of non-fixation and non-standardisation of language. Sheer cramming can never fix a long text in the way writing can. If this is really the case, then, how was the large corpus of Vedic literature composed and preserved in the non-literate oral phase of history of civilisation in Indian subcontinent? What played the role of fixation of language in oral phase if not writing?

1.2 WAY OUT OF THE PARADOX

This interesting query is resolved only when we pay attention to the development of linguistics in ancient India. Early development in linguistics is a unique feature in the history of exact science inasmuch as the absence of writing is a unique feature in history of culture in Indian subcontinent in the period *c.* 1500 BC—*c.* 500 BC. It is precisely the need to compose and preserve long texts in the absence of writing that led to the developments in linguistics quite unparalleled in the contemporaneous world. Not only long poetic and metrical compositions were meticulously retained through the peroid using knowledge of phonetics but also prose compositions and some Sutra style formulaic compositions were preserved. Fixation of language was achieved in a different way through the developments in the exact science of phonetics, etymology, and grammar in the oral phase. This thrust of exact science towards linguistics not only helped create oral devices which made possible uncontaminated retention of long compositions but also led to the creation of abstract interest in the phenomena of language, triggering later, sophisticated theoretical developments in grammar which culminated in Panini's (600 BC—400 BC) grammar of Sanskrit called *Astadhyayi,* a fine piece of reasoning available to us from the ancient world. Other products of linguistic research which are anterior to Panini are the *Nirukta* of Yaska (*c.* 700 BC—*c.* 500 BC), a book of etymology and semantics, and *Rk-*

pratisakhya of Saunaka (*c.* 600 BC—*c.* 500 BC), a book of phonetics. In this chapter we shall describe and analyse multi-faceted developments in the exact science of language in the non-literate oral phase of Indian history in the context of a successful attempt to fix and standardise long compositions orally.

2.0 NON-LITERATE ORAL PHASE AND FIXATION OF LONG COMPOSITIONS

It needs to be emphasised that the development of exact science of language in Indian subcontinent is vitally related to the unique feature of the history of Indian subcontinent. From contemporary archaeological evidence it can be minimally concluded that writing on non-perishable material did not exist in this phase and it existed before and after this phase. Alberuni in the 10th century AD had recorded a legend, prevalent in his time, that retained memory of occurrence of such a phase in Indian history when knowledge of writing was relegated to the background : "As to the writing or alphabet of the Hindus, we have already mentioned that it once had been lost and forgotten ; that nobody cared for it, and that in consequence people became illiterate, sunken into gross ignorance, and entirely estranged from science. But then Vyasa, the son of Parasara, rediscovered their alphabet of fifty letters by an inspiration of God".[4] He was referring to the traditional belief that legendary Vyasa was responsible for the collection of Veda-s and the composition of *Mahabharata* and the eighteen Purana-s. Whatever may be exegetical context of the legend the fact remains that Vyasa was believed to have revived writing which had become obscure and fallen out of use in some phase of history. At least the memory of occurrence of such a phase was retained even till the time of Alberuni.

In the light of epigraphical and archaeological evidence the legend seems to contain a reified version of the historical fact of the occurrence of the Sruti phase, though, as we would demonstrate in this chapter, the conjecture that 'people became

4. Pp. 171-172, *Alberuni's India* ed. and translated with notes and indices by Edward C. Sachau. New Delhi : Munshiram Manoharlal Reprint 1983.

entirely enstranged from science' is not quite tenable. Besides, the introduction of Brahmi and Kharosthi scripts was not merely the rediscovery of hitherto lost alphabets.

That there existed a legend, embodying a reified version of Sruti phase, is of enormous significance. It implies continuity in the complex historical change. Even if the knowledge of Harappan script did not survive through the non-literate phase, as is indicated by current archaeological information, at least the memory of the existence of writing in Harappan phase seems to have survived.

This opens the difficult question regarding the nature of interface between the Harappan writing and the composition or fixing of a long text like *Rgveda* or even some early portion of it or even some long Harappan composition. This is a doubly difficult problem. On the one hand, it is connected with the issue of the spectacular decline of Harappa, a change which is itself yet to be fully understood, and on the other hand, it is connected with the nature of Harappan script, a script which has offered insurmountable resistance for its decipherment. Despite these difficulties and the complexity of historical process which is involved it is possible to draw some conclusion regarding the relation between Harappan script and long texts.

2.1. CONJECTURE ON THE INABILITY OF HARAPPAN SCRIPT TO FIX LONG COMPOSITION

Script was one cultural feature which must have played extensive and essential role in the Harappan civilisation that had a temporal spread of about a millenium. Large number of inscribed seals (about 2500) are found made of steatite, clay, faience, ivory and little copper plates. One feature of the graphs in the inscription is that they do not seem to change significantly over the long span of about a millenium. This constancy is significant and implies that the function of the script was settled and did not require much modification and change in the script.

Another noticeable feature of the Harappan script, as it survives for us, was the brevity or the cryptic nature of text on individual inscription. The average length of the text being just 5 graphemes (signs), the total number of texts is very large

(2906 texts have been compiled by Mahadevan) implying that the sample of extant inscriptions can authentically represent Harappan use of script, the maximum length of a text or rather the longest text being only 26 graphemes long in three lines. Maximum number of lines is 7 in a text if all sides of inscribed objects are counted as one text and maximum number of lines on one side is three only. The length of a line of text varies from one grapheme to 14 graphemes, the longest line of 14 graphemes occurs in two identical texts. There is a notable absence of a continuous text, not to speak of a long one.

The cryptic nature of the Harappan text stands out as a striking and unique feature of the script and makes it different from any known script that had a spread over such a vast civilisation. This hints at the possibility of structural constraint implicit in the script itself which could have made it incapable of representing continuous long text. This seems more so given the fact that in the entire life span of the script spread over a millenium the graphemes (signs) did not undergo any noticeable change. Our positing of the incapability does not diminish the pervasive use value of the script in the Harappan civilisation and the extensive role it played therein. It merely indicates possibility, supported by two unique features of the script, that the domain of applicability could have excluded representation of long continuous composition. In other words, the semantic elaboration and fine differentiation that continuous long compositions demand could not be tamed by the Harappan script which only fixed cryptic texts.

Our conjecture is that possibly there was some formal constraint dictated by the nature of Harappan script which made it incapable of adopting itself to the need of continuous long compositions. Though the unknown nature and function of the yet undeciphered[5] Harappan script stands as a challenging prob-

5. Many decipherment attempts and claims have been made. Mostly attempts are based on comparison with other known ancient scripts and some assumed context of meanings signified by the text. There are also attempts based on statistical cogency (coherence) of symbols. But none of the attempts still today has been able to give consistent and complete account of Harappan script.

lem our observation is independent of it. It is also independent of whichever natural language or symbolic system the script represented. According to our conjecture, even if Harappan script was coextensive (e.g. in late Harappan phase) with the composition of the first kernel of *Rgveda* it would have been incapable of fixing long composition like the *Rgveda*. Notice that this observation is independent of the complex issue of relation of Indo-Aryan language with Harappan culture and can be held without assuming or positing details of the complex interface of Harappan culture and Vedic compositions. Vedic hymns are long continuous compositions which are semantically elaborate and linguistically rich. The unchanging Harappan script would have been structurally redundant to fix the Vedic composition. This would mean eventual redundancy of the script for performance of cultural role (different from the Harappan) demanded by the long Vedic composition. Hence even if the knowledge of script survived till the time when long compositions became culturally significant it would have had no role to play. Thus only the memory of script survived. Besides, the cultural significance of the auditory aspect of hymn composition and recitation would have meant a thrust towards orally fixing the long continuous composition like the *Rgveda*. The knowledge of the redundancy of the earlier script for this purpose would have added towards this thrust if our conjecture about the nature of Harappan script is true.

2.2. BEGINNING OF LONG VEDIC COMPOSITIONS AND ORAL APPROACH TO FIX THEM

The central corpus of Vedic composition, the *Rgveda*, is supposed to have been composed and compiled between about 1500 BC and 1000 BC, a period which is somewhat contemporaneous and immediately follows the decline of Harappan culture. This period is an early part of Sruti phase of Indian history. In fact the entire corpus of Vedic literature is heterogeneous[6] which includes diverse kind of literature com-

6. The corpus of Vedic literature is vast as it comes to us today. Traditional stratification of this corpus is the 4 Vedas : *Rk*, *Yajus*, *Sama* and *Atharva*, the numerous *Brahmana*-s, *Aran*-

posed in Sanskrit over a long period. The *Rgveda,* which is a compilation of hymns, is undoubtedly the earliest, archaic and pivotal body of Vedic literature.

The hymns which are compiled in the *Rgveda* were composed individually or in the groups and at different occasions. What prompted the compilation of these hymns is an open question. The compilation perhaps followed the decline of Harappa and hence must have been done orally. The oral compilation of Rgvedic hymns was called *Arsi Samhita.*[7] *Shamita,*[8] though in general means 'put together', 'collection' or union, was used in a specific connotation for designating the collection of hymns. The recitation and incantation of *Arsi Samhita* was called *samhita-patha.* More specifically, the con-

yaka-s, *Srautasutras* and *Grhyasutras* which are associated to different Vedas and some compilations out of them called the Upanisad-s. Further the later Vedic literature is classified in terms of six Vedanga-s : Vyakarana, Siksa, Nirukta, Kalpa, Chandas and Jyotisa. In the period of concern much of the literature described above was composed. Legendary Rsi Krsna Dvaipayana is supposed to have divided the original Veda into 4 portions according to their ritual significance. Interestingly *Atharvaveda* compiles that portion of *mantra*-s which is not related to Yajna or rituals. In this corpus *Rgveda* is regarded as the most archaic, *Yajurveda* being the creation of later period and also with different thematic inclination. Similarly *Atharvaveda* is a still late compilation with altogether different thematic inclination not connected with rituals.

7. The term *samhita* is used in *Rgveda* [1.168.6] and *Taittriya Samhita* (1.5.6.2) in the sense of 'put together', 'collection' or 'union'. Later literature also extensively used this term with general connotation of put together but it acquires a technical connotation of 'joining' and 'union' of sounds, words etc. *Arsi Samhita* means compilations of legendary Rsi-s and is a term which designated *Rgveda* [see Uvata's commentary on *Rkpratisakhya* 2.2].
8. Other technical connotations of the term *samhita* are given in *Taittriya Pratisakhya* [24-1-4] in terms of Pada-Samhita, Varna-Samhita, Aksara-Samhita ad Anga-Samhita. These are unions of words, syllables, letters and numbers respectively. In *Vajasaneyi Pratisakhya* [1-157-158] even collection of Pada-s has been referred to as Samhita. All these connotations imply a unified meaning of 'union' or collection'.

tinuous recitation of the compilation of hymns was called *samhita-patha*.[9] The practice of continuous recitation[10] was perhaps called for because of the need of incantation as well as because of convenience in memorising. In non-literate oral tradition *samhita-patha* was memorised and was handed down from generation to generation orally.

Memorising the *Rgveda Samhita* was not an easy task, it being a very long compilation running into 1,53,826 number of words.[11] Such a long composition if unwritten would change drastically within a dozen generations. Even if the whole composition is memorised and orally taught to the next generation, within a couple of centuries, the pronunciation, the order of hymns and even the structure would have changed. This did not happen and a meticulously fixed text of the *Rgveda Samhita* was transmitted through many many generations. Oral transmission was meticulous enough to the extent of preserving each and every syllable as if the syllables have been imprinted on immutable rocks. This gets established when late medieval manuscripts of the *Rgveda Samhita* are compared with the orally transmitted text, say, the one, among many, which is retained to this date by the Nambudri Brahmins of Kerala. The fixing of long texts in an oral milieu in the absence of writing definitely requires more than the mere skill of memorising. It requires designing of safeguards against possible contamination of all kinds. Besides developing and cultivating skill of memorising, fixing of text requires method of fixing sounds so that sounds do not get altered or changed with time. This can only be done if sounds are somehow objectively defined,

9. Samhita-patha was also called Nirbhuja Samhita which means compilation without ends (or arms) or simply continuous recitable compilation. Nirbhuja Samhita is used first in *Aitareya Aranyaka* [3-1-3].
10. See Appendix section A 3 for the example of continuous recitation. Not only poetic and metrical hymns were recited orally but also prose compositions like *Yajurveda-samhita*. *Pratisakhya* of the *Yajurveda* called *Vajasaneyi Pratisakhya* [1.158] says that *samhita* is union of syllables which can be said in one breath. That is, in Samhita-patha of prose discontinuity or pause for taking breath is accepted only.
11. p. 14 Introduction of V.K. Varma to *Rkpratisakhya*.

say, with help of detailed pronouncement procedures. In fact, a whole body of knowledge of the physiology of sound production was developed in Sruti phase to safeguard against phonemic slippage and contamination. Similarly, fixing of long text requires fixing of word-order of the hymns. This requires designing method that can check whether word-order has been disturbed or not. The counter-checking methods should be independent of mere memorising though no doubt they are designed primarily to supplement memorising. This was indeed a difficult task as it required invention and development of techniques of fixed representation which is non-graphic and yet immutable enough. Like invention of fixing long texts graphically (through writing ideograms or alphabets) requires graphic formalism, the invention of fixing long texts orally requires phonetic formalism. Such a formalism indeed was developed in the Sruti phase when a genuine need was felt to fix long texts of *Samhita*-s without the help of writing. It was in this context that the inquiry into language began in ancient India in the period which archaeologists call 'dark age',—the period between the decline of Harappan culture and the rise of Magadhan culture.

Orally fixing texts of the *Rgveda Samhita* and other *Samhita*-s was a unique achievement. Even till today right from about 1000 BC there exists the tradition of orally transmitting the text of *Samhita* without taking recourse to writing.[12] Tho-

12. J.F. Staal. *Nambudri Veda Recitation.* Hague : Mouton and and Co., 1961 deals with the Veda recitation tradition of the Numbudri Brahmins of Kerala. Some 35% of Numbudri Brahmins can recite *Rgveda,* 50% *Yajurveda* and about 1/8% *Samaveda* according to J.F. Staal. Various recensions of oral transmission of *Samhita*-s of Veda are dealt in the text like *Caranavyuha* and its commentaries.
Nambudries of Kerala have two *Rgveda* schools, one at Tirunavaji (near Kottakkal) and other at Trssivapevuv (dist. Trichur). Third school of little importance is in Kurumbrahad Taluk. *Atharveveda Samhita* in oral form perhaps exists only in a few villages of Gujarat and a few villages of Orissa. *Samaveda Samhita* can be found in Tamil Nadu. The Kauthuma Sakha is widespread but the other Sakha is near extinct. Taittiriya Sakha of *Krsna Yajurveda* can be found near

ugh today writing is available, still some archaic tradition lingers on and stands as a living testimony to the possibility of fixing long texts orally.

2.3. INSTITUTIONALISATION OF PRIESTLY COMMUNES FOR THE PURPOSE

At the time when writing was absent the process of compilation and retention was institutionalised. It was in the numerous priestly communes, known as *sakha*-s, that the *samhita-patha* was kept alive, modified and transmitted. At any point of time there were many *sakha*-s. According to the grammarian Patanjali (*c.* 200 BC), there were no less then one thousand one hundred and thirty *sakha*-s.[13] Institution of *sakha*-s was the nucleus where the fixation of *samhita-patha* was achieved. With time *sakha*-s got bifurcated into *carana*-s and lineages of various *sakha*-s even got merged. Today only *Samhita*-s of thirteen *sakha*-s are available, out of which two are incomplete.[14] The difference between the *Samhita*-s of various *sakha*-s were in terms of pronunciation, and minor differences in the contents of the hymns. The major difference in *sakha*-s was in terms of other literature associated with the *sakha*-s like *Brahmana*-s, *Aranyaka*-s, *Upanisad*-s, *Srautasutra*-s etc. For example, the Kanva *sakha* of *Sukla Yajurveda Samhita* has one hundred and eleven *mantra*-s more than the *Samhita* of Madhyandina *sakha*. *Satapatha Brahmana* of Kanva *sakha*

Madras and Rajamahendravaram (Rajahmundry) in Andhra Pradesh. *Sukla Yajurveda* can be found in Karnataka region. Madhyandina Sakha exists near Mysore and Kanvas in a few villages of Tiruchurapalli and Tancavur district of Tamil Nadu. See. V. Raghavan, 'Present Position of Vedic Chanting and its Future'. Bulletin of the Institute of Traditional Culture, pp. 48-69, Madras 1957.

13. The *Vyakarana-Mahabhasya* of Patanjali ed. by F. Kielhorn revised by K.V. Abhyankar Vol. 1 1962 Poona. According to Patanjali (in 200 B.C.) there were 21 Sakhas of *Rgveda Samhita;* 101 of *Yajurveda Samhita;* 1000 of *Samaveda Samhita;* and 9 of *Atharvaveda Samhita.*
14. Today *Samhita*-s of thirteen Sakhas have been identified and published.

has 104 chapters, but of Madhyandina *sakha* has only 100.[15] Such differences are the legacy from the complex inter-institutional divergences and convergences in the non-literate phase of Indian history. It was in that phase of history that institutions of *sakha*-s came up and fixed the texts of *Samhita*-s orally. We have put this phase of history and the unique phenomena of orally fixing long texts and the institutionalisation of the same through *sakha*-s in the background of the institutionalisation of the same through *sakha*-s in the background of the incapability of Harappan script because of its cryptic nature.

3.0. STANDARDISATION AND FIXING OF VEDIC TEXTS AND THE EXACT SCIENCE OF LANGUAGE

It was in the priestly communes,[16] that the compilation of hymns attained final standardisation. The institution of priestly communes must have precipitated when the socio-cultural need to rigorously standardise or make uniform the collection of hymns was felt. The overall process of standardisation of the *Rgveda Samhita* must have been a long drawn process beginning somewhere in the early phase of Sruti period. But the formation of *sakha*-s must have taken place in the middle of Sruti phase. The major contribution of the institution of *sakha* was not merely final standardisation but orally fixing the standardised text for uncontaminated transmission through the ages. Actual process of final standardisation of compiled hymns and orally fixing the text must have gone hand in hand as

15. See p. 5-6 of Introduction of V.K. Varma in *Vajasaneyi Pratisakhya*.
16. To call Sakha-s priestly communes needs clarification. Sakha-s were the settlements of groups of people dedicated to the cause of standardisation and fixation of *Rgveda Samhita*. Here was a group of people or community tied together with a bond of determinate cause. It is in that sense that such settlement can be called commune. As the hymns were associated with the rituals we have called it priestly communes. These priestly communes in all likelihood would not be economically self-sufficient in the nomadic milieu of the time ; hence the term commune might create some misunderstanding as commune is supposed to be economically self-sufficient. Priestly communes would have been partially supported by the surplus appropriated by the rulers.

the two were concurrent functions of the institution of *sakha*-s. The historical complexity of the process is evident from the formation of various *sakha*-s of *Rksamhita* at an early time and later formation of numerous *upasakha*-s and their further partitioning into *carana*-s.[17]

The significance of the institution of *sakha*-s was that the well-knit community dedicated to the sole task of fixing the standardised *Rksamhita* had come into being. The continuity of this community through the ages was an important pre-requisite for orally fixing the *Samhita*. Orally fixing the text, as we had remarked earlier, involved more than sheer cramming, it involved critical and creative investigation into various facets of language and it involved invention of oral devices which were not just an aid to memory but also ensured control over the natural process of contamination. This indeed was a much complex task compared to fixing of text by writing it down. Non-availability of appropriate script had left little choice but to explore ways of orally fixing the text. Sustained exploration in this direction stirred multifaceted investigations into various aspects of the phenomena of language compared to which invention and refinement of script did not involve much sustained effort. It is only a stable institution of *sakha,* with the community dedicated to the task of orally preserving the *Samhita* that could have made possible the success of such a momentous task.

3.1. PROBLEM OF STRUCTURISATION OF *RGVEDA*: ORIGIN OF INDICES, CONCORDANCES AND LEXICONS

Since *Rgveda* is a long text cramming without understanding the whole structure of the text would be extremely difficult. Even for standardisation explicit formulation of the structure of the whole text is a must. In fact, the first requirement for fixing the text orally is to clearly formulate and understand the

17. Out of 21 Sakha-s which grammarian Patanjali had recorded the following five were considerably significant—(1) Sakala, (2) Baskala, (3) Asvalayana, (4) Sankhayana and (5) Mandukayana. Though the *Samhita*-s of only Sakala and Sankhayana are available today but some of the other literature associated with the Veda like *Brahmana, Aranyaka, Grhyasutra* etc. of other Sakha-s are available today.

overall structure of the text. This is essential for easy reference and arrangement of the memorised text. Also explicit structurisation would have helped fix the order of sections of hymns in the text. Such a structure of *Samhita* could have been formulated, say, in terms of ordering the major themes covered by hymns or ordering in accordance with the deities addressed by the hymns or ordering in accordance with the purpose for which incantations of hymns is suggested or even ordering in accordance with the lineages which were responsible for composing the hymns. Indices of these kinds were necessary to understand the structure of text as a whole. Structurisation of *Samhita* was a complex historical process and discovering or imposing some order was not an easy affair. Neat theme-wise, deity-wise, purpose-wise or lineage-wise order did not exist in the *Rksamhita.* Still concordances could be made. Such corcordances and indices were indeed orally prepared and were called *Anukramani*-s.[18] This was a beginning of dictionaries (lexicons), concordances and indices in the antiquity of Indian history. To enhance understanding even the list of important words in *Samhita* or lexicons were prepared and were called *Nighantu.*[19] But the *Anukramani*-s were themselves a fairly long texts and were again memorised because they could not have been written. Though the various indices of the *Rksamhita* would have helped fix the order of the text partially, they could not have made easy the understanding of the structure of text as a whole. These indices could not have made structurisation explicit as the

18. Various *anukramani*-s from the Sruti phase have been referred to in the literature. The most important one related to *Rgveda Samhita* is Katyayana's *Rgveda Sarvanukramani. Brahaddevata* is another text which is a kind of encyclopaedia of deities etc. There is concordance which gives list of metres, this is Venkatamadhava's *Chandonukramani.*

19. *Nighantu-s* are very archaic list of important words. The great Indian book of etymology, the *Nirukta* of Yaska is a commentary on one of these *Nighantu*-s. The lists which are inherited from the Sruti phase are many indeed. There were lists of roots of words called Dhatupatha. The list of technical grammatical terms was called Ganapatha etc. But the most archaic lists are of course *Nighantu-s.* The tradition of *Nighantu-s* continued well into modern times.

structurisation should be brief and clear ; on the contrary, the indices besides being long and untidy could not have been exhaustive and neat. In fact, no single criterion could be employed to give an exhaustive and neat index which could have helped fix the order of hymns in the *Samhita.* This meant theorising on the structure was a difficult problem. In fact, an important difference among the various *sakha*-s was on the issue of structuring of *Rksamhita.*

3.2. THEORY OF METRES DEVELOPED TO ARTICULATE STRUCTURE OF RKSAMHITA

Rgvedic hymns were poetic compositions having metrical form and hence were memorised in the rhythmic mode. The metrical nature of the hymns must have made the task of memorising easy. In fact the phenomenon of metrical memorising and incantations was so imposing culturally that the grammarian Panini referred to the Veda and the Vedic language as *chandas*,[20] which he distinguished from the everyday language, *laukika.* Though the term *chandas* at a time prior to Panini, had acquired a technical meaning of poetic metre[21] but through its association with Veda its derivative term *chandasa* was, in the later periods, even used in the sense of archaic.[22] The rhythmic utterance of hymns was such a unique and imposing feature that the derivation from the term for poetic metre itself came to signify Veda.[23]

20. Panini iv. 3.71. Etymologically the word comes from the root *chand* or *chad* which means 'to cover'. *Satapatha Brahmana* 8.5.2 gives this etymology. Also *Nirukta* 7.12 gives similar etymology.
21. 'Chandasika' meant one who is familar with the Veda and 'Chandasiya' meant one who is familiar with the metrical science. [*Srutibodha* 19]. See Monier Williams, *Sanskrit English Dictonary,* p. 404-405. The two senses of Veda and metre were amalgameted. Later a famous book on poetic metres by Pingala was called *Chandah Sastra.*
22. *Sarvadarsana Samgraha* vi. 11. See also Williams 405.
23. This is not to say that ancient Indian metrical theory did not bother about metres of secular poetry. Half of Pingala's (*c.* 200 B.C.) *Chandhasutra* deals with *laukika chandah* (secular metre) and the other half deals with *vaidika chandah* (metres in Veda).

As we had remarked earlier the metrical aspect was central to the memorising and incantation of *Samhita*. Study of the metrical aspect would have helped in the classification of hymns in metres. Ordering of the metre employed would have also helped fix the order of hymns in the *Samhita*. The study of metrical aspect of hymns in fact played the most important role in fixing the order of hymns, though of course with the help of additional criteria of deities (*devata*) addressed by hymns and *rsi* families who supposedly composed or initially compiled the section of hymns.[24] Metric aspect of poetic composition changes with history and varies with cultural geography. Currency of metre in poetic style of any culture at any time is related to the practice of poetic composition and hence it changes with time and place. To determine the metre of some poetic composition is a difficult problem. *Rgveda Samhita* being a repository of hymns compiled from numerous traditions the problem of determining metre was indeed complex. In the middle of Sruti phase concept of syllable quantity[25] (time taken to pronounce a syllable and relation of its

Later *Prakrata Pingala* deals extensively with metres of compositions in various languages different from Sanskrit. The point being argued here is that the term *chandas* was an archaic signifier of Vedic language because metric and rythmic recitation was uniquely and intensely associated with memorising and ritual performance of Samhitapatha.

24. For example the *mantra* in the Appendix belongs to a section of the Tenth book of *Rgveda*. This section has 23 *mantra*-s and is composed in Anustubh metre (has 32 syllables in 4 lines or *pada*-s). And the section is in praise of *Osadhayah* (herbs and medicine) and belongs to the lineage of Atharvano Bhisag. Not every section or Sukta as it is called is composed in one type of metre, nor every section is devoted to the praise of same deity, nor is every section composed or compiled by one lineage. For example 92 Sukta or section of first book has 18 *mantras* 1-4 are in Jagati metre, 5-12 are in Tristubh metres and 13-18 are in Usnik metre. 1-15 *mantras* are in praise of Usas (morning) and 16-18 are in praise of Asvins.

25. Notion of syllable quantity is outlined in *Rkpratisakhya* 1.17-34. *Suklayajuh Pratisakhya* 1.55-61, *Taittiriya Pratisakhya* 1.31-37 and *Paniniya Siksa* 4.20. *Sambhu Siksa* has 45 *Slokas* devoted to the theory of temporal quantity of syllable. *Kalanirnaya Siksa* and its commentary called *Kalanirnaya Dipika* by Muktisvaracarya is

number with the rhythm), was evolved and on this basis various metres were classified.[26] Classification of metres employed in *Rksamhita* gave rise to the concept of fundamental metres and the derived metres.[27] Consistent and exhaustive classification of metres posited a fascinating logical problem as some

entirely devoted to the theory of temporal quantity of syllable. These texts deal with theoretical aspect of quantity and on it are based all works in Indian tradition which are related to prosody and rhythmic aspect of music. In fact syllable quantity becomes minimal unit of time in Indian astronomy. See *Aryabhatiyam* 2.2 and *Surya-siddhanta* 1.11 b. *Rkpratisakhya* 13.50 says that time taken to pronounce short vowel=Bluejay's chirp; For long vowel=crow's croak and very long vowel=peacock's cry. These three units are in ratio 1 : 2 : 3 and are fundamental to the theory of quantity.

26. The first detailed classification of metre is in *Rkpratisakhya* chapters 16, 17 and 18. The numerical arrangement of syllables forms the back-bone of classification as the first principle of classification is number of syllables. In *Sarvanukramani* [12-6] Katyayana defines metre as a result of numbers of syllables. The next principle is the number of lines and the distribution of syllables in these lines. Further finer classification was done on the basis of coherence with adjustment of accent order, metres, meaning and order of short and long syllable. In *Chandahsutra* of Pingala classification is done on the basis of additional concept of 'gana' which is a unit of three syllables and he did away with the Vedic concept of 'pada' or line. He uses permutation and combination of three long or short syllables of 'gana' to define all metres. Halayudha in his commentary on *Chandahsutra* constructs diagram of 'Meruprastara' (Pyramidical diagram) which is akin to Pascal's Triangle of coefficients of Binomial expansion.

27. *Rkpratisakhya* 17, 19, 168, 16.80, 16.89-90, gives list of 26 *chandas* which are defined for the sake of numerical elegance beginning with 4 syllable metre to 104 syllable metre by ascending in the step of 4 syllables. Each of 26 metres has four variants 1 syllable less (Nicrt), 1 syllable more (Bhurika), two syllables less (Virat) and two syllables more (Svarat). This way metre from 2 syllables to 106 syllables are classified.

In *Rgveda* first five and the last seven metres are not found (e.g. 2 to 22 syllabic metre and 78 to 106 syllable metre). Only 14 primary metres are found. About half of *Rksamhita* is Tristubh (11+11+11), 1|4 is Gayatri (8+8+8), 1|7 is Jagati (12+12+12+12) metre. So 4|5 of Samhita is just these three primary classes of metres. Then there are some which are called mixed hymns could be classified under more then one head.[28] In this keen enquiry was the birth of the discipline of Chandah-sastra or science of prosody sometime in the middle of Sruti phase. The detailed and exhaustive classification of metres of Rgvedic hymns was a most decisive way of determining the structure and thus fixing the incantation as well as helping memorising of *Samhita*. In fact, classification of hymns in the *Rksamhita* critically depended on the theory of metre evolved for the purpose. It is important to understand the concepts of *rc* (*mantra* or hymn), *pada* (line) and *avasana* (pause) to understand the classification structure of *Rgveda Samhita* and these concepts are intimately related to the theory of metre.

Socio-culturally *Rksamhita* was intrinsically related to the ritual aspect of pastoral world. *Rc*[29] or *mantra* was a mini-

metres (*misrita chandas*) as they have uneven lines. For example Viradrupa Tristubh (11+11+11+8), Jyotismati tristubh (12+8+12+12), Mahabrhati tristubh (12+8+8+8), Yavamadhya tristubh (8+8+12+8) and Abhisarini Tristubh (10+10+12+12). *Rksamhita* has metres of one line (RV 10.20.1) to 8 lines (RV 1.133.6 & 1.127.6). Even 12+12 is called Dvipadagayatri as it is equivalent to 8+8+8=24, the original Gayatri, in terms of total number of syllables. The classification becomes more complex when we note that in *Rksamhita* line with 4 syllables [RV 8.46.15] to the line with 16 syllables [RV 2.22.1] exist. To uniquely define every metre was an extremely tedious task. A text which gives list of various *chandah*-s is Venkatamadhava's *Chandonukramani*.

28. Major theoretical problem is how to divide Samhitapatha into *pada* or lines of metres. What determines how many syllables make a line of metre ? Classification into metres of the hymns will crucially depend on the solution of this question. According to Jaimini [*Jaimini sutra* 2.1.35] *pada* is obtained from meaning. Also cf. *Nidana Sutra's* commentator Tataprasad [p. 2]. *Rk pratisakhya* [17.25] gives three principles for deciding the question (1) *Prayah* (contiguity) : division of line in accordance with the neighbouring mantras of the same section or Sukta. For example RV 1.61.12 is 10 syllable line, so it should be called Vairaja line, but since in rest of the Sukta Tristubh metre is prevalent, hence *parva* is turned into *paruva* to make it 11 syllable line ; (2) *Artha* (meaning) : *pada* is that collection of syllables which completes meaning and (3) *Vrta* : In accordance

mum unit in *Samhita* which was of significance for the community in general. Units smaller than *rc* were of value only to the community in *sakha*-s. *Rc* was complete in meaning and signified invocation or praise of a particular action, instrument, theme or deity.[30] *Rc* included in the appendix for illustration is in praise of medicinal herbs that save men and domestic animals from disease and death. *Rc*-s are made out of meaningful words and contain several lines or *pada* (e.g. *rc* included in appendix has four *pada*-s), and embodies metrical aspect [e.g. *rc* in appendix is in *anustubh* (8+8+8+8 metre)] and has definite accnt.[31] Recitation of *rc* is punctuated by at least one pause or *avasana*[32] primarily to catch breath but also because of reasons of prosody. Rules for pause in the *Samhita* recitation were determined in terms of the number of lines or *pada* in the *rc*. To exhaust the domain of application of rules several exceptional cases were singled out in terms of exceptional rules (*apavada*).[33] Concept of *rc, pada*

with rules of long and short syllables. An interesting logical problem arises because application of these three criteria might give conflicting results in some cases. For example, in RV 10.73.7 3rd and 4th *pada*-s illustrate conflict between *prayah* and *artha*. Similarly in RV 8.44.23 3rd pada illustrates conflict of *Artha* and *Vrta*. For this situation *Rkpratisakhya* (17.26) gives a meta-rule that in the case of conflict earlier criterion is applicable. Besides, still finer criterion of accent is brought in. For example, *Rkpratisakhya* 17.27 says that except for *ukara* no *anudatta* accent can come in the beginning of the *pada* in the whole of *Rgveda*. Using fine distinctions and metalogical method a situation was avoided when same hymn gets classified under two metres.

29. *Rc* is what hymns were called by the *Rgveda* itself, 1.164.39 and 10.71.11 etc. The term *rc* is exclusively used for the hymns of *Rgveda* and among *Pratisakhya*-s only *Rkpratisakhya* uses the term to refer to hymns and that too abunduntly e.g. 8.9, 16.19, 31, 44, 63, 73, 78 and 17.3 etc. Interestingly according to *Jaiminiya-sutra* 2.1.35 '*Rk* is that whose lines or *pada* are decided in accordance with meaning'.
30. As Sayanacarya says : *arcyate prasasyate anaya devevisesah kriyavisesas tatsadhanaviseso va iti rksabdavyutpattiriti*|*Rgbhasyabhumika*.
31. These in fact are the characterisations of *rc*. According to Sayanacarya [*Rgbhasyabhumika* p. 71], metrical *mantras* which embody meaning and lines are called *rc-s*. And Visnumitra in

and *avasana* and the characterisation of *rc*-s according to metres made *Rksamhita* more or less technically well structured and made possible neat articulation of this structure. *Rc* is quite small a unit and *Rksamhita* has about 10552 of them. *Rksamhita* is divided into groups of *rc*-s which are called *sukta*-s. *Sukta* grouping is done according to three criteria : (1) lineage which was responsible for initial compilation of the bundle of *rc*-s, e.g. the first *sukta* of Rgveda belongs to Visvamitra lineage, (2) theme around which the bundle of *rc*-s is composed, e.g. the first *sukta,* is Agni and (3) metre in which the bundle of *rc*-s is composed, for example, the first *sukta* is composed in Gayatri metre. Most of the 1028 *sukta-s*[34] of *Rksamhita* are characterised by unique triplet marker based on the three criteria, namely, lineage, theme and metre. There are many *sukta*-s where this characterisation becomes cumbersome like the 67th *sukta* of 8th Book,[35] which contains 32 *rc*-s characterised by 8 lineages, themes and 4 metres. Finally these *sukta*-s are aggregated into 85 *anuvaka*-s and further into Ten Mandala-s or Books. This completes the structure of *Rksamhita.*[36]

3.3. THEORY OF PRONUNCIATION DEVELOPED TO FIX INCANTATION OF SAMHITA

Explicit articulation of the structure of *Samhita* not only was of pedagogical interest to the students and teachers in *sakha*-s but also helped fix the text of *Samhita* to some extent. At

his *Vargadvaya Vrtti* (p. 13) and which is included in the *Rkpratisakhya* edition says, limited syllables, lines and *ardharcas* make *rc*-s.

32. 12 *sutras* of *Rkpratisakhya* 18.46-57, deal in the rules of pause or *avasana* in *Rksamhita* incantations, *Vaj. Prat* 3.31, 4.22 and *Atharva Prat.* 14 also deal with it.
33. Rules for 3 line to 8 line *rc* are given in *Rk. Prat.* 18.47-52 respectively. And exceptions are given in *Rk. Prat.* 18.54-57.
34. The number of *rc*-s and *sukta*-s is taken from table on p. 767 of *Rksamhita* ed. by V. S. Satvalekar. Paradi : *Sarvanukramani* says there are 85 *anuvaka*-s, 1029 *sukta*-s, 10580¼ *rc*-s and 153836 in *Rksamhita.* Varma (1972) p. 4.
35. *Rgveda* 8.67.1-32.
36. See ed. of *Rgvedasamhita* and Varma (1972), p. 43.

least it standardised to a large extent order of Books, order of *sukta*-s etc. But standardising the general order is not sufficient to orally fix the text as it cannot ensure checking of insertion or deletion of *rc*-s, words and sounds etc. Fixation of text orally does not merely involve fixing order of hymns, which undoubtedly has to be preserved ; but also fixing of actual performance of incantation. Since the significance of hymns was in actual incantations it was this actual incantation which was to be transmitted through the ages and for that purpose fixed. This required that the details of utterance itself, that is, sound, accent, tempo, and even musical aspect of incantation, besides metre and position of pause, be fixed as well. Fixing of these details was in fact ach'eved in the Sruti phase. This involved detailed study of phonemes, the explicit and minimal units of utterance, and syllablisation of human speech. This effort led to development of theory of physiology of utterance or theory of the physiological basis of phonetics. We will later elaborate on this interesting development. Three kinds of accents[37] in *Samhita* were related to rising (*aroha*) and falling (*avaroha*) tone, which was related to particular physiological effort[38] needed to produce them. Eight kinds of *svarita* (circumflex) accent were distinguished,[39] phenomena of jerk in accent,[40] conjunction of accent,[41] faults

37. Whole third chapter and parts of 12th, 15th and 17th chapters of *Rkpratisakhya* deal with the theory of accent evolved for the purpose of fixing accent in Samhita. The three kinds of accent were Udatta (acute), Anudatta (grave) and Svarita (cicumflex). In the writing of *Rgveda Samhita* no sign is put for accute accent, '—' sign is put below to indicate grave accent and '|' sign is put above to indicate circumflex accent.
38. Three distinct efforts to produce three accents were called *ayama, visrambha* and *aksepa* respectively by Uvata's commentary on *Rk. Prat.* 3.1.
39. There was some amount of controversy regarding nature of Svarita as it was supposed to be half acute and half grave ; see *Rk-Prat.* 34-13 and Uvata's commentary. Three primary divisions of Svarita were suggested and three-fold two-fold, three-fold further sub-divisions were suggested in these *sutras*.
40. *Rkpratisakhya* 3.34 and Uvata's commentary on it. In this section are covered sudden shifts in accents.
41. When two words join phonetically, that is, the last letter of first and the first letter of second word coalesce the phenomenon is

of accent articulation[42] were studied, and rules related to application of accent on *samhita-patha* were formulated. Such a study in detail helped to standardise the text of *Rksamhita* in details, upto each and every syllable.

Even the tempo of utterance was divided into three modes, *vilambita* (slow), *madhyama* (intermediate) and *druta* (quick).[43] Intermediate mode was suggested for actual performance; quick one for the rehearsal and the slow for teaching.[44]

With the theoretical apparatus developed to grasp metrical aspect, sounds, accents and tempo of incantation it become possible to standardise and fix oral recitation of *Rksamhita*. Details of pronunciation[45] were made explicit so that incantation

called Sandhi (euphony). Not only end sounds but also the accent undergoes euphony. *Rkpratisakhya* 3.11-16 gives rules of conjunction of accents. Interestingly 3.13 is an exceptional rule (*apavada*) of rule 3.11. 3.11 states that acute and grave coalesce into acute whereas 3.13 states that result is circumflex in *ksipra* euphony and *abhinihita* euphony and *praslista* euphony of *ikara*-s.

42. Five standard faults are listed in *Rk. Prat.* 3.29-33—Interestingly there is a fault called Kampadosa (3.31) which according to commantator Uvata is found in 11th century Vaidika-s of south India. This seems to be related to the contemporary south Indian musical tradition which banks on Kampana.

43. *Rkpratisakhya* 13-46-50 deal with the tempo of utterance. The classification is made in 13-46. 13-8-50 defines the relation between the two by relating them to the unit of time. If in slow mode one *matra* is pronounced in intermediate 2 and in quick 3 will be pronounced.

44. *Rkpratisakhya* (3.9).

45. *Paniniya Siksa* 52 brings this point home—*mantro hinah* etc. Hymns free of care about accent and letters embody illusion and are unable to enshrine intended meaning. On the other hand they can destroy Yajamana by becoming weapon (*vagvajra*). Example is the mistake in accent of 'Indrasatru'. Example of 'Indra-satru' is classic illustration of mistake and is often presented when introducing the science of accent. In the incantation of hymn 'Indrasatrurvardhasva' the word Indrasatru' is either made by Bahuvrihi, as of, 'Indrah satrur yasya' or Sasthisamasa as of 'Indrasya satruh : In the first case 'I' of 'Indra...' will have high accent and in the second 'U' of...'tru'...In a *yaina* organised by *asura* Tvasta to obtain son Vrtra, who will kill Indra, the mistaken pronunciation of high accent 'I' in 'Indra'...resulted in Indra becoming fatal enemy of Vrtra and killing him.

itself is fixed, not merely the text of the *Rgveda Samhita,* which was standardised by articulating structure of the text. The practice of continuous recitation was called *samhitapatha,* which is compared to emptying a vessel of its liquid, again and again. In *samhita-patha* sounds were recited continuously except for the pause which was necessitated by the need to catch breath and also by the reasons of prosody or elegant incantation. In appendix we have written down *samhita-patha* of one *rc.* It is evident that all syllables are continuously recited except for the pause in the middle and at the end which is indicated today with sign of *virama.* The whole *Rksamhita* was memorised in this continuous fashion with well-set points of pause. The continuous recitation mode was also adopted at the time of incantation. In fact it was this continuous recitation mode which was the actual *Rgveda* in Sruti phase and not the printed text we have today where *rc*-s are written with modern indicators of punctuations.

3.4. SIGNIFICANCE OF MEANING FOR FIXATION AND THE IMPORTANCE OF WORDS

Memorising *samhita-patha* was quite a task. It was important that the meanings of songs and hymns be understood to avoid mindless cramming which can easily lead to contamination. Understanding meaning entailed that meanings of words used in songs and hymns be understood. Yaska, the author of the first ever book on etymology, the *Nirukta,* calls those persons blockheads and bearers of burden only, who having learnt Veda do not understand the meanings involved.[46] And he further says 'whatever is learnt without being understood is called mere cramming ; like dry logs of wood on an extinguishing fire, it can never illuminate.'[47]

46. *Nirukta* 1.18. Also quoted in Visnumitra's *Vargadvaya Vrtti* p. 8 on *Rkpratisakhya.*
47. *Nirukta* 1.18 This passage is also quoted by Patanjali in *Mahabhasya* i.1.1 p. 12 and Sayana's commentary on *Mantra Brahmana.* Durgacarya in his commentary on the above passage says—"A person who commits Vedic texts to memory without understanding is comparable to an ass bearing a load of sandalwood, who perceives its weight but not its fragrance."

The list of important and recurring words used in the *Samhita* were listed in the form of *Nighantu*-s, and the tradition of explaining their meanings also was in existence much before Yaska (*c.* 700-500 B.C.) Undoubtedly such exercises would have played a significant role in enabling to develop grasp over the key words. In fact each and every word used in *Rk-samhita* needed to be understood to intelligently memorise it. To segregate words out of continuous recitation was itself a problem. This can only be done by referring to currency of words in day to day language. The realisation that the word is a unit of human language is evident even in *Rgveda* (I.164. 45), where words are said to be of four kinds.[48]

3.5. INVENTION OF THE DEVICE OF *PADAPATHA* AND DEVELOPMENT OF THE THEORY OF *SAMDHI* OF SOUNDS AND ACCENTS

The continuous recitation, Samhitapatha, was divided into words and the recitation of words called Padapatha was devised.[49] Padapatha was word by word recitation of the continuous recitation. The third section of the appendix illustrates Padapatha of a *rc*. In word by word recitation pause is given after every word to clearly distinguish it. Word by word recitation was not just a slow form of continuous recitation or the distinction between the two is not just more or less frequent occurrence of pause. The two recitations had different sounds. For example, *brahmanastam*.....of continuous recitation becomes *'brah-*

48. Later texts give elaboration on this cryptic statement from *Rgveda* ; *Nirukta* 1.1 and 1.2 and Durgacarya's commentary on it gives these four genera of words as : *nama* (nouns), *akhyata* (verb), *upasarga* (preposition) and *nipata* (particle). Nature of these genera and their relation is further elaborated in *Nirukta.* Another ancient book *Rkpratisakhya* 12.17 to 12.26 deals extensively with it.
49. Padapatha is also called 'Asamhita' which means 'not joined'. *Vajasaneyi Pratisakhya* 1.156. 'when words are separated out it is 'Asamhita'. Padapatha is also called 'Pratrnna' in *Aitareya Aranyaka* 3.1.3, when syllables are pronounced pure without combination it is called 'pratrnna' ; 'pronouncing of two words with two clear and pure syllables without euphony is called Padapatha of 'pratrnna'.'

manah. tam.' in word by word recitation. The sound 's' of continuous recitation is replaced by sound 'h' in the word to word recitation. The difference in the sounds of the two recitations is natural. When distinct words are spoken in a way that on time gap is left in between pronunciation of these words then sometimes the sounds of words undergo modification. In other words, the end sound of one word and the initial sound of second word are to be pronounced in close proximity, sometimes these two sounds react and produce a new sound or undergo modification. This phenomenon is known as euphony or *samdhi.*[50] In fact, technically, Samhitapatha is defined in *Rkpratisakhya* as that 'when union of word-ending and word-beginning sounds is accomplished without leaving any time gap'.[51]

It is not necessary that end-sounds of words have to undergo modification (*vikara*) whenever pronounced without time gap. Most of the time in fact sounds do not undergo modification in the day to day language usage. Later the ancient Indian grammarians defined the phenomenon of euphony or *samdhi* only when modification (*vikara*) occurs because of conjunction. Contrary to this definition of *samdhi* the earlier phonetical theory associated with Samhitapatha and Padapatha maintained that any word brought together phonetically implies *samdhi* even if there is no modification.[52] This understanding of *samdhi* was called for because of the use of the term for the study of the construction of Samhitapatha from Padapatha.

50. The term 'Samdhi' in the sense of 'union' is used in *Rgveda* 8.1.12. Later the term is used explicitly in a technical sense of conjunction of syllables. *Vajasaneyi Pratisakhya* 3.3 clearly defines it. *Rkpratisakhya* uses the term frequently 2.34, 4.41, 4.78, 7.1 and 10.18. *Rktantra* 94, 96, 97, 111 and 283 uses term '*sandhaya*' for it whereas term 'Samkara' is used in this sense in *Vajasaneyi Pratisakhya* 1.1. *Atharva Pratisakhya* and *Caturadhyayi* do not employ the term at all.
51. *Rkpratisakhya* 2.2.
52. Several types of Samdhi-s propounded in *Rkpratisakhya* do not admit of any modification (*vikara*). Another Samdhi called *anuloma anvaksara* in *Rkpratisakhya* 2-8, also is one in which no modification takes place. Besides there are many exception rules (*apavada-sutra*) and counter-rules (*nipatanasutra*) dealing with Samdhi where no modification occurs.

The scholars in the *sakha*-s took theoretical interest in the phenomenon of *samdhi* because of the need to reconstruct and hence fix Samhitapatha on the basis of Padapatha. In this enterprise *samdhi* was explicitly used in the sense of bringing together words of Padapatha thus forming Samhitapatha. Now whether modification occurs or not the domain of *samdhi* is universal. In *Yajnavalkya Siksa*[53] in fact it is clearly stated that *samdhi* is of four kinds ; (1) *lopa*,[54] where elision of syllable is involved, (2) *agama*,[55] where a syllable gets inserted, (3) *vikara*,[56] where syllables get modified, and (4) *prakrtibhava*,[57] where sounds do not change. The grammarian Panini was to regard absence of *samdhi* as *prakrtibhava* in contrast to its being one kind of *samdhi*.

An extremely intricate and challenging study of the phenomenon of euphony was made in the *sakha*-s. To fix Samhitapatha with the help of Padapatha, first thing that was needed was to study in depth syllables and letters which occur at the end and the beginning of the words.[58] Such detailed study

53. *Yajnavalkyasiksa*, p. 83. There are four kinds of *samdhi*-s, *lopa, agama, vikara* and *prakrtibhava.*
54. 'Lopa' means 'to vanish' [from root *lup*]. It was first used in technical sense in *Nirukta* 2.1 *Rkpratisakhya* 4.80 etc. *Vajasaneyi Pratisakhya* 1.141 defines it as zero occurrence of letter, whereas in *Taittiriya Pratisakhya* it is defined as destruction of letter. When any letter which occurs in Padapatha does not occur in Samhitapatha it is explained by the rules of *lopa* kind of Samdhi.
55. Agama in the technical sense of insertion of letter is used in *Rkpratisakhya* (2.31). The term *upajana* is used in the same sense in *Rkpratisakhya* (4.84). It gives two kinds of *samdhi* under the title Agama.
56. This kind of Samdhi is what the later rules of Samdhi in grammatical literature deal with. The difference in sounds in Pada- and Samdhi-patha are generally explained through this kind of Samdhi.
57. Prakrtibhava means that which remains in the natural form. *Vajasaneyi Pratisakhya* L.80. The term 'Prakrti' is used in *Rk Prat.* 2.51, '*prakrtya*' is used in *Taittiriya Pratisakhya* 9.16, *Vaj. Prat.* 3.11. To indicate word endings which do not admit of any change or modification in sound *Rk. Prat.* 2.52 suggests use of *iti* in *Padapatha.*
58. Results of such an inquiry are presented earliest in 12th chapter of *Rkpratisakhya* 12.1 to 16.

of letters not occurring in the beginning,[59] letters not occurring at the end,[60] and letters not occurring together was made.[61] This was an extensive empirical exercise. Formation of *samdhi* rules could have been based only on such empirical exploration as locus of the phenomenon of euphony lies in the word-endings and word-beginnings, though of course in certain cases sounds in the middle also change. Successful formulation of *samdhi* rules was an essential pre-requisite for putting Padapatha to use. In fact, it can be said that the most significant theoretical endeavour was necessitated by the desire to reconstruct Samhitapatha from the Padapatha and the *Pratisakhya*[62] class of literature was devoted to just this problem. Even from the logical point of view the problem posited an intricate challenge and was handled by devicing three-tier rules structure.[63] We have dealt with the logical aspect elsewhere. The theory of *samdhi*[64] which evolved out of the inquity is im-

59. *Rk. Prat.* 12.1 gives list of 25 letters which do not come at the end of word. The only vowel which does not come at the end is 'ri'. The rest of 24 are consonants.
60. *Rk. Prat.* 12.2 gives list of 13 letters which do not occur in the beginning of words, 'ri', and 'li' are the two vowels among them. Varma (1972) p. 119 says that letter 'dh' also does not occur in the beginning but is not noted by the author of *Rk-pratisakhya* and Uvata who commented on it.
61. *Rk. Prat.* 12.3 to 12.35 deals with letters which do not occur together. For example 'c', cerebral 't' and dental 't' according to rule 12.3.
62. *Rk. Prat.* Chapters 2, 4, 5, 7, 8 and 9 are entirely devoted to formulating rules of euphony. Even other chapters deal with the material related to euphony like the 12th chapter quoted above.
63. The three tier rule structure is the universal rule (*samanya-sutra*) applicable universally, the exceptional rule (*apavada-sutra*) applicable in a small portion of the domain of universal rule but gives exception to the universal rule, and the counter-rule (*nipatana-sutra*) gives exception to universal as well as exceptional rule.
64. The complexity and elaborateness of theory of Samdhi is evident from the kinds of euphonies delineated and the intricate relation between these in the three-tier logical structure. The major *samdhi*-s are (1) Svarasamdhi (combination of vowels), (2) Svaravyanjanasamdhi (vowel-consonant combination), it is called Anuloma anvaksara-samdhi (combination according to succession

pressive and formed the basis of later morphological and syntactic investigations of etymologists and grammarians. An extremely novel aspect of the investigation into re-creating of Samhitapatha from Padapatha was the theory of *samdhi* of accents. Even the accents undergo modification when distinct words are brought together to recreate *samhita*.[65] Our example in the appendix illustrates difference in accent between various *patha*-s or recitations.

Such investigation into the phenomenon of euphony related to the sounds as well as accent made possible non-ambiguous re-creation of the original *Rksamhita* from the word by word recitation. The device of word by word recitation not only helped fix word-order and meanings of hymns but also accents and syllables. The importance and significance of Padapatha in fixing *samhita* cannot be over-emphasised. Also development of theory of *samdhi* in the history of linguistic ideas is intimately tied to the development of the device of word by word recitation and re-creation of continuous recitation from it. And the development of the device of Padapatha was solely related to the need for orally fixing Samhitapatha. There was no other reason to develop the device of Padapatha.

3.6. INVENTION OF THE DEVICE OF *KRAMAPATHA* AND INTERNALISATION OF THE KNOWLEDGE BODY OF *SAMDHI* IN RECITATION STRATEGY

In the oral tradition merely memorising Samhitapatha and Padapatha would not be sufficient. Without thoroughly knowing the body of knowledge related to euphony of sound and accent memorising Padapatha would not help much. Though this knowledge was evolved, the proper application of it had to be ensured also. An oral device to even ensure this was also developed and it was called Kramapatha.[66] In Kramapatha,

of syllables) ; (3) Vyanjanasvara-samdhi (combination of consonant and vowel)—it is called Pratiloma anvaksara Samdhi (combination contrary to the succession of syllables) ; euphony of consonant and consonant.

65. *Rksamhita* 3-11 to 16 gives nine rules of combination of accents.
66. There are indications of dispute from the *Rkpratisakhya*, about archaichood of Kramapatha as well as its utility and as to who

recitation is done with two words taken together as a unit. The first two words are recited, then the second word of the first pair is taken together with the third word to form a second pair, then the second word of second pair is united with the fourth word to make a third pair, this way pairs are made till half of *rc* is finished.[67] It is only then the pause is taken. See section A3 of appendix for the illustration of *karma* recitation. The *krama* recitation stands in between *samhita* recitation and *pada* recitation. It retains property of *samhita* recitation in so far as it displays modification of sounds and accent because of joining of two adjacent words, at the same time it retains word-endings of *pada* recitation. The oral device of *krama* recitation in a way internalises the knowledge-body of *samdhi*. Besides, *krama* recitation reinforces fixation of word-order, accent in *samhita* recitation, sounds of syllables, and the sounds of word-ending. It in fact helps demonstrate both sides of a bare word along with the result of words joining together through *samdhi rules*.[68]

The device of *krama* recitation was a subject of various disputes in the ancient time as we learn from *Rkpratisakhya* (11.61 to 65). According to it (11.64), "Only the *krama* recitation which was originally propounded is right but not the many kinds as propounded by numerous scholars," the originally propounded being the one devised by Babhravya and the one we have described and given in the appendix, section A3. There could be many variants of *krama* recitation in the early times. At least later many variants and deviants did exist and we have illustrated nine of them in the appendix, sections A 7.1 to A 7.9. In these recitations, as it is obvious from the appendix, word-order and pronunciation details are so juggled

invented it. *Rk.prat.* 11.65 to 11.71. It is said that Babhravya, son of Babhru, first propounded this device for the benefit of students.

67. *Rkpratisakhya* 10.2. 'Beginning with two words, take the second word and join it with the next, this way continue till the half *rc* is complete'. Similarly in *Rkpratisakhya* 11.1 : 'without losing any part of continuous recitation when two words and their union is pronounced at the same time it is called *karma* recitation.' See also *Vajasaneyi Pratisakhya* 4.183.

68. *Rkpratisakhya* 11.64.

in a very many pre-determined ways so that the *samhita* recitation becomes entirely fixed. Many of these recitations even employ pronouncement of words in reverse order. The art of well-ordered jugglery with the words of *pada* recitation using knowledge-body of *samdhi* of sounds and accents to fix Samhitapatha was well cultivated in Sruti period. Undoubtedly most of these variants and deviants of *krama* recitation were devised later as the time passed into a phase of history where *Rksamhita* got alienated from its socio-cultural roots, and stopped growing as a living tradition. Need for further rigidification of oral recitation modes to keep alive fixed Samhitapatha perhaps led to the devising of complex deviants of *kramapatha*. Even till today we find experts who have at the tip of their tongue complexly knitted recitations which help reconstruct and preserve uncontaminated transmission of *samhita* recitation through the ages. For example, an expert from Andhra Pradesh who died in January, 1968 could recite whole of *Krsna Yajurveda Samhita* in reverse, literally in reverse, from the end to the beginning word by word.[69]

3.7. SCIENTIFIC STRATEGY FOR ORALLY FIXING *RGVEDA* BECAME MODEL FOR FIXING OTHER LONG TEXTS

The tradition of the three recitations, that is, *samhita, pada* and *karma*, to a great extent successfully fixed the text of *Rksamhita* with an incredible accuracy. The additional variants of *krama* recitation if they were in use in Sruti period would have helped further consolidating the task of fixing. The invention of these devices replace the need of writing for the fixing of the long text. Oral fixation in fact even led to fixation of pronunciation of sounds and accents which graphic fixation could never have achieved at that time. The oral integrity of the incantations itself was fixed, and was preserved through the heat and dust of centuries. Most significant in the enterprise of orally fixing long text of *Rksamhita* was the development of the knowledge-body of phonetics which involved construction of the theory of syllablisation, theory of the nature of words and theory of euphony of sounds and

69. News item related to his death and obituary was published in English daily from New Delhi, Patriot, January.

accents, theory of accents, theory of prosody and detailed exhaustive accounts of relation of *samhita, pada* and *krama* reciations. The knowledge-body of phonetic formed a backbone of later celebrated works on linguistics in ancient India. Not only the knowledge-body became a back-bone of future linguistic research but the entire strategy to orally fix *Rgveda-samhita* became a model for fixing other long compositions and compilations. Only a little deviation here or there was required to adopt the *Rksamhita* fixation strategy model for fixing long compositions involving different literary styles. *Samaveda Samhita* required development of the original fixation model to fix also musical aspect to uniquely central to *saman* chants.[70] Similarly *Yajurveda Samhita* required extention of original model to fix even prose style text without the help of theory of metre. Fixation of *Atharvaveda Samhita* hardly required any change from the original model of *Rksamhita* fixation.

3.8. FIXING OF *SAMAVEDA SAMHITA* AND FAILURE TO FIX MUSICAL ASPECT OF *SAMANS*

On the basis of the model of orally fixed *Rksamhita* later even the texts of various other *Samhita*-s were standardised and fixed in the second half of the Sruti phase. *Samaveda Samhita* is almost entirely derived out of *Rksamhita,*[71] out of 1810 stanzas 1735 are borrowed from *Rksamhita,* mostly from Books VIII and IX and not at all from Book X.[72] Even the remaining 75 are derived from *Yajur* and *Atharva Samhita*-s. *Sama-*

70. If we ignore repeated stanzas out of 1549 some 1474 are borrowed from *Rksamhita*. *Samasamhita* is divided into two broad sections, Purvarcika or Arcika (858 verses) and Uttararcika (1225 verses).
71. This fact is often used to argue that the 10th *mandala* of *Rksamhita* is a later addition.
72. Samavedins employed several devices to adopt Rk hymns to music: (a) Vikara (changing pronunciation of Sigh words) (b) Vislesana (splitting the words apart), (c) Vikarsana (inserting lengthened or *pluta* vowels between split-up syllables of a word, (d) Abhyasa (repetition), (e) Virama (splitting the succeeding word and joining its first syllable to the first syllable of the preceding words before introducing a pause between the words), and (f) Stobha (insertion of syllable not presented in the *Rksamhita*). Sukumari Bhattacharji (1984), p. 161.

veda version of Rg hymns was considerably different as the hymns were sung and not flatly recited. *Samaveda* perhaps epitomised standardisation of the practice of singing hymns or rather it represented appropriation of Rg hymns in the musical practices of the time.[73] In the tradition of *Samasamhita* the process of accentuation and tonic aspect of chant became significant and at the same time complex. It perhaps needed additional development in the theory of accent and the musical scales to fix the musical aspect of the chants. Though we have later evidences of the theoretical developments in this direction yet such detailed theory was not developed in the early times. The result is that we hardly know about the *saman* music today,[74] whereas details of accent of *Rksamhita* are so meticulously preserved. Somehow devices used for the retention of *Rksamhita* or rather they were modelled on it *Samasamhita* needed additional devices to fix even musical details other than fixing sound, word-order etc. Why could not the musical richness be also fixed orally in the Sruti phase is a challenging and open problem.

3.9. FIXING *YAJURVEDA SAMHITA* AND INVENTION OF COMPLEXLY KNITTED RECITATION STRATEGY

Yajurveda is a prose composition very different from the Rg-vedic poetic composition. In fact two *Yajurveda Samhita* texts

73. *Naradiya Siksa* gives detailed theory of accent and music for melody. It gives seven tone scale, 21 semitone *sruti*-s in an octava notion of family of melodies. Bharatamuni's *Natyasastra* also deals extensively with theory of music. See for details Swami Prajnanananda, *Historical Development of Indian Music*, Calcutta, 1973.

74. "... for centuries the large bulk of Rgvedic texts were handed down by oral teaching with incredible accuracy, but...there is absolute uncertainty about the intervals of the *Samaveda* chants ...accentuation in the course of time has totally changed its characters...." B. Faddegon, *Studies on the Samaveda*, Amsterdam, 1951, part i p. 11. However, details of ancient chants have been worked on by a Samaveda Brahmin. Laksmana Shankara Bhatta. *The Ancient Mode of Singing Samagana*, Poona, 1939.

are found in various recensions.[75] *Yajurveda* countains many new liturgical compositions and about one-sixth of it is straight away or in mutilated form borrowed from *Rgveda*. Fixing of the prose text in oral tradition is an added strain compared to the fixing of poetic text. The introduction of various variants and deviants of *kramapatha* perhaps are related to the need of uncontaminated relation of prose text. In fact, *Taittiriya Pratisakhya*[76] belonging to *Yajurveda* amply cites Jatapatha which introduces reverse recitations, (see appendix A5) and

75. The two texts are *Krsnayajurveda* and *Suklayajurveda* The various *sakha*-s of *Sukla Yajurveda* of which we know were :—(1) Karva, (2) Madhyandina, (3) Sapeya, (4) Tapayaniya, (5) Kaplia, (6) Paundravatsa, (7) Avatika, (8) Paramavatika, (9) Parasarya, (10) Vaidheya, (11) Vaineya, (12) Audheya, (13) Golava (14) Baijava and (15) Katyayaniya Vajasaneyi Samhita, the text of Suklayajurveda is found today in two recensions, Kanva prevalent in Maharastra and Madhyandina prevalent in north and south India. In this text expositing prose of *Brahmana* is not mixed up with the *mantra*-s which are muttered in the sacrifice. In *Krsnayajurveda* they are mixed. We know some details of its five or six recensions though Samhita of only three have survived completely and one incompletely. These various recensions were : Taittiriya, most famous belonged to U.P., M.P. Rajasthan and north Gujarat ; Maitrayaniya, belonged to west coast in Gujrat and land between Vindhya ranges and Narmada river ; Katha, belonged to Punjab region and its text stands in between black and white Yajurveda ; Kapisthala, belonged to Punjab, Hariyana and is only incompletely available.

76. *Taittiriya Samhita* of black *Yajurveda* and *Vajasaneyi samhita* of white *Yajurveda* keep quoted *rc*-s in their original form. The rest of the recensions largely quote *rc*-s partially. *Taittiriya Pratisakhya* 3.1, 5.33, 8.8, 12, 16, 35, 9.22, 10.9, 10, 13, 11.9, 16, 17, 12.7 and 20.2 into Jatapatha of *Yajurveda*. In fact Kramapatha gets hardly cited (23.20, 24.5, 24.5). According to Whitney who edited this *Pratisakhya* these appear to be later additions. According to the commentators, interpretation rules 8.12, to 35 deal with cases that arise only in Jatapatha. Whitney even remarks that the term *vikrama* (in the sense of Krama-vikrti) signifies Jatapatha in rules 23.20 and 24.5. Whitney concludes that the weight of evidence, upon the whole, is decided in favour of the assumption that the peculiar Jata combinations were had in view.

hence introduces hordes of euphonic combinations not involved in *Samhita* or Kramapatha-s. It is in this *Pratisakhya* that the Jatapatha is first indicated and referred to unlike *Rkpratisakhya, Atharvapratisakhya* or *Rktantra* (of *Samaveda*) where no reference to Jatapatha is found. But most of the Rgvedic investigation into oral devices was centred around metric composition, much of the theoretical apparatus developed for the purpose and the key concepts were related to the metric aspect of recitation. To adopt oral devices of the fixation strategy of *Rgveda* was not possible straight away. The place of pause in the recitation had to be defined separately and then also the group of words called *pada*. This was in fact achieved by the traditions which were responsible for orally fixing prose text of *Yajurveda* by formulating separate rules for the position of pause in the recitation.

Vajasaneyi Pratisakhya of Katyayana belonging to the *Sukla Yajurveda* tradition defines *samhita* as union of letters which are pronounced in one breath.[77] This way what is decided because of metrical reasons in *Rksamhita* is decided somewhat bluntly by the normal length of breath. Separate approach to the place of pause was developed for the recitations of Yajus. With this addition whole of the apparatus of the strategy of orally fixing *Rksamhita* could be adopted. Fixing of *Yajurveda* must have naturally led to more detailed investigation of words and euphony as Jatapatha throws in many new euphonic combinations because of its employing a reverse order. Though added empirical investigation was needed no further theoretical development was entailed because of the attempt to orally fix prose text of *Yajurveda Samhita*. This is amply demonstrated from the fact that *Rkpratisakhya* is most extensive and comprehensive whereas *Vajasaneyi Pratisakhya* and *Taittiriya Pratisakhya* are smaller and largely dependent on it with no significantly new conceptual additions, the only difference being that if *rc*-s were metrically recited, Yajus were muttered.

77. *Vajasaneyi Pratisakhya* 1.158.

3.10. FIXING OF THE TEXT OF *ATHARVAVEDA SAMHITA*

Atharvaveda[78] also repeats many Rgvedic hymns but has a large corpus of its own hymns, charms, spells etc., some of them dating back to remote antiquity. The standardisation of *Atharvavada* is undoubtedly much later than *Rgveda* though it is considerably different in subject-matter from the rest of the three Vedas and perhaps there remained a doubt in early days if it is to be considered as a Vedic *samhita*. Though *Atharvaveda Pratisakhya* is an important work, it does not introduce any radical change in the theoretical apparatus developed by *Rkpratisakhya*. Only distinctive element for fixing *Atharvaveda Samhita* was some of the meaningless charms in it but they posited minor problems of memorising meaningless euphonic formations.

3.11. LINGUISTICS : A UNIQUE FEATURE OF THE EXACT SCIENCE IN INDIA

We can safely conclude that the body of knowledge developed to orally fix *Rksamhita* in the absence of writing proved to be on the whole adequate for fixing other texts orally. This knowledge-body enabled accomplishment of marvellous task of fixing pronunciation (even accent) and recitation (incantation), thus making possible uncontaminated retention through the ages. The model of orally fixing *Rksamhita* remained the dominant model for this tack throughout the Sruti phase. By the end of Sruti phase resurgence in prose literature also did not require a change in this model. Later Sutra style compositions were anyway short and did not posit any significant problem for oral fixation.

From the viewpoint of the history of scientific ideas the most significant is the successful enquiry into various facets of language which the enterprise of orally fixing the text led to. This enquiry led to the laying of foundations of phonetics. The

78. Nine *sakha*-s of *Atharvaveda* are known to have existed : Saunaka, Tanda, Manda, Paippalada, Jajala, Jalade, Cavanavaidya, Brahmavada, and Vedadarsa. *Samhita*-s of two are available today, (1) Saunaka and (2) Paippalada Sakha. The latter's *Samhita* was recently discovered in a manuscript form in some village of Orissa by D. M. Bhattacharya.

developments in phonetics created bases for later morphological analysis of words. Not only full-fledged theory of etymology got founded in the later phase of Sruti period but also syntax of the language became a subject of enquiry. This led to creation of the finest piece of reasoning from the ancient world, that is, the *Astadhyayi* of Panini,[79] which gives virtually an unambiguous grammar of Sanskrit.

What was most significant was that the spirited milieu, for the deep enquiry into the phenomena of language, got created in the later phase of non-literate period in the history of Indian subcontinent precisely because of the need to fix long text without availability of graphic means. And the success of the enquiry into various facets of language made it a unique feature of the history of exact science in India as distinct from the intellectual endeavour of other civilisations.

There is a need to pursue in detail the reconstruction of the content of various aspect of the linguistic knowledge developed in India. As an aid to this endeavour we will give a survey of the literary sources from the Indian past which can help reconstruct the detailed contents of the exact science of language developed in ancient period in India.

4.0. ANCIENT LITERATURE DEALING WITH THE KNOWLEDGE-BODY OF LINGUISTICS

The non-literate phase (*c.* 1500 B.C. to *c.* 500 B.C.) in the history of Indian subcontinent paradoxically witnessed a surge of literary activity or rather was also the phase of spurt in unprecedented literary activity. As we have seen, the kernel of linguistic knowledge (largely phonological) was developed in the period to fix the text of *Rksamhita.* Later, in the same phase, it was further developed to fix even other *Samhita*-s, the Sama, Yajus and Atharva. The *Samhita*-s and their associated literature, which were composed in the period that included various *Brahmana*-s, *Aranyaka*-s and *Upanisads* were

79. The logical rigour of Panini can only be compared with another masterpiece from the ancient world, the *Elements* of Euclid.

all handed down orally to later periods. Even the kernel of linguistic knowledge must have been handed down orally. We do not have a single integrated composition in which linguistic knowledge was codified. In a way it represents complexity of the process of knowledge generation and dispersion.

There are indications that linguistic knowledge related to fixing of pronunciation, recitation etc. did exist before 700 BC in a codified form. Yaska, in *Nirukta,* dated variously between 700 BC-500 BC, indicates existence of many phonetic treatises belonging to different branches of Vedic *sakha*-s (*caranani*).[80] Durgacarya, the 13th century AD commentator of *Nirukta,* equates these treatises with *Pratisakhya*-s.[81] Vishnumitra equates them with *Rkpratisakhya.* An 11th cent. AD commentary on *Rkpratisakhya* by Uvata was called *parsada vyakhya* or simply *Bhasya.*[82] The several medieval commentaries believed that *Pratisakhya*-s belonged to each institution of *sakha* and several others believed that they were common to many *sakha*-s but related to particular *Samhita*-s.[83]

To-day we have extant six *Pratisakhya* texts belonging to four Vedas. These are, in accordance with their relative chro-

80. *Nirukta* 1.17 says 'words are regarded as fundamental in the linguistic treatises of all the groups of Vedic recensions.'
81. Durgacarya in his commentary says that '*parsada*' or *Pratisakhya* were those treatises which dealt with division into words, Samhitapatha, Kramapatha, Padapatha and accent of the text of particular branch of Veda. Visnumitra in his *Vargadvayavrtti* on *Rkpratisakhya,* Sloka 1.
82. Another later commentary is called *Parsadavrtti* which is still unpublished, see preface pp. 47-50, M. D. Shastri. *The Rgveda Pratisakhya,* Vol. 1. Even Visnumitra, another later commentator on the first few *sutra*-s of *Rkpratisakhya* calls Pratisakhya as Parsada.
83. Anantabhatta (16th-17th century)—commenting on *Vajasaneyi Pratisakhya* 1.1 gives etymological exposition of the term as belonging to every Sakha. Bhatta in *Tantra Vartika* 5.13 thinks that Pratisakhya belongs to every *sakha.* Jnanendra Saraswati on the *Siddhantakaumudi,* Panini IV. 3.59 quotes Madhava holding this view.

nology :[84] (1) *Rkpratisakhya*[85] by Saunaka ; it belonged to Saisiriya branch of the extant Sakala-sakha of *Rksamhita ;* this is an only *Pratisakhya* which deals with metrical aspect and is the oldest as well as the longest ; (2) *Taittiriya Pratisakhya*[86] belonged to Taittiriya *sakha* of *Krsna Yajurveda Sam-*

84. S. Varma (1961) p. 12-12 gives several arguments that the *Pratisakhya* belonged to group of Sakhas. The basic arguments are that (1) we do not find as many *Pratisakhya*-s as Sakha-s (2) *Pratisakhya* cites scholars belonging to several Sakhas and (3) rules given in a particular *Pratisakhya* are applicable to many Sakha-s of a particular Veda. Besides he cites Vaidikabharana's commentary on *Taittiriya Pratisakhya* 4.11 in his favour. There it is argued that *Rkpratisakhya* belongs to Sakala as well as Bhaikala Sakha of *Rgveda,* and examples quoted in *Taittiriya Pratisakhya* pp. 184-185 cannot be traced to the extant *Taittiriya Samhita* ; so they must have belonged to extinct Sakha-s of the *Samhita.* The same arguments are advanced by Gopala Yajva in Vaidikabharana as is quoted in Varma (1972) p. 10. See Siddheshwar Varma CSPOIA pp. 20-28 for detailed discussions on chronology. He uses three criteria (1) grammatical terminology, (2) style, and (3) authorities cited.
85. Four commentaries on this *Pratisakhya* are famous : (1) *Parsada Vyakhyana* or *Bhasya* by Uvata written in 11th century AD., This *Bhasya* is included in the copy of *Rkpratisakhya* which we are referring to, (2) *Parsada Vrtti* is intimately related to Uvata's *Bhasya* and has not yet been published, (3) *Vargadvayavrtti* by Visnumitra, which comments on first 10 slokas of *Rkpratisakhya* on which Uvata's commentary is not available. This portion is included in the edition we are referring to. It has been reported that this *Vrtti* on all 18 chapters of *Rkpratisakhya* is available in manuscript form at Deccan College library. This is unpublished, (4) *Bhasya* by Pasupatinatha Sastri. This has been published but is extensively dependent on Uvata's *Bhasya.*
86. Three commentaries on *Taittiriya Pratisakhya* are famous, (1) *Padakrama Sadhana* of Mahiseya is the oldest and smallest ; (2) *Tribhasyaratna* of Somayajva. Claims to have utilised commentaries of Vararuci, Atreya and Mahiseya. It has been included in Whitney's edition of *Taittiriya Pratisakhya* which we have used, (3) *Vaidikabharana* of Gopalayajva is the most recent and refutes many assertions and interpretations of Somayajva's commentary. It is also the largest commentary.
87. It has two famous commentaries (1) Bhasya of Uvata (11th Cent. AD) also called *Matrmoda.* This is the most ex-

hita ; it is an only *Pratisakhya* which deals with Jatapatha ; (3) *Vajasaneyi Pratisakhya*[87] ascribed to Katyayana belonged to the Madhyandina *sakha* of the *Sukla Yajurveda* ; it uses artificial meaningless technical terms not employed in other *Pratisakhya*-s ; (4) *Saunikiya Caturadhyayi* is a *Pratisakhya* of *Atharvaveda* ; it uniquely employs the device of *gana*[88] (or group of words to which rules refer with the first word) to formulate rules free from complications ; (5) *Rktantra* is ascribed to Sakatayana and is a *Pratisakhya* related to the Kauthuma *sakha* of *Atharvaveda* ; (6) *Atharvaveda Pratisakhya*[89] in another treatise belonging to *Atharvaveda.* It is the smallest *Pratisakhya* and employ no technical term and cites only one authority, namely Sakalya.

None of these six *Pratisakhya*-s can be identified with *Parsada* to which Yaska refers, as the oldest among them, *Rkpratisakhya,* cities Yaska.[90] Hence at least in their present form they could not have been codified before Yaska. *Taittiriya Pratisakhya* seems to be earlier than Patanjali's *Mahabhasya* whose date has been accepted as around 150 B.C.[91] The *Tait-*

tensive commentary and is included in V. K. Varma's edition of *Vajasaneyi Pratisakhya* ; (2) *Padarthaprakasa* of Anantabhatta (16th-17th cent. AD). He has also refuted Uvata's interpretations at certain points. Even this commentary is included in V. K. Varma's edition of *Vajasaneyi Pratisakhya.*

88. Instead of listing the words over which the rule is applicable it only designates the set of these words (*gana*) by the first letter of the set followed by etc., '*adi*'. See *Atharva Pratisakhya* 1.34, 65, 85. This method is extensively put to use by Panini for the sake of brevity in his grammer *Astadhyayi. Vajasaneyi Pratisakhya* 5.38 employs it once but no other *Pratisakhya* employs it.

89. This is different from Whitney's edition and translation of *Atharvaveda Pratisakhya* which is Saunakiya Caturadhyayi. This is edited by Vishva Bandhu Shastri and published by Punjab University. This has also been published by Dr. Suryakanta.

90. *Rkpratisakhya* 17.42. 'According to Yaska no one time (*pada*) *rc* other than 10 syllable one is to be found in *Rgveda.*'

91. Whole passage in Patanjali's commentary on Panini 1.2 29-30 dealing with high and low accent has been taken from *Taittiriya Pratisakhya* 22.9-10. See Siddheswar Varma CSPOIG p. 21.

tiriya Pratisakhya[92] is later than Panini as it refers to monotonic recitation as opinioned by predecessors and Panini [1.2. 34] gives this opinion. The date of Panini is variously put between 600 B.C. to 400 B.C. Much of knowledge contained in *Pratisakhya*-s specially related to rules of euphony and phonology is plainly presumed by Panini for constructing his grammar. Absence of any influence of Panini on *Rkpratisakhya* unlike other *Pratisakhya*-s is the reason that most scholars believe *Rkpratisakhya* of Saunaka to be prior to Panini.[93] In that case the codification of *Rkpratisakhya* close to its present form can be dated in between Yaska and Panini, that is, 700 BC-400 BC or 600 BC-500 BC, rest of the *Pratisakhya*-s being codified later than Panini.

Another important fact to note is that *Rkpratisakhya*[94] is a metrical composition unlike prose composition of Yaska's *Nirukta* and Sutra style composition of Panini's *Astadhyayi.* The subject-matter of *Pratisakhya*-s exclusively deals with how to construct *samhita-patha* from *padapatha*[95]. Latter *Taittiriya Pratisakhya* also introduces Jatapatha where Krama (ascend-

92. *Vaidikhabharana* on *Taittiriya Pratisakhya* 15.9 See Siddheswar Varma CSPOIG p. 14.
93. V. K. Varma RPEP p. 89, Indra PEPSAA p. 3, Siddheswar Varma CSPOIG pp. 21-28.
94. *Rkpratisakhya* has 529 stanzas and they are in *anustubh, tristubh* and *jagati* metres. Jagati metre is less compared to the other two, the 11th chapter is wholly composed in Jagati. In 4 stanzas 9-9 syllable lines are there, in one 10-10 syllables line and in another 6-6 syllable lines are found.
    There exists a recension of *Rkpratisakhya* in Sutra style but it is decidedly later. In one such manuscript 1067 sutra-s are there. See V. K. Varma RPEP p. 23. Different styles of *Nirukta, Astadhyayi* and *Rkpratisakhya* show that *Rkpratisakhya* style of metre is older and originated with *Rgveda* itself. Prose style of *Nirukta* originated with *Yajurveda* and the *Astadhyayi* style is contemporaneous with Sutra styles of *Srauta-sutra*-s and could be called ingenious to Panini.
95. It is *Atharvaveda Pratisakhya* which in its first *sutra* formulates this objective of *Pratisakhya*-s : 'of the four kinds of words —viz. noun, verb, preposition and particle—the qualities exhibited in euphonic combination and in the state of disconnected vocables are here made the subject of treatment.'

ing with pairs of words) as well as Vikrama (descending with pairs of words) is dealt with. For this purpose *Pratisakhya*-s give details of syllablisation (*varnoccarana*), details of correct pronunciation and mistakes in pronunciation, details of words, their phonetic behaviour, details of euphony, details of metres (only *Rkpratisakhya*), details of accent, details of handling *kramapatha* and in general, instruction in teaching the Veda. *Pratisakhya*-s do not contain anything other than the phonological knowledge developed for fixing long texts orally. This is most true about *Rkpratisakhya*[96] which deals exclusively with phonological knowledge needed to retain intact Samhitapatha by its construction from Padapatha and Kramapatha. And it deals with nothing more. It is a committed text.

Composition of *Rkpratisakhaya* amounted to fixation of *Rg-veda Samhita* in the form which is available to us today. Its composition amounted to scrutinising each and every syllable of *Rksamhita* for details of its composition, accent, pronunciation, word-construction, line-construction, and metre construction. *Rkpratisakhya* itself must have taken generations to be constructed. It quotes 28 authorities[97] 78 times implying a long heritage of reflections on the subject-matter of the *Pratisakhya.* Many scholarly viewpoints and scholarly opinions and debates of the ancients have been dealt in it. It employs about 281 technical terms[98] in laying down the phonological knowledge-body. Total number of technical terms[99] in all the *Pra-*

96. *Rkpratisakhya* has 3 books with six chapters each with every chapter divided into several *varga*-s having generally five stanzas each. First chapter deals with definitions and rules of interpretation of the text : 2nd and 3rd deal with accents and their euphony ; 1st, 6th, 13th and 14th chapters deal with syllables and pronunciation ; 12th chapter deals with words ; 2nd, 4th, 5th, 7th, 8th and 9th chapters deal with euphony ; 16th, 17th and 18th chapters deal with metres ; 10th and 11th chapters deal with *kramapatha,* and 15th chapter deals with general instructions for learning Veda.
97. P. 957 of the edition of *Rkpratisakhya* gives list of names of authorities quoted.
98. 945—956 of *Rkpratisakhya.*
99. According to counting done by Indra PPPSAA p. 6 out of a total of 302 technical terms, *Rkpratisakhya* employs 281, *Taittiriya Pratisakhya* employs 75, *Vajasaneyi Pratisakhya* employs 81.

tisakhya-s are about 302 implying that most of the conceptual apparatus of the subject-matter of *Pratisakhya*-s is laid down in *Rkpratisakhya.* For the developments of extensive knowledge-body, codified in *Rkpratisakhya,* it must have taken at least two or three centuries which takes back the rudimentary prototype *Rkpratisakhya* to about 1000 BC—900 BC, i.e. about the time when oral fixation of *Rksamhita* is being achieved. Saunaka, to whom authorship of *Rkpratisakhya* is ascribed perhaps was an intellectual lineage rather than a person, which was intimately involved with the generation, retention and codification of the phonological knowledge-body. To Saunaka is also ascribed : (1) *Arsanukramani,* concordance of Veda, (2) *Chandonukramani,* concordance of metres, (3) *Devatanukramani,* concordance of deities, (4) *Anuvakanukramani,* indexing method of Samhita, (5) *Suktanukramani,* another indexing scheme of *Rksamhita,* (6) *Rgvidhana* dealing with arrangement of Rgvedic *mantra*-s or stanzas, (7) *Padavidhana,* dealing with formation and arrangement of lines in the stanzas of *Rksamhita,* (8) *Brhaddevata,* commentary on Rgvedic deities, (9) *Rgveda Pratisakhya,* phonology associated with *Rksamhita,* and (10) *Saunaka Smrti.*

The above-mentioned *Anukramani* texts [concordances (1), (2), (3), (4) and (5)] give detailed structure of *Rksamhita* in terms of its divisions into various sections, the Rsi-s with whom stanzas are associated, metres to which various stanzas are classified, deities with which stanzas are associated and the name and number of stanzas, poems (*sukta*), sections (*anuvaka*) and books (*mandala*). With the help of

Caturadhyayi employs 80, and *Rktantra* employs 56. In *Taittiriya Pratisakhya* 66 terms are same as *Rkprat.* and 9 are new, in *Vajasaneyi Pratisakhya* 73 are from *Rkpratisakhya* and 8 are new. *Caturadhyayi* has 71 same and 9 new, *Rktantra* has 51 same and 5 new. Some of these terms are expicitly defined and their properties stated, but some are employed without explicit elaboration on them. *Rkpratisakhya* defines 245, *Taittiriya Pratisakhya* defines 58, *Vajasaneyi Pratisakhya* 68, *Caturadhyayi* 39, *Rktantra* 44, whereas undefined terms used were in *Rhpratisakhya* 36, in *Taittiriya Pratisakhya* 17, in *Vajasaneyi Pratisakhya* 13 in *Caturadhyayi* 41, in *Rktantra* 12 and in *Atharvapratisakhya* 23.

these the structure of *Rksamhita* was standardised. The Vidhana texts (6) and (7) standardise the text microscopically giving a detailed commentary on stanzas and lines. *Brhaddevata* gives a study on Rgvedic deities and also deals with some theoretical aspects of linguistics. This all goes in favour of Saunaka being an intellectual lineage associated with the redaction of the tradition of oral fixation of *Rksamhita* and the knowledge-body of phonetics evolved for the purpose is contained in Saunaka's *Rkpratisakhya* whose kernel goes back to about 1000-900 BC, though the presently available codification is pre-Panini and post-Yaska.

The *Pratisakhya*-s of *Yajurveda* also seem to have their kernel before Panini and somewhat before Yaska as they seem to be internally related to the oral fixation of prose literature of *Yajurveda* which must have happened between 1000 BC-800 BC. The knowledge-body of phonetics available in the *Pratisakhya* is not self-sufficient as they presume without elaboration many phonological facts. For example, *Pratisakhya*-s do not enumerate sounds but begin the texts with statements like 'the first eight are monopthongs' and 'the next four are dipthongs'.[100] This shows that enumeration of sounds was based on some other text as order of syllables was presumed. Besides, several general aspects of phonetics were not covered by *Pratisakhya*-s. Today there are extant several treatises dealing with these aspects of phonetics in a general way, these are called *Siksa*-s. In fact Siddheswar Varma claims to have known sixtyfive of them and have personally examined fifty of them.[101] But according to his informed opinion, most of them though invariably ascribed to ancient authorship, were codified in their present form in the early medieval period and some as late as 15th century AD.[102] He does not rule out that much

100. *Rkpratisakhya* 1.1 ff.
101. Siddheswar Varma CSPOIG p. 29.
102. Siddheswar Varma CSPOIG pp. 28-54 deals with chronology of Siksa class of literature ; the arguments are based on the internal evidences of the present form of Siksas, which undoubtedly have undergone many changes and permutations. Siddheswar Varma being conscious of this leaves the possibility open that Siksas could have belonged to much earlier a period somewhat contemporareous with *Pratisakhya*-s.

of the knowledge-body contained in *Siksa*-s belonged to the time contemporaneous with the *Pratisakhya-s*. His conclusion —after examining several of them from phonological viewpoint—is that 'many of them (*Siksa*-s) contained a number of very valuable and striking phonetic observations not available in the *Pratisakhya-s*.' *Vajasaneyi-pratisakhya* (1.29) which begins the section on the letters and syllables opens with a *sutra* —'now (the aphorisms) as prescribed by *Siksa*-s".[104] *Taittiriya Upanisad*,[105] *Mundaka Upanisad*[106] and Panini[107] speak of phonetics as *Siksa*. *Siksa* traditionally is regarded as one of the six auxilliary disciplines for the study of Veda-s, which were : (1) *Vyakarana*, grammar, (2) *Siksa*, instructions on pronunciation and phonetics, (3) *Nirukta*, etymology, (4) *Kalpa*, (5) *Chandas*, metrics, and, (6) *Jyotisa*, astronomy. Though this is certainly a later exegetical arrangement, but it reflects popularity of the term *siksa* for designating discipline of phonetics. Even *Rkpratisakhya* calls itself a *siksa* which is completer, not easily available, part of the Vedic study and in accordance with ancient teachings.[108] This it says at a place which deals with mistakes in pronunciation of letters. So the term *siksa* designated on the one hand loosely the discipline of phonetics and on the other a particular class of texts.

Siksa class of texts deal with details of letters and syllables, quantity of syllables, details of accents etc. These issues are only summarily dealt with in the *Pratisakhya* literature. From the study of the knowledge-body presented in *Siksa*-s it can be said that the knowledge-body in *Pratisakhya*-s is in a sense dependent on *Siksa*. As *Atharva Pratisakhya*[109] says, the origination of accent is not seen in Pada or in Samhita. Pada and Samhita being the major botheration of *Pratisakhya* the detailed science of accent, quantity and syllable is left out for

103. Siddheswar Varma CSPOIG p. 28-29.
104. *Vajasaneyi Pratisakhya* 1.29.
105. *Taittiriya Upanisad* 1.2.
106. *Mundaka Upanisad* 1.2.1 speaks of Siksa, Kalpa and Vyakarana (grammar) as three subjects of learning.
107. Panini 4.2.61 enumerates five subjects (1) *krama*, (2) *pada*, (3) *siksa*, (4) *mimamsa* and (5) *saman*.
108. *Rkpratisakhya* 14.68, 69.
109. *Atharvaveda Pratisakhya* 4.109.

29

Siksa. On the other hand, it can be said that *Siksa*-s did not bother about phonetic issues related to *padapatha* etc., in that sense their kernel could even go back to times even before *Pratisakhya.* Out of 65 *Siksa*-s investigated by Siddheswar Varma it could not be established which one is archaic or contemporaneous with *Pratisakhya,* but many a times some mention of phonetic facts leads to the conclusion that it must have been observed in the period of early Sanskrit or the period of proto-*Pratisakhya.*[110] So as far as the knowledge-body of phonetics evolved in the oral phase is concerned *Pratisakhya*-s and *Siksa*-s provide authentic sources. *Rkpratisakhya* being the oldest and most comprehensive is an ideal source. Other *Pratisakhya*-s also add significant insights in reconstructing the knowledge-body, but care is needed as these *Pratisakhya*-s were codified later and some of them in literate period. *Siksa*-s also are sources of many significant insights leading in the reconstruction of the knowledge-body of phonetics. But *Siksa*-s need to be handled carefully and critically because they were codified much later in the literate phase.

Phonetics is not an only aspect of the critical reflection on language caused by an attempt to fix long texts orally in the non-literate phase of history of Indian subcontinent. There was also a tradition of study of meaning of key terms in the *mantra-s.* This tradition of enquiry into the root meaning of words led to development of the theory of etymology. The words were morphologically analysed in terms of roots and suffixes. The meaning of a word was derived from the meaning of the root which itself was made fully clear by analysing verbal use of the root.[111] This tradition matured in the oral phase and got codified in the *Nirukta*[112] of Yaska which is dated 700 BC—500 BC. The *Nirukta*[113] is a commentary on

110. Siddheswar Varma.

111. See Lakshamana Swarup NN pp. 57-71 and Siddheswar Varma's *Etymologies of Yaska.* for the detailed treatment of the theory of etymology employed by Yaska. For its philosophical significance see Navjyoti Singh (1984).

112. For detailed discussion of chronology, see Lakshamana Swarup NM pp. 53-54.

113. There are extant two commentaries on *Nirukta* : (1) commentary of Skandasvami, and (2) commentary of Durgacarya, a thirteenth century A.D. scholar.

the ancient list of key words called *Nighantu*.[114] In the commentary the words are morphologically analysed in accordance with the knowledge of phonetics. This work on etymology is an earliest prose literature on an exact science subject. Its prose style makes it invaluable as far as the wealth of material in it on ancient debates and discussions is concerned. Yaska cites thirty-one authorities prior to him in about 123 number of times and out of which thirteen etymologists of great antiquity can be singled out.[115] The tradition of etymologists did go back into the thick of oral phase. Even *Brahmana* and other Vedic literature have etymologies scattered here and there.[116] But *Nirukta* is an only pivotal authentic source of exact science of etymology in the oral tradition.

As far as the theoretical issues of etymology go, the references from various later grammatical works as well as phonological works like *Pratisakhya* and *Siksa*-s are of help. The *Brhaddevata* has many interesting insights into theoretical issues involved in early debates for recreating ancient theory of etymology.

The two aspects of ancient linguistic efforts, phonology and etymology, in the tail end of Sruti phase gave rise to development of grammar. The development of grammar of Sanskrit culminated in a momentous achievement of Panini, who is dated between 600 BC to 400 BC. Panini's *Astadhyayi* is written in Sutra style where each and every syllable has non-neglectable significance. Order of cryptic rules, *sutra*, and even the order of syllables in it are of significance. The background of the body of knowledge of phonetics was a pre-requisite for the development of *Astadhyayi*. The grammar of Panini

114. It is in five chapters, the first three called Naighantukakanda deal with synonyms, fourth called Nigamakanda deals with homonyms and the last called Daivatakanda deals with deities. The first three chapters have discernible order ; first deals with objects and physical things, second with man, and third with abstract qualities.

115. See index given on p. 247 of English translation of *Nirukta*.

116. Fatah Singh, *The Vedic Etymology. A critical evaluation of the science of Etymology as found in Vedic literature.* Sanskrita Sadan-kota 1952, gives listing of etymologies of 833 words attempted in the Vedic literature.

has a central body of about 4000 rules and has several appendices. These appendices are definitely drawn from the past. The history of roots or *dhatupatha* is certainly a legacy of tradition of etymology. Similarly, *unadipatha* giving lists of affixes played an important role in the constitution of grammar. Several formal and logical devices were developed by Panini to make possible writing of a non-ambiguous grammar of Sanskrit both Vaidika as well as Laukika. Many of these devices were derived from early phonetical literature. Even the appendix, *ganapatha* (list of words undergoing similar grammatical transformation) was modelled on the listing of words given in *Pratisakhya* and *Siksa* literature in the body of rules.

5.0. CONCLUSION

In conclusion we would say that the knowledge-body of linguistics which got developed in Sruti phase had several facets. The three distinct facets were the following : (1) Studies of incantation : This led to development of various dimensions of phonology. These studies were closely related to standardisation and fixing of incantations and recitation. *Pratisakhya* class of literature and *Siksa* class of literature are useful in reconstructing phonological knowledge-body. *Rkpratisakhya* (600-500 BC) of Saunaka is the most significant and important text for the study of this aspect ; (2) Studies in meaning : These studies though evolved in the context of meanings of *mantra*-s and the key words were not related to the incantation aspect directly. These studies banked on Vedic as well as ordinary language. Discovery of the root and affix morphology of the words led to establishment of exact scieice of etymology. Yaska's *Nirukta* (700-500 B.C.), a commentary on *Nighantu* (an ancient list of words) is a central source for the reconstruction of the knowledge-body of etymology, semantics and associated philology. *Brhaddevata* (800 B.C—400B.C.) of Saunaka also helps in the reconstruction ; (3) Studies in syntax : This direction of inquiry developed towords the end of Sruti phase and presupposed the above two directions of inquiry. Or, in other words, it represented development over the shortcoming of the other two traditions of inquiry. The development of the grammar of Sanskrit distanced linguistic studies from the *Samhita*-s and incantation. For grammar ordinary usage as well as Vedic textual usage were significant. Panini's *Astadhyayi* (600 B.C.-400 B.C.) foreshadowed and eclipsed all earlier grammatical inquiries and provides with complete knowledge-body of the grammar of Sanskrit. The three aspects of the knowledge-body cannot be dealt in the space of the chapter.

There is a need to re-create the most fascinating aspect of the knowledge system of linguistics and cover significant ancient debates which punctuated the actual development of the knowledge system.

APPENDICES

CORRECTIONS TO APPENDIX I

1. Page 459, line 3 : square sign to be put at the end of the first "first bracket".

2. Page 461, fig 4, B within the square O′P′SR to be read as B′.

3. Page 464., fig 8 : the letter S to be put on the the right-hand corner of the figure on the left and G to be put in the left-hand upper corner of the figure on the right.

APPENDIX—I

Basic Geometrical Propositions in the Śulva-sūtras

SUBINOY RAY

In the first chapter of his *Śulva-sūtra*, Baudhāyana has enunciated some mathematical propositions mainly geometrical in nature. A comprehensive study of these *sūtra*-s has already been made by eminent scholars like Burk, Thibaut, Dutta and others. Largely following their lead we propose to discuss here the main mathematical contents of the text. We shall first give the methods of construction as mentioned in the text and then add the proofs for these.

1. PROPOSITION 1

Construction of a square the length of whose side is given or specified. Three methods of construction are given by Baudhāyana.

Method I (*BSS* i. 22-28)

Following the directions of the *sūtra*-s the method of construction is as follows :

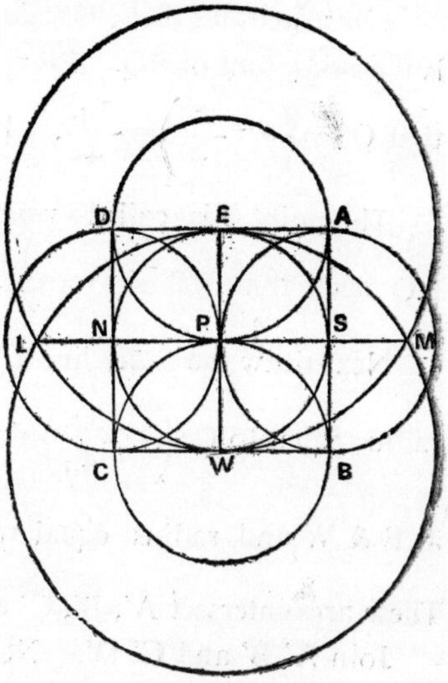

Figure 1

Construction : Let EW=the length of the side of the square. (In Śulva terminology, this line is called the *prācī*, i.e. the line drawn joining the east and west points and going through the centre of the *vedi*). Let P be the middle point of EW. With P as centre and PE (=PW) as radius, draw a circle touching E & W. Next, two circles are drawn with E and W as centres and EW as radius. These two circles intersect at L & M. Join LM. It evidently passes through P and intersects the first circle at S and N. Next, four circles are drawn with centres at E, N, W & S and radius equal to PE ($\frac{1}{2}$ of the length of the side of the square). The four circles intersect at A, B, C & D. ABCD is the required square.

The proof is evident and needs no further elaboration. This is obviously a most ancient method of construction of a square and is purely a geometrical one.

Method II

Another method of construction of a square with given side (*BSS* i. 29-35).

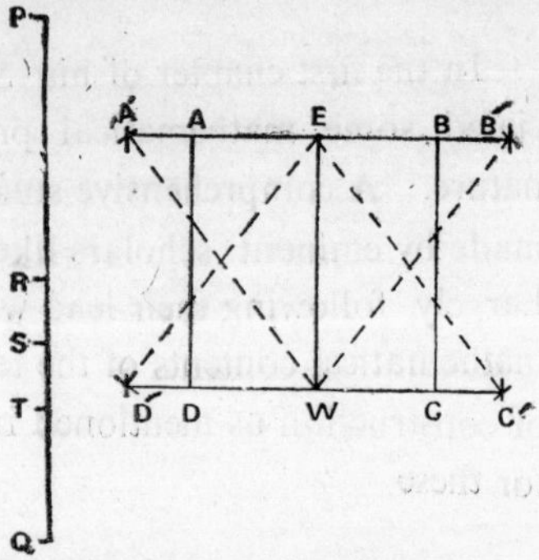

Figures 2A & 2B

Construction : Let PQ$=2a$, a being the side of the square. R is the mid-point of PQ. PR$=$RQ$=a$. S is a point on RQ such that QS$=\left(a-\frac{a}{4}\right)=\frac{3a}{4}$. PS$=\left(a+\frac{a}{4}\right)=\frac{5a}{4}$.

The point S is called *nyañchana* mark. T is another point on RQ such that RT$=$TQ$=\frac{a}{2}$.

Next draw the *prācī* line EW$=a$. With centres at E & W and radius equal to QS$\left(=\frac{3a}{4}\right)$ draw four arcs. Again with centres at E & W and radius equal to PS$\left(=\frac{5a}{4}\right)$ draw four arcs. These arcs intersect A′, B′, C′ & D′.

Join A′ B′ and C′ D′. Next with centres at A′, B′, C′ & D′ and radius equal to ST$\left(\text{RS}=\frac{a}{4}\right)$ draw arcs intersecting the line A′ B′ and C′ D′ at A, B and C, D respectively. A B C D is the required square.

Proof : Here $EA' = EB' = WD' = WC' = \frac{3a}{4}$ and

$$ED' = EC' = WA' = WB' = \frac{5a}{4}.$$

Then $\left(\frac{5a}{4}\right)^2 = \left(\frac{3a}{4}\right)^2 + (a)^2$, *i.e.* $(WB')^2 = (EB')^2 + (EW)^2$.

Therefore, EB′ is perpendicular to EW. Similarly for other triangles. Thus A′B′ passes through E. Similarly C′D′ passes through W.

Therefore $B'B = A'A = C'C = D'D = \frac{a}{4}$ and

$$AE = EB = WC = WD = \left(\frac{3a}{4} - \frac{a}{4}\right) = \frac{a}{2}.$$

In this method of construction of a square Baudhāyana has used the result of what is usually known as the Pythagorean theorem, though as applied to the specific example of the triangle whose sides are 3, 4 & 5.

Method III

Construction of a square with given side (*BSS* i. 42-44)

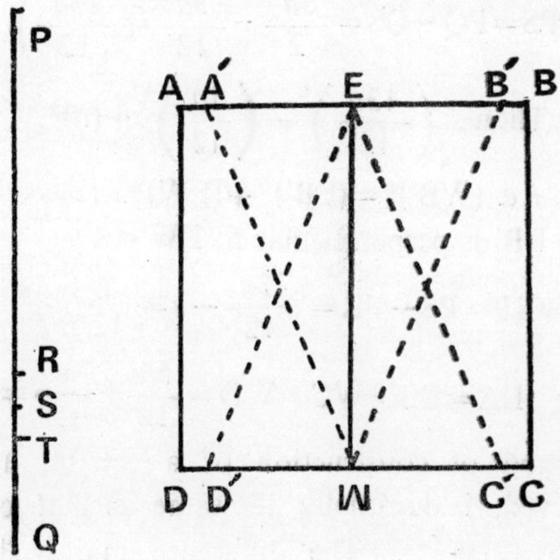

Fig. 3A and 3B

Construction : Let $PQ = a + \frac{a}{2}$, a being the side of the square. R is a point on PQ such that $QR=\frac{1}{3}\left(\frac{3a}{2}\right)=\frac{a}{2}$.

Take another point S on QR such that

$$QS=\left(\frac{a}{2}-\frac{1}{6}\cdot\frac{a}{2}\right)=\frac{a}{2}-\frac{a}{12}=\frac{5a}{12}.$$

The point S is called the *nayñchana* mark. T is another point on QR such that RS=ST. Next draw the *prācī* line $EW=a$. With centres at E & W and radius equal to QS $(=5a/12)$ draw arcs on both sides of E & W. Next with the same centres and radius equal to PS draw arcs intersecting the former arcs at A′, B′, C′ and D′. Join A′ B′ and C′ D′. Next with centres at A′, B′, C′ and D′, and radius equal to ST draw arcs intersecting the lines A′ B′ and C′ D′ produced at A, B, C & D respectively.

ABCD is the required square.

Proof: Here $EA'=EB'=WC'=WD'=QS=\frac{5a}{12}$

and $EC'=ED'=WB'=WA'=PS=\frac{13a}{12}$.

$$\left(PS=PQ-QS=\frac{3a}{2}-\frac{5a}{12}=\frac{13a}{12}\right).$$

Then $\left(\frac{13a}{12}\right)^2=\left(\frac{5a}{12}\right)^2+(a)^2$,

i.e. $(WB')^2=(EB')^2+(EW)^2$.

Therefore EB′ is perpendicular to EW etc.

Also $RS=ST=PS-PR=\frac{13a}{12}-a=\frac{a}{12}$.

Therefore, $EA=EB=WC=WD=\frac{5a}{12}+\frac{a}{12}=\frac{a}{2}$.

This method of construction of a square, as Thibaut has rightly observed, is essentially the same as that of the second method (*BSS* i. 29-35). The difference mainly lies in the fact that Baudhāyana in this method used right-angled triangle whose sides are 5, 12 & 13 while in method II he used the right-angled triangle whose sides are 3, 4 & 5.

2. PROPOSITION 2

To construct a rectangle of given length and breadth. (*BSS* i. 36·40)

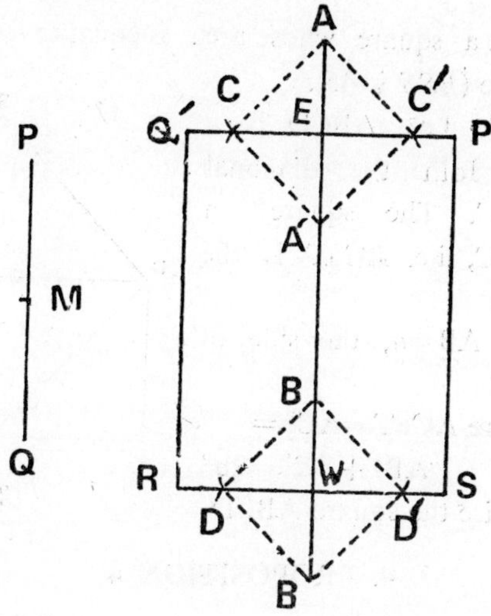

Fig. 4

Construction : First draw the *prācī* line EW$=a$, a being the length of the rectangle. Let PQ$=b$, the breadth of the rectangle. Let M be the middle-point of PQ and let A,A′ and B,B′ be points on both sides of E and W such that EA=EA′=WB =WB′. With centres at A & A′ and radius equal to PM$=\frac{b}{2}$ draw arcs on both sides intersecting at C and C′. Join CC′. Next with centre at E and radius equal to PM$\left(=\frac{b}{2}\right)$draw arcs intersecting the line CC′ at P′ and Q′. Similar cunstructions are to be made with respect to the points B, B′ and W. P′Q′RS is the required rectangle.

Proof : CAA′ is an isosceles triangle, E being the middle point of the base AA′. Therefore, CE is perpendicular to AA′. Similarly C′E is perpendicular to AA′.

Therefore, CEC′ is one straight line ; and EP′=EQ′$=\frac{b}{2}$.

Therefore, $P'Q' = P'E + EQ' = \frac{b}{2} + \frac{b}{2} = b.$

3. PROPOSITION 3

To construct a square whose area is equal to twice the area of a given square (*BSS* i. 45).

Construction : Let ABCD be a given square. Join the diagonal (*dvikaraṇī*) AC. The square on the diagonal AC, i e. ACST is the required square.

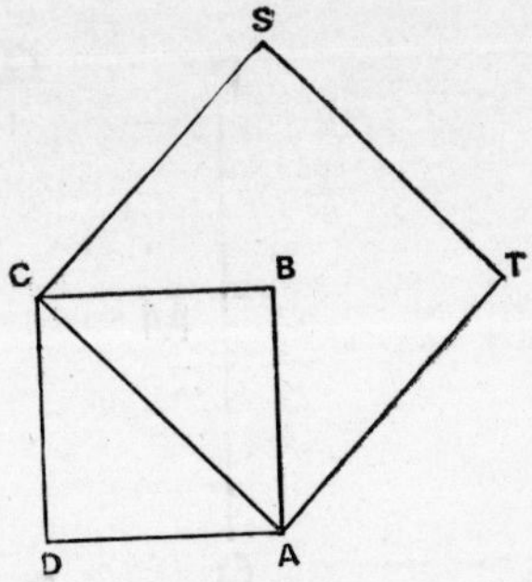

Fig. 5

Proof : Let AB $= a$, the side of the square.

$$\text{Square ACST} = AC^2 = AB^2 + BC^2 = 2a^2$$
$$= \text{twice the square ABCD}$$

4. PROPOSITION 4

To construct a square whose area is three times the area of a given square (*BSS* i. 45).

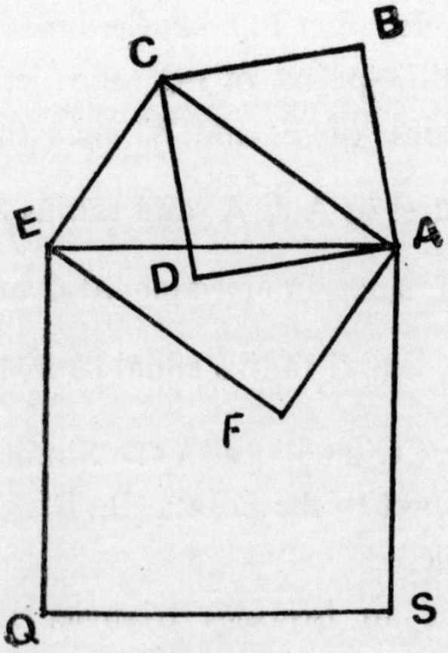

Fig. 6

Construction : Let ABCD be the given square. Join the diagonal AC. With AC as length and AB (side of the square) as breadth, construct a rectangle ACEF. Join the diagonal AE. The square on the diagonal AE, i.e. AEQS, is the required square.

Proof : Let AB=BC=a, the side of the given square.

Therefore its area=a^2, AC=$\sqrt{2}\ a$.

$$\text{Therefore}\quad AE^2=AC^2+EC^2=AC^2+AB^2\quad (AB=AD=EC)$$
$$=2a^2+a^2=3\ a^2.$$

Baudhāyana in his *sūtra* i. 47 further explains that a square whose side is $\sqrt{3}\ a$ [area three times the area of a given square with side a] can be equally divided into nine small squares by drawing lines dividing the sides of the square. The area of such a small square is 1/3rd of the area of the given square of side a which (small square) is also 1/9th of the area of the square whose side is $\sqrt{3}\ a$.

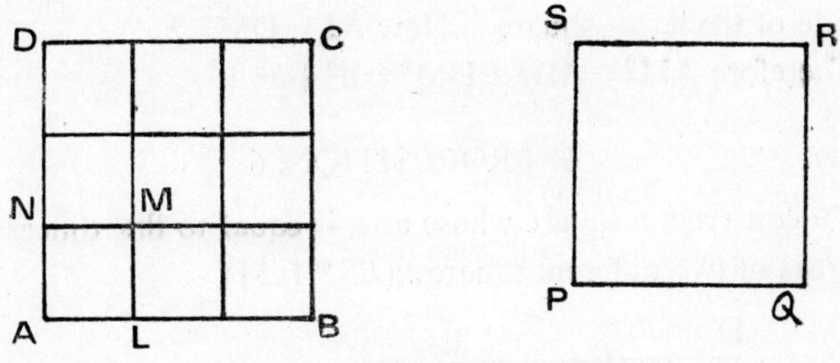

Fig. 7

Let PQRS be a square whose side is a. ABCD is the square whose side is $\sqrt{3}\ a$.

$$\text{Thus, } AB=\sqrt{3}\,a,\ AB^2=3a^2,\ AL=\frac{1}{3}.\sqrt{3}.\ a=\frac{a}{\sqrt{3}}.$$

$$\text{Hence area ALMN}=\frac{a^2}{3}=\frac{1}{9}\ (3a^2).$$

5. PROPOSITION 5

To construct a square whose area is equal to the sum of the areas of two different squares. (*BSS* i. 50)

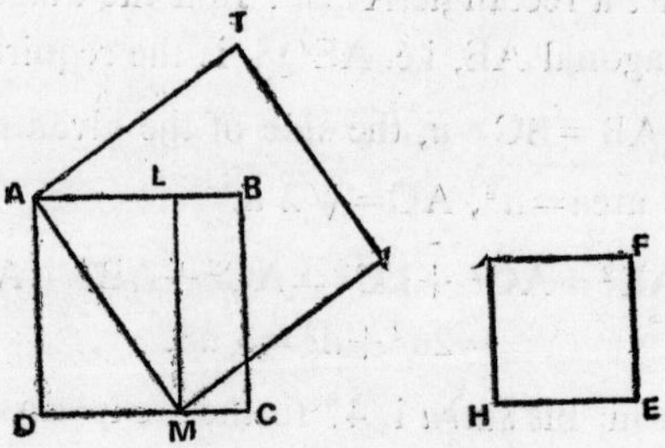

Fig. 8

Construction : Let ABCD be the larger and EFGH the smaller squares. From AB and DC cut off AL and DM respectively, each equal to EF. Join LM. The rectangle ALMD is formed. Join the diagonal AM. The square on the diagonal AM, i.e. the square AMST is the required square.

Proof : Let EF=b, the side of the smaller square and AB=a, the side of the larger square. Now AL=DM=b.

Therefore $AM^2 = AD^2 + DM^2 = a^2 + b^2$

6. PROPOSITION 6

To construct a square whose area is equal to the difference of the areas of two different squares. (*BSS* i. 51)

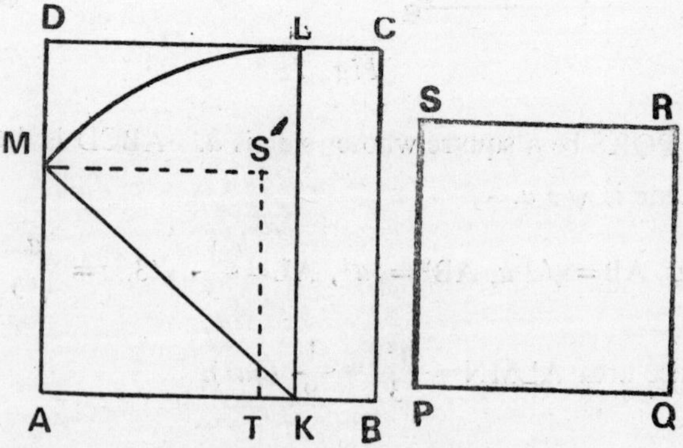

Fig. 9

Construction : Let ABCD be the larger square, and PQRS the smaller one. Let a=side of the smaller square PQRS and b=side of the larger square ABCD. Cut off AK and DL, each equal to a from AB and DC respectively. Join KL. Now the rectangle AKLD is formed whose sides are a and b. With K as centre and radius equal to KL, draw an arc meeting AD at M. Join KM. Then the square on the side AM, i.e. the square AMS′T, is the required square.

Proof : KL=KM=b. In the right-angled triangle AKM, $AM^2 = KM^2 - AK^2 = b^2 - a^2$.

7. PROPOSITION 7

To transform a square into a rectangle (whose length is equal to the diagonal of the square). (*BSS* i. 52)

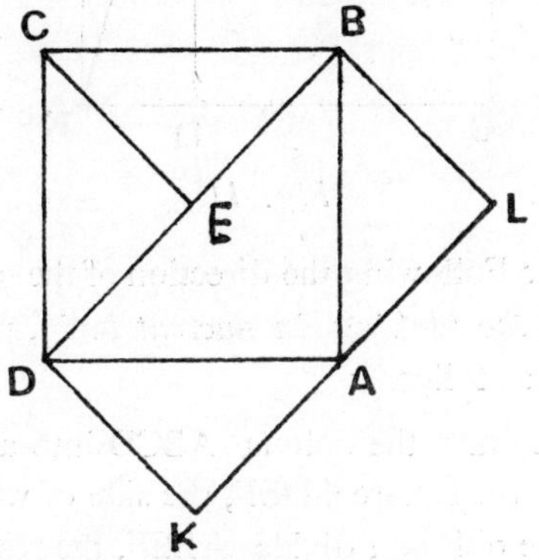

Fig. 10

Construction : Let ABCD be a square. Join the diagonal BD. Divide the right-angled triangle BCD into two small right-angled triangles by drawing a perpendicular from C on the side BD. Next join two triangles, DCE and BCE, to the two other sides AB and AD of the square, their new positions being the triangles ABL and ADK. BLKD is the required rectangle.

Proof : Rectangle BLKD = triangle ABD + triangle ALB + triangle ADK
= triangle ABD + triangle BCE + triangle DCE
= square ABCD.

8. PROPOSITION 8

To draw a square equal in area of a given rectangle (*BSS* i. 54).

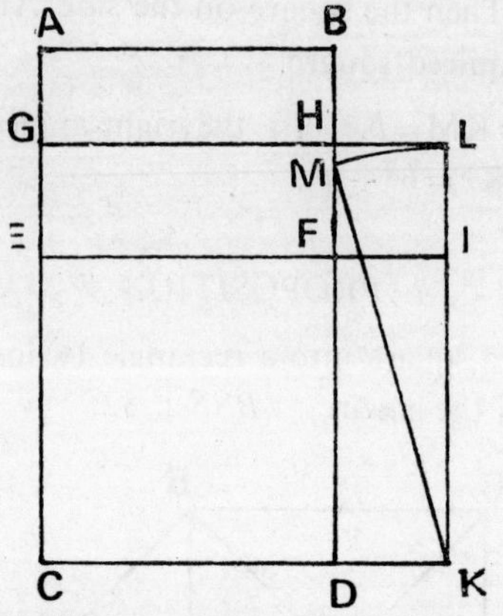

Fig. 11

Construction : Following the direction of the *sūtra*, G. Thibaut (*Mathematics in the Making in Ancient India*, p. 25) has stated the construction as follows :

If you wish to turn the oblong ABCD into a square, cut off from the oblong the square CDEF, the side of which is equal to the breadth of the oblong ; divide ABEF, the rest of the oblong, into two parts, ABGH and GHEF. Take ABGH and place it into the position DFIK ; fill up the empty place in the corner by the small square FHLI ; then deduct by *sama-caturasra-nirhāra* (i.e. according to a special method of deducting one square from another) the small square FHLI from the large square GLKC. The square you get by this deduction will be equal to the oblong ABCD.

Proof : Rectangle ABCD = square CKLG − square FILH.

[By the method of construction as stated in proposition 6 : *BSS* i. 51, draw a square, i.e. with centre at K and radius equal to KL, draw an arc meeting DH at M. Join KM. Square on the side DM is the required square.]

$DM^2 = KM^2 - KD^2 = KL^2 - HL^2$
$=$ Square CKLG $-$ Square FILH
$=$ Rectangle ABCD.

9. PROPOSITION 9

To draw an isosceles trapezium of given side equal in area of a given square or a rectangle, the given side of the trapezium being less than the side of the square (*BSS* i. 55).

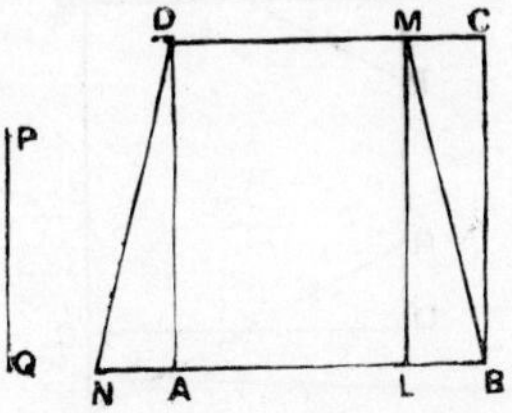

Fig. 12

Construction : Let ABCD be a given square. Let PQ be the given side of the trapezium (PQ being less than AB). With centers at A and D and radius equal to PQ, draw two arcs intersecting AB at L and DC at M. Join LM. Join also the diagonal MB of the rectangle LBCM. Then take the right-angled triangle MCB and place it after inverting on the side AD, the new position of the triangle MCB being triangle AND. DMBN is the required trapezium.

Proof : Triangle LMB = triangle MCB = triangle AND.

Now, the trapezium NBMD = triangle AND + rectangle ALMD + triangle BLM
= triangle MBC + rectangle ALMD + triangle BLM
= Square ABCD

We may mention here that in the context of constructing a falcon-shaped altar, a method of transformation of a square (or a rectangle) into a trapezium is already suggested in *Śatapatha Brāhmaṇa.* (x. 2.1.4), which reads as follows :

"He contracts (the right wing) inside on both sides by just four finger-breadths *(aṅgula)* and expands it outside on both sides by four finger-breadths : he thus expands it by just as much as he contracts it ; and thus, indeed, he neither exceeds (its proper size) nor does he make it too small. In the same way in regard to the tail, and in the same way in regard to the left wing."

Mathematically, it may be explained as follows :

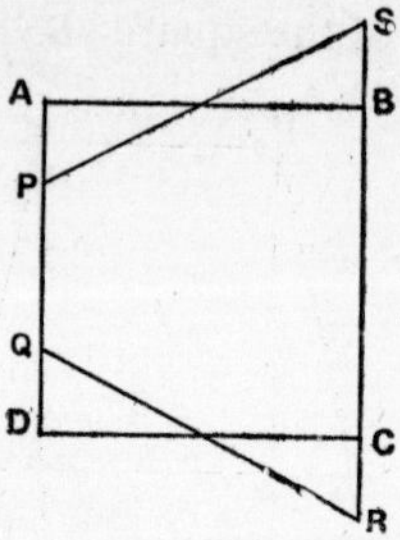

Fig. 13

Let the right wing be in the form of a rectangle ABCD.

Take AP=DQ=4 *aṅgulas.* Produce BC both sides such that BS=CR=4 *aṅgulas.* Join PS & QR. PQRS is the trapezium whose area is equal to the area of the giv n rectangle ABCD.

(See Eggeling *ŚB* iii. 419 [*SBE* 49] for his diagram)

10. PROPOSITION 10

To construct a triangle *(prauga)* whose area is equal to the area of a given square (or rectangle). (*BSS* i. 56).

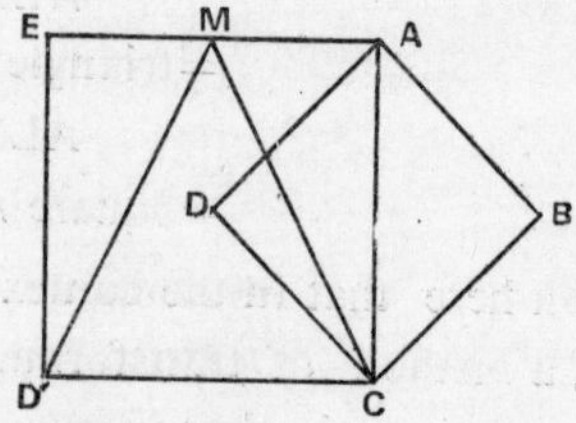

Fig. 14

Construction : Let ABCD be a square with side a. Join the diagonal AC. With AC ($=\sqrt{2}a$) as side, a square ACD′E is constructed, whose area is evidently twice the area of the given square. Considering AE as the eastern side, let M be its middle point. Join MD′ and MC. The triangle MD′C is the required one.

Proof : Area of the square ABCD $= a^2$

Now, $AC = \sqrt{2}\, a$.

The height of the triangle $MD'C = \sqrt{2}\, a$.

Area of the triangle $MD'C = \frac{1}{2} . \sqrt{2}a . \sqrt{2}a$

$= a^2$.

Note : If instead of a square a rectangle be given, first transform the rectangle into a square by the method described in *sūtra BSS* i. 54. Next follow the method described above.

11. PROPOSITION 11

To construct a double *prauga* (double triangle, i.e. a figure formed by two triangles joined with their bases, or, in other words, a rhombus) whose area is equal to the area of a given square (or a rectangle) (*BSS* i. 57).

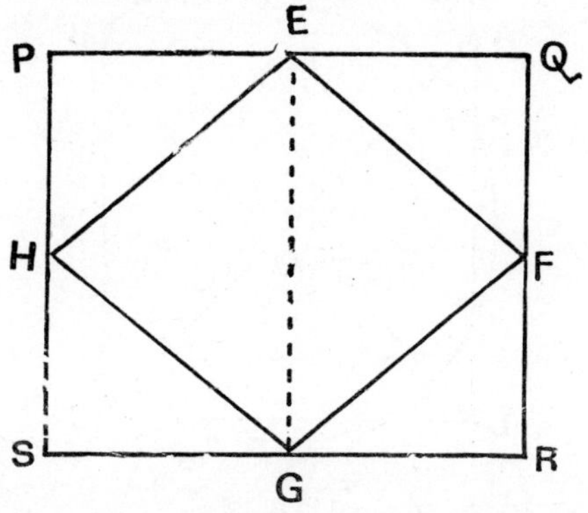

Fig. 15

Construction : Let ABCD be a given square with side *a*. First construct a square whose area is twice the area of the given square by the method described in *sūtra BSS* i. 45. Next transform this square into a rectangle by the method described in *BSS* i. 52. Let PQRS be the transformed rectangle, whose area is $2a^2$. Let PQ be the eastern side. Let E, F, G and H be the middle points of the sides PQ, QR, RS and SP respectively. Join EF, FG, GH and HE. EFGH is the required rhombus.

Proof : Rhombus EFGH=triangle EHG+triangle GHE.

$=\frac{1}{2}$[Rectangle EQRG+Rectangle EGSP].

$=\frac{1}{2}$ Rectangle PQRS$=\frac{1}{2}.\ 2a^2=a^2$.

Note : If instead of a square a rectangle is given, first construct a rectangle whose area is twice the area of the given rectangle by placing the two rectangles side by side. Next follow the above method.

12. PROPOSITION 12

To construct a circle equal in area of a given square (*BSS* i. 58).

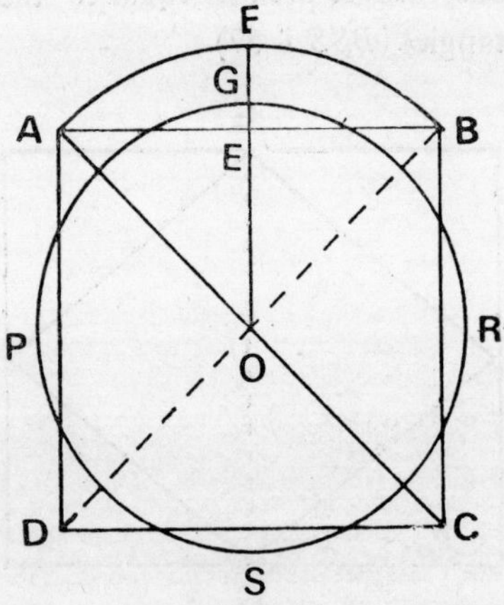

Fig. 16

Construction : Let ABCD be a given square with side *a*. Join the diagonal AC. Let O be the middle point of AC, with O as centre and OA as radius draw an arc AFB. Let E be the middle point of AB. Join OE and produce it to meet the arc at F. Take EG=1/3 EF. Next with O as centre and OG as radius draw a circle PGRS which is the required one.

Let us find the area of the circle :

Radius of the circle $r = OG = OE + EG$

$$= \tfrac{1}{2}a + \tfrac{1}{3}\,EF$$

$$= \tfrac{1}{2}a + \tfrac{1}{3}\,(OF - OE)$$

$$= \frac{1}{2}\,a + \frac{1}{3}\left(\frac{\sqrt{2}}{2}a - \frac{1}{2}\,a\right)$$

$$[OF = OA = \tfrac{1}{2}\,AC = \tfrac{1}{2}\,\sqrt{2}\,a]$$

$$= a.\left(\frac{1}{2} + \frac{1}{3\sqrt{2}} - \frac{1}{6}\right)$$

$$= a.\left(\frac{3\sqrt{2} + 2 - \sqrt{2}}{6\sqrt{2}}\right)$$

$$= a\,\frac{2(\sqrt{2}+1)}{6\sqrt{2}}$$

$$= a\,\frac{(\sqrt{2}+1)}{3\sqrt{2}}$$

If we take the present day value of $\sqrt{2} = 1{\cdot}4142135...$ and $\pi = \frac{22}{7} = 3{\cdot}142857...$

the area of the circle becomes

$$\pi\, r^2 = \frac{22}{7}.\ a^2\left(\frac{\sqrt{2}+1}{3\sqrt{2}}\right)^2$$

$$= a^2.\frac{22}{7}\left(\frac{3+2\sqrt{2}}{18}\right)$$

$$= a^2.\frac{11}{63}\,(3 + 2\times 1{\cdot}4142135...)$$

$$= a^2.\frac{11}{63}\,(3 + 2{\cdot}828427...)$$

$$= a^2.\frac{11}{63} \times 5{\cdot}828427...$$

$$= a^2 \times \frac{64{\cdot}112697\,..}{63}$$

$$= a^2 \times 1{\cdot}017661857...,$$

which is rather a close approximation to a^2—the area of the given square.

From this, the value of π may be ascertained as follows.

Area of the square = Area of the circle

$$i.e.\ a^2 = \pi r^2 = \pi a^2 \left(\frac{\sqrt{2}+1}{3\sqrt{2}}\right)^2$$

$$\therefore \quad \pi = \frac{18}{3+2\sqrt{2}} = \frac{18}{5{\cdot}828427} = 3{\cdot}08831182,$$

which is not very accurate in comparison with the modern value of $\pi = 3{\cdot}142857$.

13. PROPOSITION 13

To transform a circle into a square (*BSS* i. 59).

Following the direction of the *sūtra* we find that if d be the diameter of the given circle and a the side of the transformed square, then :

$$a = \frac{7d}{8} + \left[\frac{d}{8} - \left\{\frac{28d}{8.29} + \left(\frac{d}{8.29.6} - \frac{d}{8.29.6.8}\right)\right\}\right]$$

$$= d\left[\frac{7}{8} + \left(\frac{1}{8} - \frac{28}{8.29}\right) - \frac{1}{8.29.6} + \frac{1}{8.29.6.8}\right]$$

$$= d\left[\frac{7}{8} + \frac{1}{8.29} - \frac{1}{8.29.6} + \frac{1}{8.29.6.8}\right]$$

$$= d\left[\frac{7.29.6.8 + 6.8 - 8 + 1}{8{\cdot}29.6.8}\right]$$

$$= d\left[\frac{9744 + 48 - 7}{11136}\right]$$

$$= d \times \frac{9785}{11136}$$

$$= d \times 0{\cdot}87868\ldots$$

Now the area of the given circle is $\pi\left(\frac{d}{2}\right) = \frac{\pi}{4} d^2$.

Considering the modern value of $\pi = 3.142857$, we get the area of the circle $= \frac{\pi}{4} d^2 = \frac{22}{7} \cdot \frac{1}{4} \cdot d^2$

$$= \frac{11}{14} \cdot d^2 = d^2 \times 0{\cdot}7857.$$

Also the area of the transformed square $=a^2$

$$=(0{\cdot}87868...)^2\times d^2$$
$$=d^2\times 0{\cdot}772078...$$

From this also we can find a possible value of π. Thus :

$$\frac{\pi}{4}d^2=d^2\times 0{\cdot}772078$$

$$\therefore\quad \pi=4\times 0{\cdot}772078=3{\cdot}088321$$

It is interesting to note that in both the cases the value of π is more or less the same. So considering these identical values of π and this *sūtra* (*BSS* i. 59) very closely, Thibaut has rightly suggested that it is nothing but the reverse of the rule for turning a square into a circle (*BSS* i. 58).

14. PROPOSITION 14

Alternative method of turning a circle into a square (*BSS* i. 60).

According to the *sūtra*, if ***d*** be the diameter of the given circle and *a* be the *gross* side of the transformed circle, then

$$a=d-\frac{2}{15}d=d.\frac{13}{15}$$

Let us find the value of π in this case :

Area of the given circle $=\pi\frac{d^2}{4}$

Area of the transformed square $=a^2=d^2\times\left(\frac{13}{15}\right)^2$

$$\therefore\quad \pi\frac{d^2}{4}=d^2.\left(\frac{13}{15}\right)^2$$

$$\therefore\quad \pi=4\times\frac{169}{225}=4\times 0{\cdot}7511=3{\cdot}0044$$

This value of π is evidently less accurate than the previous value of π obtained from *sūtra*-s i. 58 and 59.

The rationale of this method has been obtained by B.B. Datta in his *Science of the Śulbas*, p. 146 as follows :

That is to say $2a = d - \frac{2}{15}.d$

or, $a = r - \frac{2}{15}.r$

Datta has taken d=diameter of the given circle

r=radius of the given circle

$2a$=side of the transformed square

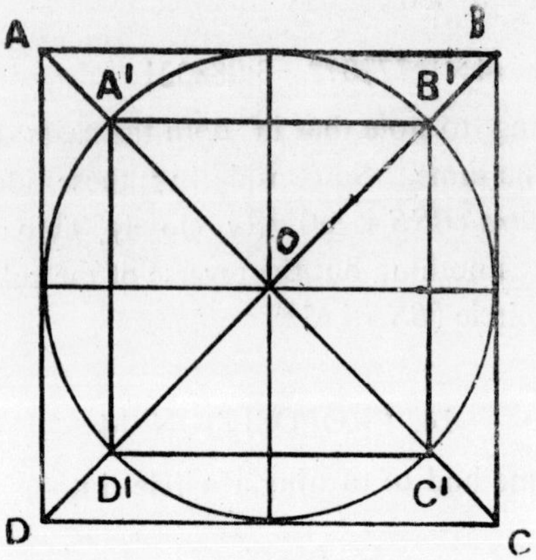

Fig. 17

The rationale of this formula seems to be :

Draw the square ABCD circumscribing the circle and draw also the square A′B′C′D′ inscribed within it. Then apparently the area of the circle will be smaller than the area of the square ABCD ($=4r^2$) and greater than the area of the square A′B′C′D′ ($=2r^2$), that is, $4r^2 >$ area of the circle $> 2r^2$.

An obvious, approximation will be :

$$\text{Area of the circle} = \frac{4r^2 + 2r^2}{2} = 3r^2$$

If $2a$ denote a side of the square equivalent in area of the circle, we shall have approximately

$$4a^2 = 3r^2 \text{ or } a = \frac{\sqrt{3}}{2}.\ r$$

But it will be shown later on (Datta pp. 194-5)

$\sqrt{3} = 1 + \frac{2}{3} + \frac{1}{15} = \frac{26}{15}$ upto the second order of approximation.

Therefore, $a=\frac{13}{15}r.$

$$=r-\frac{2}{15}r.$$

$$[\text{i.e. } 2a=d-\tfrac{2}{15}d]$$

15. APPROXIMATE VALUE OF $\sqrt{2}$ *(BSS. i. 61)*

Baudhāyana's *sūtra* i. 61 states that if d be the side of a square whose area is twice the area of a square with side a then,

$$d=a+\frac{a}{3}+\frac{a}{3.4}-\frac{a}{3.4.34}.$$

$$\text{or, } \frac{d}{a}=1+\frac{1}{3}+\frac{1}{3.4}-\frac{1}{3.4.34}.$$

Now, it has already been stated in *sūtra* i. 45, that the square on the diagonal of a given square with side a will produce twice the area of the given square. Thus this rule gives us a relation between the diagonal and the side of a square. If we take 1 unit as the side of the square, we get,

$$\sqrt{2}=1+\frac{1}{3}+\frac{1}{3.4}-\frac{1}{3.4.34}$$

$$=\frac{3.4.34+4.34+34-1}{3.4.34}$$

$$=\frac{408+136+33}{408}$$

$$=\frac{577}{408}=1.414215686.$$

This value of $\sqrt{2}$ is a very close approximation to the modern value of $\sqrt{2}=1{\cdot}414213562...$

From this it may not be out of place to mention that the ancient Indian mathematicians calculated the value of $\sqrt{2}$ to a remarkable degree of accuracy; but they did not mention the method by which this result was obtained. Both Thibaut and Datta gave two independent hypotheses by which this formula to calculate the value of $\sqrt{2}$ has been arrived at by the Śulvakāras.

1. *G. Thibaut :*

"The question arises : how did Baudhāyana or Āpastamba or whoever may have the merit of the first investigation, find this value ? Certainly they were not able to extract the square root of 2 to six places of decimals ; if they had been able to do so, they would have arrived at a still greater degree of accuracy. I suppose they arrived at their result by the following method which accounts for the exact degree of accuracy they reached.

"Endeavouring to discover a square the side and diagonal of which might be expressed in integral numbers they began by assuming 2 as the measure of a square's side. Squaring 2 and doubling the result they got the square of the diagonal, in this case=8. Then they tried to arrange 8, let us say again, 8 peobles, in a square ; as we should say, they tried to extract the square root of 8. Being unsuccessful in this attempt, they tried the next number, taking 3 for the side of a square ; but 18 yielded a square root no more than 8 had done. They proceeded in consequence to 4, 5 etc. Undoubtedly they arrived soon at the conclusion that they would never find exactly what they wanted, and had to be contented with an approximation. The object was now to single out a case in which the number expressing the square of the diagonal approached as closely as possible to a real square number. I subjoin a list in which the numbers in the first column express the side of the squares which they subsequently tried, those in the second column the square of the diagonal, those in the third the nearest square number.

1	2	1	11	242	256
2	8	9	12	288	289
3	18	16	13	338	324
4	32	36	14	392	400
5	50	49	15	450	441
6	72	64	16	512	529
7	98	100	17	578	576
8	128	121	18	648	625
9	162	169	19	722	729
10	200	196	20	800	784

"How far the *sūtrakāra*-s went in their experiments we are of course unable to say ; the list upto 20 suffices for our purposes.

"Three cases occur in which the number expressing the square of the diagonal of a square differs only by 1 from a square number ; 8-9 ; 50-49 ; 288-289 ; the last case being the most favourable, as it involves the largest numbers. The diagonal of the square, the side of which was equal to twelve, was very little shorter than seventeen ($\sqrt{289}=17$). Would it then not be possible to reduce 17 in such a way as to render the square of the reduced number equal or almost equal to 288 ?

"Suppose they drew a square the side of which was 17 *pada*-s long, and divided it into $17 \times 17 = 289$ small squares. If the side of the square could now be shortened by so much, that its area would contain not 289 but only 288 such small squares, then the measure of the side would be the exact measure of the diagonal of the square, the side of which is equal to 12 ($12^2+12^2=288$). When the side of the square is shortened a little, the consequence is that from two sides of the square a stripe is cut off ; therefore a piece of that length had to be cut off from the side that the area of the two stripes would be equal to one of the 289 small squares. Now, as the square is composed of 17×17 squares, one of the two stripes cuts off a part of 17 small squares and the other likewise of 17, both together of 34 and since these 34 cut-off pieces are to be equal to one of the squares, the length of the piece to be cut off from the side is fixed thereby : it must be the 34th part of the side of one of the 289 small squares.

"The 34th part of 34 small squares being cut off, one whole small square would be cut off and the area of the large square reduced exactly to 288 small squares ; if it were not for one unavoidable circumstance. The two stripes which are cut off from two sides of the squares, let us say the east side and the south side, intersect or overlap each other in the south-east corner and the consequence is, that from the small square in that corner not 2/34 are cut off, but only $\frac{2}{34}-\frac{1}{34.34}$. Thence the error in the determination of the value of the *saviśeṣa* (a ıatio : diagonal of a square divided by the side of the same square $=\sqrt{2}$). When the side of a square was reduced from 17 to $16\frac{33}{34}$ the area of the square of that reduced side was not 288, but $288+\frac{1}{34.34}$ or putting it in a different way : taking 12 for the side of a square, dividing each of the 12 parts into 34 parts (altogether 408) and dividing the square into the corresponding small squares, we get $408 \times 408 = 166464$. This doubled is 332928. Then taking the *saviśeṣa*-value of $16\frac{33}{34}$ for the diagonal and dividing the square of the diagonal into the small squares just described, we get $577 \times 577 = 332929$ each small squares. The difference is slight enough.

"The relation of $16\frac{33}{34}$ to 12 was finally generalized into the rule : increase a measure by its third, this third by its own fourth less the thirty-fourth part of this fourth

$$\left(16\frac{33}{34}=12+\frac{12}{3}+\frac{12}{3.4}-\frac{12}{3.4.34}\right).$$

"The examale of the *saviśeṣa* given by commentators is indeed $16\frac{33}{34}$: 12 ; the case recommended itself by being the first in which the third part of a number and the fourth part of the third part were both whole numbers." (G. Thibaut, *Mathematics in the Making* etc. pp. 18-21).

2. *Datta.*

"But a more simple and very plausible hypothesis will be that the expression for $\sqrt{2}$ was obtained in the following way. Take two squares whose sides are of unit length. Divide the second square into three equal strips.

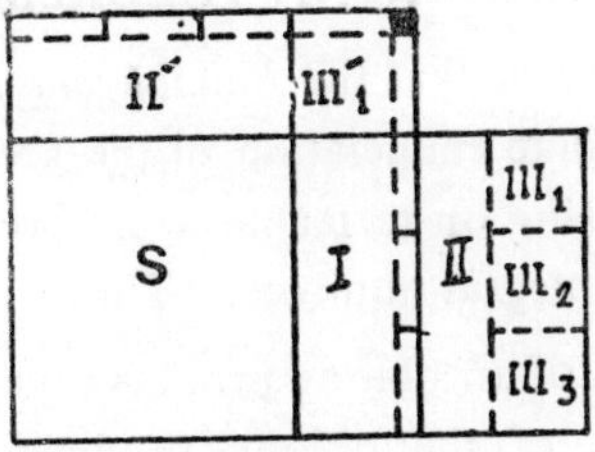

Fig. 18

"I, II and III. Sub-divide the last strip into three small squares III_1, III_2, III_3 of sides 1/3 each. Then on placing II and III_1 about the first square S in the positions II′ and III'_1 a new square will be formed. Now divide each of the portions III_2 and III_3 into four equal strips. Placing four and four of them about the square just formed, on its east and south sides, say, and introducing a small square at the south-east corner, a larger square will be formed, each side of which will be obviously equal to $1+\frac{1}{3}+\frac{1}{3.4}$.

"Now this square is clearly larger than the two original squares by an amount $\left(\frac{1}{3.4}\right)^2$, the area of the small square introduced at the corner.

"So to get equivalence, cut off from the either sides of the former square two thin strips. If x be the breadth of each thin strip, we must have $2x\left(1+\frac{1}{3}+\frac{1}{3.4}\right)-x^2=\left(\frac{1}{3.4}\right)^2$.

"Whence, neglecting x^2 as too small, we get

$$x=\left(\frac{1}{3.4}\right)^2 \cdot \frac{3.4}{34}=\frac{1}{3.4.34}.$$

Thus we have finally $\sqrt{2}=1+\frac{1}{3}+\frac{1}{3.4}-\frac{1}{3.4.34}$ nearly."

(Datta, *Science of the Śulbas*, pp. 192-194).

16. THE PYTHAGOREAN THEOREM SO-CALLED

It seems remarkable that the *Baudhāyana Śulva-sūtra* (i. 48-9) seems to give a definite enunciation of the so-called Pythagorean triangle before passing on to mention a few cases of this, the sides of which are integral numbers. Thus :

"The diagonal of an oblong produces by itself both the areas which the two sides of the oblong produce separately (i.e. the square of the diagonal is equal to the sum of the squares of the two sides. (i. 48) "This is seen in those oblongs the sides of which are 3 & 4, 12 & 5, 15 & 8, 7 & 24, 12 & 35, 15 & 36. (i.49)". (Tr. Thibaut).

APPENDIX II

Some Observations Relating to the Lunar Asterism Kṛttikā

RAMATOSH SARKAR

Śatapatha Brāhmaṇa, II. 1.2.1 to II. 1.2.5, as expounded by Sāyaṇa and rendered into English by Eggeling, reads as follows :

"He may set up the two fires under the Kṛttikās ; for they, the Kṛttikās, are doubtless Agni's asterism, so that if he sets up his fires under Agni's asterism, (he will bring about) a correspondence (between his fires and the asterism) : for this reason he may set up his fires under the Kṛttikās.

"Moreover, the other lunar asterisms (consist of), one, two, three, or four (stars), so that the Kṛttikās are the most numerous (of asterisms) : hence he thereby obtains an abundance. For this reason he may set up his fires under the Kṛttikās.

"And again, they do not move away from the eastern quarter, whilst the other asterisms do move from the eastern quarter. Thus his (two fires) are established in the eastern quarter : for this reason he may set up his fires under the Kṛttikās.

"On the other hand (it is argued) why he should not set up the fires under the Kṛttikās. Originally, namely, the latter were the wives of the Bears (ṛkṣa) ; for the seven Ṛṣis were in former times called the Ṛkṣas (bears). They were, however, precluded from intercourse (with their husbands), for the latter, the seven Ṛṣis, rise in the north, and they (the Kṛttikās) in the east. Now it is a misfortune for one to be precluded from intercourse (with his wife) he should therefore not set up his fires under the Kṛttikās, lest he should thereby be precluded from intercourse.

"But he may nevertheless set up (his fire under the Kṛttikās) ; for Agni doubtless is their mate, and it is with Agni that they have intercourse : for this reason he may set up (the fire under the Kṛttikās)."

It would appear that the text is to some extent elliptical—with one or more words needed here and there to complete the construction of sentences or their meanings. The probable words, that were left out, have been suggested within parentheses by Eggeling possibly following Sāyaṇa ; but even with them, the passage does not admit of easy or straightforward comprehension.

That is perhaps partly due to our ignorance of the rites referred to and their purpose. Partly it is due to the myths or allegories interwoven into the text.

Yet some astronomical statements seem to stand out clearly from the "theological twaddle", and examining them in the light of modern astronomical knowledge, we can ascribe some meanings to them and thence arrive at some conclusions without much doubt.

Following the time-honoured practice initiated by outstanding Indologists, we can take the Kṛttikās to mean *Eta* Tauri and the stars in its immediate vicinity, forming the 'open cluster' Pleiades. It has six stars visible to the naked eye without much effort and some more under conditions of good visibility. That explains its being described as 'the most numerous'.

The seven Ṛṣis are without doubt stars that are even today, all over India, referred to as 'Saptarṣi Maṇḍala'. They are the stars of Ursa Major designated by *Alpha*, *Beta*, *Gamma*, *Delta*, *Epsilon*, *Zeta* and *Eta*. These stars are now all very much to the north of the stars of Kṛttikā—so much in fact that the slow rate at which the relevant coordinates are known to undergo change cannot lead to the possibility that the relative positions were quantitatively different even several thousand years ago. To be somewhat more specific, *Alpha* Ursae Majoris and *Eta* Tauri have now—in 1985—declinations of about 61° 49′ 46″ and 24° 3′ 40″ respectively and the difference is now dwindling at the rate of about 30″ per year ; the celestial latitude of the former is now about 49°40′46″ and that of the latter about 4°2′58″ and the difference is decreasing at the negligible rate of about 0″·200 annually. In other words, just as the stars of Saptarṣi are all to the north of Kṛttikās today, so were they several thousand years ago. That explains the statement : "...the seven Ṛṣis rise in the north, and they (the Kṛttikās) in the east."

But some explanation is still called for. What may be plausibly implied by saying, about some stars, that they rise in the east ? All stars rise in the east or, in other words, seem to do so (due to earth's rotation from west to east), unless they are 'circumpolar' (in which case, the question of rising does not arise at all). Why then single out Kṛttikās ?

The only answer seems to be that here, with reference to Kṛttikās, the word 'east' has not been used loosely to mean just any part of the horizon stretching from the 'north point' to the 'south point' *via* the 'east point', but precisely the 'east point' itself or points very close thereto. In fact, in the present context it should have the second alternative meaning, because Kṛttikās refer to a star-cluster and not just a single star.

But does any of the Kṛttikās—say, *Eta* Tauri—really rise at the east point ? No, it does not today. Did it do at any time in the past ? Let us see.

The point on the 'celestial sphere' called the 'vernal equinox' or anything situated there has necessarily to rise exactly at the east point. The distance from the vernal equinox measured along the 'ecliptic' eastwards is 'celestial longitude' and the vernal equinox itself is slowly moving in the opposite direction (due to what is known as 'precession of the equinoxes'). As a result, the celestial longitude of any star (barring the sun whose behaviour-pattern is different) keeps on systematically increasing (till it attains the highest permissible value of 360°, when the value is considered to be 0° once again and the whole cycle starts afresh). The rate of increase is always small but it is not uniform—it slightly varies. In 1900 A.D. the rate was 50″·256 per year. In 2000 A.D. the figure would be 50″·279 and in 2000 B.C. it was 49″·390. Let us take the mean of these two last mentioned values i.e., for our historical purpose, let the rate be 49″·835 per year. Now, in 1985, the celestial longitude of *Eta* Tauri is about 59°47′24″. But 59°47′·24″÷49″·835=4319·1331=4319 approximately. Therefore we conclude that about 4319 years back from now, its celestial longitude must have been zero and therefore it would more or less coincide with the vernal equinox. So it could rise at the east point or almost there.

The expressions 'could', 'more or less', and 'almost', used here are necessary, but they pose no serious problem. They are used to admit the fact that, even when the celestial longitude vanishes, the celestial latitude may not. In other words, even when a star is neither ahead of nor behind the vernal equinox in the eastward journey across the sky, it may still be on the northern side or the southern side. In fact, for *Eta* Tauri, the celestial latitude is now about 4°2′58″ and because this co-ordinate can change only very minutely for any star (for *Eta* Tauri it is now increasing at the rate of only about 0″·377 per year, i.e. about 1° per 9549 years), it could have been only very slightly different for *Eta* Tauri 4319 years back i e. in 2334 B.C. In other words, (supposing, as one may, the rate of change not to have varied significantly) in 2334 B.C. *Eta* Tauri would rise not exactly at the east point but at a point a little to its north—say, 3° to 4° away.

A deviation of 3°—4° is arguably so small as to have been ignored by the ancient observers. Another way out (a better one, if one chooses to be a stickler for precision) is provided by the fact that Kṛttikās comprise several stars, of which *Eta* Tauri happens to be just one. Some other star of the group, more to the south than *Eta* Tauri—e.g. Merope—might fit the description more exactly, and there is nothing in the text to indicate that it was *Eta* Tauri that was particularly chosen for consideration and not any other star.

The upshot of all this is that *circa* 2334 B.C., 'Kṛttikās rise in the east' was a particularly significant statement to make—more significant, in fact, than at any other earlier time in the history of human civilization. [As the cycle due to precession of the equinoxes takes about 26,000 years to complete, the previous occurrences of the phenomenon under consideration took place around (2334+26,000) B.C., (2334+52,000) B.C. and so on].

References

1. S B E
2 *The Indian Astronomical Ephemeris for the year* 1985 :
Published by India Meteorological Dept., Govt. of India ; 1985.
3. *Report of the Calendar Reform Committee* ;
Published by the Council of Scientific and Industrial Research, Govt. of India ; 1955.

APPENDIX III

Some Observations Relating to the Longest Day of the Year

RAMATOSH SARKAR

Vedāṅga Jyotiṣa, the oldest Sanskrit composition relating specifically to astronomy that Indologists have come to know about, is not generally considered to be a very commendable work. It is very concise and, by and large, quite obscure too. Even knowledgable scholars find it almost unintelligible or even meaningless in many places. Nevertheless, there are a few statements in it that are not difficult to understand and that can be straightway put to test and examined for hidden information —direct or indirect.

Vedāṅga Jyotiṣa indicates in verse 8 that the greatest duration of day-time is 18 *muhūrta*-s, and the shortest is 12. Let us take this statement for scrutiny.

As is well-known, we have alternation of day and night because of the rotation of our earth. The earth—which with a very fair measure of accuracy may be taken as a sphere—is turning about one of its diameters, called its 'axis, and by dint of its gravitational pull forces us to take part in that motion. So we are also moving and, in the process, we are sometimes turned more or less towards the sun and sometimes away from it. In the latter case, the body of the earth stands in the way of our viewing the sun : we cannot possibly see the sun and it is night-time. When no such obstruction is created by the body of the earth, it is day-time. Now, the rate at which the earth is rotating is practically a constant : the time taken for a full turn with reference to the sun, called a 'solar day', is therefore almost of constant length. It consists of 24 (solar) hours, each hour comprising 60 (solar) minutes and each minute in its turn comprises 60 (solar) seconds. (The term 'solar' is used to distinguish from another set of similar quantities—'sideral day', 'sidereal hour' etc.

Since we will not have occasions here to use the other set, we shall henceforth be simply using the words 'day', 'hour' etc without the qualifying adjective 'solar'.)

From what has been said it will be clear why the sum-total of day-time and night-time should always be 24 hours. One might expect each of these parts to have the same duration. But one knows from experience that it is not so.

The fact of the matter is that, as the earth is rotating about its axis, it is also revolving round the sun. The path it follows in the course of revolution lies on a plane—the 'ecliptic plane'. If the axis of rotation had been perpendicular to the ecliptic plane, day and night would indeed have been of equal duration all over the earth, throughout the 'year' (i.e. throughout the time taken by the earth to complete a turn round the sun). But it so happens that the axis is inclined to the perpendicular at an angle whose measure slightly fluctuates at a slow rate and has a mean value of about 23° 17'·5 (the extreme values being about 21°59' and 24° 36'). As a result, during a certain period in the year, the northern half of the earth is a little bent towards the sun and therefore enjoys a little more of the solar energy, and some months later it is a little tilted away from the sun and consequently gets a smaller share of sun's rays. The peaks of these periods are called the 'northern solstice' and the 'southern solstice'. During these two periods, the effects are just opposite in the southern hemisphere of the earth. In between the solstices, we have two 'equinoxes'—the 'vernal equinox' and the 'autumnal equinox'—when day and night are of equal duration throughout the earth.

Now, how much out of 24 hours would be day-time (or, for that matter, night-time) depends on the day of the year. In the northern hemisphere, the largest quota for day-time comes exactly on the northern solsticial day and in the southern hemisphere, it is just the opposite. The quantum is furthermore dictated by the place of observation, i.e. where on the surface of the earth the observer is situated—how much away he is from the earth's equator or mid-region. In other words, it depends on the observer's geographical latitude. For instance, in high northern latitudes (of about $+68°$ or more) day-time may extend to the whole of 24 hours and similarly extreme southern latitudes (of about $-68°$ or less) may experience 24 hours of night-time at a stretch. On the equator (i.e. on latitude 0°), however, day and night cannot but be of equal duration.

Vedāṅga Jyotiṣa uses *muhūrta* to measure time and indicates that the longest day-time is 18 *muhūrta*-s in length and the shortest is 12 ; a *muhūrta*, means the 30th part of a day, i.e. the 4/5th part of an hour or 48 minutes. *Vedāṅga Jyotiṣa* unquestionably refers to the northern hemisphere. But which place exactly, or even roughly, it relates to we do not know. The *Vedāṅga Jyotiṣa* is silent on the point. This silence does not redound to the credit of its composer or composers. Probably it betrays ignorance of the fact that the phenomenon cannot have global significance—it is only local in character. But this shortcoming, whatever its cause, is helpful to Indologists—it gives some clue in the matter of geographical identification. In fact, it clearly indicates (but for possible variation in longitude) the place on the earth's surface that must have been associated with the observation recorded in the *Vedāṅga Jyotiṣa* in some way. To be more explicit, it specifies the latitude of the place—not necessarily the place where the composition of *Vedāṅga Jyotiṣa* was undertaken but at least the place whence the relevant observation (as to the length of the day-time) was made.

Let us examine the issue from different angles by admitting different possible meanings of 'sun-rise' and 'sun-set'. (A little knowledge of spherical astronomy will be pre-supposed).

I. *Without correction for refractional error*

Ordinarily, with reference to any heavenly body, it is said to 'rise' when it just shows up wholly on the eastern horizon ; it is said to 'set' when it has just dropped wholly out of sight, going beneath the western horizon

If the 'declination' of the heavenly body is *delta* and the latitude of the place of observation is *phi* (both measured northwards), then (by calculating its 'hour-angle' both at the time of its rise and set) one can write for its duration of stay above the horizon in terms of hours, the expression $\frac{2}{15^\circ}\cos^{-1}(-\tan\phi\tan\delta)^\circ$, where the angle denoted by the inverse circular function is assumed to be an angle in degrees. Now, in the case of any star of the night-sky, this expression will more or less hold good throughout the year. For, the declination of a star during a year remains almost the same. In the case of the sun, however, the declination changes from day to day—it either waxes or wanes albeit slowly : the sun's declination systematically fluctuates during the course of a year. For the longest day-time in the northern hemisphere, we must put, in the mathematical expression just cited, the greatest possible value of *delta*, i.e. the sun's declination on the day of the northern solstice. This declination now-a-days measures about 23°27′ ; but this extreme value of solar declination is again subject to some slow periodic change—pendulating from 21°59′ to 24°36′.

(a) *Greatest solar declination=23°27′*

Since we do not know for certain when the observations might have been made and therefore what precisely was the value of sun's extreme northern declination obtaining then, we may first of all, as a test case, without any prejudice whatsoever, calculate on the basis of the current value i.e. 23°27′.

In this case, since 18 *muhūrta*-s=14·4 hours and 15° correspond to an hour (*vide* any book on spherical astronomy), we can write $\frac{2}{15^\circ}\cos^{-1}(-\tan\phi\ \tan 23^\circ 27')^\circ = 14{\cdot}4$ (taking the angle expressed in terms of $\cos^{-1}$ in degree measure)

$$\text{or, } -\tan\phi\tan 23^\circ 27' = \cos 108^\circ$$

$$\text{or, } \tan\phi = \frac{30902}{43378} = {\cdot}7124$$

$$\text{or, } \phi^\circ = 35{\cdot}5^\circ \text{ nearly......(1a).}$$

N.B. As a matter of curiosity, one may be tempted to make use of the other bit of information supplied side by side in *Vedāṅga Jyotiṣa*, viz. that the shortest day is 12 *muhūrta*-s, i.e. 9·6 hours in length. But the value of *phi* that results from that approach turns out to be the same (as indeed it should be). For, in this case, the least declination for the sun is to be taken, which of necessity would be $-23°27'$. Then the relevant equation is :

$$\frac{2}{15°}\cos^{-1}(\tan 23°27' \tan\phi)° = 9{\cdot}6$$

$$\text{or, } \tan 23°27' \tan\phi = \cos 72°$$

$$\text{or, } \tan\phi = \frac{30902}{43378} = {\cdot}7124 \text{ or, } \phi° = 35°{\cdot}5 \text{ nearly.}$$

(b) *Greatest solar declination=23°17′·5*

We next take the greatest solar declination to be numerically equal to 23°17′·5 which perhaps is the safest value to take while calculating for a year whose calendrical position (i.e. which year A.D. or B.C. it is) is not definitely known. Because, 23°17′·5 is the mean of the different values that the quantity may possibly have.

Here the equation becomes

$$\frac{2}{15°}\{\cos^{-1}(-\tan\phi \tan 23°17'{\cdot}5)\}° = 14{\cdot}4$$

$$\text{which yields } \phi° = 35°{\cdot}66 \text{ nearly.}$$

II. *With correction for refractional error*

The preceding arguments and calculations are open to some objections. The question of refraction has been by-passed.

Our earth is surrounded by a rather dense layer of atmosphere extending upto a certain height. As a result, light from the heavenly bodies, unless they are directly overhead, get inwardly bent before reaching our eyes and therefore we see them at some elevations above their actual positions. The effect is greatest near the horizon—amounting to about 34′. In consequence of horizontal refraction, a heavenly body is seen just on the horizon when in reality it is 34′ below it. It applies equally to both the occasions of rising and setting. So the period of time, during which the heavenly body is seen, is a little longer than what it should be because of the two additions at the beginning and at the end of its journey across the sky above.

What has just been said, is true of course for the sun as well. Therefore, if the day-time appears to be h hours say, then the real day-time (measured from the actual sun-rise to actual sun-set) would be sometime less. Modern astronomers know this and they accordingly modify their statements. But it would be the height of improbability to assume that the ancient people knew about this error in their astronomical observations. Therefore, it would be in the fitness of things to subject the observations of *Vedāṅga Jyotiṣa* to refractional correction.

The equation in this case assumes the form

$$\frac{2}{15^{\circ}}\{\cos^{-1}(-\tan\phi\ \tan\delta)\}^{\circ}+2\times\frac{34}{60}\times\frac{1}{15}\times\frac{1}{\sqrt{\cos(\phi+\delta)\ \cos(\phi-\delta)}}=14{\cdot}4.$$

Let us, as before, first put 23°27′ for *delta* in this equation and try to solve it for *phi*.

The solution as obtained by means of computers turns out to be $\phi^{\circ}=34^{\circ}{\cdot}3$ nearly (2a).

If on the other hand we make use of the mean value of *delta*, *viz*. 23°17′·5 the computer gives $\phi^{\circ}=34^{\circ}{\cdot}5$ nearly (2b).

III. *Taking the sun's upper limb for the sun itself*

There is another question to be considered. The sun, unlike a star of the nocturnal sky, does not appear to be a point of light. Therefore, the question of its rising or setting is a little tricky. The sun looks like a disc whose angular diameter is about 32′. Naturally, therefore, the whole of the solar body does not appear on the horizon or drop below it simultaneously : sun-rise or sun-set is not an instantaneous phenomenon.

What, then, people ordinarily do ? There are two distinct procedures. One is to choose any specific part of the sun. Usually, it is the centre of the solar disc that is chosen. When this chosen part of the sun rises or sets, the sun (as a whole) is considered to have risen or set. The other alternative is to choose what is called the 'upper limb' of the sun and treat it as representative of the sun. But in this case, the upper limb does not signify the same part of the sun throughout : in the eastern sky the point of the sun first to appear on the horizon is referred to as the upper limb, while in the west the last point to disappear is given that appellation. It would immediately appear that it is rather an improper stand to take ; but it so happens that it is by far the more popular stand : to people in general, the day begins as soon as a bit of the sun is seen ; as long as the last bit of it is seen, the day is still on.

If we adopt the first procedure, which in fact is the more rational of the two, we are *de facto* treating the sun as a star of the night-sky and therefore the equations and their solutions of II hold *in toto.*

If on the other hand we assume that the observers of *Vedāṅga Jyotiṣa* took the usual or popular view, then to them the day-time got unduly lengthened. If we want to correct this error, our equation becomes $\frac{2}{15^\circ}\{\cos^{-1}(-\tan\phi\ \tan\delta)\}^\circ +$

$$2\times\frac{34+16}{15}\times\frac{1}{60}\times\frac{1}{\sqrt{\cos(\phi+\delta)\cos(\phi-\delta)}}=14{\cdot}4.$$

Replacing *delta* by 23°27′ and then solving this eqnation for *phi* by means of computers, one gets $\phi^\circ=33^\circ{\cdot}8$ nearly (3a).

If, on the other hand, *delta* is equated with 23°17′·5, the computer-solution turns out to be $\phi^\circ=34^\circ$ nearly. (3b).

IV. *Admitting correction for dip of horizon*

There still remains one point to consider. It is the altitude i.e. the height from the sea level of the observer. This affects observations – only fractionally for small heights, but considerably for larger heights. In other words, depending on the height, the sun is seen by the observer even when it is some distance below his proper horizon. As a result of this also daytime gets stretched at both ends. And, for strict accuracy, correction in this regard is also called for.

The equation in this case is $\frac{2}{15^\circ}\{\cos^{-1}(-\tan\phi\ \tan\delta)\}^\circ+2\times$

$$\frac{34+16}{15} \times \frac{1}{60} \times \frac{1}{\sqrt{\cos(\phi+\delta)\cos(\phi-\delta)}} +$$

$$\frac{2}{15} \times \frac{1}{60} \times \frac{1}{60} \times \frac{\sqrt{h} \times 60}{\sqrt{\cos(\phi+\delta)\cos(\phi-\delta)}} = 14{\cdot}4.$$

where h is the height of the place of the observer (above sea-level) in feet.

This equation has also been solved by means of computers. The different values that have been assigned to h are 300, 600, 1050, 1500, 2250, 3000, 4500, 6000, 9000, 12,000 ft. successively. The choice of these values has not been made entirely arbitrarily. It has been dirctated by the high probability that the place could be somewhere in the northern or north-western parts of India or a region nearby. Referring to a dependable atlas, in its relief maps, the aforesaid altitudes have been selected.

If *delta* is taken to be 23°27′ the equation yields, through computers $\phi°=33°{\cdot}2$, 32°·9, 32°·7, 32°·5, 32°·2, 31°·9, 31°·5, 31°·1, 30°·6 and 30° nearly. ...(4a).

If *delta* be identified with 23°17′·5, one gets for the same set of alternative values of h, $\phi°=33°{\cdot}4$, 33°·1, 32°·9, 32°·6, 32°·4, 32°·1, 31°·7, 31°·3, 30°·7 and 30°·2 nearly...(4b).

Since there exists no clue whatsoever to guide in the matter of choice of the value of h, the author (of this appendix), even though he took pains to obtain them, is sceptical about the usefulness of solutions (4a) and (4b). He has also his doubts about (1a) and (1b), because he believes that the ancient observers could not have had the knowledge of refraction and therefore to that extent their observations, as recorded, should be regarded as faulty.

But, in the opinion of the present author, solutions (2a)—(2b) and (3a)—(3b) are likely to prove useful to researchers who are trying to locate the sites associated (directly or indirectly) with the observation recorded in the *Vedāṅga Jyotiṣa*.

The author is indebted to A. Bandyopadhyay, Director, Positional Astronomy Centre (Indian Meteorological Department), New Alipore, Calcutta and to Sanjay Biswas, Associate Professor, Department of Mechanical Engineering, Indian Institute of Science, Bangalore, for the help that he received in getting some of the values of *phi* through the computers at their disposal.

References :

1. *Vedāṅga Jyotiṣa.*
2. *Text-Book on Spherical Astronomy* by W.M. Smart ; Published by Vikas Publishing House Pvt. Ltd., New Delhi, Bombay etc ; 1979 (Reprinted by permission of Cambridge University Press).
3. *Report of the Calendar Reform Committee* ; Published by the Council of Scientific and Industrial Research, Govt. of India ; 1955.
4. *The Readers' Digest Great World Atlas* ; 1962.

APPENDIX IV

The Asterisms

APURBA KUMAR CHAKRAVARTY

A system of 27 asterisms, called *nakṣatras*, plays an important role in Indian astronomy and calendrical science. The present convention is that the ecliptic is divided into 27 equal parts, each 13° 20′ long, from one initial point. The arcs are called *nakṣatras*. Again, a bright or prominent star of each such arc is called *yogatārā* bearing the name of that arc : e.g. Kṛttikā as a *nakṣatra* means an arc division and as a *yogatārā* it means the star *Eta* Tauri. There are three exceptions : the *yogatārās* Uttara-Āṣāḍhā, Śravaṇā and Dhaniṣṭhā lie marginally outside the respective arc divisions.

The origin of this asterism system can be traced to an early period. The names of these *nakṣatras* appear in Vedic literature ; but there these names denote stars—either single star or star-groups and not arc divisions. The *Taittirīya*, *Kāṭhaka* and *Maitrāyaṇī Saṃhitās*, the 19th Book of the *Atharvaveda*—each has given a list of 27 *nakṣatras* (the last two have included an additional star, Abhijit, or α Lyrae making a total of 28 *nakṣatras*). All these lists always begin with the name Kṛttikās (the star group Pleiades), and the order of names of the *nakṣatras* are more or less the same. Some of the names of the *nakṣatras* changed in different hands, but in all texts they are single star or star-groups.

In some cases physical descriptions of the stars are given from which they can be identified, and some of these descriptions have astronomical significance which we shall now consider.

The Kṛttikās are described as consisting of seven stars (*Taitt. Saṃhitā* iv.4.5.1), many stars (*Śatapatha Brāhmaṇa* ii. 1.2.1-4) and accordingly the Kṛttikās have been identified with the

Pleiades. The *Śatapatha Br.* further states that the Kṛttikās rise in the east and the seven sages (the Great Bear) rise in the north.

Let us take this statement on its face value. If the Kṛttikās do not shift from the east at rising, it must be on the equator. We suppose that at the time of observation of this phenomenon, the central star of Pleiades, *Eta* Tauri had zero declination so that the Kṛttikās were seen to rise in the east.

Now, in 1967, the latitude of *Eta* Tauri=4°2′51″ N, (unaffected by precession) and its longitude=59° 31′53″. Also, obliquity epsilon=24° 1′ 34″ in 3000 B.C. and the average rate of precession between 3000 B.C. and 1967 AD=49″·7 per year.

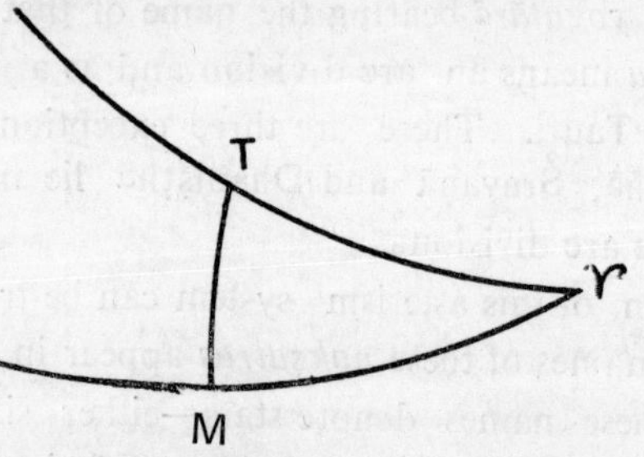

In the diagram, let γ be the vernal equinox, T be position of *Eta* Tauri on equator, TM is secondary to the ecliptic such that TM=latitude (N) of *Eta* Tauri=4° 2′ 51″ (proper motion of the star and other variations being neglected as they will not substantially affect our result). From triangle TMγ, sin γM $=\frac{\tan TM}{\tan \varepsilon}=\cdot 15871$ or $\gamma M=9°8'$. Hence total precession till 1967=59°31′53″+9°8′=68°39′53″ or approximately 68°40′. At the average rate of precession, the epoch comes to about 3000 B.C., far before the generally accepted date of beginnings of Vedic civilisation in India. Hence we are inclined to believe that the tradition of Kṛttikās rising in the east is based on observation earlier than the birth of Vedic literature. But somehow this tradition was handed down to the compilers of *Brāhmaṇas* and *Saṃhitās*, who also recorded it without any verification. Our guess is that the original source from where this tradition reached the Vedic literature was held in so high esteem that nobody questioned the validity or truth of the statement, i.e. its authority was beyond any question and the tradition was recorded as a token of respect to the authority of the source. This means that the authors of the Vedic literature were indifferent to actual observation, though this does not explain why the tradition became so authoritative for them.

There is, of course, a second view. In this opinion, the Kṛttikās have been named first in all these lists because of some preferential reason, and the reason is that the Kṛttikās at that age were on the equinoctial colure i.e. vernal equinox was at the Kṛttikās. Of course there is no express or explicit statement to this effect in Vedic literature ; a few verses can only be pressed to make out such a meaning. However, proceeding as before and taking *Eta* Tauri on the equinoctial colure, we get the epoch at nearly 2400 B.C., which is also earlier than Vedic period. But in that case, its shift from the east point at rising becomes so pronounced (about 5 degrees) that it cannot escape any sky-watcher's notice. We cannot correlate the statement "Kṛttikās rise in the east" and the supposition that the Kṛttikās were at the vernal equinox. If there was any preferential reason for naming Kṛttikās first in the list, it may be that the Kṛttikās rise in the east or that it contains many members or any such specific property of the asterism.

Bentley has given another interpretation of this statement about Kṛttikās, which we shall consider later. We shall see that there are indeed some evidences, but not stated in terms of Kṛttikās, which relate the astronomical observations to the epoch 2400 B.C.

As traditionally considered a "limb of Vedic studies" *(vedāṅga)*, comes down to us a small text, the Vedāṅga Jyotiṣa. Therein the winter solstice has been placed at Dhaniṣṭhā (*Beta* Delphini) which refers to the epoch 1400 B.C. But the period of actual composition of the text is generally viewed by the modern scholars as *circa* 5th or 6th century B.C.

In this text, the *nakṣatras* are arc divisions only and they are no longer stars or star-groups. The lunar zodiac is divided into 27 equal parts commencing from Dhaniṣṭhā each 13° 20′ long. Lunar positions, *ayanas* etc. are always given in terms of fractions of these divisions. *Nakṣatras* lost their original meaning of stars. In this new scheme of *nakṣatras*, vernal equinox falls at 10° of the Bharaṇī *nakṣatra* arc, i.e. 3° 20′ to the west of the beginning of the Kṛttikā *nakṣatra* arc.

Bentley's assumption is that at the time of *Śatapatha Brāhmaṇa*, vernal equinox coincided with this beginning point of Kṛttikā arc division, and since then, upto the epoch of *Vedāṅga Jyotiṣa*, there has been a precession of 3° 20′ which occurs in 240 years. Accordingly, Bentley arrives at 1700 B.C. as the date of composition of the *Śat. Br.*

But there are difficulties in accepting this view. The main purport of the text, that the Kṛttikās rise in the east, does not hold true in this assumption ; besides *nakṣatras* are stars and not arc divisions in the *Brāhmaṇas*. We are nowhere told in Vedic literature the specific purpose for which the asterism system was developed or devised, but there are indications that measures of year, month and fixation of auspicious dates and dates for ritual performance were related to many of these stars. Expressions like "Citrā full moon day", or "day of Phalgunīs" etc. indicate a relationship of day with one or other stars. Although the exact purport of "Citrā full moon" is not clear, it may mean the day when the full moon disc is visible near the star Citrā (Spica) in the night sky or the full moon day in the month of Caitra ; yet an attempt of calendarisation of the asterisms is discernible here. It is not known whether the system was devised as a framework for calendar or once the system was ready at hand, its services were used for calendrical purposes. But, undoubtedly, the calendrical scheme was based on this system of asterisms.

Again, the *Maitrī. Up.* (VI. 14) says that the sun's southern course starts from the beginning of Maghā and ends in the middle of Śraviṣṭhā. Very clearly, Maghā and Śraviṣṭhā (=Dhaniṣṭhā) here are arc divisions and not stars. Now, in the Vedāṅga

Jyotiṣa scale, summer solstice falls at middle of Aśleṣā, and this point is at 4° east of the star *Epsilon* Hydrae. Also from middle of Aśleṣā to the beginning of Maghā it is 6° 40′. Assuming this 6° 40′ to be the total precession between the records of *Maitrī Upaniṣad* and *Vedāṅga Jyotiṣa* the epoch of this tradition roughly corresponds to 1800 B.C.

But surely the *Maitrī. Up.* did not use the scale of *Vedāṅga Jyotiṣa,* which was based on a formulation of some 400 years later, just as an inscription written in European language cannot be read in terms of B.C. years, when the very concept of B.C. years did not come into being. The text must have used its own zodiac or one formulated before it.

We take this zodiac to have commenced from the star Maghā (Regulus) as its beginning point. This initial point cannot be vernal equinox or winter solstice because then we get an epoch roughly of 8000 B.C. or earlier, when there was no human civilisation at all, and hence no question of any astronomical calculation. Hence it was the summer solstice. The total precession between this text and *Vedāṅga Jyotiṣa* then becomes 13° 29′= arc between *Epsilon* Hydrae and Regulus (17° 29′) minus 4°. The epoch of the observation recorded in the text then corresponds to about 2400 B.C., a period when the text itself was not composed.

We now get a highly interesting result. In the scale of this text, the vernal equinox was at *Eta* Tauri (as it is 89° 50′ behind Maghā). Here we get an indirect reference to the fact that vernal equinox occurs at Kṛttikās, though no explicit reference to vernal equinox occurring at Kṛttikās is found in so many words in any of the texts under consideration.

We thus see that the transformation of *nakṣatras* from stars to arc divisions and calendarisation of the system were complete around 2400 B.C. We are inclined to believe that this tradition was also handed down to Aryan India, which without questioning the authority, recorded the tradition in Vedic literature.

The obvious presumption is that the practice of determining the nearness of the moon to one or another star by eye-estimation was not considered a reliable method. To avoid any dispute, the stars were replaced by arc-divisions and the eye-estimation method was replaced by a computational method using moon's mean motion.

We can make a rough estimate of the region from where the above traditions were observationally correct. The star Svātī (*Alpha* Bootes) has been described as an outcaste (for being far away from other stars) traversing the northern sky (Vedic Index, vol. i.p. 417). The declination of Svātī works out to be 45° 3′ in 3000 B.C. using the standard transformation formula, and the second to it in declination was Abhijit with declination 43° 38′. As Svātī had rising and setting, the latitude of the place of observation, must be after some marginal allowances, less than 40°. It may be noted here that the declinations of these two stars, by 2400 B.C., were 42° 29′ and 42° 4′ resp. i.e. they were equally northern stars. It is likely that Abhijit was recognised as an additional star around this period, although it was not assigned any arc division. Thus by 2400 B C. the Svātī could no longer be regarded as an outcaste.

Again, the statement that the seven sages rise in the north seems to bear an inner meaning. By 3000 B.C. *Alpha* Draconis was the pole star (within tolerable limit of accuracy) and the stars *Epsilon* and *Zeta* Ursae Majoris were circumpolar almost throughout India, and *Beta* Ursae Majoris had the highest co-declination among the seven sages. If we assume that the purport of the text was that all the seven sages were then nearly circumpolar or that *Beta* Ursae Majoris was then seen to touch the northern horizon at rising, then the latitude of the place should be, after marginal allowances, greater than 30° ; in other words, the traditions recorded in Vedic literature were observationally correct from a latitude belt of 30° to 40° N.

We refer to another tradition in favour of this assumption.

The *Vedāṅga Jyotiṣa* has measured the length of the longest day as 14 hours 24 minutes. The unit used in the text is 18 *muhūrtas* where 30 *muhūrta*s equal 24 hours. A rough estimate of the latitude of the place of observation can be made thus :

Half of the longest day=7 hours 12 minutes=108° ; declination *delta* of sun on longest day=*epsilon*=23° 51′42″ (in 1400 B.C.). Hence, from the formula cos H=−tan *phi* tan *delta*, we get *phi*=35°. But this result is unreliable, as effect of refraction has not been considered here. The Indian system is that day begins when the sun is just visible on the horizon. Now, horizontal refraction is 35′, and sun's angular radius is 16′. Hence, at apparent beginning of day, the zenith distance of the sun's centre will be 90°+35′+16′=90°51′. Now consider the spherical triangle *PZS* where *S* is centre of the sun at apparent rising, *P* and *Z* are Pole and Zenith, hour angle *H* is 108°. Then we get, using cosine rule on triangle *PZS*, *phi*=33° 12′ which we take as 33°. We neglect the dip because any height favourable to the development of an advanced civilisation will affect our result only very slightly. Thus the tradition of *Vedāṅga Jyotiṣa* also conforms to our estimation of the latitude belt.

We referred earlier to an opinion that the Kṛttikās were in the equinoctial colure when the *nakṣatra* list was originally formulated. We have also stated that under such an assumption Kṛttikās shift by over 5° from east at rising. If we place Kṛttikās on the equinoctial colure, declination of *Eta* Tauri works out to be 4° 26′ and hence, its shift from east at rising as seen from a latitude of 33° works out to be 5° 17′.

We have so long considered only such astronomical references which are stated in explicit terms and our finding is that the convention of asterism system was developed before the birth of Aryan civilisation in India and that it occurred somewhere in a latitude belt of 30° to 40° N. This was recorded in Vedic literature as unquestionable truth.

The third and final reformation of asterism system was done in 6th century A.D., after which it assumed its present form, where *nakṣatras* appear both as arc divisions and stars, but neither as single star only nor as star-groups. But more of this in our subsequent volumes.

APPENDIX V

Further Note on The Kṛttikās

RAMATOSH SARKAR

The period of time arrived at by calculations, with regard to Kṛttikās rising in the east, by Chakravarty and by myself has not been quite the same (Vide Appendices II and IV). Of course we have both given only approximate dates and the two suggested years both fall in the third millennium B.C. ; but they are still separated by about 600 years—Chakravarty's being earlier than mine. To account for this disagreement, I have the following words to say :

The discrepancy stems from the following facts.

My colleague Chakravarty has chosen to concentrate exclusively on the star *Eta* Tauri (which indeed is supposed to be the *yogatārā* of Kṛttikās) ; but I have kept the interpretation a little more open (as the text, according to me, does not warrant the rigid view).

Chakravarty has opted to go as far back in time as necessary to get *Eta* Tauri on the celestial equator, whereas I have chosen just to get that star as close to vernal equinox as possible. For Chakravarty the declination of *Eta* Tauri should be zero and for me its celestial longitude should be zero and celestial latitude either zero or nearly so. (As a result, Chakravarty had to admit a celestial longitude of about 9° and I had to contend with a declination of about 3-4°.)

The rationale behind my procedure was two-fold. First, I wanted to go back into the past only as far as I considered to be imperative. Secondly, because ancient Indian texts enlist nakṣatras with Kṛttikā heading the list, I wanted Kṛttikā to be as near the vernal equinox as possible, so that it would have a special significance for calendrical purpose.

The two of us had worked independently without any prior

consultation and then compared our results. We decided, in consultation with Professor Chattopadhyaya, not to discard any of the two slightly different viewpoints in favour of the other. Because we thought that unwittingly we had arrived at two extreme limits that possibly delineate the period of time we were looking for.

APPENDIX VI

Some remarks on Brij Bhusan Vij's paper on linear standard in the Indus Civilisation.

APURBA KUMAR CHAKRAVARTY

Sri Brij Bhusan Vij has attempted to throw some new light on the Indus Valley Civilisation (in FIC. pp. 153 ff.).

This paper relates as mainly to the so called great bath of Mohenjodaro ; from a consideration of the dimensions of this bath the writer claims to have found out the purpose of its construction. Although we appreciate the labour and time devoted by the writer in his researches, yet we are unable to accept his views.

The dimensions of the bath are, as quoted by the writer, length 12·124 metres, breadth 7·01 metres, base diagonal (not skew-diagonal) 14 02 metres.

Consider a right-angled triangle ABC, right-angled at B where the hypotenuse AC is double of one side BC. Then, by Pythagoras' theorem, $4a^2=a^2+c^2$ or $c=a\sqrt{3}$.

Sri Vij's opinion is that a hypotenuse of length 14·02 metres and breadth 7·01 metres were chosen in advance so as to obtain a third side of length $\sqrt{3}\times 7{\cdot}01$ metres$=12{\cdot}124$ (*sic*) metres, the actual length of the bath. Thus the actual purpose of construction of the bath was to give a physical demonstration of the number $\sqrt{3}$, and for this purpose, the Pythagoras' theorem was used. Thereby the antiquity of Pythagoras' theorem is sought to be traced to Mohenjodaro.

We totally fail to understand this deduction. Which side or which dimension of the bath does demonstrate this number $\sqrt{3}$? What purpose does this demonstration serve ? This assumption pre-supposes a knowledge of the real number system and with that knowledge an irrational number is represented by a point on the real number axis and not demonstrated by masonry works. Further also, how can we be certain that the breadth and diagonal were chosen in advance so as to obtain a desired length ? If one constructs a rectangle knowing or even without knowing the Pythagoras' theorem, the diagonal will always have its geometrical value.

The height of the bath is 3 metres and so a skew-diagonal works out to be $\sqrt{14{\cdot}02^2+9}=14{\cdot}3373$ metres. Thus the angle between the base diagonal and skew-diagonal works out to be

$$\theta=\sin^{-1}\frac{3}{14{\cdot}3373}=\frac{\pi}{18}\ (sic).$$

Thus the bath demonstrates sub-multiple angles also.

The walls of these ruins, some 5000 years old, cannot have such sharp edges as to be measurable with the accuracy of 3 places of decimal in metres. Authorities connected with the Indus Valley excavations have generally given these dimensions in rounded off figures.

The writer also thinks that the bath was used as an observatory. We do not expect the inhabitants of Mohenjodaro to have used telescopes and hence do not expect broken glass or quartz lenses in the ruins. However, the writer mentions nothing whatsoever about the bath itself, which can even remotely suggest any device for making measurement relevant for astronomical observations. Again, for watching the night sky, an elevated place (like a tower) is more suitable than a depressed place (like the floor of the bath).

We admire the knowledge of mensuration and geometry the people of Mohenjodaro displayed in the planning of the town ; we also admit that they must have had some working knowledge of astronomy as an agricultural civilisation cannot thrive without the knowledge of the cycle of seasons. But we are unable to find out the reason as to why the construction of a big-sized cistern became necessary to demonstrate the knowledge of such subjects.

APPENDIX VII

NAVJYOTI SINGH

Illustration of Various Kinds of Recitations of Rgveda Which Were Devised to Preserve Long Compositions Orally[1].

A. 1 We will take a particular hymn (*rc*) say x. 97.22, from the *Rgveda* for illustrating various ways of recitation. Ninety-seventh *sukta* of the tenth book (*mandala*) is a song in praise of medicinal herbs and plants according to *Brhaddevata*[2]. In *Arsanukramani*[3] it is said that this song was composed by Bhisaj who belonged to the lineage of Atharvan. This song comprises of 23 *mantra*-s. All these are in *anustubh* metre (8, 8, 8, 8), each having four lines (*pada*) with eight syllables each. Hence this *sukta* is characterised by : authorship : Atharvano Bhisaj, theme : Osadhayah, and metre : Anustubh.

In this song Soma is described as a king of plants and herbs. The kingdom of herbs is ascribed the power of driving away disease from men and their domesticated animals (20 *rc*). A physician is described as a king among the crowd of men as he/she chases diseases with the help of power of herbs (6th *rc*). We choose the twenty-second *rc* of this song in praise of medicinal herbs for our illustration.

A. 2 In the metric style this *rc* can be written as :

Oṣadhayaḥ Saṃ Vadante Somena Saha Rājñā/
Yasmai Kṛṇoti Brāhmaṇas Taṃ Rājan Pārayāmasi//

'With Soma as their supreme Lord the plants hold conversation and say :

---

1. The appendix is largely based on pp. 792-808 of *Rgveda Samhita*, ed. by Pt. S.D. Satavalekava. Paradi (Gujrat) Svadhyaya Mandala, 4th edition. 1982.
2. *Brhaddevata* vii. 154. It further says : In (its) employment this (hymn) of Bhisaj is applicable to the cure of consumption (*yaksma*).
3. *Arsanukramani* X-45. Even *Brhaddevata* assigns this song to Bhisaj.

O King, we save from death the man whose cure a Brahmana undertakes'[4].

The three kinds of accents are written by marking '/' above syllable for circumflex accent (*svarita*), '—' below syllable for grave accent (*anudatta*), and leaving unmarked acute accent (*udatta*), since most syllables in *Rgveda* are pronounced in acute accent. Not that this *rc* has four equal-sized lines or *pada* as is the case with ordinary *anustubh* metre.

A. 3 Samhitapatha : The continuous recitation of this *rc* can be represented as—

Osadhayahsamvadantesomenasaharajna/Yasmaikrnotibrahmanastamrajanparayamasi/

In Samhitapatha all the syllables or sounds are recited continuously except for the pause which is necessary for breathing and is decided for the reasons of prosody. The place of pause is indicated with long dark line '/'.

A. 4 Padapatha : The word by word recitation of the *rc* is—

Osadhayah/Sam/Vadante/Somena/Saha/Rajna/
Yasmai/Krnoti/Brahmanah/Tam/Rajan/Parayamasi//

This *rc* consists of twelve words which are recited with a pause after each word. So there are twelve pauses in Padapatha instead of two in Samhitapatha.

Note that the accents in Padapatha are somewhat different from Samhitapatha. Also some sounds are different like... *brahmanastam*...becomes...*Brahmanah/Tam*. .

We have also given number markings to all the 12 words which constitute this *rc* for the sake of convenience. Padapatha helps fix words used in Samhitapatha.

A. 5 Kramapatha : The order of words and accent is fixed by this recitation—

Osadhayahsam/Samvadante/Vadantesomena/
Somenasaha/Saharajna/Rajnetirajna//

---

4. English translation is taken from Ralph T.N. Griffith's *Hymns of Rgveda* ed. by J.L. Shastri. Delhi Motilal Banarsidass Reprint. 1977.

Yasmaikrnoti/Krnotibrahmanah/Brahmanastam/
Tamrajan/Rajanparayamasi/Parayamasitiparayamasi//

In this recitation pause is given after continuously pronouncing two words. Each word occurs twice in the Kramapatha, e.g. '1,2/ 2,3/3,4...', 2 occurs twice.

'//' symbol helps us relate to Samhitapatha but it does not mean longer pause. The compounds 6.6 and 12.12 with the intermediate 'iti' help relate to the position of pause in Samhitapatha. The important function of Kramapatha is that it fixes accent to a large extent and also the order of words by showing their relation with neighbouring words. In a way Kramapatha has features of both, it is partly Samhitapatha and partly Padapatha.

A. 6 Pancasandhipatha : This recitation rigidifies more the Kramapatha. It represents word-serialisation of every couplet in five-fold way, that is, *krama* (1,2), *utkrama* (2,2), *vyutkrama* (2,1), *abhikrama* (1,1) and *sankrama* (1,2). In the number representation we have adopted it goes like—
1,2/ 2,2/ 2,1/ 1,1/ 1,2/ 2,3/ 3,3/ 3,2/ 2,2/ 2,3/ 3,4/ 4,4/ 4,3/
3,3/ 3,4/ 4,5/ 5,5/ 5,4/ 4,4/ 4,5/ 5,6/ 6,6/ 6,5/ 5,5/ 5,6/ 6,6/.

A. 7.1 Jatapatha : This recitation is further consolidation of Kramapatha with a view to fully fixing accent and syllables of words and the Kramapatha itself. In this *patha* basic unit is doublets, but pause is not allowed after each doublet, but only when three doublets have been pronounced. The comma is used to indicate that the *sandhi* or euphonic combination does not occur. The *rc* will be recited as :

1.2, 2.1, 1.2/
2.3, 3.2, 2.3/
3.4, 4.3, 3.4/
4.5, 5.4, 4.5/
5.6, 6.5, 5.6/6 iti 6/
7.8, 8.7, 7.8/
8.9, 9.8, 8.9/
9.10, 10.9, 9.10/
10.11, 11.10, 10.11/

11.12, 12.11, 11.12/12 iti 12/

Jatapatha is one of the eight variations of Kramapatha. The other seven are called Malapatha, Sikhapatha, Rekhapatha, Dhvajapatha, Dandapatha, Rathapatha and Ghanapatha. We will illustrate them with number notations. For the details of accent etc. reference 1 will have to be consulted.

A. 7.2 Mala : There are two kinds of *mala*.

(a) Kramamalapatha : It has to be read from left to right. The gap has been shown to illustrate the design and also because if two persons recite one after another, one person can recite the column on the left and the other on the right. This means one person should go from 1 to 6 whereas the other person should recite in opposite direction. This way two people can recite one after another Kramamalapatha.

1,2/	6,6/
2,3/	6,5/
3,4/	5,4/
4,5/	4,3/
5,6/	3,2/
6,6/	2,1/

(b) Puspamalapatha : This is almost similar to Jatapatha, except that pause is there after every doublet and at the place of pause in Jatapatha 'iti' is said before that pause. Besides in doublets euphonic combination is universally observed so that 'Brahmanah tam' becomes 'Brahmanastam' and 'Rajan tam' becomes 'Rajamstam' etc. The recitation according to the order of words is—

1,2/ 2,1/ 1,2/ iti/
2,3/ 3,2/ 2,3/ iti/
3,4/ 4,3/ 3,4/ iti/
4,5/ 5,4/ 4,5/ iti/
5,6/ 6,5/ 5,6/ iti/
7,8/ 8,7/ 7,8/ iti/

8,9/ 9,8/ 8,9/ iti/
9,10/ 10,9/ 9,10/ iti/
10,11/ 11,10/ 10,11/ iti/
11,12/ 12,11/ 11,12/ iti/
12 iti 12/

A. 7.3 Sikhapatha : In the Kramapatha the first word of the next line is added to the first line to turn it into Sikhapatha. According to the word-order it goes like—

1,2-2,1-1,2-3/
2,3-3,2-2,3-4/
3,4-4,3-3,4-5/
4,5-5,4-4,5-6/
5,6-6,5-5,6/
6,6/
7,8-8,7-7,8-9/
8,9-9,8-8,9-10/
9,10-10,9-9,10-11/
10,11-11,10-10,11-12/
11,12-12,11-11,12/
12,12/

The sign '-' in the above recitation is equivalent to what comma stands for in Kramapatha, so that no euphonic combination is allowed.

A. 7.4 Rekhapatha : It can be done in two ways. One involving only half of the *rc* (*ardharca*) and other involving whole *rc*.

(a) First part of *rc* :
1,2/ 2,1/ 1,2/
2,3,4/4,3,2/ 2,3/
3,4,5,6/ 6,5,4,3/ 3,4/ 4,5/ 5,6/ 6 iti 6/
Second part of *rc* :
7,8/ 8,7/ 7,8/
8,9,10/ 10,9,8/ 8,9/
9,10,11,12/ 12,11,10,9/ 9,10/10,11/ 11,12/12 iti 12/

(b) Involving whole *rc*-s
1,2/ 2,1/ 1,2/
2,3,4/ 4,3,2/ 2,3/

3,4,5,6/ 6,5,4,3/ 3,4/
4,5,6,7,8/ 8,7,6,5,4/ 4,5/
5,6,7,8,9,10/ 10,9,8,7,6,5/5,6/
6,7,8,9,10,11,12/ 12,11,10,9,8,7,6/ 7,8/ 8,9/
9,10/ 10,11/ 11,12/ 12 iti 12/

A. 7.5 Dhvajapatha : Doublets are recited alternatively from forward recitation and inverse recitation :

1,2/	12 iti 12/
2,3/	11,12/
3,4/	10,11/
4,5/	9,10/
5,6/	8,9/
6 iti 6/	7,8/
7,8/	6 iti 6/
8,9/	5,6/
9,10/	4,5/
10,11/	3,4/
11,12/	2,3/
12 iti 12/	1,2/

A. 7.6 Dandapatha : It is designed for half *rc* (*ardharca*) as a unit.
For first *ardharca*
1,2/ 2,1/
1,2/ 2,3/ 3,2,1/
1,2/ 2,3/ 3,4/ 4,3,2,1/
1,2/ 2,3/ 3,4/ 4,5/ 5,4,3,2,1/
1,2/ 2,3/ 3,4/ 4,5/ 5,6/ 6,5,4,3,2,1/
1,2/ 2, 3/ 3,4/ 4,5/ 5,6/ 6 iti 6/

For the second half of *rc* :
7,8/ 8,7/
7,8/ 8,9/ 9,8,7/
7,8/ 8,9/ 9,10/ 10,9,8,7/
7,8/ 8,9/ 9,10/ 10,11/ 11,10,9,8,7/
7,8/ 8,9/ 9,10/ 10,11/ 11,12/ 12,11,10,9,8,7/
7,8/ 8,9/ 9,10/ 10,11/ 11/ 12/ 12 iti 12/

A. 7.7 Rathapatha : There are three primary Rathapatha-s involving one *rc*. But they can be done by taking 2, 3 or 4 *rc*-s together. The three kinds are :

(a) Dvicakraratha

(1) 1,2/7,8/
2,1/8,7/

(2) 1,2/7,8/
2,3/8,9/
3,2,1/9,8,7/

(3) 1,2/7,8/
2,3/8,9/
3,4/9,10/
4,3,2,1/10,9,8,7/

(4) 1,2/7,8/
2,3/8,9/
3,4/9,10/
4,5/9,10,11/
5,4,3,2/11,10,9,8,7/
1,2/7,8/
2,3/8,9/
3,4/9,10/
4,5/10,11/
5,6/11,12/
6 iti 6/12 iti 12/

Similar recitation can be designed for the next half of *rc*.

(b) Tricakraratha : This is valid for the *rc*-s which have three *pada*-s or multiple of three *pada*-s. The *rc* which we have chosen has four *pada*-s ; hence this recitation cannot be illustrated.

(c) Catuscakraratha : This is valid for *rc*-s having four or multiple of four *pada*-s. For the *rc* we have taken in the appendix this recitation has following word-order :

(1) 1,2/4,5/7,8/10,11/
2,1/5,4/8,7/11,10/

(2) 1,2/4,5/,7,8/10,11/
2,3/5,6/8,9/11,12/
3,2,1/6,5,4/9,8,7/12,11,10/
1,2/2,3/3 iti 3/
4,5/5,6/ 6 iti 6/
7,8/8,9/9 iti 9/
10,11/11,12/12 iti 12/

A. 7.8 Ghanapatha :

It is of two kinds

(a) 6 iti 6/5,6/4,5/3,4/2,3/1,2—2,3/3,4/4,5/5,6/6 iti 6/
12 iti 12/11,12/10,11/9,10/8,9/7,8—8,9/9,10/10,11/
11, 12/12 iti 12/

(b) 1,2—2,1—1,2,3—3,2,1—1,2,3/
2,3—3,2—2,3,4—4,3,2—2,3,4/
3,4—4,3—3,4,5—`,4,3—3,4,5/
4,5—5,4—4,5,6—6,5, —4,5,6/
5,6—6,5—5,6/
6 iti 6/
7,8—8,7—7,8,9—9,8,7—7,8,9/
8,9—9,8—8,9,10—10,9,8—8,9,10/
9,10—10,9—9,10,11—11,10,9—9,10,11/
10,11—11,10—10,11,12—12,11,10—10,11,12
11,12—12,11—11,12/
11 iti 12/

BIBLIOGRAPHY

I. JOURNALS, GENERAL REFERENCE WORKS AND ANTHOLOGIES

ABORI — *Annals of the Bhandarkar Oriental Research Institute,* Poona.

ACI — *Ancient Cities of The Indus,* Possehl, G.L. (ed.), New Delhi 1979.

AI — *Ancient India,* Bulletin of the Archaeological Survey of India, New Delhi.

Am An — *American Anthropologist,* Beloit.

An Pa — *Ancient Pakistan,* Peshwar.

Antiquity — A Quarterly Review of Archaeology, Cambridge, England.

ARASI — *Annual Report of the Archaeological Survey of India,* Delhi.

BSOAS — *Bulletin of the School of Oriental and African Studies,* London.

CF — *Cultural Forum,* New Delhi.

CHI — *The Cambridge History of India,* Vol. I, Ancient India, Edited by E.J. Rapson, Cambridge, 1922.

CR — *Calcutta Review,* Calcutta.

EIP — *Essays in Indian Proto-history.* Agrawal, D.P. and Chakrabarti, D.K. (eds.), New Delhi 1979.

ERE — *Encyclopaedia of Religion and Ethics,* James Hastings (ed.).

EW — *East and West,* Rome.

Expedition — Bulletin of the University Museum, University of Pennsylvania.

FIC — *Frontiers of The Indus Civilization* : (Sir Mortimer Wheeler Commemoration Volume), Lal, B.B. and Gupta, S.P. (eds.), Delhi, 1984.

HC — *Harappan Civilization : A Contemporary Perspective,* Possehl, G.L. (ed.), Delhi, 1982.

HR — *History of Religions,* University of Chicago.

HS — *History and Society,* (Essays in Honour of Niharranjan Ray) Edited by D.P. Chattopadhyaya, Calcutta, 1978.

HT — *A History of Technology,* Charles Singer, E.J. Holmyard and A.R. Hall (eds.), 7 vols. Oxford. 1954—

IA-NP — Indian Archaeology—New Perspectives, Possehl, G.L. (ed.).

IAR — *Indian Archaeology—A Review*, New Delhi.

IC — *Indian Culture,* Calcutta.

IHQ — *The Indian Historical Quarterly,* Calcutta.

IHR — *The Indian Historical Review,* Journal of the Indian Council of Historical Research, New Delhi.

IJHS — *Indian Journal of History of Science*, Indian National Science Academy, New Delhi.

ILN — *The Illustrated London News.*

Ind. Stud. — *Indische Studien*, herausgegeben von A. Weber.

IP — Indian Prehistory. Ed. Misra V. N. and Mate M.S., Poona 1965.

ISPP — *Indian Studies : Past and Present*, Calcutta.

IR — *Indian Review,* Madras.

JAIH — *Journal of Ancient Indian History.*

JESHO — *Journal of the Economic and Social History of the Orient,* Leiden.

JIH — *Journal of Indian History,* Trivandrum.

JRAS — *Journal of the Royal Asiatic Society of Great Britain and Ireland.*

JUB — *Journal of University of Bombay.*

LK — *Lalit Kala,* A Journal of Oriental Art, Chiefly Indian, New Delhi.

MASI — *Memoirs of the Archaeological Survey of India,* New Delhi.

ME — *Man and Environments.*

MIC — *Mohenjo-daro and Indus Civilization,* Marshall, J. (ed.) 3 Vols. London, 1931.

Novitates — *American Museum Novitates,* New York.

OC — Transactions (Verhandlungen, Acts) of Congresses of Orientalists.

OLS — *Oriental and Linguistic Studies,* New York.

PIHC — *Proceedings of the Indian History Congress.*

Puratattva — Indian Archaeological Society, New Delhi.

RA — *Rationalist Annual,* London.

RIA — *Radiocarbon and Indian Archaeology,* Agrawal, D.P. and Ghosh, A. (eds.), Bombay, 1973.

S — *Science,* The American Association for Advancement of Science, Lanchaster.

SA — *Scientific American,* Pub. by W.H. Freeman and Co., Sanfrancisco, California.

SBE — *Sacred Books of the East,* Max Muller, F. (ed), First published 1882, Reprinted Delhi, 1972.

SC — *Science and Culture* : A monthly Journal of Natural and Cultural Science. Calcutta, 1935.

Science in USSR — USSR Academy of Sciences. An Illustrated Science and Information Journal, Established 1981, Published in Russian and English.

SHSI — *Studies in the History of Science in India,* 2 Vols. Chattopadhyaya D.P. (ed), New Delhi, 1982.

SHSIA — *Symposium on the History of Sciences in India, Abstracts of Papers,* New Delhi, Oct. 1968.

SSP — *Social Science Probings,* Sharma, R.S. *et al.* (ed.)

ST — *Science Today,* Ed. Surendr Jha, Bombay.

The Pandit Varanasi

VIJ — *Vishveshvarananda Indological Journal,* Hosiarpur.

II. OTHER BOOKS AND ARTICLES

Adams, R.M. *The Evolution of Urban Society*: Chicago, 1966.

Agrawal, D.P. *Archaeology of India,* Scandinavian Institute of Asian Studies, Copenhagen K. 1982.

— "Harappan Chronology; a re-examination of the evidence", *Studies in Prehistory,* Ed. by Sen, D. and Ghosh, A.K. 139-48, Calcutta 1966.

— "Harappan Culture; new evidence for a shorter chronology", in *Science,* No. 3609 : 950-2, 1964.

— "Metal Technology of the Harappans", in FIC.

Alberuni India, Tr. by E.C. Sachau, Reprint, New Delhi, 1983.

Allchin, B. and R. *The Birth of Indian Civilization.* Penguin, 1968.

— *The Rise of Civilization in India and Pakistan,* New Delhi, 1983.

Allchin, F.R. *Antecedents of the Indus Civilization*, London 1980.

— "The legacy of the Indus Civilization", in *HC*.

Ansari, Z.D. and Dhavalikar M. "New Light on the Prehistoric Cultures of Central India", in *World .Archaeology*, Vol. 2, 337-46, 1971.

Ansari, Z.D. and Dhavalikar M. "New Light on the Prehistoric Cultures of Central India", in *World Archaeology*, Vol. 2, 337-46, 1971.

Aravamuthan, T.G. "Archaeological discoveries in India", in *Antiquity*, Vol. 2 : 83-5. 1928.

— *Progress of Archaeology in India from 1764 to 1952 (and) 1953 to 1956*. Delhi, 1969.

Asthana, Shashi, "Indus-Mesopotamia trade : Nature of Trade and Structural Analysis of Operative System", in *EIP*.

Atharvaveda Tr. W.D. Whitney, 2 vols. Reprint, new Delhi, 1962.

Banerjee, N.R. *The Iron Age in India*. Delhi, 1965.

Banerji, R.D. "Exploration and Research, Western Circle, Sind, Mohenjo-daro", in *ARASI*, 1922-23, 102-4.

Basham, A.L. *Aspects of Ancient Indian Culture*, Bombay, 1966.

— "Recent Work on the Indus Civilization", in *BSOAS*, Vol. 13 : 140-45, London, 1949/51.

Bernal, J.D. *Science in History*, 4 Volumes (First Pub. 1954), Revised edition, Penguin, 1969.

Bhandarkar, D.R. *Some Aspects of Ancient Indian Culture*, Madras, 1940.

Bhardwaj, H.C. "Aspects of Early Iron Technology in India" in *RIA*.

— "Some Aspects of Ancient Indian Metallurgy" in *SHSIA* : 85, New Delhi, Oct, 1968.

Bhatta L.S. *The Ancient Mode of Singing Samagana,* Poona, 1939.

Bhattacharji Sukumari *Literature in the Vedic Age,* 2 vols., Calcutta, 1984-86.

Bhattacharya, Chaya, "Some means of Transport of the Harappan Culture—their technical development" in *SHSIA* : 98-9, Delhi, 1968.

Bhattacharyya, Sitesh Chandra (ed.), *Vedanga Jyotisam,* (with Bengali translation), Calcutta, 1974.

Bisht, R.S. "Excavation at Banawalli : 1974-77", in *HC*.

Bongard-Levin, G. and Vigasin, A. *The Image of India* : The Study of Ancient Indian Civilization in the USSR, Moscow, 1984.

Bose, D.M, Sen S.N. and Subbarayappa B.V, *A Concise History of Science in India,* New Delhi 1971.

Braidwood, Robert J, "The Agricultural Revolution", in *SA*, Vol. 203, No. 3, 130-48. Sanfrancisco, California, 1960.

Brown, Burton, T. *Excavation at Azarbaijan,* London 1951.

Burrow, T. "On the Significance of the Term arma-armaka in Early Sanskrit literature" in *JIH,* Vol. 41, 1963.

Calder, C.C. *in An Outline of Field Science in India,* Edited by S.L. Hora, Calcutta, Indian Science Congress, 1937.

Caldwell, J.R. and Shahmirzadi, S.M. *Tal-i-Iblis* : *The Kerman range and the beginning of Smelting,* Springfield, Illinois State Museum, 1966.

Casal, J.M. "Fouilles de Mundigak" ; *Memories de la Delegation Archeologique Francaise en Afganistan* 17, Paris, 1961.

Chakrabarti, D.K. "The Aryan Hypothesis in Indian Archaeology", in *ISPP* Vol. 9, No. 4 : 343-58, Calcutta, 1968.

— "Harappan Chronology" in *JAIH,* Vol. 1 Nos. 1-2 ; 78-82, 1967-68.

— "Origin of the Indus Civilization : Theories and Problem", in *FIC*.

— "Size of the Harappa Settlements" in *EIP*.

— "Some Theoretical Aspects of Early Indian Urban Growth", in *Puratattva,* No. 7, 1974. New Delhi.

Chanda, Rama Prasad, "The Indus Valley in the Vedic Period", in *MASI,* No. 31, Delhi 1926.

— "Survival of the Prehistoric Civilization of The Indus Valley", in *MASI* No. 41, Delhi 1929.

— "Harappa" in *SC*, Vol. 6, 377-81, Calcutta, 1941.

Chatterjee, B.K. "The Date and Character of Indus Civilization" in *JBRS,* Vol. 42, Nos. 3-4, 389-95, 1956.

— "Mohenjo-daro Civilization" in *CR,* Vol. 139 : 121-6 ; Vol. 141 ; 252-60 ; Vol. 144 : 127-33, 1956-57.

— "Mohenjo-daro and Vedic Civilization", in *IR,* Vol. 107, 408-11, 1956.

— "Mohenjo-daro Civilization" in *Kalyana Kalpataru* Vol. 20 : 465-8, 1966.

— "Mohenjo-daro—an Aryan Colonization of Mesopotamia", in *VIJ*, Vol. 3 : 111-16, 1965.

Chattopadhyaya, Debiprasad *Bharatiya Darsana* (in Bengali), Calcutta 1960. Reprinted 1980.

— *Lokayata : A Study in Ancient Indian Materialism,* New Delhi, July 1959, 6th Reprint 1985.

— *Science and Society in Ancient India,* Calcutta, 1977, Reprinted 1979.

— *What is Living and What is Dead in Indian Philosophy,* New Delhi, 1976.

Chattopadhyaya, D. (Ed.) *Nyaya,* Complete English Translation of *Nyaya Sutra* with Vatsyayana's Commentary by M. Gangopadhyaya with Introduction by editor, Calcutta, 1982.

Chattopadhyaya, K.C. *Studies in Vedic and Indo-Iranian Religion and Literature,* Vol. II, ed. V.N. Misra, Varanasi, 1978. 41-50.

Childe, V. Gordon *Man Makes Himself.* London (First Pub. 1936). 1951 Reprint.

— *New Light on the Most Ancient East,* New York, Rev. ed. 1954.

— "The Indus Civilization", in *Antiquity.* Vol. 11 : 351. 1937.

— "The Urban Revolution", Reprinted in *ACI.*

— *What Happened in History* (Penguin, First Pub. 1942) 1976 Reprint.

Chitawala, Y.M. "The Problem of Class Structures in the Indus Civilization", in *FIC.*

Coghlan, H.H., *Viking Fund Publications in Anthropology,* 28.1960.

Cornford, F.M. *From Religion to Philosophy,* Cambridge, 1912.

Cowell, E.B. (ed.), *The Jatakas,* 6 Vols. 1st Published 1895, Reprinted Delhi, 1973.

Cunningham, Sir Alexander, *The Ancient Geography of India,* London, 1871.

— "Harappa", in *Archaeological Survey of India, Report,* 1872-73 : 105--8, Delhi 1875.

Dales, G.F. "The Decline of Harappans" in *ACI.*

— "The Mythical Massacre at Mohenjo-daro" in *ACI.*

— "New Investigations at Mohenjo-daro" in *ACI.*

— "Sex and Stone at Mohenjo-daro" in *FIC*.

Datta, Bibhutibhusan, *The Science of the Sulba* : A study in *Early Hindu Geometry,* Calcutta, 1932.

Datta B.B and Singh A.N, *History of Hindu Mathematics,* Lahore 1935 and 1938.

Dey, Nandalal, *The Geographical Dictionary of Ancient and Mediaeval India,* Delhi, 1979.

Dikshit, K.N. *Prehistoric Civilization of the Indus Valley,* Madras, 1967 (2nd edn.)

— "Hulas and the Late Harappan Complex in Western Uttar Pradesh", in *HC*.

— "Late Harappa in Northern India", in *FIC*.

— "The Late Harappan Cultures in India" in *EIP*.

Dikshit, Sankar Balakrishna *Bharatiya Jyotish Sastra,* (History of Indian Astronomy), English Translation (Translated by Prof. R.V. Vaidya) Part II (History of Astronomy during the Siddhantic and Modern Periods) Delhi 1981.

Diringer, D. *The Alphabet ; a Key to the History of Mankind* (with foreword by Sir Ellis M. Mins), New York. 1958.

Dupree, L., "Deh Morasi Ghundai : A Chalcolithic Site in South Central Afghanistan", *Archaeological Papers of the Museum of Natural History,* 50, Pt. 2. New York, 1963.

Dutt, Binode Behari *Town Planning in Ancient India,* Calcutta and Simla, 1925.

Engels, F. *The Origin of the Family, Private Property and the State,* Moscow 1952 edn.

Faddegon B. *Studies on the Samaveda,* Pt. i, Amsterdam 1951.

Fairservis, W.A. Jr. *The Roots of Ancient India* : The

Archaeology of Early Indian Civilization, Macmillan, New York, 1971, (2nd edn. Chicago : 1975).

— "The Harappan Civilization : New evidence and More theory", in *ACI*.

"The Origin, Character and Decline of an Early Civilization", in *ACI*.

Forbes, R.T. *Studies in Ancient Technology,* 9 ; Leiden, 1964.

Farrington, Benjamin, *Greek Science* : Its Meaning for us. Penguin, 1963 (First Pub. 1944).

— *Head and Hand in Ancient Greece* : For Studies in the Social Relations of Thought, London, 1947.

— *Science in Antiquity*, London 1936.

— *The Civilization of Greece and Rome,* London, 1938.

Filliozat, J. *The Classical Doctrine of Indian Medicine,* Delhi, 1964.

Gaur, R.C. *Excavation at Atranjikhera* : Early civilization of the Upper Ganga Basin, Delhi, 1983.

Ghosh, A(Ed.) *Archaeological Remains, Monuments and Museums,* 2 Vols, New Delhi 1964.

Ghosh, A. *The City in Early Historical India,* Simla 1973.

— "Deurbanization of the Harappan Civilization", in *HC*.

Gopal, Lalanji "The Date of the Krsi-Parasara", in *JIH*, Golden Jubilee Number, pp. 151-168, 1973.

Gordon, D.H. *The Pre- Historic Background of Indian Culture,* Bombay 1958.

Gordon, D.H. and Gordon M.E. "Survivals of the Indus Culture" in *JRASB* Vol. 6 : 61-72, 1940.

Gupta, S.P. "Two Urbanizations in India : A side

Study in Their Social Structure," in *Puratattva,* No. 7, 53-60, 1974.

Hawkes, Jacquetta and Woolley, Leonard, *History of Mankind*: Cultural and Scientific Development. Vol. I: Pre-history and the Beginnings of Civilization. (Part One: Prehistory by Jacquetta Hawkes, pp. 3-358, and Part Two: The Beginning of Civilization, pp. 359-865). George Allen and Unwin, UNESCO, 1963.

Heras, H. "Mohenjo-daro the most Important Archaeological Site in India", in *JIH*, Vol. 16, 1-12, 1937.

— "Mohenjo-daro, The People and The Land" in *IC,* Vol. 3 n.p., 1937.

Jaiswal, O.P. "The Indus Culture and Its legacy", in *PIHC,* Vol. 10 : 113-18, 1968.

Jarrig, Jean-Francois, and Richard H. Medow, "The Antecedents of Civilization in the Indus Valley", in *SA*, Aug. Vol. 243 No. 2 : 122-33, 1980.

Jayaswal, K.P. "The Punch-Marked Coins : A survival of The Indus Civilization", in *JRAS*, 720 ff. 1935.

Joshi, J.P. "The Nature of Settlement of Surkota-da", in *EIP.*

Joshi, J.P. and Madhubala, "Manda, a Harappan Site in Jammu and Kashmir" in *HC.*

Kashikar, C.G. "Agnicayana : Extension of Vedic-Aryan Rituals", in *ABORI,* Vol. LXII, Pts. 1-4, 121-133, 1981.

Katyayana, *Rgveda Sarvanukramani,* Ed. A.A. Macdonell, Oxford, 1886.

Keith, A.B. *The Religion and Philosophy of the Veda and Upanishads.* Harvard Oriental Series, Vol. 31. First edition 1925. Indian Reprint 1970.

— *The Veda of The Black Yajus School*; entitled *Taittiriya Samhita.* H.O.S.; XIX. Originally Published Harvard : 1941, Reprinted New Delhi 1967.

Kennedy, K.A.R. "Trauma and Disease in the Ancient Harappans" in *FIC*, 425-436.

Kinner Wilson, J.V. "The case for Accountancy", in *FIC,* 173-178.

Kosambi, D.D. *An Introduction To The Study of Indian History,* Bombay, 1956.

— *The Culture and Civilization of Ancient India,* London, 1965.

Lahiri, Bela, *Indigenous States of Northern India* (C. 200 B.C. to 320 A.D.), Calcutta University, 1974.

Lal, B.B. "Excavation at Hastinapur and Other Explorations", in *Al* Nos. 9 and 10, 1955.

— "Expeditions and Excavation Since Independence," in *CF*, Dec. 1961.

— "History of Technology in Ancient and Mediaeval India," in *SHSIA*, 87-88, 1968.

— "Kalibangan and Indus Civilization" in *EIP*.

— "The Role of Bhagwanpura as a bridge between Certain Stage of the Indus and the Ganges Civilization", in *HC*.

Lal, B.B. and Thapar, B.K. "Excavation at Kalibangan : New Light on the Indus Civilization", in *CF*, Vol. 9 No. 4, 78-88, 1967.

Lamberg-Karlovsky, C.C. and Sabloff, J.A. *Ancient Civilizations : The Near East and Mesoamerica,* California, 1937.

— "An Idea or Pot-luck", in *FIC,* 347-351.

— "Archaeology and Metallurgical Technology in Prehistoric Afganistan, India and Pakistan" in *Am An*, Vol. 69; 145-62, 1967.

Law, B.C. *Geography of Early Buddhism,* Calcutta 1973.

— *Historical Geography of Ancient India,* Paris, 1954.

Mackay, E.J.H. *Early Indus Civilization,* Revised and enlarged edn. New Delhi, 1976.

— *Further Excavations at Mohenjodaro,* 2 Vols. Delhi, 1938.

MacDonald, Sir. M. *Mohenjo-daro : Interim Report on Investigations of Deterioration of Brick Work Through Dampness and Salinity,* Hyderabad 1962.

Macdonell, A.A. *The Vedic Mythology,* Strassburg 1897, Reprinted 1963, Varanasi.

Macdonell, A.A. and Keith, A.B. *Vedic Index of Names and Subjects,* 2 Vols. London, 1912, Reprinted, Delhi 1928.

Madhavacarya, *Sarvadarsanasamgraha,* Poona, 1928.

Mahadevan I, *The Indus Script : Text, Concordance and Tables,* MASI, No. 77, 1977.

Mainkar, V.B., "Metrology in the Indus Civilization" in *FIC,* 141-151.

Majumdar, N.G. "Amri" in *MASI,* No. 48, 1934.

— "Excavation at Jhukar" in *ARASI,* 76-82, 1927-28.

— "Pre-historic and Proto-historic Civilization", in *Revealing India's Past,* Sir John Cumming (ed.) 91-116, London, The Indian Society, 1939.

Malik, S.C. (ed.) *Indian Civilization : The Formative Period,* Simla, Indian Institute of Advanced Study, 1966.

— "Harappan Social and Political Life" in *FIC,* 201-209.

Manchanda, O., *A Study of the Harappan Pottery,* Oriental Publishers, New Delhi, 1972.

Marga Bandhu, C. "Early Transport Vehicles from Ganga Valley", in *Ranga-Valli : Recent Research in Indology,* Narasimha Murthy, A.V. and Gururaja Rao, (eds.) Delhi, 1983.

Marshall, Sir John (ed.), *Mohenjodaro and The Indus Civilization,* 3 Vols., London, 1931.

— *Taxila,* 2 Vols. Cambridge, 1951.

Marshall, J., Mackay, E.J.H., and R.B. Dayaram Sahni "Mohenjo-daro" in *ARASI,* 72-98, 1925-26, Delhi 1927.

Michaels Axel, *A Comprehensive Sulvasutra Word Index.* Wiesbaden : Steiner, 1983.

Max Muller F, *A History of Ancient Sanskrit Literature,* London 1899.

McNairn Barbara, *The Method and Theory of V. Gordon Childe,* Edinburgh 1980.

Misra, V.N. "Climate, a Factor in the Rise and Fall of The Indus Civilization", in *FIC,* 461-485.

Monier-Williams M, *Sanskrit English Dictionary,* Reprint, New Delhi 1963.

Mookerji, Radha Kumud, *Indian Shipping* : A History of the Sea-Borne Trade and Maritime Activity of the Indians from the Earliest Times. Bombay, Calcutta, etc., 1957 (2nd edn.).

— *Ancient Indian Education* (Brahmanical and Buddhist), 3rd edn. Delhi, 1960.

Moran, H.A. *The Alphabet and The Ancient Calendar Science , Astrological Elements in the Origin of The Alphabet* (with a foreward by D. Diringer), California, 1953 (Litho-printed).

Mughal, Mohammad Rafique, *The Early Harappan Period in the Greater Indus Valley and Northern Baluchistan,* Philadelphia, 1970.

Needham, Joseph, *Science and Civilization in China,* several volumes published so far, Cambridge, 1954—.

— *The Grand Titration* : Science and Society in East and West, (First Pub. 1969), 2nd impression, 1979, London.

Nilson, M.P. *Primitive Time Reckoning,* C.W.K. Gleerup, Lund, 1920.

Pande, B.M. and Ramachandran, K.S. *Bibliography of the Harappan Culture,* Henry Field, (ed.), Field Research Projects, Coconut Grove, 1971.

Pandey, M.S. "Potteries in the Brahmanical Literature (Upto 2nd century B.C.)", in *PAI,* 155-160.

Pandya, Sumanben, "Kotado—A Major Harappan Urban Settlement in the Greater Rann of Kutch", in *IANP.*

Patanjali, *Vyakarana Mahabhasya,* Ed. F. Kielhorn, Vol. i, Poona 1962.

Petrie, Sir Flinders, *Prehistoric Egypt*, London, 1920.

Piggot, Stuart, *Prehistoric India,* Penguin, 1950.

— *The Dawn of Civilization,* London and New York, 1961.

— *Some Ancient Cities of India*, London, 1945.

Pillai, L.D.S.K. *Indian Ephemeris,* Vol. I, pt. I.

Pingala, *Chandahsutra,* Ed. Sitanath Bhattacharyya, Calcutta 1935.

Przyluski, Jean, "On the Origin of the Aryan Word Istaka", in *IHQ* Vol. vii, No. 4, 1931.

Radhakrishnan, S., *Indian Philosophy,* Vol. I, London, 1923.

— *Idealist View of Life,* London 1937.

Raikes, R.L. "The End of The Ancient Cities of The Indus" in *ACI,* 297-306.

— "The Mohenjo-daro Floods" in *Antiquity,* Vol. 39 : 196-203, 1965.

— "The Mohenjo-daro Floods—riposte" in *Antiquity,* Vol. 41, 309-10, 1967.

Rao, S.R. "Lothal", in *MASI,* No. 78. Delhi, 1979.

— *Lothal and The Indus Civilization,* Asia Pub. House, New York, 1973.

— Indus Cities : Planning for *Perfection*, in *ST*, June, 1982.

— "New Light on the Post-Urban (Late-Harappan) Phase of the Indus Civilization in India" in *HC*.

Ray, Prafulla Chandra, *Autobiography of a Bengali Chemist,* Calcutta 1958, edn. (Originally published with the title *Life and Experiences of a Bengali Chemist,* 1932-35).

— *History of Hindu Chemistry,* First Published—Vol. I, 1902-3 ; Vol. II, 1909 by The Bengal Chemical and Pharmaceutical Works Limited, Calcutta.

Ray, Priyadaranjan, (ed.) *History of Chemistry in Ancient and Mediaeval India* Calcutta, 1956.

Renou, Louis, *Religions of Ancient India,* London, 1953.

Rgveda Samhita, Vaidic Samshodhana Mandala, Poona, 1933.

Roy, T.N. *The Ganges Civilization,* New Delhi, 1983.

Saha, Meghnad, *Meghnad Racana Sankalan,* (in Bengali), Calcutta, 1965.

— *Report of the Calendar Reform Committee,* New Delhi 1955.

Sankalia, H.D. *Aspects of Indian History and Archaeology,* Delhi 1977.

— *Indian Archaeology Today,* Bombay, 1962.

— *Pre-History and Proto-History of India and Pakistan,* 2nd edn., Poona 1974.

— *Some Aspects of Prehistoric Techonology in India,* New Delhi, Indian National Science Academy, 1970.

Sarkar, B.K. *Hindu Achievements in the Exact Science,* Longmans Green, New York and Calcutta, 1918.

Sarup, Lakshman, *The Nighantu and the Nirukta,* Delhi 1984

Satyaprakash, Dr., and Sharma, Pt. Ram Swarup (eds.) *Baudhayana Sulbasutram* : With Sanskrit Commentary by Dwarakanath Yajvan and English translation and critical notes by Prof. G. Thibaut. New Delhi, 1968.

Saunaka, *Rgveda Pratisakhya,* Ed. V.K. Varma, Varanasi 1970.

Seal, B.N. *The Positive Sciences of the Ancient Hindus ;* London, 1915, Reprinted New Delhi 1958.

Sen, S.N. and Bag, A.K. *The Sulbasutras,* Text, English Translation and Commentary, New Delhi 1983.

Sharma, G.R. *The Excavation at Kausambi* (1957-59), Allahabad, 1960.

Sharma, R.K. (ed.), *Indian Archaeology* : *New Perspectives,* Agam Kala Prakashan, Delhi, 1982.

Sharma, Ram Sharan, *Material Culture and Social Formations in Ancient India,* New Delhi, 1983.

Sharma, Y.D. *Archaeological Remains Monuments and Museums,* 2 vols. New Delhi 1964.

— "Past Patterns in Living as Unfolded by Excavations at Rupar", in *Lalitkala* Nos. 1-2, 1955-56.

Shendge, Malati J. *The Civilized Demons* : *The Harappans in Rgveda,* New Delhi, 1977.

"Harappan and Rigvedic Inter-relations", in *Puratattva,* No. 8, 1975-1976.

Shukla, Kripa Shankar and Sarma, K.V. *Aryabhatiya of Aryabhata,* Indian National Science Academy, New Delhi, 1976.

Sing, H.N. *History and Archaeology of Black-And-Red-Ware* (Chalcolithic Period), Delhi, 1982.

Singer, Charles, *A Short History of Scientific Ideas To* 1900, Oxford, 1959, Reprinted 1965.

Singh Fatah, *The Vedic Etymology,* Sanskrita Sadan-kota 1952.

Singh, Shivaji, "Vedic Literature on Pottery", in *PAI,* 301-313.

Sinha, B.P. *Potteries in Ancient India,* Patna, 1969.

Sinha, B.P. and Narain, S.A., *Pataliputra Excavation,* 1955-56, Bihar 1970.

Sircar, D.C. *Cosmography and Geography in Early Indian Literature*, Calcutta, 1967.

Smith, D.E., *History of Mathematics,* 2 Vols. New York, 1923, Reprinted Dover Pub. Inc. 1951.

Spate, O.H.K. *India and Pakistan*, London, 1957.

Srinivasan, Saradha, *Mensuration in Ancient India,* Delhi, 1979.

Srivastava, K.M. "The Myth of Aryan Invasion of Harappan Town", in *FIC,* 437-443.

Staal J.F., *Nambudri Veda Recitation,* Mouton and Co, Hague, 1961.

Stebbing, E.P., *The Forests of India*, Vol. 1. The Bodley Head, 1922.

Steward, J. *Theory of Culture Change,* Univ. of Illinois Press. Urbana, 111. 1955.

Subbarao, B. *The Personality of India,* 2nd Edn., Poona, 1958.

Swami Prajnanananda, *Historical Development of Indian Music*, Calcutta 1973.

Thapar, B.K. "Comments on the Indus Civilization : Its origins, authors, extent and chronology by A. Ghosh", in *IP*.

— "Kalibangan : A Harappan Metropolis Beyond the Indus Valley", in *ACI,* 196-202.

— "Six Decades of the Indus Studies" in *FIC*, 1-25.

Taittiriya Samhita, Eng. tr. by A.B. Keith, Reprint, pts. i-ii, New Delhi 1967.

Thibaut, G. *Mathematics in the Making in Ancient India* Ed. by Chattopadhyaya, D, Calcutta 1984.

Thomson, G. *An Essay on Religion*, London, 1950 edn.

— *Studies in Ancient Greek Society,* Vol. I, London, 1949; Vol. II, (The First Philosophers), London, 1955.

Tilak, B.G. *The Orion : or Researches into the Antiquity of the Vedas,* First Published, Oct. 1893, Fifth Edn. Poona, 1972.

Tripathi, Vibha *The Painted Grey Ware : An Iron Age Culture of Northern India* (Foreword by A. Ghosh), Delhi, 1976.

— "Introduction of Iron in India—A Chronological Perspective", in *RCIA*.

Tylecote, R.F. *Metallurgy in Archaeology,* London, 1962.

Varma, K.C. "The Iron Age, The Veda and the Historical Urbanization" in *IANP*.

Varma S, *The Etymologies of Yaska,* Hoshiarpur 1953.

Vats, M.S. *Excavation at Harappa,* 2 Vols. Delhi, 1940.

— "Every Day Life in the Indus Civilization", in *The World of the Past,* Hawkes, Jacquetta, (ed.) Vol. 2. 240-241, 1963.

— "Indus Valley Civilization", in *The Cultural Heritage of India.* Vol. 1: 110-28, 1958.

Vij, Brij-Bhusan, "Linear Standard in the Indus Civilization", in *FIC* 153-156.

Wheeler, M. *Archaeology From the Earth*, Oxford, 1954.

— *Civilization of the Indus Valley and Beyond,* London, 1966.

— *Early India and Pakistan,* London, 1959. Rev. edn. 1968.

— *Five Thousand Years of Pakistan : An Archaeological Outline,* London 1950.

— *The Indus Civilization* (First Published in 1953), 3rd ed. Cambridge, 1979 (Reprint).

Winternitz M. *A History of Indian Literature,* Vol. I (Translated from the original German, by Mrs. S. Ketkar), Calcutta, 1927.

Woolley, Leonard, *History of Mankind : Culture and Scientific Development,* Vol. I, Part. 2 (pp. 3539-865), The Beginnings of Civilization, Unesco, 1963.

INDEX

A

Abbasid 36
Abhijit 256
abhinihita 427n
Abhisāriṇī Triṣṭubh 423n
Abhyankar, K.V. 416n
Abhyāsa 436n
Abul-Qasim Sa'id bin 'Abdur-Rahman bin Muhammad bin Sa'id al-Andalusi 46n
abutilon Indicum 280
Academic des Sciences 3
ācaryate praśasyate anayā devaviśeṣaḥ kriyāviśeṣas tatsādhanaviśeṣo vā iti ṛkśabdavyutpattiriti Ṛkbhāṣyabhūmikā 424n
Adhikārma 131
adhipatnī 180
Adhyardhā 191, 193, 194
Adhvaryu-s 137, 144, 145
ādi 444n
aditi 381
Āditya 182, 264, 387, 390
Afghanistan 87, 316
āgama 431
agni 119, 121, 145, 151, 170, 178, 182, 185, 186, 187, 198, 211, 257 385, 387, 388, 390, 301, 397
agnicayana 129, 130, 138, 139, 140, 141
agnicit 174
agninā apacat 130
agninā pacāni 130
Agnirahasya 138
agram 211
Agrawal, D.P. 73, 240, 274, 330 n 331n. 354, 355, 356, 357
Agarwala, V.S. 131, 132, 362, 403
agri 211
Ahar 325
āhavanīya 161, 162, 164, 165. 166, 257
ahi 368
Ahicchatra 124, 125
āhuti 397
Aitareya Āraṇyaka 414n, 429n
ajāloma 134
ākhyāta 429n
Akkad 66
akṣa 158
Akṣapāda 33, 34
akṣayastomīya 179
Akṣṇya 202
akṣṇya-rajju 202
akṣṇaye 159
ākṣepa 426n
Akṣara-Saṃhitā 413n
akṣaya 179
Alāja 152, 153
Alājaciti 153
Alamgirpur 126
al-Battani 44
(Al) al-Biruni 4, 50, 252, 409

Alexander 63
Al Forabi 36
Almagest 50n.
Alexandria 36, 38
Al-kindi 36
Allchin, B and R. 65n. 72, 84n. 86, 87, 92, 93n. 97, 98, 100, 101, 102, 139n., 221, 275n., 276, 279, 289, 293n., 294, 295, 296, 298, 300, 302, 305, 346, 354, 357, 358, 373, 374n
Allchin, F.R. 94n., 283, 284n
al-manazil 251, 252
Al Razi (Rhazes) 36
al-Tusi 50n
āma 130
Amri 97, 101
Aṃśa 201, 394
āmnāya 379
Anantabhaṭṭa 442n, 444n
Anatolia 316
Aṅgasaṃhitā 413n
Aṅgirasa-s 384, 388, 389
aṅgula-s 142, 158, 159, 161, 166, 185, 186, 191, 192, 193, 194, 195, 196, 198, 200, 201, 230, 231, 232
anna 381
anṛta 385, 388
aṇu-s 158, 231
anudātta 424n., 426n
Anuloma anvakṣara-saṃdhi 430n, 432n
anuvāka-s 425, 447
Anuvākānukramaṇī 447
Anukramaṇī 419, 447
Anuṣṭubh 181, 421n, 424, 445n
anvahārya-pacana 161
anyad vyavacchinatti 26
apāka 129
apāna 179
apānabhṛt 179
Āpastamba 123, 145, 150, 156, 165, 190, 198
Āpastamba Śrautasūtra 133
Āpastamba Śulva-sūtra 156 198, 199, 219
apasya 180
apavāda 424, 427n
apavāda-sūtra 430n, 432n
apsujit 368
ara-s 187
Arab 41, 44
Āraṇyakas 62, 129, 156, 407n, 412n, 416, 418n, 441
aratni 158, 166. 198, 231
Ārcajyotiṣam 266
Archimedes 38
Arcikā 436n
ardha-adhyardhā 193
ardharca 425n.
ardhyā 192, 193
Aristotle 11, 12, 30, 36
Arjuna 264
Ārjunis 264
Arkwright 19
arma 131, 132, 133
arma-iṣṭakā 135
armaka 131, 133
arma-kapāla 133, 134, 135
arma-kapālāni 133, 134
Armenia 316
arṇas 369
āroha 426

Ārṣânukramaṇî 447
Ārṣī Saṃhitā 413
artha 423n, 424n
Arthaśāstra 158, 230, 231
Āryabhaṭa 167, 231
Āryabhaṭa I 219
Āryabhaṭīya 219, 220, 231n, 422
Āryāvarta (or Madhyadeśa) 272
asa 386
asaṃhita 429n
asapatnya 179
Asia 36, 42, 43, 48, 52
Aśleṣā 267
aśmanmayī 360, 364
Asmar 61
Aśoka 117
aśru 211
Assurbanipal (Ashur-baniapli) 259
Asṭādhyāyī 408, 441, 444n, 445, 451, 453
Aṣṭamī 194
Asthana, Sashi 58, 60, 358n
Asuras 265, 387, 427n
aśva 211
Āśvalāyana 418n
Aśvinī 179
Aśvins 179, 365, 388, 390, 421
āsye 379
Atharvan 375
Atharvāṇo Bhiṣag 421n
Atharvaveda 140, 249, 256, 260, 384, 403, 47n., 412n., 413n, 440
Atharvaveda Prātiśākhya 425n, 431n., 439, 440, 444, 445n, 447n, 449
Atharvaveda Saṃhitā 415n, 416n, 436, 439, 440, 441
atha yad armakapālāni bhavanti armād evainaṃ tat pṛthivyāḥ saṃbharati 133
Atichandas 181
ātman 177, 198, 199
Atranjikhera 124, 129, 375, 377
Ātreya 443n
Audheya 438n
autumnal equinoxes 250
avama 397
avāroha 426
avasāna 423, 424, 425
Avestā 249, 386
āyāma 426n
ayana 269
ayas 364, 376
āyasī 360
Āyur-veda 34
āyuṣya 180
Adams, R.M. 80

B

Bâbhravya 434
Babhru 434 n.
Babiru 60
Babylonia 56, 58, 60, 114, 271, 272
Bacon, Francis 1, 16, 39
Bag, A. K. 160, 403
Bahrein islands 61, 243
bāhu 158
Baijava 438n.
Bajra 111
Bakshali (village) 218

Bakshali Manuscript 218
Balarāma 249
Baluchistan 276, 279
Banerjee, H.N. 376
Banerjee N. R. 304, 376
Banerjee, R. D. 55
Banwali 102, 106, 107, 111
basileus 396
Baskala 418n.
Baudhāyana 123, 145, 150, 156, 157, 164, 190, 191, 192, 203.
Baudhāyana Śulva-Sūtra 130, 143n, 158, 161, 163, 166, 167, 185, 186, 187, 190, 191, 192, 193n, 194, 201, 220, 231
Bāveru 60
Bāveru Jātaka 60
Beck, Horace 302, 307
Bernal J. D. 2, 11, 12n, 13, 15, 17, 28, 33, 41, 42n, 43, 44n, 54, 68, 69, 77n, 115, 130, 169
Beriman 238
Berthelot M. 3, 45
Besnagar (Vidiśā) 124
Bhagwanpura 109, 126, 127, 128, 129
bhaga 338, 394
Bhaikāla Śākhā 443n
Bhardvāja 365
Bharatas 364
Bhāratīya-darśana (Bengali) 33n
Bharatamuni 437
Bhāskara II 49n, 167, 219
Bhāṣya 442, 443n
Bhatnagar, S. 22
Bhatta, Laksmana Shankara 437n
Bhattacharji, Sukumari 436n
Bhattacharya, D. H. 440n
Bhattacharyya, A. K. 3n
Bhattacharyya, Ramkrishna 49n
Bhattacharyya, Satish Chandra 266n, 270
bheda-s 186, 187, 190
Bhūrika 422n
bhūyaskṛt 179
Bhūtārma 131
Bijayanand 50n
Bihar 327
Biot 252, 253, 256
Bisht 103n, 107n
Black 19
Black and Red-Ware 140n, 277, 283, 285, 374, 375
Bloch, N. A. 50n
Bluejay 422
Bolan 111
Bose, Satyendranath 5
Bongard Levin G. 84n., 250
Bose 406n.
Bose, D. M. 52
Bose, Rajsekhar 5n.
Boyle 7, 10, 15
Brahmā 249
Brāhmaṇa 47, 197, 205, 207, 208, 209, 210, 211, 249, 253, 254, 255, 256, 267, 271, 363, 375, 407n., 412n., 416, 418, 438n., 441, 451
Brāhmaṇa 62, 65, 130, 137, 141, 142, 145, 147, 149, 154, 156, 157, 162, 172, 176, 177, 196, 204, 207
brāhmaṇaḥ tam 429-30
brāhmaṇastam 429

Brahmavāda 440n.
Brahmins 163, 207
Brahma-sūtra 175, 176
Brazil 400
Brevik 315
Bṛbhu 365
Bṛhadāraṇyaka Upaniṣad 210, 212, 213
Bṛhaddevatā 419n., 447, 448, 451, 452
Bṛhatsaṃhitā 267
bṛhatī 179, 181
Bronze Age 61, 66
Brown, Burton T. 319, 327
Brunswing, R.H. 62n.
Buck 382
Buckle 36
Buddha 124
Buddhism 9
Buhler 60, 117, 218, 365
Burgess, E. 4, 258
Burk A. 59, 147, 360, 167
Burnell, A. C. 147
Burrow, T. 131, 132, 133, 134, 249
Butler, Samuel 358

C

Caitriyayāna 181
Cakṣuścit 175
Calder, C. C. 333n
Caldwell, J. R. 317n
Caliphs Mansur 36
Cambay 304
Caraka 32
Caraka-saṃhitā 31, 34
caraṇa 416, 418
Caraṇākṣa 33
caraṇāni 442
Caraṇavyulha 415n
cariṣṇva 364
Carlyle 25
carṣaṇayaḥ 383
carṣaṇi 383
cas 383
cāṣā 383
Casal, J. M. 101n
Caturadhyāyī 430n, 447n
caturasra 166, 202
Celts 63
chad 420n
Chakrabarti, D. K. 61n, 67n, 79n, 86n, 87, 88, 89, 90n, 94, 96, 97n, 100, 110f, 283n, 284
Chakravarty, Apurva Kumar 247, 258
Chalcolithic 66
chand 420n.
chanda 153, 163, 423n.
Chandas 413n, 420, 421n, 422n, 449
Chanda, R. P. 64, 217, 363
Chāndogya Upaniṣad (Ch.Up) 116n.
Chandaḥ Śāstra 420n., 423
Chandaś-citi 120, 153
chandaścit 174, 175, 176
Chāndasika 420n
Chāndasīya 420n
Chandaḥsūtra 420n., 422n.
Chandonukramaṇī 419n., 423n., 447
Chanhu-daro 90, 291, 299, 305, 306, 307, 330, 359

Charsada 124
Chatley 252
Chattopadhyaya, B.K 64n.
Chattopadhyaya, D. 5n., 9n., 20n., 21n., 29n., 33n., 47n., 110, 177n., 298n., 299n., 383n.
Chattopadhyaya, K.C. 64
Chautang 355, 356
Childe, Gordon V 55, 56, 64n., 65, 66, 69, 71, 72, 77, 78, 79, 80, 81n., 82, 83, 84, 85, 88n, 90, 91, 96, 113, 114, 118, 130, 169, 217, 223, 241, 242, 291, 293n., 327, 332, 334, 336, 339, 353, 360, 376
China 41, 42, 43, 44, 48, 52, 53, 251, 252, 348, 349, 372, 407
Chirand 124
Christopher Wren 7
Chu Kro-Chen 252
Cirakāla-śūnyagrāme bhūmau avasthitāni purātanāni 134
Cire-perdue 329, 330
Citrā 264
Citi 119, 121, 122, 151, 170, 175, 183
Clements, Geoffrey 401
Coghlan, H. H. 319, 329n
Colebrooke, H.T. 4, 6
Conford, F.M. 386
Converse, Hyla Stuntz 139, 140, 141, 142
Conway, M. 325n.
Copparis aphyllan 310
Cordier, P. 4, 5n.
Cowell 117
Cretan 65
Csoma de Koros, Alexander 4
Cunningham, Alexander 55, 299
Cyclops 10

D

Dadheri 109, 126, 127, 128
Daivatakāṇḍa 451
dakṣiṇā 146, 164, 166, 173, 204, 221, 255, 261
dakṣiṇa-agni 161, 164, 165
dakṣināyana 269, 270
Dales, G.F. 355, 361
Damb Sadaat 101, 317
daṇḍa 269
Dani, A.H. 101n.
Darius 359
"Dark Age" or "Dark Period" 71, 73, 74, 75, 173, 352
Daruyi 317
Darwin 17
Dasgupta S.N. 31n., 33, 34
Das, S.R. 269n.
Datta, Akshay Kumar 3
Datta, B.B. 144, 147, 148, 151, 152, 156, 160, 164, 166, 167, 170n., 174, 218, 219
Datta, Rasik Lal 5
deha 177
dehātmavāda 177
Deh Muras 317
Delhi 372, 375
Democritus 17
De Rerum Natura 16
Desch, C.H. 306, 318
Des Cartes 10

Des Noettes 40
Deshpande, M. N. 98, 224, 233, 240, 242, 243n. 244
devatā 421n.
Devatānukramaṇī 447
Dey, Makhanlal 5
Dharmakīrti 51
Dharmaśāstra 151, 221
Dharma-sūtra 149, 150, 171, 172
Dharma-vijaya 380
Dhar, Nilratan 5
Dhātupāṭha 419n, 452
Dheradun 136
Dhiṣṇya 167
Dignāga 29n
Dikshit, K.N. 108, 109, 310, 355
Dikshit, S.B. 268n, 269n
dīrgha-adhyardhā 193, 194
dīrgha-caturasra 159, 202
diś 379
Diyala region 61
Divodāsa 360, 364
divya 382
draviṇodā 180
Diringer, D. 251
Dṛṣadvatī 133, 356
druta 427
D'Sylva 333
duhitṛ 381
Dupree L. 317n
durga 364
Durgācārya 428n, 429n, 442, 450n
During Caspers E.C.L., 61n
dūrvā 180, 181
Dutt, Binod Behari 56, 57n, 117n
Dvarikanātha 220
Dvipadāgāyatrī 423n
Dyāvāpṛthivī 388

E

Eggeling 137, 138, 154, 183n, 205, 206n, 207, 209, 210n, 215n, 256n, 258
Egypt 37, 48, 56, 58, 71, 78, 79, 80, 81, 83f, 96, 109, 113, 114, 116, 118, 173, 216, 217, 223, 227, 238, 241, 277, 329, 339, 343, 344, 345, 346, 347, 407
Einstein, A. 400, 401
ekura 249
Elam 277
Emerson 36
Emeenesu 249
Engels, F. 39, 40n, 394, 395, 396
Epics 65
Erasistratus 38
Eta Tauri 258, 259
etā ha vai prācyai diśo na cyavante sarvāṇi ha vā anyāni nakṣatrāṇi prācyaī disaś cyavante 257
Euphratis-Tigris 61, 79, 95
Europe 36, 42, 45, 52, 53

F

Faddegon, B. 437n.
Fairservis, W.A. (Jr.) 73, 89, 94, 101n, 331
Falasifa School 36

Fallaize, E.N. 244
Farrington, B. 11, 12, 37n, 38, 39, 40n, 69, 76, 169, 344
Father Heras 249
Fick 117
Fields, P.R. 325n
Filliozat, J. 44, 45n, 386, 389n
Fleet, J.F. 158, 231, 269n
Florence 13
Forbes, R.T. 316n, 376
Friedman, A.M. 325n
fungal hypae 288

G

Galen 36, 38, 39
Galileo 83
gāma 132
gaṇa 390, 444
gaṇaka 171
Gaṇapāṭha 419n, 452
Ganges 138
Gaṇitapāda 219
Gaṇitayuktayaḥ 49n
Garbe 156, 157
gārhapatya 161, 162, 164, 165, 166, 257
gāyatrī 162, 163, 180, 181, 379, 422n, 423n, 425
gaveṣaṇa 381
gavyāt 381
gavyu 381
gavyūti 381
Gelb, I.J. 62n
Geo-metry 114, 116
German 63
Ghaggar 133, 355, 356
Ghaggar-Hakra river 87
ghee (brick) 180
Ghora 379
Ghosh, A. 73, 74n, 75n, 79n, 94, 101n, 123, 125n, 126, 140, 216, 217, 355, 374, 377n, 380
Ghosh, Jnan Chandra 5, 22
Ghosh, Partha 400, 402
Giles, P. 63n, 19
Gītā 7, 25
Glashow 402
gojāta 381, 382
Golavā 438n
gomat 381
Gopa 381
Gopati 381
Gopāla Yajvā 443n
Gossypium urboreum 288
go¸u 381
gaviṣṭi 381
grāma 132
Great Bath (at Mohenjo-daro) 234, 245, 246, 247, 347
Greco-Roman 38, 39, 40, 41
Greece, 36, 38, 48, 113
Greeks 63
Gṛhya-sūtra 149, 413n, 418n
Gudea 115
Guha, B.S. 64
Gujrat 59, 87, 111, 305
Gulati, A.N. 289
Gumla 97, 101
Gupta, R.C. 403
Gupta, S.P. 58

hala 282, 383
Halāyudha 422n.
Hall, A. R., 329n.
Haloxylon amo dendron 316
Hamid, D. A., 279, 284n, 311, 314,
Hamilton, Buchanan 5n
Haṁsamukhī (Swan-beaked) brick 195
Hanumangarh 135
Hanumangarh-Suratgarh 102, 103
Harappa 55, 57, 59, 61, 62, 64, 66, 70, 86, 89, 90, 91, 92, 95, 96, 98, 102, 108, 134, 140, 141, 224, 225, 226, 233, 235, 238, 279, 291, 294, 298, 299, 307, 317, 318, 346, 359, 362, 366, 372, 374, 410
Hargreaves 19, 361
Haryana 375
hasta 231, 264
Hastinapur 124, 125, 129, 372
Hegde, K.T.M., 282, 308, 325
Heine-Geldern 96
Hemmy, A.S., 237, 238, 239
Henry Ford 1
Herodotus 113, 129
Herophilus 38
Hindon 126
Hipparchus 44
Hissar 359
Hobbes 7
Hoernle, A.F.R. 4, 218
Holmyard, E.J. 329n
Hooke, 7
Hooper 314
Hopkins 383
Hora, S.L. 329n, 331n,
hsiu 251, 252

I

Ibn Rhshd (Averroes) 36
Ibn Sina (Avicenna) 36, 46
Ibṛahim, S.M. 46n
ila 381
indh 211
indha 211
Indra 208, 211, 264, 358, 362, 363, 365, 366, 367, 368, 369, 370, 371 387, 391, 427n, 445n, 446n
Indrasya śatruḥ 427n
indriya 211
Indus 231, 232, 233, 278, 279, 305, 315, 318, 319, 331, 354, 358, 372,
Indus-Mesopotamian Trade 60
Indus Valley 18, 61, 66, 79, 81, 87, 101, 228, 229, 238, 243, 276, 291, 315, 345, 363, 365, 366
Indus (Valley) Civilization 229, 230, 234, 235, 237, 239, 240, 241. 242, 243, 244, 245, 246, 247, 248, 256, 262, 268, 273, 274, 289, 332, 339, 345, 346, 349, 352, 353, 357, 358, 360, 362, 363, 366, 370, 379, 404
Iran 95, 277, 279, 317, 330, 359
Iron Age 333, 376
Iroquois 395
iṣā 158
Isaacs 392
iṣṭa 183
iṣṭakā 141, 178, 183, 184
Iṣṭakāpūraṇa Pariśiṣṭa 142

iti abhyupadiśanti 156, 157, 168, 172
iti daśa 194
iti ekapramāṇa vyāpāraḥ 27
iti uktam 156, 157
iti vijñāyate 156, 157
I-Tsing 4, 50

J

Jacobi, H 254, 255, 256, 260, 261, 262
(Paṇḍit) Jagannāth 37, 50n
Jagatgram 136
Jagatguru Swami Sri Bharati Krishna Tirtha Maharaja Shankaracharya 402
Jagatī 162, 163, 180, 181, 421n, 422n, 445n.
Jaimini 169, 423n
Jaimini Sūtra 423n, 424n
Jainism 20
Jajala 440n
Jalade 440n
Jalayantra 269
Jalipur 97
James Watt 19
Janaka 212
jānu 158, 185, 186
jarethām asmat vi paṇeḥ manīṣām 365
Jarrige Jean Francois 97n
Jātakas 117
Jaṭāpāṭha 438, 439, 444, 445
Jayanta 27, 34
Jayasiṃha 50n
Jay Sings 37
Jemet Nasr 227
Jhukar 359
jñāna 32, 35
Jñānendra Sarasvatī 442n
Jolly, J. 4
Joshi J. P. 126, 309
Jumna 138
Junghans 319
Jupiter 250
jus-gentuim 392
Jyotirmīmāṃsā 49n
Jyotiṣa 268, 270, 413, 449
Jyotiṣmatī 423n

K

Kachchi plain 111
Kahum 58
Kāla (Yama) 249
Kālakajña 265, 266
Kālanirṇaya Śikṣā 421n
Kalās 10
Kalibangan 58, 70, 73, 97, 98, 99, 101, 102, 103, 106, 107, 133, 134, 139, 224, 225, 229, 233, 236, 276, 277, 294, 299, 317, 346
Kālanirṇaya Dīpikā 421n
Kalpa 413n, 449
Kalpasūtra 144, 145, 149, 150, 171, 183, 214
Kāmānām 380n
Kampadoṣa 427n
Kāmya-agni 151, 185
Kaṇāda 14
Kāṇḍas 138
Kaṅka-citi 153

Kane, P.V. 123, 361
Kangle 231
Kāṇva 379, 416, 438n
Kapila 438n
Kapiṣṭhala 438n
Kar 211
Karani 167, 202, 218
Karat 211
Karavindasvāmin 219
Karma 398
Karma-kāṇḍa 176
Karmāntika 171
Kashikar, C.G. 141, 142, 221, 222,
Kāśikā 131
Kastner, M 325n
Kāśyapa Saṃhitā 109
Kāṭhaka Saṃhitā 129, 256
Kātyāyana 123, 150, 165, 419n, 422, 439, 444
Kātyāyana Śrauta-sūtra 142, 145
Kātyāyana Śulva-sūtra 170, 219
Kātyāyanīya Vājasaneyī Saṃhitā 438n
Kausambi 124, 125, 135, 136, 376
Keith, A.B. 60, 123, 179n, 180, 181, 256, 385, 386, 390, 397, 404
Kenoyer, Jonathan M. 296
Kerala 49n
Keśava's *Paddhati* 142, 222
Khadilkar, S.D. 170n
Khan, F.A. 101n
Khanaka 171
Khaṇḍanakhaṇḍakhādya 26
Khettasamika 117
Kielhorn, F 416n
Kiral bush 310
Kish 59, 61
Knorozov, Yuroj V 250
Koinos nomos 393
Kosambi, D.D. 99, 347, 348, 350, 367, 369, 370
Kot-Diji 73, 90, 97, 101, 102, 294, 317
Krama 434, 436, 445, 449n
Kramapāṭha 433, 435, 438, 442n, 446
Krama-vikṛti 438n
Kramer, S.N. 62n
kṛṣ 383
Kṛṣṇa 249
(Ṛṣi) Kṛṣṇa 413n 415
Kṛṣṇājinaloma 134
Kṛṣṇayajurveda 435, 438n, 443
kṛṣṭi 383
Kṛttikās 180, 181, 256, 257, 258, 259, 260, 262, 263, 264, 265
Kṣatra 209, 264, 265
Kṣatriyas 204, 206, 208, 209
kṣipra 427n
Kulkarni, R.P. 234, 235
Kuppana Sastri, T.S. 403
Kuru 375
Kurus 377
Kurukshetra 126
Kutch 87, 356
kuṭumba 117

L

Lagadha 266, 267
Lagash 55, 115
Lahiri Bela 136

Lahore 103, 135
Lal, B.B. 73, 98, 99n, 101n, 103n, 125n, 126, 128, 136n, 275n, 276n
Lal, B.B. and Gupta, S.P. 61n, 62n, 245
Lal, D. 333
Lallanji Gopal 376
Lamberg-Karlovsky 2, 55, 96, 317
Lambrick 100, 101, 111, 355
Lancashire 57
lāṅgala 383
Langdom 227
Lāṭyāyana Śrauta-sūtra 133
Laufer 349
laukika 420, 452
laukika chandaḥ 420n
Liebig 12
Līlāvatī 148, 218
Logic of Therapeutics 32
lokam-pṛṇa 204, 205
lokam-pṛṇa chidram-pṛṇa 204
Lokāyata 20, 21, 177
loka-vyavahāra 26
lopa 431
Lothal 58, 59, 61, 70, 73, 84n, 90, 103, 107. 108, 139, 224, 225, 226, 227, 228, 229, 231, 232, 233, 234, 235, 238, 240, 241, 243, 290, 291, 292, 293, 294, 296, 297, 299, 303, 304, 305, 307, 329, 346, 356, 357
Louis Pasteur 3, 256
Lucretius 16
Ludhiyana 126
Lyceum 38

M

Macdonell A.A. 60, 256, 367, 368, 369, 385, 389n, 390n, 397n, 404
Mackay 57, 98, 104, 105, 225, 226, 235, 238n, 243, 275, 276, 278, 279, 281, 282, 283, 284n, 286, 288, 289, 292n, 295, 297, 298, 299, 300, 301, 302, 303, 305, 306, 307, 308, 309n, 314, 317, 329n, 330n, 353, 361
Mādhava 49n, 442n,
madhyama 427
Mādhyandina 416, 417, 438n 444
Madrārma 131, 132
Madras 377
Magan 358
Mahābhārata 172, 409
Mahābhāṣya 380, 428n, 444
Mahābṛhatī triṣṭubh 423n
Mahadevan, Iravatham 84n, 248, 299n, 411
Maharashtra 129, 234
Mahā Silam 117n
Mahāvrata 129
Mahesh Yogi 400
Māhiṣeya 443n
Maine, Henry 393
Mainkar, V.B. 158, 185, 224, 226n, 227, 228, 229, 231, 232n, 233, 234n, 237, 238
Maithili 383
Maitrāyaṇa 150
Maitrāyaṇīya 438n
Maitrāyaṇī Saṃhitā 129, 256

Major Clark 55
Majumdar, N.G. 97
Makran Coast 87
Mamun 36
Man 319n
manaścit 174, 175, 176
Mānava 150
Mānava Śulva-sūtra 219
manazil 252
Manchanda, O. 275
Manda 440n
maṇḍala 397, 436n, 447
Māṇḍūkāyana 418n
manomaya citi 120
mantra-s 142, 174, 175, 413n, 416, 421n, 423, 424n, 427n, 438n, 447, 450, 452
mantro hīnaḥ 427n
Manu 9, 165, 365
mā paṇiḥbhūḥ 365
Marshall J. 55, 56, 57, 64, 65, 66, 67, 72, 94, 104n, 217, 235, 237, 240, 275, 276, 279n, 282, 284n, 286n, 288, 295n, 298n, 300n, 303n, 309, 311, 314, 315, 329n, 330n, 348, 349, 361, 362
Maruts 208, 368, 379, 390
Marx K. 16, 395, 397
Mashiz Valley 317
Mātrāmoda 443n
Matsyas 377
Mauryan Dynasty 372
Max Muller F. 162, 215, 255, 369
māyā-vāda 14, 24
mazzaloth 252
Medicie Speciali 13
Meerut 126
Mehi 317
Mehrgarh 111
Meluhha 358
Meruprastāra 422
Mesopotamia 37, 48, 55, 60, 61, 66, 71, 78, 79, 80, 81, 83, 84, 95, 96, 114, 116, 217, 223, 241, 271, 276, 277, 279, 327, 329, 339, 345, 346, 347
Metta, D. 325n
Michaels, A. 236
Milsted, J. 325n
Mill, J.S. 30
mīmāṃsā 449n
Mīmāṃsā-sūtra 169
Minayef 60
miśrita chandas 423n
Mitalhal 97
Mitra 387, 388, 389, 390
mnā kathane 379
Mohenjo-daro 55, 57, 58, 59, 62, 64, 70, 86, 88, 89, 90, 91, 92, 95, 96, 98, 103, 104, 105, 108, 134, 224, 225, 226, 227, 228, 229, 233, 234, 235, 238, 240, 241, 277, 279, 281, 290, 292, 294, 295, 296, 298, 299, 314, 317, 319, 325, 328, 329, 330, 331, 346, 347, 353, 354,
Mookerjee Radha Kumud 59, 60
Mathura 124
Moran, H.A. 251
Mṛgaśīrṣa 264
Mughal, M.R. 38, 72n, 86n, 87n, 93, 94, 97, 101n
muhūrta 269, 270, 272

Mukherjee, Jnanedra Nath 5, 22
Muktīśvarācārya 421n
Multan 103, 135
Muṇḍaka Upaniṣad 449
Mundigak 101, 317
Muttra 92

N

nābhi 187
nāḍikā 269
nagara, nagarin 129, 132, 304, 364
Naitandhava 133
Nigamakāṇḍa 451
nākasat 179, 186
Nakṣatra 180, 249, 251, 252, 253, 256, 260, 263, 264, 265
nakṣatratvam 265
Nal 317
nāma 429n
Namaśūdras 25
Nāradīya Śikṣā 437n
Nasik 124
nata (bent) 200
Nāṭyaśāstra 437n
Navagraha 249
Navārma 132
Navdatoli 124
navyaḥ jāyatam ṛtam 397
Needham, Joseph 42, 43, 44, 48, 52, 53n, 54, 69, 169, 249, 251, 252, 259, 348, 349, 391, 392, 393, 398
Neolithic 66
neo-Malthusians 1
Nevasa 124
Newton I. 10
New York 58
Nidāna Sūtra 423n
Nigam, J.S. 277
Nighaṇṭu 379n, 381, 419, 429, 451, 452
Nile 79, 114, 216
Nīlakaṇṭha 49n
nīlalohita 140
Nineveh 55, 259
nipāta 429n
nipātanasūtra 430n, 432n
Nippur 55
Nirbhuja Saṃhitā 414n
Nirmud 55
Nirukta 365, 379, 379n, 380, 407n, 408, 413, 419n, 420n, 428, 429n, 431n, 442, 445, 449, 450, 451, 452
Nitya 151
niyamena śuddha-prācyām eva udyanti 258
Niyogi 376
Northern Black Polished Ware (NBP) 123, 124, 125, 215, 374, 375, 377
Nurul Hasan 46n
nūtana 397
nyañchana 202
Nyāyamañjarī 27, 34
Nyāya School 29
Nyāya-sūtra 31, 33, 34

O

Oldberg 319
Oldenberg, H 253

Olsen, E. 325n
Ori 249
Oṣadhayaḥ 421n

P

Pada 158, 421n, 423, 424, 434, 436, 439, 444n, 449
Padakrama Sādhana 443n
Padapāṭha 429, 430, 431, 432, 433, 442n, 445, 446, 450
Padārthaprakāśa 444n
Pada-saṃhitā 413n
Pada Vidhāna 447
Pādyā 192, 193, 194
Painted Grey Ware (PGW) 74, 75n, 125, 126, 127, 128, 129, 130, 176, 222, 356, 374, 375, 376, 377
Paippalāda Śākhā 440n
pāka 130
Pakistan 66, 87
paṇa 365
pañcacoḍā 179, 186
pañcajanāḥ 383
Pañcālas 375, 377
Pañcamī 191, 192, 193, 194
Pañcasiddhāntikā 267, 268
Pañcaviṃśa Brāhmaṇa 133, 381
Pande, B.M. 67, 87
Pāṇini 131, 132, 133, 134, 408, 420, 441, 442n, 444, 445, 448, 449, 451, 452, 453
Pāṇinīya Śikṣā 421n, 427n
Paṅktis 181
Paracelsus 12, 13
Paramāvaṭika 438n
parṣada 442n, 444
Parṣada Vṛtti 443n
Parṣada Vyākhyā 442, 443n
Parāśara 409
Pārāśarya 438n
paraśāstra-kutūhalaḥ 171
Paricāyyacit 152
Parikarṣaṇa 202
parimaṇḍala 166
pariśayanam 368
Pariśiṣṭa 403
parokṣa-kāmāḥ iva hi devāḥ 210
parokṣa-priyāḥ iva hi devāḥ 210, 212, 261
Parpola, A. 62n, 84n, 248, 249, 250, 251, 253,
Pārśvamānī 202
pārthiva 382
paruva 423n
Parva 423n
Pascal 422n
Pasupatinatha Sastri 443n
Pāṭaliputra 124, 372
Patañjali 380, 416, 418n, 428n, 444
P.A.T.B. von Hohenheim 12n
Pāṭha 443
Patiala 356
Paul Tannery 44
Pauṇḍravatsa 438n
Periano-Ghundai 87
Persia 276
Peshwar 92, 218
phala 383
phalānāṃ varṣitā 380
Phalgunīs 264

Piggott, S 92, 95, 103, 290, 293n, 362, 363
Pillai 267n
Piṅgala 420n, 421n
Pitāmaha Siddhānta 268
Pittioni 319, 325
Plato 16, 36, 341, 342, 343, 344, 385
Pleiades 258
pluta 436n
Polybius 210
Plenderleith, H.P. 314
Possehl, G.L. 72, 78n, 80n, 84n, 86n, 87n, 89n, 93n, 97n, 101n, 103n, 106n, 107n, 296n, 308n, 358, 360n, 361n, 380n
Poussin, Valle 261
prācyai na cyavante 257
pradeśa 158, 166, 198
prajā 207
Prajñānānanda 437n
prākṛtajanāḥ 177
Prabhumath 240
prajāḥ 383
Prajāpati 182, 183, 207, 264
prakrama 158, 161
Prākṛta Piṅgala 421n
prakṛtibhāva 431
prakṛtya 431n
pramā 26
Pramāṇa 32, 33, 202
pramāṇa-dvaividhyakhaṇḍanam 27
prāṇa 179
prāṇabhṛt 179
praśliṣṭa 427n
prastāra-s 186
prastāra-citi 186
prastha 269
Pratiloma-anvakṣara Saṃdhi 433n
Prātiśākhya 407n, 414n, 424n, 425n, 432, 438n, 439, 442, 443, 444, 445. 446, 447 448, 449, 450. 451, 452
pratyakṣa-dviṣaḥ 212, 261
prauga 121, 152, 202
praugaciti 153
pravṛtti-sāmarthyāt arthavat pramāṇam 26
prayaḥ 424n
pṛtha 158
pṛthivī 360, 381
Przyluski, J. 178
Psalm 392
Ptolemy 36, 39, 50n
pūḥ 211
Punarvasu 267
Punjab 125, 132, 279, 290, 331, 360, 363, 375
Purandara 360, 366, 367, 371
Pur 360
Pura 364, 367
Purāṇas 9, 15, 409
Puroha 366
Purus 364
Purukutsa 364
Puruṣa 121, 122, 159, 185, 191, 194, 195, 196, 198, 199
Puruṣa-sūkta 9n, 207
Puruṣṇi 364
Purvarcikā 436n

Purvarcikā 436n
Pūrva Mīmāṃsā 169
pūrvapakṣa 176
Pūṣan 365

Q

Quetta Valley 73

R

Radhakrishnan, S. 46, 47, 385
Raghavan, V. 416n.
Rahman, A. 45n.
Rahman Dheri, 101
Raikes, R.L. 331, 355
Rājāmahendravarma 416n.
Rājanya 161, 163
Rajasthan 327, 355, 356, 375
Rajghat 124
rajju 118, 177
rajju-gahaka (rope-holder) 117
rajju-gahaka-amacca 116, 117
rajjuka 117
Rāma 219
Rāmacandra 7, 21
Ramachandran K., S., 67
Ramachandran, T. N. 64, 87, 136, 379, 380
Rāmānuja 175, 176
Rāmāyaṇa 171
Rana Ghundai 317
Rangpur 74, 294, 325
Rann 356
Rao, S.R. 58, 61, 62, 74n, 84n, 90, 103n, 107n, 108, 139n, 216, 225, 227, 228, 232, 234, 238, 239, 240, 290, 292, 293, 296, 300, 302, 303, 304, 305, 306, 307, 378, 379, 380
' as-al-Qala 61
Ras Sharma 61
Rathacakraciti 153
rayi 381
Raymond, 37
Ray, P.C. 2, 3, 4, 5, 6, 7, 8, 9, 10, 11, 12, 13, 14, 15, 17, 18, 19, 20, 21, 22, 23, 24, 25, 26, 27, 35, 36, 37, 45, 169. 213
Ray, Priyadaranjan 23, 24
Red Ochre-Washed Ware 125
Renou, Louis 208, 255
rex 396
Ṛc-s 423, 424, 425, 426, 438n, 444n
Richard H. Meadow 97n
Ṛgveda 59, 75, 116, 141, 163, 168, 178, 249, 255, 266, 272, 273, 292, 360, 363, 364, 365, 366, 367, 368, 369, 370, 371, 380, 381, 382, 383, 384, 385, 386, 387, 389n, 390, 391, 392, 393, 394, 395, 396, 397, 400, 404, 405. 407n, 410, 412, 413, 415n, 421n, 422n, 428, 429, 430n, 438, 439, 440, 443n, 444n, 445n
Ṛgveda-saṃhitā 373, 378, 382, 383, 414, 415, 416n, 417, 419n, 421, 423, 425n, 426n, 428, 436, 446
Ṛgvidhāna 447
Ṛju-lekha 202
ṛk 378, 397, 407n, 412n
Ṛksaṃhitā 418, 419, 420, 422, 423, 425, 428, 433, 435, 436, 437, 439,

440, 441, 443, 447, 448
riṇag rodhāṃsi kṛtrimāṇi 369
Ṛṣi 365
Ṛṣi-s 399, 421, 447
ṛta 384, 385, 386, 387, 388, 389, 390, 391, 392, 393, 394, 395, 396, 397, 398
ṛta jātaḥ pūrvīḥ 390
ṛtasya dhenāḥ ānayanta sa-śrutaḥ 387
ṛtasya gopā 389
ṛtasya yonau 388
ṛtena yaḥ ṛta-jātaḥ vivavṛdhe rājā devaḥ ṛtambṛhat 390
ṛtavān 390
ṛtvij 172
ṛtavya 180, 206
rodhas 369
Roger Bacon 1, 17
Rohiṇī 264
Romans 63
Rome 36, 154, 372
Ropar 356
Roth 397
Roy, Amita 298, 299
Roy T.N. 109n
Rudra 249
Rupar 299
Russell 47

S

Sabacan-Himayaritic 216
Śabara 169
Sabar Khan, M 46n
Sabarmati 59
sabhā 394
Sablof, J.A. 96n
Sachau Edward C 409n
Saha, M. N. 2, 3, 4n, 5, 8, 17, 18, 19, 20, 36, 37, 63n, 399
Saharanpur 87
Sahni, Daya Ram 55
Sahni, M.R. 354
Śākala 418n, 443n
Śākaṭāyana 444
Śākhā 416, 417, 418, 420, 424, 425, 431, 438n, 440n, 442, 444
Salam 402
samacaturasra 159n, 202
sāman 436, 437, 449n
sāmānya-sūtra 432n
Samarkand 50n
Sāmasaṃhitā 436n, 437
sam āyus 182
Sambara 360, 364
Sambhava 32
Śambhu Śikṣā 421n
Sandhi 430, 431, 432, 433, 434, 435
Saṃhitā 62, 139, 254, 363, 413, 414n, 415, 416, 417, 418, 419, 420, 421, 422n, 423, 424, 426, 429, 433, 434, 435, 436, 438n, 439, 440, 441, 442, 443n, 447, 449, 452
Saṃhitā pāṭha 413, 414, 416, 421n, 423n, 427, 428, 429, 430, 431, 432, 433, 435, 442n, 445, 446
samiti 394
Śaṃkara 14, 20, 24, 33, 116n, 175, 176, 177, 212n, 430n
Sāṃkhya 7
Sāman 407n, 412n, 415n, 436-37, 439

śamyā 158
saṃyāna 179
Saṃyānīs 179, 182,
saṃvāda 26
saṃvādijñāna 26
Sāmaveda Brāhmaṇas 139, 141
Sanaullah 301, 317n 319
Sanchi 372
sandhāya 430n
Śāṇḍilya 138
Sangmeister 319
Sañjīvārma 131
Sankalia, H.D. 99, 101, 275, 284, 294, 310
Sāṅkhyāyana 418n
Santānāntara-siddhi 51n,
sapāda 192
sapeya 438n
Saptarṣih—Ursa Major or Wain 257
Saptavidha agni 198
Śaradi 360, 364
Sarai khola 97
Sārarathacakraciti 187
Sarasvatī 133, 356
śarkarā 134
Sarkar, B. K. 27n, 52
Sarkar, Pulin Behari 5
Sarma, K. V. 49n
Sarup, L. 379, 450n
Sarvadarśana Saṃgraha 420
Sarvānukramaṇī 422n, 425n
Suśruta-saṃhitā 45
Śāstra 171
śāstrabuddhyā vibhāgajñaḥ para-śāstra-kutūhalaḥ/śilpibhyaḥ sthaptibhyaḥ ca ādadīta matīḥ sadā 170
Śāstrān'ara 34
śāstrāntarābhyāsāt 34
Sastri, N 169
Sastri Vishva Bandhu 444n
Sastri, T. S. Kuppana 267
śatabhuji 360, 364
Śatapatha Brāhmana 130, 137, 138, 139, 141, 142, 143, 149, 162, 163, 168, 176, 183, 184, 204, 206, 207, 208n, 209, 210, 211n, 214, 215, 256, 258, 259, 262, 263, 264, 265, 407n, 416, 420n
Satvabkar V. S. 425n
Śaunaka 409, 440n, 443, 445, 447, 448, 452
Śaunaka Smṛti 447
Saurashtra 87, 240, 279, 285
savana 269
saviśeṣa 202
Sāyaṇa 116, 134, 257, 364, 365, 371, 379, 380, 381, 384, 397, 424n, 428n
sayuj 179
Saxon Pirates 372
Schilpp P. A. 400
Seal, B. N. 21, 22, 23, 26, 27, 28, 29, 30, 31, 32, 33, 34, 35, 49
Sen, S. N. 50n, 52, 160
Sen, S. N. and Bag, A. K. 196, 197
Sengupta, P. C. 257n, 258
Shaffer, Jim G. 101n
Shahmirzadi S. M. 317n
Sharma, G. R. 125, 126. 135, 136
Sharma, R. S. 64n, 91, 117, 118, 119, 123, 126, 127n, 129, 130, 255,

374, 375, 376, 377, 381, 382, 383
Sharma, Y. D. 124, 136
Shatrana 356
Sherwani, H. K. 46n
Shikarpur 89
Shivramkrishna 333
Shukla and Sharma 231n
Siah-Damb 317
Siddhānta-samrāj 50n
Śikṣā 407n, 413n, 448, 449, 450, 451, 452
śilājatu 314
Śilpakara 171
Śilpī 171
Sind 50n, 83, 97, 100, 101, 108, 278, 279, 290, 291, 293, 331, 363
Singer, C. 113, 114n, 329n
Singhbhum 327
Singh, Fatah 451n
Singh, H. N. 140n, 285
Singh, Shivaji 134
Sinha, B. P. 134
Sinha, K. K. 125n, 126, 376
śirā 383
śiras 264
Sirhind 356
Siswal 97
sita 383
Śivadāsa 219
Skanda 249
Skandasvāmī 450n
Ślokas 421n
śmaśānaciti 153
Smith, D. E. 113, 167n, 243
Smith, Vincent 56
Smṛtis 15
Soma 129, 208, 371, 390
Somākara 268
Somayajvā 443n
Sonpur 124
Śrāddha 19
Śrauta, 149, 207, 407n
Śrautasūtra 142, 149, 150, 222, 413n, 416, 445n
Śraviṣṭhā (Delphini) 267
śreṣṭha 264
Śrīdhara 219
Śrīharṣa 26
Srinivasan 116n
Srivastava, K. M. 362, 370, 371
śrotracit 175
Śruti, 378, 409, 410, 415, 417n, 419n, 421, 428, 435, 437n, 440, 441, 451, 452
Śrutibodha, 420n
Staal, Frits, 140, 141, 415
Stcherbatsky, Th. 4, 27n, 51
Stebbing, E. P. 331n
Stein, Aurel 101n, 102
Sthapati 171, 172
Stobha 436n
Strato 38
Subbara, B. 73, 94
Subbarayappa, B. V. 52
Sūdas 364
Śūdra 206
Sukkur 294, 295
Śuklayajurveda 416, 438n, 439, 444
sūkta 421n, 425, 447
Sūktānukramaṇī 447
Śūla-pādyā 193
Śulva (Śulba) 144, 146, 147, 148,

151, 156, 155, 162, 163, 164, 166, 167, 168, 173, 177, 182, 190, 195, 196, 197, 201, 203, 213, 214, 215, 217, 219, 220, 230, 231, 232, 233, 235, 236, 237
Śulva-sūtra (Śulbha-sūtra) 112, 118, 119, 121, 122, 123, 129, 130, 131, 143, 144, 145, 148, 149, 150, 154, 155, 156, 157, 168, 170, 174n, 182, 183, 184, 197, 201, 203, 209 213, 214, 215, 218, 219n, 220, 230, 232, 236, 253
Sumer 109, 407
Sumerian 65, 66
sūnu 383
Surathgarh 135
Surkotada 90, 97
Suryakānta 444n
Suśruta 9
Sutkagendor 357
Sutlej 356
Sūtra 133, 177, 198, 218, 363, 403, 425n, 426n, 442n, 445n, 449, 451,
Sūtrakāra 200
svakīya-mukhe 379
Svarasandhi 432n
Svaravyañjana-sandhi 432n
Svarita 425
śyena 198, 199, 200
Śyenaciti 126, 135, 136, 153
Syrians 41
svabhāva 398
Svarāṭ 422
Svayamātṛṇṇa 179
Switzerland 400

T

Tagore, R. 19
Taittirīya Saṃhitā 129, 130, 134, 135, 137, 149, 152, 153, 156, 168, 172, 174, 176, 178, 180, 181, 256, 371, 413n, 438n, 443n
Taittirīya Upaniṣad 449
Tal-i-iblis 317
Tāṇḍa 440n
Tāṇḍya 204, 205
tanu 183
tao 386
Taoist 386
Tāpayanīya 438n
Tarakai Qila 111
tastabhānaḥ 369
Tataprasad 423n
tathā adhyardhāyāḥ 193
tattvanirṇaya 33
Taxila 92
Teicher 392
Thapar B. K. 55, 72n, 101n, 102, 103. 106, 276n
Theophrastus 38
Thibaut, G. 121, 144, 145, 147, 155, 158, 160, 161n, 167, 168, 169, 170, 175n, 177, 185n, 186, 190, 191n, 193, 194, 197, 198, 200, 202, 218, 219, 231, 253n, 267, 268, 272, 273
Thomson F. C. 319
Thomson, G. 69, 335, 338, 340n, 341, 394
Tibet 48
Tigris 79, 95

tila-s 158, 194, 196
Tilak B. G. 254, 255, 256, 260, 262
Tilaurkot 124
Tiryak 202
tiryaṅ-maṇi 202
tridhā baddho vṛṣabho roravīti maho-devo martyān-āviveśa 380
Trī-karaṇī 202
Tṛtīya-karaṇī 202
Tripathi V. 374, 376, 377n
Tripiṭaka 375
Tripuri 124
triṣṭubh, 162, 163, 180, 181, 421n, 422n, 445n
tṛtīya-prakārābhāvaṃ ca sūcayati 26
Tso-chuan 349
Turner James, 288
Turvasas 364
Tylecote R. F. 328n

U

Ubhayataḥ-prauga 152, 153, 202
Ubhayī 193, 194
Udabhārṣīt 211
Udātta 426n
Udumbara 211
Uilsson 249
Ujjain 124, 304
Ukha 211
Ukthya 379
Ulugh-Beg 50n
Ulūkhala 211
Umayyad Caliphates 36
Uṇādipāṭha 452
Upadeśana 379
Upajana 431n
Upaniṣads 7, 62, 65, 179, 177, 255, 375, 398, 404, 412n, 416, 441
Upaśākhā 418
Upasarga 429n
Ur 55, 58, 59, 61, 95, 358
Urta 386
Uru 211
Urukara 211
Urvī 360
Uṣas 388, 421n
Uṣṇihs 181
Uṣṇik 421n
Utkha 211, 388
Ut-khan 211
Uttarāṣāḍhā 267
Uttarāyaṇa 269
Uttarayuga 158
Uttaravedī 129
Uvaṭa 413n, 426n, 432n, 442, 443n, 444n

V

Vāc 381
Vādhryasva 364
Vādhula 150
Vādhulasūtra 133
Vāgvajra 427n
Vāgbhaṭa 9
Vāhīka country 132
Vaidheya 438n
Vaidika 452
Vaidikābharaṇa 443n, 445n
Vaidika Chandaḥ 423n
Vaikurumtn (Tamil) 249
Vaineya 438n
Vairāja 423n

Vaiśālī 124
Vaiśeṣika 15
Vaiśeṣika-sūtra 379
Vaiśya 163, 206, 209
Vājasaneyin 137, 138
Vākcit 175
Vakrapakṣa 194
Vakrapakṣa-śyena, 190, 198
Vakrapakṣa-śyena citi, 190, 197,
Vālakhilya 179
Vāmabhṛt 180
Vaṃśa 215
Vapra 369
Varāha 150
Varāhamihira 267, 268
Vararuci 443n
Varatra 383
Vardhakī 171
Varga 167, 218, 446n
Varma, S. 443n, 444n, 445n, 448, 449n, 450
Varma, V. K. 414n, 417n, 425n, 432n, 443n, 444n, 445n
Varṇoccāraṇa 446
Varuṇa 208, 386, 387, 388, 389, 390, 397
Vāstuvidyā 172
Vats, M. S. 225, 317n, 362
Vātsyāyana 26
Vayas 179
Vayasyā 179
Vāyu 179
Veda 15, 56, 176, 262, 271, 398, 399, 400, 401, 403, 404, 406, 409, 412n, 413n, 418n, 420, 421n, 428, 442, 446, 447
Vedādarśa 440n
Vedāṅga 271, 413n
Vedāṅga Jyotiṣa 266, 267, 268, 269, 270, 271, 272
Vedānta 14, 24, 47
Vedānta-sūtras 14
Veṅkaṭamādhava 419n, 423n
Vernal equinoxes 250, 256
Vibhāgajñaḥ 170
Videha 212
Vigasin, A. 250
Vihavya 180
Vikarṇī 179
Vījagaṇita 148
Vij, Brij Bhusan 245
Vijñāna 32, 35
Vijñāyate, 157, 160, 161, 166, 168, 172
Vikāra 430, 431, 436n
Vikarṣaṇa 436n
Vikrama 438n
Vilambita 427
Vinaṣṭagrāma 132
Viradrūpa Triṣṭubh 423n
Virāj 179
Virāma 428, 436n
Virāṭ 422n
Viśākhā 180
Visaṃvādi jñāna 26
Viṣkambha 202
Viśleṣaṇa 436n
Viṣṇumitra 424n, 428n, 442, 443n
Viśrambha 426n
Viśvadevas 388, 391
Viśvajyotis 180
Vrata 385

Vṛdhra 202
Vṛṣabha 380
Vṛṣṭiḥ 383
Vṛṣṭisāni 179
Vṛta 423n, 424n
Vṛtrahan, 367
Vṛtra 211, 367, 368, 369, 370, 427n
Vṛtti 443n
Vyākaraṇa 413n, 449
Vyāma 159
Vyañjanasvara-saṃdhi 433n
Vyarṇa 133
Vyāsa 409
Vyuṣṭi 179
Vyāyāma 159, 166

W

Weber A. 4, 138, 253
Weinberg 402
Weinstock 252
Wertime, T. A. 317n
Wheeler, M. 95, 96, 126n, 129, 305, 331n, 353, 357, 359, 360, 362, 363, 370, 371, 376
Whitney, W.D. 253n, 258, 267, 272, 397, 438n, 443n, 444n
William, B. 103
William Jones 63
Winternitz, M. 149, 254n, 261, 381, 385
Woolley, L. 55, 372, 373

Y

Yadus 364
Yajamāna 120, 152, 153, 173, 204, 221, 427n
Yajña 387, 388, 389, 397, 413n, 427n
Yajñasena 181
Yājñavalkya 137, 138, 210, 212, 213
Yajurveda 135, 137, 139, 140, 141, 142, 143, 148, 149, 153, 157, 176, 178, 183, 196, 197, 214, 255, 260, 266, 267, 271, 407n, 412n, 413n, 414n, 416n, 436, 437, 438, 439, 441, 445n, 448
Yajus 255, 375, 407n
Yājuṣa-jyotiṣam 266
Yajuṣmatī 205
Yallaya 231
Yamunā 126
Yāska 365, 379, 380, 407n, 408, 419n, 428, 429, 442, 444, 445, 448, 450, 451, 452
Yaśodhara 7, 8, 21, 23
Yavamadhyā 231
Yavamadhyā Triṣṭubh 423n
yuga 158

Z

Zekda 308
Zide Arlene, R.K. 84n
Zilsel, E. 11, 393, 398